D1565851

Linear Systems in Communication and Control

Linear Systems in Communication and Control

Dean K. Frederick and A. Bruce Carlson

Systems Engineering Division
Rensselaer Polytechnic Institute

JOHN WILEY AND SONS, INC.

New York · London · Sydney · Toronto

Library of Congress Catalogue Card Number: 71-155118

ISBN 0-471-27721-5

Printed in the United States of America

10 9 8 7 6 5 4 3

To our wives and children:

Phyllis Frederick, Freddy, Ann, David, and Steven
Patricia Carlson, Kendra, Kyle, and Kristen

Preface

This textbook is an introduction to systems analysis with emphasis on linear, time-invariant systems and applications to communication and automatic control. It gives a unified presentation of basic analytic methods at a level appropriate for college seniors and provides a foundation for further study of communication, control, or systems engineering in general.

We have adopted this unified approach because of our firm belief that it is educationally more efficient and desirable to develop the various analytic techniques in a way that takes advantage of their common base and inherent similarities. Not only is the student spared unnecessary duplication but he is also more likely to gain a conceptual understanding that will prove valuable in the future, especially as the areas of communication and control become increasingly entwined in practice. Our conviction that communication and control should be taught together in an introductory course is reinforced by the mutual benefits we have personally found in team teaching and coauthorship; we hope to convey some of that spirit to the users of this book.

The material is not intended exclusively for electrical engineers. Therefore, the assumed background is limited to ordinary differential equations, the algebra of complex numbers, matrix algebra, elementary mechanics, and basic circuit theory. No prior knowledge of transforms is required, and the few necessary elements of complex-variable theory are covered by an appendix.

The specific topics and organization are given in the Contents and are amplified in Section 1.4. But we would point out here certain salient features, some of which are new to an undergraduate textbook.

1. State variables and related concepts are introduced in Chapter 2 and are woven throughout the text; however, we have refrained from the aspects that entail advanced matrix theory.

2. Fourier and Laplace transforms (Chapters 6 to 8) are discussed only after a

review of classical methods (Chapter 3) and a careful development of time-domain analysis based on convolution (Chapter 4).

3. Chapter 4 also contains a derivation of the superposition integral from first principles, impulses presented in a fashion consistent with distribution theory, and a time-domain approach to transfer functions.

4. Signal analysis is covered in Chapter 5 using the geometric interpretations of signal space, with signal approximations viewed as vector projections.

5. The several analytic tools for feedback systems (Chapters 9 to 11) are illustrated by repeated application to a satellite control problem.

6. Postponing discrete signals and systems to Chapters 12 and 13 makes possible a compact presentation by analogy with the continuous case. These chapters include the discrete Fourier transform as well as the z-transform.

Since 1967 we have taught the contents of the first eleven chapters in a three-credit, one-semester course at Rensselaer. The pace is brisk, and the enrollment has been primarily electrical engineering seniors from the upper half of their class. Most of our students have followed up this course with design-oriented electives in communication and control, where the material of the last two chapters is covered.

A less demanding course, possibly at the junior-year level, may be devised by selected omission of some of the following units: Sections 3.5, 8.4, and 9.4; Sections 5.2, 5.3, 5.4, and 12.4; Sections 7.5 and 12.2; Chapters 10 and 11; Section 12.3; Chapter 13; Chapters 12 and 13. On the other hand, when supplemented by laboratory or computer simulation work, the text would be sufficient for a year-long course.

Problems for the student are at the end of each chapter; some of them have been designed for computer solution and are so indicated. An instructor's manual is available from the publisher on request.

Dean K. Frederick
A. Bruce Carlson

February 1971

Acknowledgments

As with any undertaking of this nature, there are many to whom we are indebted for guidance, assistance, encouragement, and tolerance. Professor Dean Arden, chairman of the Electrical Engineering Curriculum at Rensselaer, was instrumental in establishing the philosophy of the course from which this text evolved, and his approach to signal analysis is used in Chapter 5. For reviewing all or parts of the manuscript, we thank Professors Mac Van Valkenburg and John Thomas of Princeton, William Perkins of Illinois, Paul DeRusso of Rensselaer, and Isador Cogan of Drexel. Mrs. Joan Hayner cheerfully carried the brunt of the typing chores, and Carl Pavarini and Donald Van Luvanee helped in the tedious work of error detection. We also thank our colleagues in the Systems Engineering Division, the Rensselaer administration, and the editorial staff of John Wiley & Sons. Finally, the dedication attempts to express the depth of our gratitude for support on the home front throughout the several years this project was in the making.

D.K.F.
A.B.C.

Contents

Linear Systems in Communication and Control

1

Introduction

The main subject of this text is the *analysis of linear time-invariant systems*, particularly as related to electrical communication and automatic control. Although we shall speak primarily in the language of electrical engineering, much of the material also has applications in mechanical, chemical, and aeronautical engineering. Indeed, linear systems analysis is only a part of the newly emerged discipline known as *systems engineering*, which transcends the traditional fields and has relevance to many areas of investigation, including some not usually associated with engineering.

This chapter introduces the basic ideas of systems and systems engineering, and discusses the particular role of linear systems analysis in the systems approach.

1.1 SYSTEMS ENGINEERING

A definition of systems engineering follows quite naturally from a definition of the term *system:*

> A system is an organized collection of interacting units – possibly including men as well as machines – designed to achieve some specified objective or set of objectives by manipulation and control of materials, energy, and information.

Our definition is very general, since there are many kinds of systems, e.g., power systems, telephone systems, information processing systems, transportation systems, economic systems, political systems, etc.

Common to all of these are the two distinguishing features of a system: *1*. it consists of smaller units, variously called the system elements, components, or subsystems; and *2*. the function performed by the system as a totality is made possible only by organizing and coordinating the functions of the individual elements. In short, the system is a whole that truly does exceed the sum of its parts.

However, not all systems fall within the domain of systems engineering as the term is used here. The deciding factor is whether the system in question is based fundamentally in engineering technology. Accordingly, we state that:

> *Systems engineering* is the practice of engineering uniquely associated with systems as a whole, concentrating on the overall function of the system rather than the specific details of its constituent parts.

The need for a systems approach and professionals functioning as systems engineers is readily appreciated by recognizing that the design of a modern system, even one that is physically small, involves a wide variety of technological specializations. Consequently, no single engineer could have the detailed knowledge or the time to do the complete job by himself; this requires a *team* of experts, one or more for each of the different subsystems.

Coordinating the team effort is the role of the systems engineer, and systems engineering is the organizing force, the "glue" that binds together the enormous and diverse work required to produce a complex, sophisticated system. The systems engineer is charged with the overview of the entire system. He looks into the details of subsystems and components only to the extent that they affect overall system performance, reliability, and cost. Thus we can say that systems engineering is a *philosophy of approach* to complex engineering problems, an orderly way of assessing the requirements on the system and following through to see that those requirements are met by the final design.

There are two significant characteristics of the systems approach. First, at the outset the problem is cast into quantitative form by defining requirements and criteria. (This, of course, is not unique to systems engineering, since a precise statement of the problem is half the battle in any engineering task.) Second, throughout the design process one continually checks the work against these requirements, thereby detecting major flaws before implementation rather than after. Final testing should thus take the form of minor "debugging" only.

While the systems approach seems quite logical, history shows that many systems have been built by trial-and-error methods wherein one leaps immediately to an implementation, be it ever so crude, and patches it up to suit. Such seat-of-the-pants engineering may work well enough for small problems, but for space-age systems – involving a tremendous investment of time and money, to say nothing of the possible risk of human lives – there must be a high degree of confidence in the design long before it is committed to hardware realization. This is true especially if, as is sometimes the case, final testing comes only with actual use of the system and it is too late to "go back to the old drawing board."

Having gained some feeling for systems engineering, we can now state the relationship of this text to the systems approach. The chapters that follow discuss linear systems, their mathematical models, analysis, simulation, and evaluation. Hence, excluding the design phase per se, we are dealing with the central part of the systems approach for the special case of *linear, time-invariant* systems. Although it may seem that this is a very small slice of systems engineering, it is nevertheless a very important one because an understanding of linear systems analysis forms the best preparation for other, more difficult analysis problems. Moreover, since design is essentially analysis done backwards, the study of analysis is the first steppingstone toward design.

Along this same line of thought, it should be pointed out that a systems engineer seldom conceives an approach without taking account of possible devices for the final implementation. He would be very foolish indeed to develop mathematical specifications that are beyond the realm of practical achievement. Accordingly, the designer must have some familiarity with what can and cannot be achieved by actual devices. It is from an apprenticeship in analysis that this important familiarity is gained.

1.2 SYSTEMS AND SIGNALS

The evolution of systems engineering was strongly spurred by electrical engineers concerned with communication and automatic control. And many of today's large-scale systems heavily involve these functions. Therefore, it is appropriate to introduce specific system concepts in the context of communication and control.

Communication systems

As a familiar example of a communication system, consider Figure 1.1, which purports to show the major steps in a television broadcast of the evening news. The image of the announcer is picked up by the TV camera and converted to an electrical waveform called the video signal;

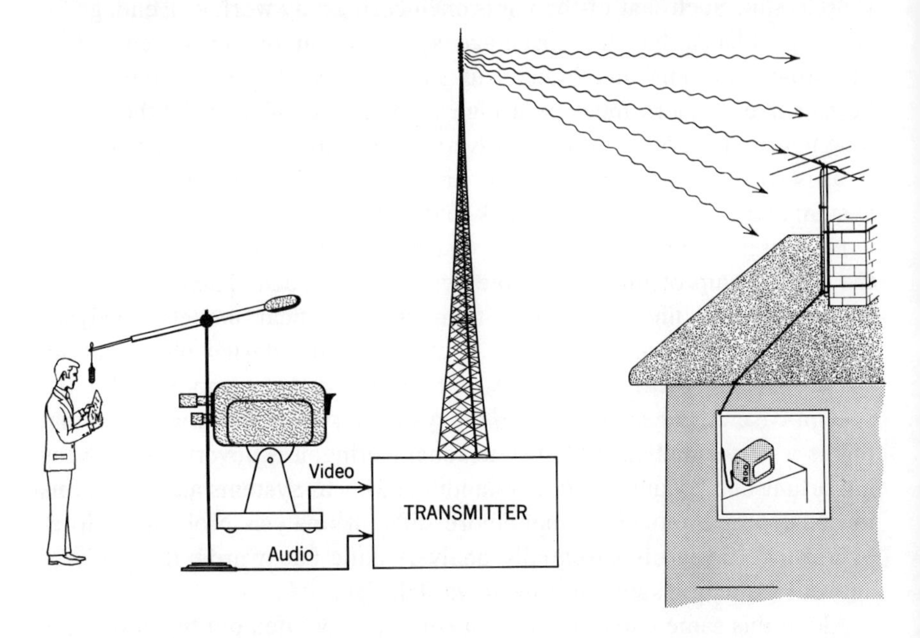

Figure 1.1. A television broadcast.

similarly, the microphone converts sound waves to an electrical audio signal. Both audio and video are fed into a transmitter which further operates on them to produce a new signal, still containing the sound and picture information but having properties better suited to transmission by radio-wave propagation. A home antenna at some distance from the transmitter then extracts energy from the broadcast signal – which by now has suffered some changes in the course of its travel – and delivers this received signal to a receiver: the TV set. The receiver performs essentially the reverse of the transmitter operations, that is, it separates the audio and video signals and applies them to the loudspeaker and picture tube, thereby recreating sound and image.

Because this example has all the essential ingredients, we can now draw a generalized block diagram for any electrical communication system (Figure 1.2). The information source – sound waves, image, data, and the like – is passed through a transducer to give an electrical *input signal* $x(t)$, which the *transmitter* then modifies to prepare for transmission. The *transmission medium*, which may take the form of wire pairs, coaxial

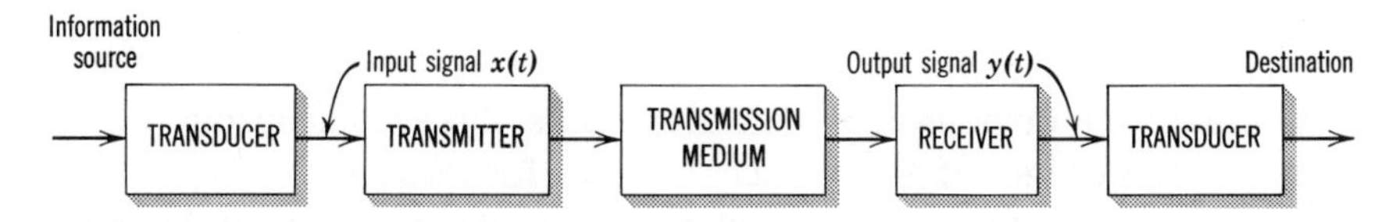

Figure 1.2. Elements of a communication system.

cable, or electromagnetic radiation, provides the electrical link between transmitter and receiver. The *receiver* operates on the received signal, giving an electrical *output signal* $y(t)$, which the output transducer converts to the final desired form. Notice that Figure 1.2 has a *cascade* or *open-loop* configuration in that the signals flow directly from left to right, passing successively through various system blocks but not doubling back. Hence, the output of any one block depends only on its input and not on the effect of subsequent units.

Generally, the objective of a communication system is to make the output signal $y(t)$ as faithful a reproduction of the input $x(t)$ as possible, subject to numerous technical and economic constraints. (However, the criteria for what constitutes an "acceptable" reproduction is somewhat ill-defined when human users are involved. Just look at a color TV picture that has been adjusted by somebody else!) Complicating this, the communications engineer is faced with equipment imperfections, over which he has some control, and various disturbances in the form of noise, interference, and changing transmission conditions, over which he has little control.

At first glance, a cure for some of these design problems might appear to be possible by comparing the output signal with the input, and using the resultant *error signal* as a correction factor. This approach would give a *closed-loop* or *feedback* configuration as depicted in Figure 1.3. Unfortunately, feedback from destination to source is seldom practical in a communication system since another communication system going in the reverse direction would be required to establish the closed-loop path.

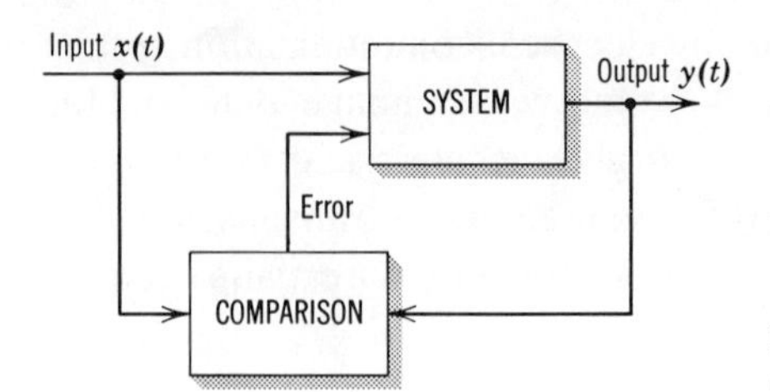

Figure 1.3. A closed-loop configuration for error compensation.

And aside from obvious economic drawbacks, this second system would suffer the same kinds of disturbances as in the forward path.

Because of the distance involved, the physical remoteness of the input and output, communication systems usually have an overall cascade structure and other techniques are used to reduce disturbances without depending on feedback around the entire system. These other techniques generally fall in the domain of *modulation* and *coding*. Thus, for example, frequency-modulation (FM) broadcasting was developed for its relative immunity from atmospheric static and noise, as compared with the older and simpler amplitude modulation (AM).

Control systems

Turning to control systems, where distance between input and output generally is not a factor, we find closed-loop configurations extensively employed. In fact, it is feedback that puts the *automatic* into automatic control.

Figure 1.4 illustrates this concept for a simple strip-steel finishing mill, where steel strip is passed through a set of rolls to obtain a desired thickness. A hydraulic unit provides the compressive force of the rolls and the resulting thickness is measured by a sensor such as a beta-ray counter. Under manual control, an operator watches the thickness readout meter and adjusts the hydraulic pressure according to a strategy based on his past experience and, perhaps, dependent on the way he happens to feel that day. Alternately, the system can be automated by electrically or mechanically comparing the thickness signal with a suitable reference signal and generating therefrom a control signal for the hydraulic unit.

Generalizing from this example we get the block diagram representation of a reasonably arbitrary feedback control system shown in Figure 1.5. The block marked *plant* or, equivalently, *process* is that part of the system which performs the useful work and needs to be controlled. To

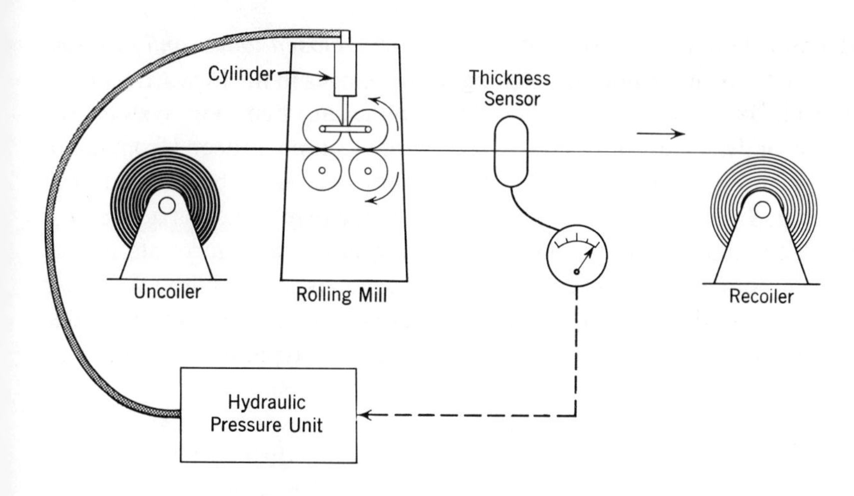

Figure 1.4. A rolling mill with thickness control.

affect this control, certain of the plant's output variables are monitored by appropriate sensors to generate signals that are compared with their desired values or *references*. The differences are passed to the *controller*, which dictates the "control strategy," i.e., the type of corrective action that must be applied to the plant. The *actuators* are those devices that can be manipulated so as to make the control signals affect the operation of the process; commonly, they govern the flow of energy or raw materials to the plant. Although this diagram indicates single signals flowing between blocks, actual systems generally involve a number of signals going into and out of each of the units.

Figure 1.5 applies both to manual and automatic operation since, in the manual case, the operator performs the functions of the comparator

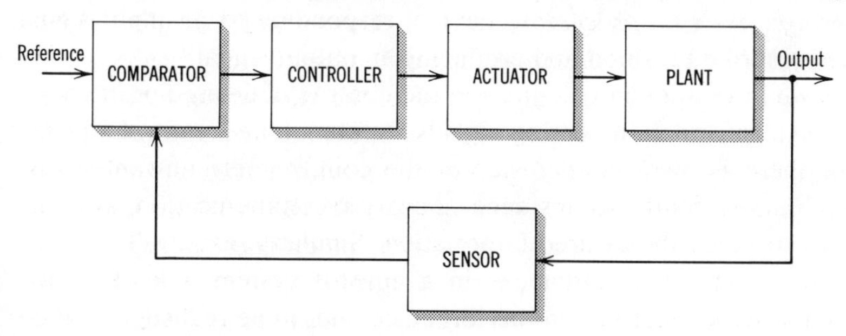

Figure 1.5. Elements of a feedback control system.

and controller. But in either case we have a *closed-loop configuration*, and variations in the output resulting from changes in the forward path are fed back for corrective action. Thus, unlike an open-loop system, the analysis or design of a feedback system cannot simply proceed from left to right, from input to output, because the output actually becomes part of the input. It is the presence or absence of feedback that, as a rule, distinguishes control versus communication systems, and accounts for the fact that somewhat different analysis techniques are required for these two areas. On the other hand, comparing Figures 1.2 and 1.5 reveals one very important similarity in the generalized block diagrams, as discussed below.

Signals

Both communication and control systems consist of blocks or subsystems that are excited by an input, respond to that input in some fashion, and produce a corresponding response at the output. Therefore, each block is characterized by an input–output or cause-and-effect relationship. By convention the inputs and outputs, or excitations and responses, are called *signals*, with the broad-sense interpretation that *a signal is any physical quantity that varies with time*. In a communication system, excluding the transducers, all signals are time-varying electrical quantities – voltage, current, etc. – which is the common notion of a signal. In a control system, however, a signal may be the pressure in a hydraulic line, the thickness of a steel strip, or, even more abstractly, the reference mark on a meter or the setting on a control dial. In any case, with our broader interpretation of signals, one can say quite simply that a *subsystem* is a *signal-processing* device, responding to an input signal in some prescribed fashion and producing an output signal.

So far we have spoken of signals in the context of desired or intended signals, e.g., information-bearing signals, reference and control signals, etc. But there is also another side of the coin, namely unwanted *disturbance signals*. Static, for instance, appears in communications systems and tends to mask the desired information. Similarly, externally caused perturbations act as disturbances in a control system. Clearly, such undesired signals affect system performance and, to be realistic, must be considered as part of the system.

Inherent limitations of physical systems

Many potential obstacles stand between a system design and its physical realization. Among these are *feasibility* constraints such as cost, complexity, state-of-the-art technological limitations, and so forth. But of more fundamental importance are the *inherent limitations* of physical systems themselves. Feasibility constraints have cures in theory, although they may not be practical; the inherent physical limitations, however, are dictated solely by Mother Nature and fall beyond the engineer's control. Let us briefly consider the problems imposed by system dynamics, time delay, and noise.

System dynamics. A system is said to be *instantaneous* if the response at any time depends only on the value of the excitation at that instant and not on past or future values. The output of a *dynamic* system, in contrast, generally depends on the entire past history of the input, and so a dynamic system may be thought of as having *memory.* This memory effect arises primarily because of *stored energy* in the system which cannot be altered or dissipated instantaneously. Thus, for example, if the excitation is turned off, the stored energy within the system continues to produce an output.

Real systems always have energy-storage elements – inductance and capacitance in electrical systems, inertia and compliance in mechanical systems, etc. – and therefore *physical systems are dynamic systems.* The dynamic nature means, for example, that a control system cannot respond immediately to a change in the reference signal, while for communication systems there will be some maximum rate above which information cannot be transmitted successfully. Another way of expressing this concept is to say that real systems have a finite upper frequency-response limit or a *finite bandwidth.*

Time delay. In addition to and independent of dynamic "sluggishness," physical systems also have time delay. This means that some nonzero amount of time must elapse before the output makes *any* change in response to a change at the input. Figure 1.6 illustrates the difference between typical dynamic response and pure time-delayed response for a simple off-on-off excitation.

Time delay is obviously present in communication systems because electrical signals travel at a finite velocity and therefore require time to

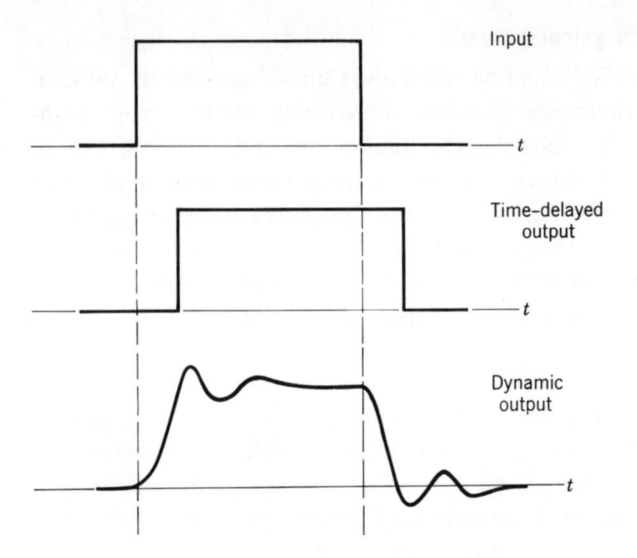

Figure 1.6. Time-delayed and dynamic response.

travel from transmitter to receiver. Fortunately, except for transcontinental and extraterrestrial transmission, this delay is usually negligible.

Another more serious source of time delay occurs in feedback systems when sensors cannot be placed directly at the point where the variables they are supposed to measure first appear; e.g., the thickness gauge in Figure 1.4 must be located some distance downstream of the rollers. Hence, there is a delay between the time when the output actually changes and when this change is sensed and acted upon. Meanwhile, if the output changes in the opposite direction, then the corrective action will be the reverse of what is needed. Taken to the extreme where output variations and corrections are exactly out of phase, the system may begin to oscillate and become *unstable*. Stability, in fact, is a major concern of the control system designer. Potential system instability has its roots in both the dynamic nature and the time delay of physical systems.

Noise. Noise is the term commonly used for the unpredictable or *random* signals that are always present in physical systems. Somewhat surprisingly, the inevitability of noise stems from the fact that real systems must operate at temperatures above absolute zero. Kinetic theory then tells us that all particles – electrons, molecules, etc. – will have thermal energy manifested as thermal agitation or random motion. Clearly, a randomly moving electron produces a random electrical signal; hence, there will always be noise in an electrical system. Likewise, no matter what type of system is considered, random particle motion will result in "noise" of one kind or another.

Historically, noise has been a serious factor only in communication systems. And at that, only when the received signal is very weak so as to be of the same order of magnitude as the electrical noise in the receiver. Under this circumstance, special techniques are required to distinguish the desired signal from the contaminating noise that tends to mask it, and complete noise elimination is impossible. Lately, however, noise is also becoming a problem in the more sophisticated control systems using highly sensitive and delicate sensors. This can lead to the unfortunate situation where the system attempts to control the spurious noise variations picked up by the sensor along with the actual output variations.

Of the three inherent limitations we have enumerated, the first two will receive extensive consideration in later chapters. The third one, noise, will be touched upon but not in any depth. The reason for this omission is that the study of systems analysis with random signals involves probability theory and statistics, and is best tackled after one has mastered systems analysis with nonrandom signals.

1.3 THE TOOLS OF THE TRADE

The ultimate task of systems engineering is getting the job done, bringing into fruition a working system that meets the stated requirements and does so in a practical fashion. This means that the engineer must be a very realistic type of person, having a firm grasp of the principles of physics coupled with sound economic judgment and perseverance. It also means that *mathematical tools* play a central role.

The reason for the importance of mathematics goes back to our earlier observation that sophisticated systems cannot be designed by seat-of-the-pants methods. A system must be conceived and refined on paper, not in the flesh. The design must be quantitatively expressed, and evaluated in a quantitative fashion. The only vehicle for such work is the language and techniques of mathematics.

Mathematical models: the "black box" system

Since a systems engineer is not as concerned with the internal workings as he is with overall performance, the entire system can be viewed as a signal processor, taking the basic input or reference signals – plus disturbances – and producing the final output signals. This leads to the abstract viewpoint illustrated in Figure 1.7, a "black box" system representation. It emphasizes the overall system characteristics, expressed mathematically, and says nothing about what is inside the box.

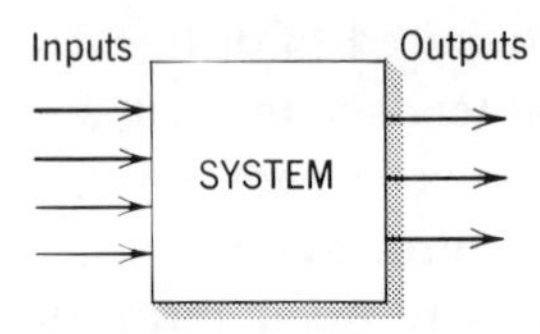

Figure 1.7. The black-box system representation.

Eventually, of course, implementation must be dealt with in terms of hardware. But implementation comes only after the overall specifications of the system have been delineated and tested to insure that the conceptual design meets the requirements. And, although we shall from time to time consider some illustrative physical components, the primary emphasis in this text is on systems and subsystems viewed as black boxes having input-output characteristics described by mathematical models.

Because systems engineering relies so heavily on such models, the reader must clearly understand from the start the distinction between models and actual systems. Specifically, *the model is not the system*; rather, it is a mathematical idealization that represents selected aspects of the system's behavior over a limited range of operation. If the model is a good one, we can predict from it, to a reasonable degree of accuracy, how the system will perform under expected conditions. However, no model can exactly represent a real system under all possible circumstances. Golomb (1968)† expressed these points another way:

> Basically, the model is like a *map*. It may obscure the complexities of the terrain, but it provides a simple enough picture to be grasped at a glance and is helpful in plotting a route from one point to another. But you will never strike oil by drilling through the map.

†Throughout the text, references will be indicated in this fashion, by author's last name and year of publication. Complete citations are listed under "References."

To fully appreciate why system models always have limited validity, one should actually go through the labor of modeling a system from its physical description, as will be illustrated in the next chapter. For the moment, however, it can be pointed out that models are generally derived from physical "laws," and these laws themselves are only mathematical approximations of the real world. Newton's laws of motion, for example, are valid only when relativistic effects can be ignored; Ohm's law applies only to ideal linear resistors; and so on. Thus, at the very basis, our models must be approximations to reality.

In addition, mathematical models can seldom include all the physical phenomena present in a system. Attempting to do so would result in a hopelessly complex mass of detail prohibiting useful manipulation and study. Obviously, further simplifying approximations are called for to obtain a model in practical form. These approximations frequently are based on the types of signals expected, rather than on the nature of the system alone. As a result, one system may have several different models depending on the intended application. When dealing with a transistor amplifier, for example, we might use the small-signal transistor model, the large-signal model, the high-frequency model, etc.

By the same kind of reasoning, it follows that the *signals* used in systems analysis and evaluation are also mathematical models or *representations*, not the exact physical variable. Indeed, whenever we write an expression, say for $x(t)$, we are actually making an approximation or idealization. Thus, signal representations are subject to the same limitations as the system models.

Mathematics and engineering

If systems engineering is a highly mathematical discipline, it is still first and foremost an *engineering* discipline. Putting the matter in proper perspective, the systems engineer uses mathematics as a tool, a means to an end. He must not, at the one extreme, become a slave to mathematical manipulation nor should he, at the other extreme, be reluctant to invoke advanced mathematical ideas if they seem helpful in a given problem – providing, of course, that he understands their limitations.

The final test is good engineering judgment and common sense. Using a powerful and complicated mathematical technique on a simple problem – using a piledriver where a tackhammer will do – is nearly as bad a practice as attempting a difficult problem with inadequate preparation.

Similarly, when mathematical results defy physical intuition, one must carefully examine the situation to determine whether the mathematics has been improperly applied or the intuitive reasoning is wrong. Thus, in this text we shall give equal weight to mathematical results and physical interpretation, the one complementing the other.

Several different kinds of analytic tools will be developed because systems engineering involves several different kinds of problems. Insofar as possible, the theoretical underpinnings of a given technique are discussed, at least to the point of the limitations, for without that knowledge the tool is useless. Nonetheless, a balance must be struck between rigor and vigor. After all, engineers are engineers, not professional mathematicians. When certain facts are stated without proof, saying "it can be shown that . . . ," we are not begging the issue but rather, in the interests of getting to the main point, we are relying on the fact that mathematicians have already provided rigorous justification.

Experimentation, simulation, and numerical methods

Along with a wide variety of mathematical tools, the systems engineer must have at his disposal the traditional engineering techniques of experimentation, simulation, and numerical analysis.

Suppose, for example, that a system model results in a formal expression for the output signal as a function of the input, but the expression cannot be solved conveniently. One way around this problem is to use *numerical methods*, expedited perhaps by a digital computer. Truly, as the reader is probably well aware, the design and evaluation of today's complex systems would be unthinkable without the help of high-speed computers. Furthermore, it is now not unusual to find computers actually built in as components of these systems. Obviously, digital computation and systems engineering go hand in hand. Therefore, although numerical methods and computer programming must be the subjects of other courses, we shall at least indicate their role in systems analysis and set up a few typical problems for numerical solution.

Another type of computer, the analog computer, is also used by systems engineers. Analog computers have the advantage that they can be made to *simulate* a system on a continuous basis – as distinguished from digital simulation which yields only a sampled and quantized approximation. Hence, for a system or subsystem of not extreme complexity, the performance can be evaluated under various conditions by simulation on the analog machine using an appropriate input signal generator. Also, since the analog-computer simulation can be carried out in "real time," which often cannot be done with present digital computers, it is practical to incorporate some of the actual system hardware into an analog simulation for increased realism. Using a simulation of the system, one can modify and improve a design experimentally with relative ease simply by adjusting machine parameters and repeating the solution. Moreover, simulation diagrams themselves offer further insight into system properties, and therefore will receive attention in later chapters.

Finally, it should not be necessary to argue for the importance of experimentation in systems engineering. During the analysis phase, for instance, the designer may have a specific component in mind and will conduct experiments to obtain a suitable model of that component for evaluation in the context of the entire system. And when the design is implemented, with actual devices replacing black boxes, laboratory measurements provide the only true assurance that the devices live up to the specifications of their mathematical counterparts.

1.4 SURVEYING THE TEXT

To give the reader a feeling for what lies ahead, and why, we briefly summarize the content of the remaining chapters, their relationships and objectives.

Chapter 2 develops some examples of system models to the point where the more abstract concepts and terminology of systems engineering can be appreciated, including the notion of the state of a system. The special properties of fixed linear systems are then explored using the new notation. Based on this material, the reader should be able to construct suitable models of simple physical systems, select state variables, and test the model for linearity and time invariance.

Methods of analyzing fixed linear systems, given their models, are introduced in Chapters 3 and 4 – by brute-force solution of differential equations in Chapter 3, and by the superposition integral in Chapter 4. Supporting topics encountered in the course of this work include system modes and stability, simulation diagrams, formulation of state equations, convolution, impulses, and the impulse response and transfer function of a system. Upon completion of these chapters the reader should be able to start from a system described by a differential equation and do the following things: identify the modes and classify the stability; write expressions for the complete response, the impulse response, and the transfer function; convert the model to a simulation diagram or a set of state equations; and, in the case of integrable expressions, carry out the convolution to obtain the response for a given input.

As the counterpart to Chapter 2, Chapter 5 deals with the mathematical representation of signals. The objective is to provide means for characterizing signals and expressing them as linear combinations of elementary functions, notably exponential functions. The viewpoint of treating functions as vectors in a signal space is introduced and applied to signal approximations and the theory of Fourier analysis. Although this background theory is supplementary and not strictly required for subsequent chapters, an understanding of it is helpful at our level and is essential for most advanced work in the field.

Under appropriate conditions, it is found that systems analysis can be greatly expedited by using transfer functions and the related frequency-domain approach. That approach is taken in Chapter 6 to handle periodic steady-state problems, with the aid of the Fourier series. It is broadened in Chapter 7, via the Fourier transform, and applied to some of the basic tasks of signal transmission, i.e., communication systems. The reader will learn the conditions under which frequency-domain analysis is valid and useful, the interpretation of signals and systems in the frequency domain, and how to set up frequency-domain solutions.

While the Fourier series and transform usually are sufficient for the study of steady-state response and open-loop signal transmission systems, the study of transient response and closed-loop control systems requires somewhat different analytic tools. Accordingly, Chapter 8 presents the Laplace transform, which is best suited to working with systems having feedback where transients and stability are critical concerns. Then, using Laplace transforms and transfer functions, Chapter 9 discusses the basic elements and characteristics of feedback systems, the discussion being illustrated by a satellite-attitude control system. At this point the reader should have gained the ability to analyze the transient behavior of fixed linear systems, both open- and closed-loop types, along with an understanding of the distinguishing features of feedback systems – whether they are in the context of automatic control, electronic amplifiers, or even economics or ecology.

The next two chapters are devoted to the most commonly used techniques of control-systems engineering: the root-locus method (Chapter 10) and the frequency-response methods (Chapter 11). Both techniques are applied to the satellite-attitude control system as a practical example. From this material the reader will learn how to design a stable single-loop system, selecting a control law and choosing parameter values.

Finally, the text concludes with an introduction to discrete-time signals and systems. Sampling theory and the properties of discrete signals are surveyed in Chapter 12, from which the reader should gain an understanding of the sampling process, its values and limitations. The dynamic response of systems describable by a discrete model is studied in Chapter 13 by using the z-transform. Illustrative applications in digital filtering and sampled-data control systems are included. Both chapters draw heavily on earlier material, especially the frequency-domain concept, thereby allowing one to utilize much of what has been learned about continuous systems for an understanding of discrete systems.

2

Systems and their models

Investigating a given system usually involves two kinds of tasks: formulating a mathematical model, and evaluating system performance as predicted by the model. Most of this book is devoted to the latter aspect and assumes that a suitable model has been obtained. This chapter, however, deals primarily with the question of modeling and its implications. We begin by considering the modeling process itself, illustrated with several examples. Two of the examples are then solved to yield explicit input-output relationships. With this background, the basic concepts and notation of modern systems engineering are introduced. Finally, we discuss the general properties of fixed linear systems – the class of mathematical models that receives most attention in later chapters – and we state just why they warrant extensive study.

2.1 INTRODUCTION TO MODELING

Modeling is the process whereby the physical properties of a system are expressed in a mathematical form suited to further analysis. Bear in mind, however, that a model is at best an approximation to physical reality, a mathematical idealization representing selected aspects of the system's behavior relevant to the problem at hand.

Here we cannot develop modeling techniques in depth, for a separate book would be required to cover the myriad types of system components. But we can introduce the subject by examining some relatively simple systems whose physical laws should be familiar. As preparation for these examples, we first outline the modeling process.

The modeling process

To model a system of moderate complexity, the following steps might typically be taken.

1. Enumerate the obvious elements or components of the system and determine their individual *describing equations*. Describing equations are the physical laws associated with each component, e.g., the voltage-current equation of an electrical device or the force-displacement equation for a mechanical element. If a describing equation is not known, one must fall back on experimental techniques and curve-fitting to obtain an appropriate relationship.

2. Look for the presence of not-so-obvious physical phenomena, discarding those having negligible influence. Stray capacitance, air drag, and temperature variation are examples of effects that may or may not be important in a given situation.

3. Write down the *combining equations* relating the variables of the individual elements to each other. These combining equations – based on circuit diagrams, free-body diagrams, etc. – reflect how the various elements are interconnected.

4. Using the combining equations, eliminate all variables that are not of interest, thereby obtaining a single equation or set of equations involving the pertinent variables and parameters. This constitutes the first cut at the model and, more often than not, results in *implicit* relationships, equations that must eventually be solved to get the unknown variables expressed directly in terms of the things that are known.

5. If possible, reduce the model's complexity by invoking any reasonable assumptions or approximations. In particular, if nonlinearities are present, test the model for possible linearization.

These steps are not necessarily performed in the order given, and some steps may be repeated several times as the model is refined and improved. Each step, of course, is subject to close scrutiny to determine whether it is valid for the case at hand, particularly those steps involving simplifications. Most important, throughout the process we are guided by the dual objective of obtaining a model that accurately represents the system under expected operating conditions but, at the same time, has a minimum of mathematical complexity.

For the case of systems containing relatively few components, a model may be written down by inspection without formally going through the above steps. For the case of very large or involved systems, it is usually best to break down the system into smaller blocks or subsystems, and model each one separately – taking account of interaction between them.

Armed with these ideas, plus a knowledge of elementary physics, let us now examine three simple but illustrative examples. Additional examples are presented in the course of the chapter.

Example 2.1 An RC circuit

Electrical systems consisting only of resistors, inductors, and capacitors are, perhaps, the easiest to model. The elements usually have linear properties over a wide range, and electric circuit diagrams lead more readily to systematic analysis then do graphical representations of other types of systems. Therefore, we begin our examples with the simple RC circuit of Figure 2.1, where the input signal $x(t)$ is a voltage source and the output $y(t) = e_C(t)$ is the open-circuit voltage across the capacitor.

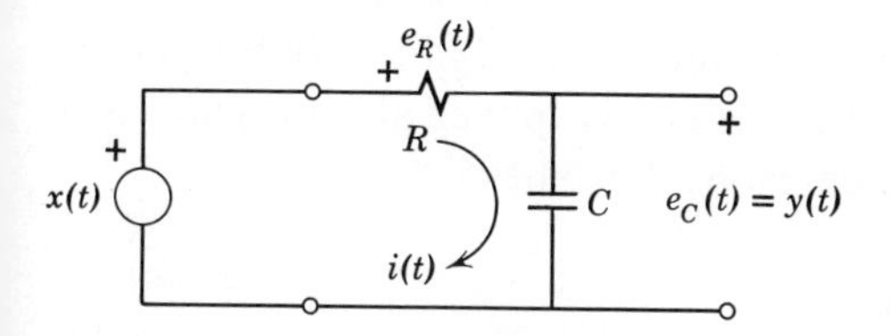

Figure 2.1. An RC circuit.

Assuming that the elements are ideal with constant parameters, the two describing equations are

$$e_R(t) = Ri(t) \qquad i(t) = C\frac{de_C(t)}{dt} = C\frac{dy(t)}{dt}$$

Note that the second equation is written in differential form rather than as the integral $e_C(t) = C^{-1} \int i(t)dt$; this practice has the advantage of leading directly to a differential equation, instead of an integro-differential equation, and will be followed in all our examples.

The appropriate combining equation is Kirchoff's voltage law. Thus

$$x(t) - e_R(t) - y(t) = 0$$

which, upon substituting for $e_R(t)$ and rearranging, gives

$$Ri(t) + y(t) = x(t)$$

Then, because the loop current is not of immediate interest, we eliminate it via the capacitor's describing equation to yield

$$RC\frac{dy(t)}{dt}+y(t)=x(t) \tag{1}$$

a first-order differential equation with constant coefficients. For convenience, and when there is no danger of misinterpretation, we shall frequently omit the independent variable t – writing y instead of $y(t)$, etc. – and indicate time derivatives by the dot convention $\dot{y} \triangleq dy(t)/dt$, $\ddot{y} \triangleq d^2y(t)/dt^2$, and so forth. In this notation (1) becomes

$$RC\dot{y}+y=x \tag{2}$$

Since, presumably, $x(t)$ is known and $y(t)$ is to be found, the above is an *implicit* relationship. Thus, the next logical step would be solving for $y(t)$ *explicitly* in terms of $x(t)$, a step we postpone until the next section. It should be observed, however, that the explicit solution of (2) will involve an unknown constant reflecting the *initial condition* of the system when the input is applied. Assuming that $x(t)$ is applied at some particular time, say $t=t_0$, the complete system model becomes

$$\begin{aligned} RC\dot{y}+y&=x \\ y(t_0)&=y_0 \end{aligned} \qquad t \geq t_0 \tag{3}$$

where y_0 is the voltage across the capacitor at time t_0.

This result is an *exact* mathematical description if the elements are *ideal.* But actually most of the modeling has already been done when we start from a circuit diagram, the diagram itself being a simplification of the physical system. For example, one can infer from Figure 2.1 that the input is expected to be a low-frequency signal. At high frequencies, say in the megahertz region, even a simple resistor requires a complicated equivalent circuit to account for stray capacitance and lead inductance, while at still higher frequencies, a distributed-parameter representation would be necessary rather than one with lumped elements. Other possible factors that apparently have been omitted from the diagram include output loading, capacitor leakage, element nonlinearities, etc. In short, our model is tailored to a somewhat restricted operating condition, albeit a common one.

Example 2.2 A high-speed vehicle

Figure 2.2 represents a high-speed vehicle having mass M and driven at a velocity $V(t)$ by a force $F(t)$. The driving force is opposed by bearing friction, proportional to velocity, and air drag, proportional to velocity

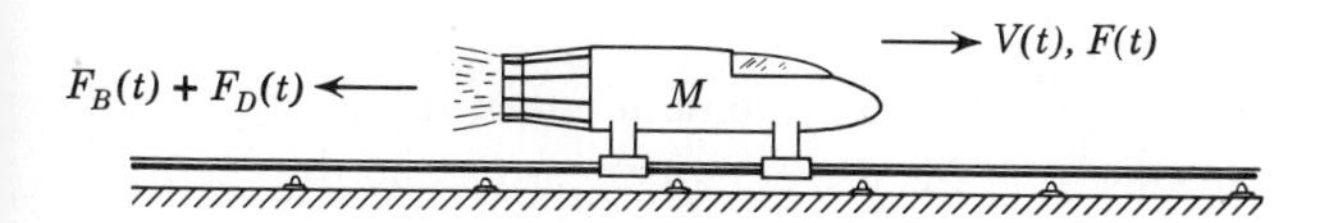

Figure 2.2. A high-speed vehicle.

squared. Beginning at time t_0 the driving force undergoes small variations so that

$$F(t) = \bar{F} + f(t) \qquad t \geq t_0$$

where $\bar{F}$ is a constant. The problem is to find the resulting velocity.

Although it is not so obvious in the figure, there are three "elements" in this system: the mass, the bearing friction, and the air drag. The corresponding describing equations for the latter two elements are the force-velocity relations

$$F_B(t) = BV(t) \qquad F_D(t) = DV^2(t)$$

and Newton's second law provides the combining equation

$$M\frac{dV}{dt} = F(t) - F_B(t) - F_D(t)$$

Substituting and rearranging gives

$$M\dot{V} + BV + DV^2 = F \tag{4}$$

Because of the V^2 term, (4) is a *nonlinear* differential equation, generally much more difficult to analyze than a linear one. Let us, therefore, look for a simplification taking advantage of the fact that the force variations $f(t)$ are "small." Specifically, since $F(t) = \bar{F} + f(t)$, it seems reasonable to write the velocity as

$$V(t) = \bar{V} + v(t) \qquad t \geq t_0$$

with $\bar{V}$ a constant. Making this substitution in (4) and noting that $\dot{V} = \dot{v}$, we have

$$M\dot{v}+B\bar{V}+Bv+D\bar{V}^2+2D\bar{V}v+Dv^2=\bar{F}+f$$

or

$$(B\bar{V}+D\bar{V}^2)+(M\dot{v}+Bv+2D\bar{V}v+Dv^2)=\bar{F}+f \tag{5}$$

where all terms not involving v or f are independent of time. This means we can write *two* equations, one for the constant terms and the other for the time variables, by the following argument.

Suppose there are no force or velocity variations, i.e., $f=v=\dot{v}=0$; then (5) becomes

$$B\bar{V}+D\bar{V}^2=\bar{F} \tag{6}$$

But all terms in this equation are constants, and therefore it must also hold true when the variations f and v are *not* zero. Accordingly, if (6) is inserted back into (5), all the time-independent terms cancel out, leaving

$$M\dot{v}+Bv+2D\bar{V}v+Dv^2=f \tag{7}$$

Clearly (6) determines the steady-state velocity in the absence of force variations, while (7) represents the velocity variations, relative to $\bar{V}$, caused by the force variations f.

At this point we make the assumption that $|Dv| \ll |B+2D\bar{V}|$ for all $t \geq t_0$, which constitutes the criterion for "small" variations. Then the nonlinear term Dv^2 can be dropped from (7), giving the *incremental* model

$$M\dot{v}+(B+2D\bar{V})v=f \qquad t \geq t_0 \tag{8}$$

which has the advantage of being a linear equation. To finish the stated problem, one solves (6) for $\bar{V}$ and substitutes the result into (8) which can then be solved for $v(t)$; then one checks to be sure that the linearizing assumption was justified. For our purposes, however, the modeling phase is completed when (8) has been obtained.

One final point should be drawn from this example, together with the previous one, namely that drastically different physical systems can have very similar or even identical mathematical models. Indeed, if we let $M=K(B+2D\bar{V})$ and $f(t)=(B+2D\bar{V})g(t)$, (8) reduces to $K\dot{v}+v=g$, whose form is identical to (2), the RC-circuit equation. Systems whose mathematical models have the same structure are said to be *analogs*. It is this observation that arouses interest in *analog computers* and *simulation* methods since, for instance, we can experimentally study the velocity variations of the high-speed vehicle system by simulating it in the form of an RC circuit. Simulation has many merits, conceptual as well as experimental, and will be considered again at a later point.

Example 2.3 A DC motor system

Unlike the prior examples, the DC motor system of Figure 2.3 involves two types of physical elements, electrical and mechanical. The motor,

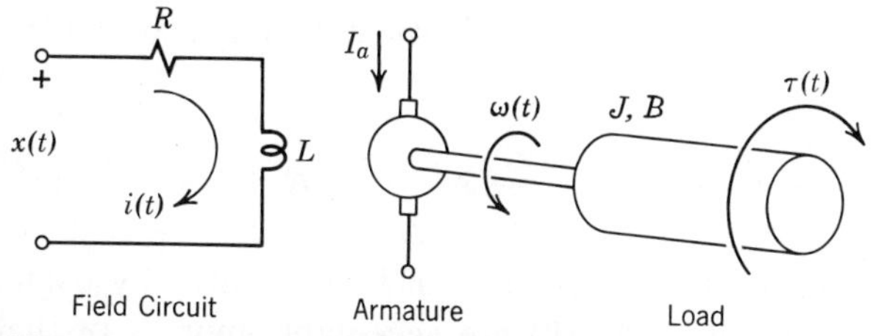

Figure 2.3. Elements of a field-controlled motor.

operating in the *field-controlled mode*, is excited by a field voltage $x(t)$ and a fixed armature current I_a. The rotating load has moment of inertia J, bearing friction with coefficient B, and is subject to a disturbance torque $\tau(t)$. This system, for instance, might represent a scanning antenna with wind loading.

Neglecting nonlinearities such as saturation, hysteresis, etc., the torque developed by the motor is proportional to I_a and the field current $i(t)$, so that

$$\tau_m = K_m I_a i = Ki$$

where $K = K_m I_a$ is constant, since I_a is constant. By inspection, the field circuit obeys

$$x = Ri + L di/dt$$

and, summing the torques,

$$\tau_m - (B\omega + \tau) = J d\omega/dt$$

where $\omega(t)$ is the angular velocity of the load.

Rewriting these equations by eliminating τ_m yields

$$\begin{aligned} L(di/dt) + Ri &= x \\ J(d\omega/dt) + B\omega &= Ki - \tau \end{aligned} \qquad t \geq t_0 \qquad (9)$$

with initial conditions $i(t_0) = i_0$ and $\omega(t_0) = \omega_0$. Incidentally, we point out that these initial conditions merely reflect the presence of *stored energy* in the system at time t_0: electromagnetic energy in the inductor, $E_L = (1/2)Li_0^2$, and kinetic energy in the rotating mass, $E_J = (1/2)J\omega_0^2$. (The energy associated with the armature current does not enter the problem simply because I_a is assumed to be constant.) The same observation about stored energy and initial conditions applies to Examples 2.1 and 2.2, as the reader can confirm.

Equation (9) expresses the output $\omega(t)$ in terms of the two inputs, $x(t)$ and $\tau(t)$, and an additional variable $i(t)$. Alternately, $i(t)$ also can be eliminated, giving

$$JL\ddot{\omega} + (BL + JR)\dot{\omega} + BR\omega = Kx - (L\dot{\tau} + R\tau) \tag{10}$$

whose derivation requires a little effort but can be expedited by using transform techniques. Both (9) and (10) are acceptable models of the system. The former, however, has several advantages, despite the fact that it is a set of simultaneous equations rather than a single equation. First, solving (9) gives us $i(t)$ as well as $\omega(t)$, and it may be important to have this additional information – excessive field current, for example, might drive the motor into saturation, thereby invalidating the model and possibly overheating the motor. Second, coupled first-order differential equations are no more difficult to solve than a single differential equation of higher order, and frequently they are easier. Third, there is little difficulty specifying the initial conditions for (9), whereas the required second initial condition for (10) is not immediately obvious. Finally, as we shall later see, the presence of $\dot{\tau}$ in (10) makes it awkward to simulate on an analog computer.

2.2 SOLVING THE MODEL: TWO EXAMPLES

The reader, no doubt, has tackled differential equations more difficult than those encountered in our modeling examples. In this section we shall obtain solutions by using the technique known as *variation of parameters*. Our purpose is to introduce the perspective of systems engineering and some new concepts that will pave the way for a more abstract discussion of systems analysis given later.

Example 2.4 Solution of a first-order system

Recall that the RC circuit in Example 2.1 led to the model $RC\dot{y} + y = x$; now, given $x(t)$ for $t \geq t_0$ and $y(t_0) = y_0$, we wish to solve for $y(t)$. To be more inclusive, let us consider the first-order differential equation

$$a_1 \dot{y} + a_0 y = bx \tag{1}$$

which includes the RC-circuit problem as a special case.

The first step toward solving (1) is to find the *complementary solution*, denoted by $y_H(t)$, from the related *homogeneous* equation

$$a_1 \dot{y} + a_0 y = 0 \tag{2}$$

which is the original *inhomogeneous* equation with the forcing function $bx(t)$ set to zero. Separating variables in (2) and integrating, we have

$$y_H(t) = Ae^{-(a_0/a_1)t} \tag{3}$$

where A incorporates the arbitrary constant of integration.

Turning to the inhomogeneous equation, the variation-of-parameters method continues by postulating that its solution has the form of $y_H(t)$ but with the constant A replaced by a function of time, say $\alpha(t)$, reflecting the effect of the forcing function. Thus, we assume the complete solution is

$$y(t) = \alpha(t)e^{pt} \qquad p = -\frac{a_0}{a_1} \tag{4}$$

with $\alpha(t)$ to be found from (1). Thus, substituting (4) for y in (1) and using the chain rule for differentiation, we get

$$a_1 \overbrace{(\dot{\alpha}e^{pt} + p\alpha e^{pt})}^{\dot{y}} + a_0 \overbrace{(\alpha e^{pt})}^{y} = bx$$

But since $p = -a_0/a_1$, this immediately simplifies to† $a_1 \dot{\alpha} e^{pt} = bx$, or

$$\frac{d\alpha(t)}{dt} = \frac{b}{a_1} e^{-pt} x(t)$$

which can be integrated to find $\alpha(t)$.

When performing this integration we should bear in mind that $x(t)$ is known only for $t \geq t_0$; whether or not it is zero for $t < t_0$ is immaterial, as

†For compactness, we continue to use p rather than $-a_0/a_1$ in the exponentials.

long as we have specified the value of $y(t)$ at t_0. Accordingly, rather than using the indefinite integral $\alpha(t) = (b/a_1) \int e^{-pt}x(t)dt$ it is better to write

$$\alpha(t) = \frac{b}{a_1}\int_{t_0}^{t} e^{-p\lambda}x(\lambda)d\lambda + \alpha(t_0) \tag{5}$$

with λ being a dummy variable of integration. By doing so, we remove from consideration all time $t < t_0$ – for which $x(t)$ is not given – and the initial condition drops out immediately as the integration constant $\alpha(t_0)$. With respect to the latter, setting $t = t_0$ in (4) shows that $y(t_0) = \alpha(t_0)e^{pt_0}$ and hence

$$\alpha(t_0) = y_0 e^{-pt_0} \tag{6}$$

Finally, inserting (5) and (6) into the assumed solution, we obtain,

$$y(t) = e^{pt}\overbrace{\left[\frac{b}{a_1}\int_{t_0}^{t} e^{-p\lambda}x(\lambda)d\lambda + \alpha(t_0)\right]}^{\alpha(t)}$$

$$= \int_{t_0}^{t} x(\lambda)\frac{b}{a_1}e^{p(t-\lambda)}d\lambda + y_0 e^{p(t-t_0)} \tag{7}$$

where $(b/a_1)e^{pt}$ has been brought into the integral – an odd but nonetheless valid step since it is independent of λ. Implicit in this result is that it applies only for $t \geq t_0$. Similarly, returning to the original RC-circuit problem (i.e., $a_1 = RC$, $a_0 = b = 1$), its solution is

$$y(t) = \int_{t_0}^{t} x(\lambda)\frac{1}{RC}e^{-(t-\lambda)/RC}d\lambda + y_0 e^{-(t-t_0)/RC} \tag{8}$$

where $p = -a_0/a_1 = -1/RC$ has been written out explicitly.

Several important observations can be made from (8) together with the original circuit – redrawn for convenience in Figure 2.4*a*. First, suppose $x(t) = 0$ for all time; then $x(\lambda) = 0$ and

$$y(t) = y_0 e^{-(t-t_0)/RC} \qquad t \geq t_0 \tag{9}$$

Physically, a voltage source $x(t)$ with zero output is simply a short circuit and, since the capacitor has an initial charge at time t_0, $y(t)$ represents the voltage across the capacitor as it discharges back through the resistor, as shown in Figure 2.4*b*. Now recall that the discharge characteristic or *natural behavior* of a simple RC circuit is an exponential time function with a time constant equal to RC; that is, if the discharge starts at $t = 0$ we would have $y(t) = y_0 e^{-t/RC}$. But in the case at hand the initial time is $t = t_0$, so the discharge starts at time t_0, as sketched in Figure 2.4*c*. Equation (9) says precisely that.

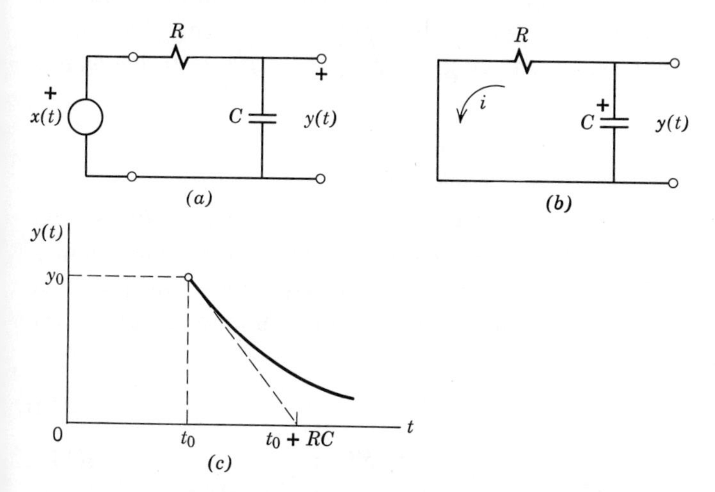

Figure 2.4. (*a*) Circuit diagram. (*b*) Natural behavior conditions, $x(t) = 0$. (*c*) Natural behavior waveform.

On the other hand, suppose the capacitor has zero charge at $t = t_0$, i.e., $y_0 = 0$. Equation (8), for this case, becomes

$$y(t) = \int_{t_0}^{t} x(\lambda) \frac{1}{RC} e^{-(t-\lambda)/RC} d\lambda \qquad t \geq t_0 \tag{10}$$

and, quite logically, $y(t)$ is determined only by $x(t)$ and the characteristic of the circuit. Furthermore, the circuit characteristic as it appears in this integral has the same kind of time dependence as the natural behavior. Underscoring this point, let us define a new time function

$$\eta(t) \triangleq \frac{1}{RC} e^{-t/RC}$$

which would be the natural behavior for an initial condition of $y_0 = 1/RC$ at $t = 0$. We then have

$$y(t) = \int_{t_0}^{t} x(\lambda) \eta(t - \lambda) d\lambda$$

where $\eta(t-\lambda)$ is $\eta(t)$ shifted or delayed by λ seconds. Notice, however, that $y(t)$ here is the response to $x(t)$ with no initial charge on the capacitor; the presence of $\eta(t-\lambda)$ in the integral accounts for the circuit characteristic, not an initial condition.

Finally, going back to (8), we see that the complete solution is the sum of two components, one due only to the input $x(t)$, the other due only to the initial condition. Systems having this property are said to be *decomposable.*

Example 2.5 Solution of a second-order system

The DC motor system of Example 2.3 is called a second-order system because its model is a second-order differential equation or, equivalently, a set of two simultaneous first-order differential equations. Since we have already solved a first-order system, let us try the same technique on the set of equations

$$L(di/dt)+Ri=x \tag{11a}$$

$$J(d\omega/dt)+B\omega=Ki-\tau \tag{11b}$$

with $i(t_0)=i_0$ and $\omega(t_0)=\omega_0$.

As it happens, the result for $i(t)$ can be set down right away by comparing (11a) with (1). Hence, substituting appropriately in (7),

$$i(t)=\int_{t_0}^{t} x(\lambda)\frac{1}{L}e^{p_1(t-\lambda)}d\lambda+i_0e^{p_1(t-t_0)} \tag{12}$$

where $p_1=-R/L$ is recognized as the familiar time constant of an RL circuit. In fact, with minor wording changes, all our remarks about (8) could be repeated here for (12).

The same solution technique applies as well to (11b) if one treats the entire right-hand side as the forcing function, i.e., letting

$$v(t)=Ki(t)-\tau(t)$$

it follows that

$$\omega(t)=\int_{t_0}^{t} v(\lambda)\frac{1}{J}e^{p_2(t-\lambda)}d\lambda+\omega_0e^{p_2(t-t_0)} \tag{13}$$

where $p_2=-B/J$.

From a formal viewpoint the problem is now complete. But a systems engineer would rather have $\omega(t)$ expressed directly in terms of the two inputs $x(t)$ and $\tau(t)$, rather than the form of (13). For this purpose, consider the integral

$$\mathcal{I}(t) = \int_{t_0}^{t} v(\lambda)\frac{1}{J}e^{p_2(t-\lambda)}d\lambda$$
$$= \int_{t_0}^{t} [Ki(\mu) - \tau(\mu)]\frac{1}{J}e^{p_2(t-\mu)}d\mu$$

which is the first part of (13) with $v(\lambda)$ written out and the dummy variable changed from λ to μ for clarity. Referring to (12), $i(\mu)$ is simply $i(t)$ with μ replacing t wherever it appears, i.e.,

$$i(\mu) = \int_{t_0}^{\mu} x(\lambda)\frac{1}{L}e^{p_1(\mu-\lambda)}d\lambda + i_0 e^{p_1(\mu-t_0)}$$

Thus, $\mathcal{I}(t)$ becomes a sum of three integrals, say

$$\mathcal{I}(t) = \mathcal{I}_1(t) + \mathcal{I}_2(t) + \mathcal{I}_3(t)$$

where

$$\mathcal{I}_1(t) = \int_{t_0}^{t}\left[\int_{t_0}^{\mu} x(\lambda)e^{p_1(\mu-\lambda)}\frac{K}{JL}e^{p_2(t-\mu)}d\lambda\right]d\mu \qquad (14a)$$

$$\mathcal{I}_2(t) = \int_{t_0}^{t}\frac{K}{J}i_0 e^{p_1(\lambda-t_0)}e^{p_2(t-\lambda)}d\lambda \qquad (14b)$$

$$\mathcal{I}_3(t) = -\int_{t_0}^{t}\tau(\lambda)\frac{1}{J}e^{p_2(t-\lambda)}d\lambda \qquad (14c)$$

Notice the importance of distinguishing the two dummy variables in (14a); but there is just one variable of integration in the other two, and we can relabel it as λ without confusion.

The integral $\mathcal{I}_3(t)$ admits of no further simplification, while the integration for $\mathcal{I}_2(t)$ is routine and yields

$$\mathcal{I}_2(t) = \frac{K}{J(p_1 - p_2)}i_0[e^{p_1(t-t_0)} - e^{p_2(t-t_0)}]$$

However, with the quantity $(p_1 - p_2)$ in the denominator, the result is valid only if $p_1 \neq p_2$ or $R/L \neq B/J$. In the rather special case when these two time constants are equal, the solution must be modified in a manner that will be covered in the next chapter.

As to $\mathcal{I}_1(t)$, one integration can be performed by taking advantage of the handy formula†

$$\int_a^b \left[\int_a^\mu w(\mu, \lambda) d\lambda \right] d\mu = \int_a^b \left[\int_\lambda^b w(\mu, \lambda) d\mu \right] d\lambda \tag{15}$$

where $w(\mu, \lambda)$ is any function of the two variables μ and λ, and a and b are constants. Applied to the case at hand, (14a) becomes

$$\begin{aligned} \mathcal{I}_1(t) &= \int_{t_0}^t \left[\int_\lambda^t x(\lambda) e^{p_1(\mu-\lambda)} \frac{K}{JL} e^{p_2(t-\mu)} d\mu \right] d\lambda \\ &= \int_{t_0}^t x(\lambda) \frac{K}{JL(p_1-p_2)} [e^{p_1(t-\lambda)} - e^{p_2(t-\lambda)}] \, d\lambda \end{aligned}$$

which, like $\mathcal{I}_2(t)$, involves both natural-behavior functions, $e^{p_1 t}$ and $e^{p_2 t}$.

Inserting these results and rearranging, the final result for $\omega(t)$ is

$$\begin{aligned} \omega(t) &= \mathcal{I}(t) + \omega_0 e^{p_2(t-t_0)} \\ &= \int_{t_0}^t x(\lambda) \frac{K}{JL(p_1-p_2)} [e^{p_1(t-\lambda)} - e^{p_2(t-\lambda)}] \, d\lambda \\ &\quad - \int_{t_0}^t \tau(\lambda) \frac{1}{J} e^{p_2(t-\lambda)} d\lambda + \omega_0 e^{p_2(t-t_0)} + \frac{K}{J(p_1-p_2)} i_0 [e^{p_1(t-t_0)} - e^{p_2(t-t_0)}] \end{aligned} \tag{16}$$

where

$$p_1 = -R/L \qquad p_2 = -B/J \qquad p_1 \neq p_2$$

Although it appears formidable at first, (16) is interpreted in a fashion similar to that of the RC-circuit solution. The first term clearly represents the effect of the input field voltage $x(t)$, while the second describes the influence of the disturbance torque $\tau(t)$. They have different forms because the two inputs enter at different points in the system – which explains why only the mechanical parameters appear in the torque integral whereas both mechanical and electrical parameters are in the voltage integral. Likewise, the last two terms are the natural behaviors due to an initial angular velocity and an initial field current, respectively.

Not to mislead the reader, we point out that this particular problem was easily handled because the first-order equations are not fully coupled:

†The reader can justify this relationship by sketching the integration boundary and differential strips in the μ-λ plane.

that is, $i(t)$ is in the equation for $\omega(t)$, (11b), but $\omega(t)$ is *not* in the equation for $i(t)$, (11a). Hence, it was possible to solve independently for $i(t)$ and then, using that result, to solve for $\omega(t)$. Had we been dealing with an *armature-controlled* motor system, for instance, a simultaneous solution would have been required. That type of problem will be treated in Chapter 3.

2.3 BASIC CONCEPTS AND NOTATION

The few examples we have given barely begin to scratch the surface of the many different system types likely to be encountered in practice. And, clearly, it would be impossible to cover all of them here. We can, however, seek out and examine the basic concepts that are common to most systems, regardless of type. This, together with the introduction of some new notation, is the purpose of the present section. Of necessity, we shall use rather abstract terms to preserve as much generality as possible. Our previous examples should help the reader in relating the abstract to the concrete.†

System equations

The time-honored black-box representation of a single-input-output (SIO) system is shown in Figure 2.5. The excitation is an input signal

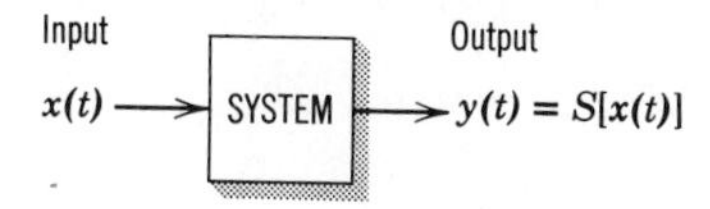

Figure 2.5. SIO system diagram.

$x(t)$, and the corresponding response is the output signal $y(t)$. Presumably, $y(t)$ depends in some fashion on $x(t)$; we shall express this symbolically as

$$y(t) = S[x(t)] \tag{1}$$

to indicate that there is a cause-and-effect relationship between excitation and response, denoted by the *function S*.

Equation (1), called an output equation, may be a sufficient model for

†Desoer and Kuh (1969) may be consulted for further discussion in the specific context of electrical circuits.

many cases, but it overlooks one important aspect: *initial conditions*. As we have seen, initial conditions in the form of stored energy within the system – e.g., a charged capacitor or a rotating mass – also contribute to the output. Thus, if a system has n independent initial conditions, say $q_1, q_2, \ldots q_n$, at $t = t_0$, then the output equation should take the form

$$y(t) = S[q_1, q_2, \ldots q_n; x(t)] \qquad t \geq t_0 \tag{2}$$

where S is now a function of $n+1$ variables, the input plus the n initial conditions. We say that the q's define the *initial state* of the system and, because there are n of them, the system is called an nth-order system.

Common though they may be, SIO systems are just a special case of the multi-input-output (MIO) system depicted by Figure 2.6. Here the

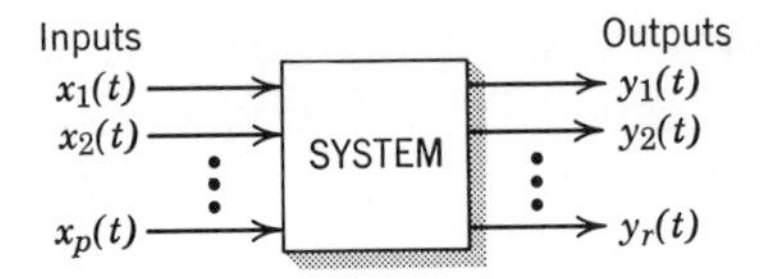

Figure 2.6. MIO system diagram.

excitation consists of p input signals, $x_1(t)$, $x_2(t), \ldots x_p(t)$, and the response is the set of r output signals, $y_1(t)$, $y_2(t), \ldots y_r(t)$. Since each output may depend on all the inputs and the initial state, we must write not one but r equations, one for each output. Thus

$$\begin{aligned} y_1(t) &= S_1[q_1, q_2, \ldots q_n; x_1(t), x_2(t), \ldots x_p(t)] \\ y_2(t) &= S_2[q_1, q_2, \ldots q_n; x_1(t), x_2(t), \ldots x_p(t)] \\ &\vdots \\ y_r(t) &= S_r[q_1, q_2, \ldots q_n; x_1(t), x_2(t), \ldots x_p(t)] \end{aligned} \tag{3}$$

where there will be a different function S_i for each $y_i(t)$. Obviously, (3) is not only tedious to write out but also is cumbersome to manipulate and difficult to interpret. In answer to this problem, many familiar words and notational forms have taken on somewhat different or restricted meanings in the context of systems engineering. Because these terms permit increased precision and compactness in discussions to come, the more important ones must be introduced here.

As implied in the first chapter, we use the term *signal* to refer to any time-varying physical quantity. The mathematical representation of a signal is, therefore, a function of t – say $v(t)$. The value of a physical variable must, of course, be a real number. Even so, $v(t)$ is allowed to be complex, if desired, providing some operation on it – usually $\text{Re}[v(t)]$ – yields an appropriate real time function. However, one thing $v(t)$ cannot be is *multivalued*, i.e., for each value of t there is only one value for $v(t)$. Thus, our usage of the term *function* will be in accord with the standard mathematical definition.

Now suppose we have to deal with several signals at once, say $v_1(t)$, $v_2(t), \ldots v_m(t)$. A convenient way to handle them is the *matrix* notation†

$$\mathbf{v}(t) \triangleq \begin{bmatrix} v_1(t) \\ v_2(t) \\ \cdot \\ \cdot \\ v_m(t) \end{bmatrix} \tag{4}$$

which is called an *m-dimensional vector*. By convention, signal vectors are defined in column format as shown. However, to save space, they are often written out as the *matrix transpose* of a row vector, denoted by a superscript T; e.g.,

$$\mathbf{v}(t) = [v_1(t) \quad v_2(t) \cdots v_m(t)]^T$$

is equivalent to (4)

There is nothing particularly exotic or mysterious about signal vectors. For the present they should be viewed as just a notational shorthand having mathematical properties identical to ordinary column matrices, and the operations of scalar multiplication and vector addition are defined in the usual fashion. Signal vectors do differ from ordinary matrices in that their elements are functions of time, generally, instead of constants. Thus, we can speak of the time derivative of $\mathbf{v}(t)$, written as

$$\dot{\mathbf{v}} \triangleq \frac{d\mathbf{v}(t)}{dt} = [\dot{v}_1 \quad \dot{v}_2 \cdots \dot{v}_m]^T \tag{5}$$

where $\dot{v}_1 = dv_1(t)/dt$, etc. If it happens that all the elements of $\mathbf{v}(t)$ are constant, then $\dot{\mathbf{v}} = \mathbf{0}$, the symbol $\mathbf{0}$ standing for

$$\mathbf{0} = [0 \quad 0 \cdots 0]^T \tag{6}$$

which is called the *zero vector*.

†Readers who are not familiar with matrix notation and manipulations should consult any of the standard texts on the subject. Dorf (1969), a self-study introduction, is recommended.

To get back to the main subject, consider an nth-order system with one input and one output. Writing the state at t_0 as a vector with constant elements

$$\mathbf{q}(t_0) \triangleq [q_1(t_0) \quad q_2(t_0) \cdots q_n(t_0)]^T \tag{7}$$

the output equation, (2), becomes

$$y(t) = S[\mathbf{q}(t_0); x(t)] \qquad t \geq t_0 \tag{8}$$

In the language of mathematics, the function S is said to transform or map the vector $\mathbf{q}(t_0)$ and the scalar $x(t)$ into the scalar $y(t)$ for each value of $t \geq t_0$.

Extending these notions to the MIO case, we define input and output signal vectors, $\mathbf{x}(t)$ and $\mathbf{y}(t)$, by

$$\begin{aligned} \mathbf{x}(t) &\triangleq [x_1(t) \quad x_2(t) \cdots x_p(t)]^T \\ \mathbf{y}(t) &\triangleq [y_1(t) \quad y_2(t) \cdots y_r(t)]^T \end{aligned} \tag{9}$$

and write the output equation as

$$\mathbf{y}(t) = \mathbf{S}[\mathbf{q}(t_0);\ \mathbf{x}(t)] \qquad t \geq t_0 \tag{10}$$

where $\mathbf{S}$, now a *vector function*, transforms vectors into vectors. Again, this is just notational convenience, (10) being a compact way of writing the set of r equations in (3).

Zero-input and zero-state response

Recall that in Section 2.2 we found that the system response was decomposable into two terms: one due only to the input, the other due only to the initial state. Although this simple form does not always hold – with a nonlinear system, for instance, there generally will be additional terms – it is nonetheless important to distinguish between these two output components.

Specifically, we define the *zero-input response* $\mathbf{y}_{zi}(t)$ as the system output when all inputs are zero but the initial state is nonzero. Formally, setting $\mathbf{x}(t) = \mathbf{0}$ in the output equation,

$$\mathbf{y}_{zi}(t) \triangleq \mathbf{S}[\mathbf{q}(t_0); \mathbf{0}] \qquad t \geq t_0 \tag{11}$$

Thus, $\mathbf{y}_{zi}(t)$ represents the response caused by the initial state, notably the internal stored energy, and clearly relates to the system's *natural behavior* or *free* response.

On the other hand, the zero-initial-state response is defined as that which results when $\mathbf{x}(t) \neq \mathbf{0}$ and $\mathbf{q}(t_0) = \mathbf{0}$. Hereafter, this concept is referred to as the *zero-state response*, symbolized by

$$\mathbf{y}_{zs}(t) \triangleq \mathbf{S}[\mathbf{0};\ \mathbf{x}(t)] \qquad t \geq t_0 \tag{12}$$

Thus, $\mathbf{y}_{zs}(t)$ is the system's *forced* response.

To be totally correct, the definition in (12) must be qualified by the additional assumption that the system is at rest or in equilibrium with zero output when both the initial state and input are zero, i.e.,

$$\mathbf{S}[\mathbf{0};\mathbf{0}] = \mathbf{0} \qquad t \geq t_0 \tag{13}$$

The vast majority of systems do satisfy (13), particularly when one is dealing with passive systems or incremental variables. For a discussion of the exceptions, along with other related definitions, see Zadeh and Desoer (1963, Chap. 2).

These concepts become most meaningful when $\mathbf{S}[\mathbf{q}(t_0);\mathbf{x}(t)]$ is decomposable for, as illustrated by Examples 2.4 and 2.5, the complete response is then

$$\mathbf{y}(t) = \mathbf{y}_{zs}(t) + \mathbf{y}_{zi}(t) \tag{14a}$$

or

$$\mathbf{S}[\mathbf{q}(t_0);\ \mathbf{x}(t)] = \mathbf{S}[\mathbf{0};\ \mathbf{x}(t)] + \mathbf{S}[\mathbf{q}(t_0);\ \mathbf{0}] \tag{14b}$$

But, to reiterate, this holds in general only for *linear* systems.

State variables

An interesting and significant feature of the output equation, (10), is that the past history, $t < t_0$, does not appear. Regardless of whether the input was zero for $t < t_0$, the output is completely determined for $t \geq t_0$ by the initial state $\mathbf{q}(t_0)$, the input $\mathbf{x}(t)$, and the system model $\mathbf{S}$. In this light, $\mathbf{q}(t_0)$ summarizes all that has gone on in the past as it affects future behavior.

Now consider some later time $t_1 > t_0$. Taking t_1 as a new initial time, it follows that $\mathbf{y}(t) = \mathbf{S}[\mathbf{q}(t_1);\mathbf{x}(t)]$ for $t \geq t_1$, where $\mathbf{q}(t_1)$ is the new initial state. Since, generally, the state has changed in going from t_0 to t_1, we can speak of a *state vector*, which is a function of time,

$$\mathbf{q}(t) \triangleq [q_1(t) \quad q_2(t) \cdots q_n(t)]^T \tag{15}$$

whose n components are known as the *state variables* of the system.

One property of the state vector $\mathbf{q}(t)$ is that, for any specific instant t, it completely describes the internal state of the system at that time. Expanding on this concept, we say that:

> The *state of a system* at any time t_0 is the minimum amount of information which, together with the system model and specification of the input for $t \geq t_0$, uniquely determines the state at any later time.

The essential aspect implied by this apparently circular definition is that $\mathbf{q}(t)$ provides just enough information, and no more, to immediately evaluate any desired system variable. Consequently, n, the number of components in the state vector, is an important system characteristic. Previously we said that that number equals the order of the system; now we say:

> The *order* of a system is defined as the number of components in the state vector.

Defining system order in this fashion merely gets the horse back in front of the cart.

Analytically, for $\mathbf{q}(t)$ to qualify as a state vector, it must satisfy an explicit *state equation*

$$\mathbf{q}(t) = \mathbf{S}_q[\mathbf{q}(t_0);\ \mathbf{x}(t)] \qquad t \geq t_0 \tag{16}$$

as well as the *output equation*, $\mathbf{y}(t) = \mathbf{S}[\mathbf{q}(t_0); \mathbf{x}(t)]$. The state function $\mathbf{S}_q$ should not be confused with the output function $\mathbf{S}$. The distinction between these two is simply that $\mathbf{S}$ gives the output signals while $\mathbf{S}_q$ gives the internal state variables, some of which may be considered outputs, but not necessarily. Taken together, they constitute a complete mathematical description of the system since any signal of interest, be it internal or an output, can be computed from one or the other. If an engineer is only concerned with a system's input-output characteristics, as in (10), then (16) need not be considered. Unfortunately, this black-box viewpoint sometimes proves disasterous, ignoring as it does the internal behavior.

The state concept is illustrated in the following example, and further developed in the next chapter. For the moment, we add that, just as a given system can have several mathematical models, there may be several different state representations. In other words, the state vector is *not unique.*

Example 2.6 A tuned electronic amplifier

Figure 2.7 shows the small-signal equivalent circuit of an electronic amplifier with a parallel-tuned (resonant) LC circuit. The input $x(t)$

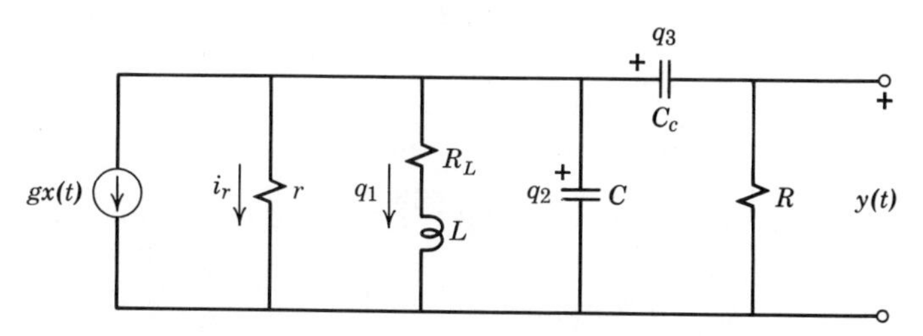

Figure 2.7. Equivalent circuit of a tuned electronic amplifier.

appears here in the controlled current source $i(t) = gx(t)$, while the output is the voltage across the resistor in the coupling network composed of R and C_c.

To select the state variables, we use the fact that they must define the initial conditions which, in turn, are related to the stored energy. Since there are three possible energies, namely

$$E_L = \tfrac{1}{2}Li_L^2 \qquad E_C = \tfrac{1}{2}Ce_C^2 \qquad E_{C_c} = \tfrac{1}{2}C_c e_{C_c}^2$$

let us try

$$q_1(t) = i_L(t) \qquad q_2(t) = e_C(t) \qquad q_3(t) = e_{C_c}(t)$$

as indicated on the diagram. A little thought should convince the reader that, given these variables, any other variable in the system can be computed directly from an *algebraic* equation.

For the combining equation, Kirchhoff's current law says that

$$gx + i_r + i_L + i_C + i_{C_c} = 0$$

where $i_r = q_2/r$, $i_L = q_1$, $i_C = C\dot{q}_2$, and $i_{C_c} = C_c\dot{q}_3$. In addition, we see that

$$q_2 = R_L q_1 + L\dot{q}_1 \qquad y = q_2 - q_3$$

and

$$C_c \dot{q}_3 = \frac{y}{R} = \frac{q_2 - q_3}{R}$$

After routine manipulation one obtains three simultaneous first-order differential equations

$$\begin{aligned} L\dot{q}_1 + R_L q_1 &= q_2 \\ C\dot{q}_2 + \frac{R+r}{Rr} q_2 &= -q_1 + \frac{1}{R} q_3 - gx \\ C_c \dot{q}_3 + \frac{1}{R} q_3 &= \frac{1}{R} q_2 \end{aligned} \tag{17}$$

This set constitutes the state equation, and the appropriate initial conditions are $q_1(t_0) = i_L(t_0)$, $q_2(t_0) = e_C(t_0)$, and $q_3(t_0) = e_{C_c}(t_0)$. The output equation, in terms of the state variables, is simply

$$y(t) = q_2(t) - q_3(t) \tag{18}$$

Two points should be made with respect to this example. First, the state-variable approach has appreciably simplified the modeling process and, in retrospect, is the same technique we used to model the DC motor system of Example 2.3. Second, the resulting state and output equations are in implicit form, rather than the explicit functional form of (16) and (10). Thus, although we assure the reader that our choice of state variables is valid, proving this assertion requires that one solve (17) and (18) explicitly for $\mathbf{q}(t)$ and $y(t)$ in terms of $\mathbf{q}(t_0)$ and $x(t)$.

2.4 FIXED LINEAR SYSTEMS

This textbook emphasizes the analysis of systems whose mathematical models are classified as being *linear* and *time-invariant*. For short, they will be called *fixed linear systems*. There are four reasons why this special class deserves extensive treatment.

1. Many common systems have models that, at least for a small signal approximation, are linear and time-invariant.

2. Frequently the properties characterizing fixed linear systems are precisely those desired for a given application.

3. From a very pragmatic viewpoint, fixed linear systems are easier to analyze than other types.

4. A firm understanding of fixed linear systems is the best preparation for tackling more difficult problems, particularly since some of the analytic tools can be applied, with modification, to nonlinear or time-varying systems.

Therefore, we shall conclude this chapter with a general discussion of fixed linear systems.

Linear systems

Defining a linear system requires first the definition of a linear function. The function $S(x)$ is said to be linear if, for any constant α,

$$S(\alpha x) = \alpha S(x)$$

and if, with $x = v + w$,

$$S(v+w) = S(v) + S(w)$$

These properties, known as *homogeneity* and *additivity*, are compactly combined in the familiar principle of *superposition*

$$S(\alpha v + \beta w) = \alpha S(v) + \beta S(w) \tag{1}$$

for any constants α and β.

Similarly, a vector function $\mathbf{S}(\mathbf{x})$ is linear if replacing $\mathbf{x}$ by

$$\alpha\mathbf{v} + \beta\mathbf{w} = [\alpha v_1 + \beta w_1 \quad \alpha v_2 + \beta w_2 \quad \cdots \;]^T$$

yields

$$\mathbf{S}(\alpha\mathbf{v} + \beta\mathbf{w}) = \alpha\mathbf{S}(\mathbf{v}) + \beta\mathbf{S}(\mathbf{w}) \tag{2}$$

where α and β are scalar constants. Interpreting this definition, we recall that a vector function represents a *set* of equations, like Eq. (3), Sect. 2.3, so (2) means that each component function $S_i(\mathbf{x}) = S_i(x_1, x_2, \ldots)$ satisfies

$$S_i(\alpha\mathbf{v} + \beta\mathbf{w}) = \alpha S_i(v_1, v_2, \ldots) + \beta S_i(w_1, w_2, \ldots)$$

Turning now to linear systems, there are two different functions to consider: the zero-state and zero-input responses. Thus, we define a *linear system* as one whose zero-state response is a linear function of the input, whose zero-input response is a linear function of the initial state, and whose complete response is decomposable. Expressed symbolically,

$$\mathbf{S}[\mathbf{0}; \alpha\mathbf{v}(t) + \beta\mathbf{w}(t)] = \alpha\mathbf{S}[\mathbf{0}; \mathbf{v}(t)] + \beta\mathbf{S}[\mathbf{0}; \mathbf{w}(t)] \tag{3a}$$

$$\mathbf{S}[\alpha\mathbf{q}(t_0) + \beta\mathbf{p}(t_0); \mathbf{0}] = \alpha\mathbf{S}[\mathbf{q}(t_0); \mathbf{0}] + \beta\mathbf{S}[\mathbf{p}(t_0); \mathbf{0}] \tag{3b}$$

$$\mathbf{S}[\mathbf{q}(t_0); \mathbf{x}(t)] = \mathbf{S}[\mathbf{0}; \mathbf{x}(t)] + \mathbf{S}[\mathbf{q}(t_0); \mathbf{0}] \tag{3c}$$

for the general MIO case.

Equation (3) is a more complete definition of linear systems than is usually encountered. Because the primary concern in most analysis problems is the zero-state response, the common definition is just the first part, (3a), which says that superposition applies to the input–output relationship; i.e., in the SIO case,

$$S[\mathbf{0}; \alpha v(t) + \beta w(t)] = \alpha S[\mathbf{0}; v(t)] + \beta S[\mathbf{0}; w(t)] \tag{4}$$

And, more often than not, a system satisfying this condition also satisfies the other two. Therefore, with a few pathological exceptions, zero-state linearity can be taken as a sufficient test.

Example 2.7

To illustrate how one tests a system's model for linearity, consider the zero-state response

$$y_{zs}(t) = S[0; x(t)] = \int_{t_0}^{t} x(\lambda)\frac{1}{RC}e^{-(t-\lambda)/RC}\,d\lambda$$

corresponding to the RC-circuit response, Eq. (8), Sect. 2.2, with the initial state $y_0 = 0$. Replacing $x(t)$ by $\alpha v(t) + \beta w(t)$ yields

$$\begin{aligned} S[0; \alpha v(t) + \beta w(t)] &= \int_{t_0}^{t} [\alpha v(\lambda) + \beta w(\lambda)]\frac{1}{RC}e^{-(t-\lambda)/RC}\,d\lambda \\ &= \alpha \int_{t_0}^{t} v(\lambda)\frac{1}{RC}e^{-(t-\lambda)/RC}\,d\lambda \\ &\quad + \beta \int_{t_0}^{t} w(\lambda)\frac{1}{RC}e^{-(t-\lambda)/RC}\,d\lambda \\ &= \alpha S[0; v(t)] + \beta S[0; w(t)] \end{aligned}$$

so $S[0; x(t)]$ satisfies (4) and, hence, this system is linear with respect to its zero-state response.

Similarly, one can show that it is linear with respect to its zero-input response and, as we have already noted, the complete response is decomposable. Therefore, the RC-circuit model meets all requirements of a linear system. The same is true for the DC motor system in Example 2.3.

Nonlinear systems and linerization

Nonlinear systems, by definition, fail to satisfy one or more parts of (3). This fact is usually obvious from the model for SIO systems; with multiple inputs, however, one must proceed rather carefully in the interpretation of (3a). A sufficient but not necessary indication of nonlinearity is when the output involves *multiplication* of two or more inputs or states.

Nonlinear systems almost always prove harder to analyze than linear systems. *Linearizing approximations*, when applicable, thus become very useful tools of the engineer. Example 2.2 illustrated one linearization method; there we wrote $V(t) = \bar{V} + v(t)$, etc., and assumed $v(t)$ to be a small or incremental variation so that nonlinear terms involving $v^2(t)$ could be neglected. Alternately, if a nonlinear function is reasonably smooth over the range in question, one can obtain a linear approximation from the first two terms of the *Taylor-series* expansion about the nominal operating point. Students of electrical engineering should be acquainted with both methods in the guise of the *small-signal* approximation.

Time invariance

Systems are further categorized as being time-invariant or time-varying. A *fixed* or *stationary* or *time-invariant* system, whether linear or nonlinear, is one whose response to a given excitation does not change shape if the excitation is applied at a different time. We express this concept mathematically by considering an arbitrary input vector $\mathbf{x}(t)$ and its *delayed* version $\mathbf{x}(t - t_d)$. As illustrated in Figure 2.8 for a scalar signal, $\mathbf{x}(t - t_d)$ is just $\mathbf{x}(t)$ displaced along the time axis by t_d units in the positive

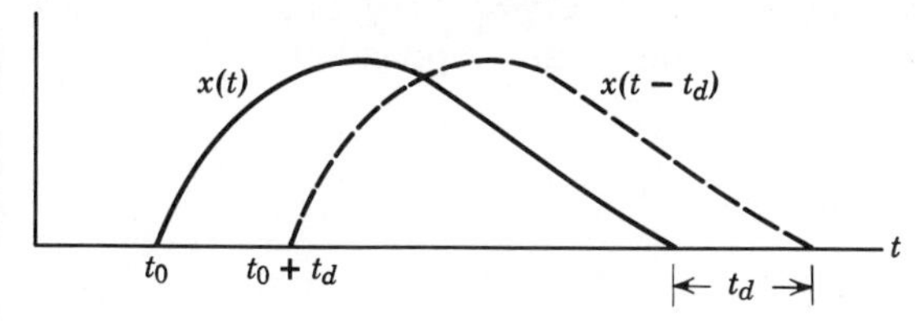

Figure 2.8. A signal and its delayed version.

direction. A fixed system is then defined to have the property that if

$$\mathbf{y}(t) = \mathbf{S}[\mathbf{q}(t_0); \mathbf{x}(t)] \qquad t \geq t_0$$

then

$$\mathbf{S}[\mathbf{q}(t_0); \mathbf{x}(t-t_d)] = \mathbf{y}(t-t_d) \qquad t \geq t_0 + t_d \tag{5}$$

so that delaying the excitation (both input and initial state) has no effect other than delaying the response a like amount.

Physically, (5) implies that the parameters of a fixed system are constant. However, if the nature of the system changes with time, we say that it is *nonstationary* or *time-varying.*

One should note that linearity and time-invariance are independent classifications; a time-varying system can be linear or nonlinear, and vice versa. In fact, it frequently happens that linearizing a fixed nonlinear system model results in a time-varying linear model, as illustrated in Problem 2.16.

Properties of fixed linear systems

Coming at last to the heart of this section, we now examine the distinguishing properties of fixed linear systems. Basically, they must satisfy the definitions (3) and (5). Concentrating on SIO systems with zero initial state we write

$$y(t) = S[x(t)]$$

where, for simplicity, $\mathbf{q}(t_0) = \mathbf{0}$ is implied. Then, if the system is fixed and linear,

$$S[\alpha v(t) + \beta w(t)] = \alpha S[v(t)] + \beta S[w(t)] \tag{6a}$$

$$S[x(t-t_d)] = y(t-t_d) \tag{6b}$$

which represent superposition and time-invariance, respectively.

To give further meaning to the rather abstract notation above, let us list the four most primative fixed linear systems, called *scalors, differentiators, integrators,* and *delayors.* Assuming the input is applied at t_0, their zero-state output equations are:

Scalor $$y(t) = Kx(t) \tag{7a}$$

Differentiator $$y(t) = \frac{dx(t)}{dt} \tag{7b}$$

Integrator $$y(t) = \int_{t_0}^{t} x(\lambda)\,d\lambda \tag{7c}$$

Delayor $$y(t) = x(t - t_D) \tag{7d}$$

where K and t_D are constants. More complex systems can be built up by taking a linear combination of these basic operations or by applying two or more operations in succession to the same input. In fact, if a system consists of a finite number of elements, each having a describing equation that is a fixed linear operation, then the complete system is fixed linear, at least with respect to the zero-state response. Excluding delayors, such systems can be modeled by linear differential equations with constant coefficients, of the general form

$$a_n \frac{d^n y}{dt^n} + a_{n-1} \frac{d^{n-1} y}{dt^{n-1}} + \cdots + a_1 \frac{dy}{dt} + a_0 y$$
$$= b_m \frac{d^m x}{dt^m} + b_{m-1} \frac{d^{m-1} x}{dt^{m-1}} + \cdots + b_1 \frac{dx}{dt} + b_0 x \tag{8}$$

We have given several example systems that, sometimes with the help of linearizing approximations, resulted in equations like this.

Not only do many systems of practical interest have models that are fixed and linear but, from the design viewpoint, the properties expressed by (6) often are highly desirable in systems engineering. Consider, for instance, a communication system whose task is to take an input signal, say the sound of a person speaking, and deliver to a distant point an output signal that reasonably approximates the input. Recognizing that there will be some inevitable time delay in transmission, we would like the output to be

$$y(t) = Cx(t - t_d) \tag{9}$$

Such a system is said to give *distortionless transmission* since the output has the same shape as the input, differing only by a scale factor C and a time delay t_d. Clearly, distortionless transmission requires a communication system whose net properties are linear and time invariant.

Or consider a control system whose output is the physical variable being controlled, say the trajectory of a rocket, and whose input is the controlling reference signal. Although one seldom can achieve perfect control in the sense of "distortionless transmission," linearity and time invariance of the system are still helpful, e.g., it would be nice if the output responds proportionally to the reference, and if a certain control signal always yields the same result, regardless of application time.

Returning to analysis, the distinguishing feature of fixed linear systems as a whole is that they are *solvable*. By this we mean that one can always find an explicit relationship for the output in terms of the input. The relationship may well be a nontrivial integral, but at least it is a formal solution. For nonlinear or time-varying systems, explicit solutions cannot, in general, be found.

The reason why fixed linear systems are easier to analyze can be seen by studying (6). For instance, suppose we know the response to two different input signals, $v(t)$ and $w(t)$; then, by superposition and time invariance, we immediately have the response to any input of the form

$$x(t) = \alpha v(t-t_1) + \beta w(t-t_2) \tag{10a}$$

namely

$$y(t) = \alpha y_v(t-t_1) + \beta y_w(t-t_2) \tag{10b}$$

where $y_v(t) = S[v(t)]$, etc., and α, β, t_1, and t_2, are constants.

Going further in this direction, it is also true of fixed linear systems that if $y(t) = S[x(t)]$ then

$$S\left[\frac{dx(t)}{dt}\right] = \frac{dy(t)}{dt} \tag{11}$$

and

$$S\left[\int_{-\infty}^{t} x(\lambda)\,d\lambda\right] = \int_{-\infty}^{t} y(\lambda)\,d\lambda \tag{12}$$

In words, the derivative of the response equals the response to the derivative of the excitation, and similarly for integration. Thus, if we know the response to a certain input, we can find the response to its derivative or integral by differentiating or integrating the original response, respectively. Having asserted this property, let us prove (11); proof of (12) follows parallel lines.

Recall the fundamental definition of differentiation:

$$\frac{dx(t)}{dt} \triangleq \lim_{\epsilon \to 0} \frac{1}{\epsilon}[x(t+\epsilon) - x(t)]$$

Then, invoking (6a) and then (6b),

$$S\left[\frac{x(t+\epsilon)}{\epsilon}-\frac{x(t)}{\epsilon}\right]=\frac{1}{\epsilon}S[x(t+\epsilon)]-\frac{1}{\epsilon}S[x(t)]$$
$$=\frac{1}{\epsilon}y(t+\epsilon)-\frac{1}{\epsilon}y(t)$$

so, taking the limit as $\epsilon \to 0$ yields

$$S\left[\frac{dx(t)}{dt}\right]=\lim_{\epsilon\to 0}\frac{1}{\epsilon}[y(t+\epsilon)-y(t)]=\frac{dy(t)}{dt}$$

which completes the proof.

Taken together, (10) to (12) mean that an extensive catalog of input-output pairs can be compiled for a particular system by first calculating the response to a few elementary input signals. Because of this fact, we shall later seek elementary signals for which the response is easily found; these turn out to be exponential time functions. Then, to obtain full advantage of this approach, we shall also seek ways of representing more complicated inputs in terms of the elementary signals; these turn out to be the Fourier series, Fourier transform, and Laplace transform. Moreover, as will be discussed in the last chapter, similar techniques are applicable for *discrete* systems – i.e., systems described by *difference equations* rather than differential equations – providing they are fixed and linear. In short, there exists a large body of highly developed mathematical tools that are eminently suited to systems analysis when linearity and time-invariance hold.

Postscript: interchanging mathematical operations

In the proof of (11) we tacitly assumed that

$$\lim_{\epsilon\to 0} S\left[\frac{x(t+\epsilon)-x(t)}{\epsilon}\right]=S\left[\lim_{\epsilon\to 0}\frac{x(t+\epsilon)-x(t)}{\epsilon}\right]$$

and in the future we shall frequently interchange the order of mathematical operations such as integration, differentiation, and limiting. While that step is allowable in most situations of interest in this text, there are occasional exceptions and one should always give the matter some thought rather than plowing blindly ahead.

Mathematical tests do exist† for determining the conditions under

†For instance, the Moore-Smith theorem; see Gleason (1966, pp. 256–257).

which interchanges are valid, but they can be applied only on a case-by-case basis and are likely to be rather complicated. A more practical approach is to look for a test of the result that is independent of the derivation. For example, we can check (11) when the specific system function $S[x(t)]$ is known simply by finding $dS[x(t)]/dt$ and comparing it with $S[dx(t)/dt]$. Note, however, that that test checks only the specific case, not the general assertion. Fortunately, the various interchange operations assumed in systems engineering have been tested exhaustively and verified – barring pathological exceptions having little engineering significance. Therefore, in the future, we shall take those steps without comment.

Problems

2.1 Modify Eq. (3), Sect. 2.1, to account for a load resistance R_L across the output terminals.

2.2 A fluid reservoir system has a tank of cross-sectional area A in which the fluid depth is $H(t)$. The input and output flow rates are $X(t)$ and $Y(t)$, in fluid volume per unit time, so $X(t) - Y(t) = A\dot{H}(t)$. The nature of the outlet valve is such that $Y(t) = CH^{1/2}(t)$, with C being a constant. Obtain a nonlinear differential equation relating X and Y, not containing H. *Answer*: $(2A/C^2)Y\dot{Y} + Y = X$.

2.3 A slack loop is maintained between two sets of drive rolls, Figure P2.1, in a certain computer tape deck. All rolls have diameter D. Under normal operating conditions $\Omega_1(t) = \Omega_2(t) = \bar{\Omega}$ and $L(t) = \bar{L}$. Starting at time t_0 there are small, independent variations in the drive speeds so that $\Omega_1(t) = \bar{\Omega} + \omega_1(t)$ and $\Omega_2(t) = \bar{\Omega} + \omega_2(t)$. Derive an expression for the resulting change in loop length $l(t)$.

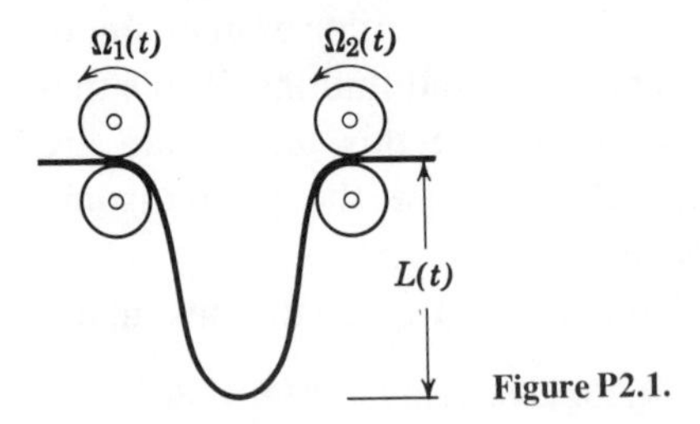

Figure P2.1.

2.4 Suppose the RC circuit of Figure 2.1 has $y(t_0) = 10$ and $x(t) = 20$ for $t \geq t_0$. Using Eq. (8), Sect. 2.2, find and sketch $y(t)$, $t \geq t_0$. Also sketch on the same plot the two components in $y(t)$ and give a physical interpretation.

2.5 Solve Eq. (8), Sect. 2.1, for $v(t)$ in terms of $f(t)$.

2.6 Referring to Eq. (7), Sect. 2.2, with $t_0 = 0$, find an input $x(t)$ such that $y(t) = y_0$ for $t \geq 0$, i.e., the output stays constant.

2.7 Apply the method of Example 2.4 to the case of $a_1\dot{y} + a_0 y = b_1\dot{x} + b_0 x$, and show that the solution is

$$y(t) = \int_{t_0}^{t} \frac{1}{a_1}\left[b_1 \frac{dx(\lambda)}{d\lambda} + b_0 x(\lambda)\right] e^{p(t-\lambda)} d\lambda + y_0 e^{p(t-t_0)}$$

where $p = -a_1/a_0$.

2.8 If the DC motor of Example 2.3 operates in the *armature-controlled* mode, the equivalent electrical circuit is as shown in Figure P2.2. The field current is held constant and the motor torque is $\tau_m = Ki_a$. Obtain a model similar to Eq. (9), Sect. 2.1, and explain why the method used in Example 2.5 will not work for solving this case.

2.9 (a) Referring to Eq. (8), Sect. 2.2, identify the initial state $\mathbf{q}(t_0)$ and the zero-input and zero-state responses. Show that Eqs. (13) and (14), Sect. 2.3, apply. (b) Obtain an expression for $y(t)$ for the circuit in Figure P2.3. Identify the initial state and show that $S[0; 0] \neq 0$.

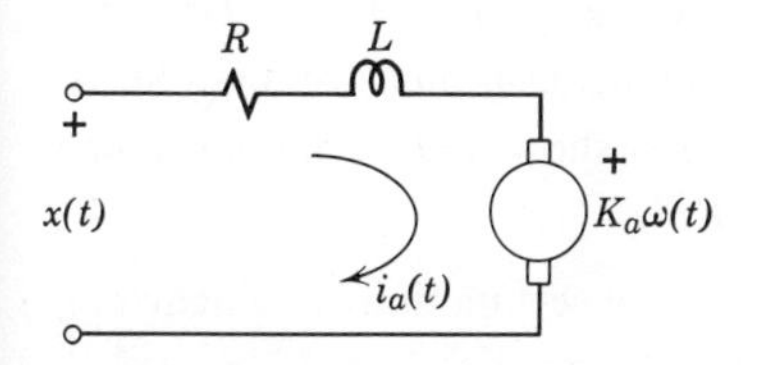

Figure P2.2.

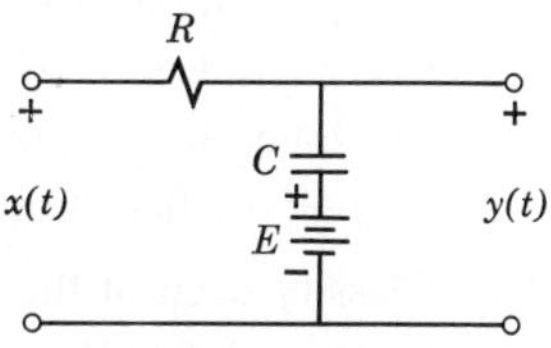

Figure P2.3.

2.10 Verify, by doing so, that *algebraic* equations can be written for all the circuit variables in terms of the state variables indicated in Figure 2.7.

2.11 Write appropriate state and output equations, similar to Eqs. (17) and (18), Sect. 2.3, for the two-input, two-output network of Figure P2.4.

2.12 Consider the system model $\dddot{y} + a_2\ddot{y} + a_1\dot{y} = x$, whose complete solution requires *three* initial conditions. However, letting $z = y$ we get the *second-order* equation $\ddot{z} + a_2\dot{z} + a_1 z = x$. Resolve this anomoly.

2.13 When subjected to an extension force $0 \leq F < F_c$, the length of a certain

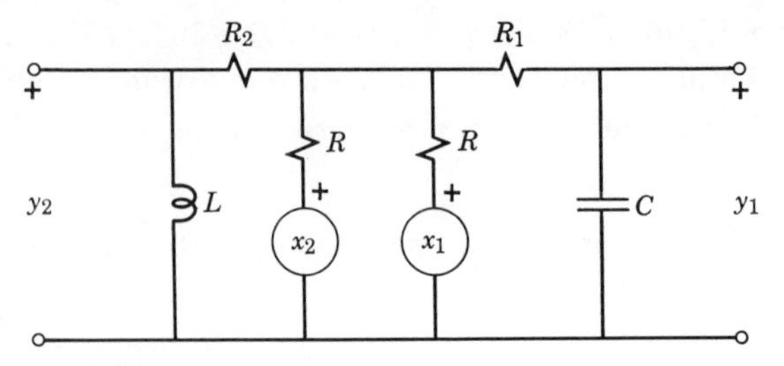

Figure P2.4.

nonlinear spring is $X = X_0[1 + \tan(\pi F/2F_c)]$ – i.e., the spring breaks if $F \geq F_c$. Write a Taylor-series expansion to obtain a linear approximation for the change in length $x(t)$ when there are small force variations $f(t)$ with respect to $\bar{F} = F_c/2$.

2.14 For the system in Problem 2.2, take $X(t) = \bar{X} + x(t)$, $H(t) = \bar{H} + h(t)$, etc., and obtain a linear incremental model relating $x(t)$ and $y(t)$ in terms of $\bar{H}$. Also state the relationships between $\bar{X}$, $\bar{H}$, and $\bar{Y}$. *Hint*: first linearize the valve equation.

2.15 When both the armature and field currents of a DC motor are allowed to vary, the torque developed is $\tau_m(t) = K_m i_a(t) i_f(t)$. Show that τ_m is not a linear function of i_a and i_f.

2.16 Referring to Example 2.2, suppose it is desired to have $V(t) = Kt$ and the force required is $\bar{F}(t)$. If the actual force is $F(t) = \bar{F}(t) + f(t)$ then $V(t) = Kt + v(t)$, where $f(t)$ and $v(t)$ are small variations. Obtain a linearized differential equation relating the latter, and show that it contains a time-varying coefficient. *Answer*: $M\dot{v} + (B + 2DKt)v = f$.

2.17 Classify each of the following zero-state response functions as linear/nonlinear and fixed/time-varying: (a) $y(t) = e^{x(t)}$; (b) $y(t) = Ad[x(t-\tau)]/dt$; (c) $y(t) = \int_{t_0}^{t} \{x(\lambda) + [dx(\lambda)/d\lambda]\} e^{t-\lambda} d\lambda$; (d) $y(t) = tx(t)$.

2.18 Prove that the primitive operators listed in Eq. (7), Sect. 2.4, are fixed and linear.

2.19 In later chapters we shall derive a zero-state response expression of the form $S[x(t)] = \int_{t_0}^{t} x(\lambda) h(t-\lambda) d\lambda$, where $h(t-\lambda)$ is arbitrary but does not involve $x(t)$. Show that this expression satisfies Eq. (6), Sect. 2.4.

2.20 Taking $x(t) = A \sin \Omega t$, $t \geq 0$, show that Eq. (10), Sect. 2.2, has the properties of Eq. (11), Sect. 2.4.

3

Systems described by differential equations

The most commonly encountered model of a dynamic system is a linear differential equation with constant coefficients. This chapter discusses the solution and interpretation of such equations, along with certain related topics.

We shall start by solving homogeneous differential equations, the solutions being viewed in the context of a system's zero-input response. Because the zero-input response generally involves exponential time functions of the form e^{st}, the properties of these system modes are also investigated. With this as preparation, we then tackle the complete system response as found from an inhomogeneous equation. Although most of this material should be primarily review for the reader, new facets especially pertinent to systems engineering will be brought out as we go along. In the interests of simplicity, these investigations will concentrate on the single-input-output case.

Also covered here are the simulation of differential equations and further discussion of state-variable equations. In the latter, the scope is broadened to include multi-input-output systems and systems with time-varying parameters.

3.1 ZERO-INPUT RESPONSE

Throughout this chapter we shall discuss systems described by the input-output model

$$\frac{d^n y}{dt^n} + a_{n-1}\frac{d^{n-1}y}{dt^{n-1}} + \cdots + a_1\frac{dy}{dt} + a_0 y$$
$$= b_m\frac{d^m x}{dt^m} + b_{m-1}\frac{d^{m-1}x}{dt^{m-1}} + \cdots + b_1\frac{dx}{dt} + b_0 x \qquad (1)$$

where x and y are functions of time, the a's and b's are constants, and, without loss of generality, we have taken the coefficient of the highest derivative of y to be unity, i.e., $a_n = 1$. As distinguished from the complete response $y(t)$, this section considers only the zero-input response $y_{zi}(t)$. But before getting into the heart of the problem, it is worthwhile to introduce some handy notational devices.

Besides the dot convention, we shall represent time derivatives by

$$v^{(k)} \triangleq \frac{d^k v(t)}{dt^k} \qquad (2)$$

which readily lends itself to writing (1) in the compact summation form

$$\sum_{i=0}^{n} a_i y^{(i)} = \sum_{k=0}^{m} b_k x^{(k)} \qquad (3)$$

Different summation indices, i and k, are used here for clarity. Also, on occasion, the entire right-hand side of (1) or (3) will be denoted as

$$F_x(t) \triangleq b_m\frac{d^m x}{dt^m} + \cdots + b_0 x = \sum_{k=0}^{m} b_k x^{(k)} \qquad (4)$$

and called the *forcing function.*

For the zero-input response, the forcing function is set to zero, by definition, and we have a set of n initial conditions corresponding to the system's state at time $t = t_0$. Since $y(t)$ is the only variable at hand, the initial state must be specified in terms of $y(t)$ and its derivatives. Hence, our task becomes that of solving for $y(t)$, $t \geq t_0$, given

$$y^{(n)} + a_{n-1}y^{(n-1)} + \cdots + a_1\dot{y} + a_0 y = 0 \qquad (5)$$

and the initial conditions

$$y(t_0) = y_0, \quad \dot{y}(t_0) = \dot{y}_0, \; \ldots \; y^{(n-1)}(t_0) = y_0^{(n-1)} \qquad (6)$$

Zero-input response of a second-order system

As preparation for the general nth-order case, let us first tackle the second-order problem

$$\ddot{y}+a_1\dot{y}+a_0y=0 \tag{7}$$

Based on the results of Section 2.2, it is plausible to assume an exponential solution of the form†

$$y(t)=Ae^{st}$$

where A and s are constants, not yet known, and hence

$$\dot{y}=sAe^{st} \qquad \ddot{y}=s^2Ae^{st}$$

Inserting these into (7) and regrouping gives $(s^2+a_1s+a_0)Ae^{st}=0$, which is satisfied if either $Ae^{st}=0$ or $s^2+a_1s+a_0=0$.

Obviously, the *trivial* solution $y=Ae^{st}=0$ does not hold much interest, so we concentrate on the other possibility

$$s^2+a_1s+a_0=0$$

This equation, a quadratic polynomial in s, can be factored into the form

$$(s-p_1)(s-p_2)=0$$

where the *roots*, p_1 and p_2, are

$$\begin{aligned} p_1 &= -\tfrac{1}{2}\left(a_1+\sqrt{a_1{}^2-4a_0}\right) \\ p_2 &= -\tfrac{1}{2}\left(a_1-\sqrt{a_1{}^2-4a_0}\right) \end{aligned} \tag{8}$$

assuming $a_1{}^2 \neq 4a_0$ so that $p_1 \neq p_2$.

Since the trial solution $y=Ae^{st}$ satisfies (7) with either $s=p_1$ or $s=p_2$, and with any value of A, it follows that the most general solution consists of a *linear combination of two exponentials*,

$$y=A_1e^{p_1t}+A_2e^{p_2t} \tag{9}$$

Confirming this point, substitute (9) into (7), as follows:

$$\overbrace{(p_1{}^2A_1e^{p_1t}+p_2{}^2A_2e^{p_2t})}^{\ddot{y}}+a_1\overbrace{(p_1A_1e^{p_1t}+p_2A_2e^{p_2t})}^{\dot{y}}+a_0\overbrace{(A_1e^{p_1t}+A_2e^{p_2t})}^{y}=0$$

or

$$(p_1{}^2+a_1p_1+a_0)A_1e^{p_1t}+(p_2{}^2+a_1p_2+a_0)A_2e^{p_2t}=0$$

†In Section 2.2 we used p in the exponential because its value was known from the start. Here we shall use s until it is evaluated.

But, from (8), $p_1^2 + a_1 p_1 + a_0 = p_2^2 + a_1 p_2 + a_0 = 0$, so (9) satisfies the homogeneous equation regardless of the values of A_1 and A_2.

We stress that either term of (9) is a valid solution, and therefore both terms must be included for completeness. Mathematically, it is known that a second-order homogeneous differential equation has two linearly independent solutions; physically, we have seen that a system with two energy-storage elements has two independent initial conditions. That is why our solution contains two constants, still to be evaluated.

Turning to this task, A_1 and A_2 are determined from the *initial conditions* $y(t_0) = y_0$ and $\dot{y}(t_0) = \dot{y}_0$. Accordingly, one sets $t = t_0$ in (9) and in its first derivative, i.e.,

$$y(t_0) = A_1 e^{p_1 t_0} + A_2 e^{p_2 t_0} = y_0$$

$$\dot{y}(t_0) = p_1 A_1 e^{p_1 t_0} + p_2 A_2 e^{p_2 t_0} = \dot{y}_0$$

or, in matrix-equation form,

$$\begin{bmatrix} 1 & 1 \\ p_1 & p_2 \end{bmatrix} \begin{bmatrix} A_1 e^{p_1 t_0} \\ A_2 e^{p_2 t_0} \end{bmatrix} = \begin{bmatrix} y_0 \\ \dot{y}_0 \end{bmatrix} \tag{10}$$

which can be solved for $A_1 e^{p_1 t_0}$ and $A_2 e^{p_2 t_0}$ via *Cramer's rule*. Specifically, introducing the determinants

$$D = \begin{vmatrix} 1 & 1 \\ p_1 & p_2 \end{vmatrix} = p_2 - p_1 \tag{11a}$$

and

$$D_1 = \begin{vmatrix} y_0 & 1 \\ \dot{y}_0 & p_2 \end{vmatrix} = p_2 y_0 - \dot{y}_0 \qquad D_2 = \begin{vmatrix} 1 & y_0 \\ p_1 & \dot{y}_0 \end{vmatrix} = \dot{y}_0 - p_1 y_0 \tag{11b}$$

we have

$$A_1 e^{p_1 t_0} = \frac{D_1}{D} = \frac{p_2 y_0 - \dot{y}_0}{p_2 - p_1} \qquad A_2 e^{p_2 t_0} = \frac{D_2}{D} = -\frac{p_1 y_0 - \dot{y}_0}{p_2 - p_1} \tag{12}$$

Finally, we put these results back into (9) to get the zero-input response

$$y_{zi}(t) = \frac{p_2 y_0 - \dot{y}_0}{p_2 - p_1} e^{p_1(t-t_0)} - \frac{p_1 y_0 - \dot{y}_0}{p_2 - p_1} e^{p_2(t-t_0)} \tag{13}$$

which, of course, only holds for $t \geq t_0$. Two comments about this result are in order here. First, the roots p_1 and p_2 must be *distinct*, i.e., $p_1 \neq p_2$ as previously assumed; otherwise, our expression for $y_{zi}(t)$ blows up. Second, the roots may turn out to be *complex* numbers; even so, because it represents a physical variable, $y_{zi}(t)$ will be real. The necessary modifications when $p_1 = p_2$ are briefly treated below, while complex roots are covered in the next section.

Repeated roots

Suppose the system parameters are such that $a_1^2 = 4a_0$ and hence there is only the one root

$$p = -a_1/2 = -\sqrt{a_0} \tag{14}$$

Since, in effect, $p_1 = p_2 = p$, this is known as the case of *repeated roots.* Mathematicians have long ago shown that the complementary solution for this case must be written as

$$y = (A_1 + A_2 t)e^{pt} \tag{15}$$

Whence, upon evaluating the constants, one obtains

$$y_{zi}(t) = [y_0 + (\dot{y}_0 - py_0)(t - t_0)]e^{p(t-t_0)} \tag{16}$$

whose verification is left as an exercise.

Zero-input response of an *n*th-order system

With the aid of some matrix algebra, the above results for a second-order system are readily extended to the nth-order case, as follows. Inserting the trial solution $y = Ae^{st}$ into the homogeneous equation, (5), yields

$$s^n + a_{n-1}s^{n-1} + \cdots + a_1 s + a_0 = 0 \tag{17}$$

called the *characteristic equation* of the system. The left-hand side, an nth-order polynomial in s, is known as the *characteristic polynomial.* In general, (17) will have n roots, which we denote by p_i, $i = 1, 2, \ldots n$, each of which satisfies the characteristic equation, i.e.,

$$s^n + a_{n-1}s^{n-1} + \cdots + a_1 s + a_0 = (s - p_1)(s - p_2) \cdots (s - p_n)$$

Assuming all the p_i are distinct (no repeated roots) the complete solution for y_{zi} takes the form

$$y_{zi}(t) = A_1 e^{p_1 t} + A_2 e^{p_2 t} + \cdots + A_n e^{p_n t} = \sum_{i=1}^{n} A_i e^{p_i t} \tag{18}$$

and the n constants A_i must be evaluated from the initial conditions.

For that purpose, differentiating (18) $n-1$ times, setting $t = t_0$, and applying Cramer's rule gives the nth-order version of (12) as

$$A_i e^{p_i t_0} = \frac{D_i}{D} \qquad i = 1, 2, \ldots n$$

where D and D_i are the determinants†

$$D = \begin{vmatrix} 1 & 1 & \cdots & 1 \\ p_1 & p_2 & \cdots & p_n \\ p_1^2 & p_2^2 & \cdots & p_n^2 \\ \vdots & \vdots & & \vdots \\ p_1^{n-1} & p_2^{n-1} & \cdots & p_n^{n-1} \end{vmatrix} \tag{19a}$$

$$D_i = \begin{vmatrix} 1 & \cdots & y_0 & \cdots & 1 \\ p_1 & \cdots & \dot{y}_0 & \cdots & p_n \\ p_1^2 & \cdots & \ddot{y}_0 & \cdots & p_n^2 \\ \vdots & & \vdots & & \vdots \\ p_1^{n-1} & \cdots & y_0^{(n-1)} & \cdots & p_n^{n-1} \end{vmatrix} \tag{19b}$$

(ith column indicated by arrow at y_0 column)

Note that D_i is just D with its ith column replaced by the initial-condition vector $[y_0 \;\; \dot{y}_0 \cdots y_0^{(n-1)}]^T$. Finally, substituting for A_i in (18), our answer is

$$y_{zi}(t) = \sum_{i=1}^{n} \frac{D_i}{D} e^{p_i(t-t_0)} \tag{20}$$

for the case of distinct roots.

If it happens that one or more of the roots are repeated, the trial solution (18) must be modified along a line similar to (15). For instance, if p_1 is repeated r times, then the first r terms of (18) are replaced by

$$(A_1 + A_2 t + A_3 t^2 + \cdots + A_r t^{r-1}) e^{p_1 t}$$

Because repeated roots are a rather special case, and notationally far more cumbersome, we carry the solution no further than this point.

†Equation (19a) is sometimes called the determinant of the *Vandermonde matrix*.

Example 3.1

As a specific application of (20), consider the third-order homogeneous equation

$$\dddot{y} + 4\ddot{y} + 3\dot{y} = 0$$

subject to the initial conditions $y_0 = 10$, $\dot{y}_0 = 110$, and $\ddot{y}_0 = -410$.

By inspection, the characteristic equation is $s^3 + 4s^2 + 3s = 0$, which readily factors into $s(s+1)(s+3) = 0$. Hence, the roots are

$$p_i = 0, -1, -3$$

and the determinant D, per (19a), becomes

$$D = \begin{vmatrix} 1 & 1 & 1 \\ p_1 & p_2 & p_3 \\ p_1^2 & p_2^2 & p_3^2 \end{vmatrix} = \begin{vmatrix} 1 & 1 & 1 \\ 0 & -1 & -3 \\ 0 & 1 & 9 \end{vmatrix} = -6$$

Using (19b) for D_1 gives

$$D_1 = \begin{vmatrix} y_0 & 1 & 1 \\ \dot{y}_0 & p_2 & p_3 \\ \ddot{y}_0 & p_2^2 & p_3^2 \end{vmatrix} = \begin{vmatrix} 10 & 1 & 1 \\ 110 & -1 & -3 \\ -410 & 1 & 9 \end{vmatrix} = -120$$

and likewise $D_2 = -240$, $D_3 = 300$.

Putting these numbers into (20) – and noting that $e^{p_1(t-t_0)} = 1$ since $p_1 = 0$ – we find that

$$y_{zi}(t) = 20 + 40e^{-(t-t_0)} - 50e^{-3(t-t_0)} \qquad t \geq t_0$$

which is plotted in Figure 3.1 versus $t - t_0$. For a quick check on this

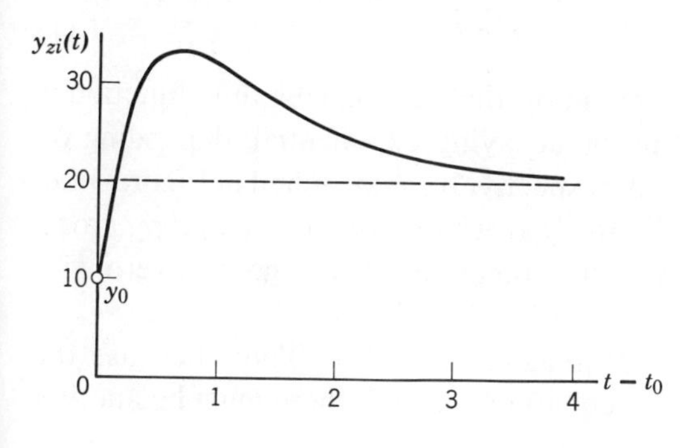

Figure 3.1.

answer one can immediately see that $y_{zi}(t_0) = 20+40-50 = 10$, in agreement with the specified value y_0. And it only takes a bit more work to check on $\dot{y}_{zi}(t_0)$ and $\ddot{y}_{zi}(t_0)$.

3.2 SYSTEM MODES AND THE s-PLANE

It should be recognized that Example 3.1, given for illustration purposes, was set up to be rather simple. In practice, however, finding $y_{zi}(t)$ may entail appreciable effort, particularly if n is large, say $n > 3$. Usually, the characteristic equation cannot be factored directly, so numerical methods are needed to calculate the roots. Then there are determinants to evaluate, and the possible further complications of repeated roots.

Nonetheless, and regardless of how large n is, the zero-input response of any system described by Eq. (1), Sect. 3.1, always consists of a sum of exponential time functions such as Ae^{pt} or At^re^{pt}. Thus, whether or not we actually solve for $y_{zi}(t)$, we can learn a great deal about these systems by examining their natural-behavior time functions, called the system *modes*.

Mode functions

If one is dealing with a physical system, the coefficients of its characteristic equation must be real quantities since they represent the system's parameters. But the roots can be real, imaginary, or complex, and each root may be expressed in the general form

$$p = \sigma + j\omega$$

where σ and ω are real while $j \triangleq \sqrt{-1}$.

Assume for the moment that p is, in fact, *real* ($\omega = 0$) and *nonrepeated*; then one of the system's modes is

$$Ae^{\sigma t} \qquad t \geq 0 \tag{1}$$

where we have adopted the convention that $t_0 = 0$. This time function is either a constant, or a growing or decaying exponential, depending on whether $\sigma = 0$, $\sigma > 0$, or $\sigma < 0$, respectively, as sketched in Figure 3.2*a*. Figure 3.2*b* illustrates the mode $Ate^{\sigma t}$ associated with a *repeated real* root. In either case, one should notice that the mode always goes to zero with increasing time if σ is negative.

Now suppose p is *complex* or imaginary ($\omega \neq 0$). Then, because the coefficients in the characteristic equation are real, there must be another

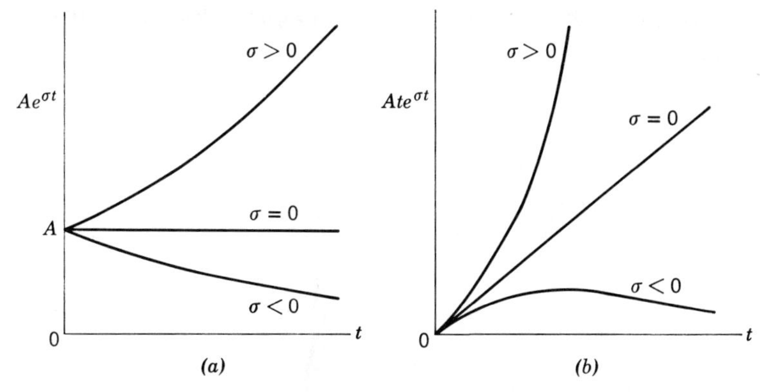

Figure 3.2. Mode functions. (*a*) Single real root. (*b*) Repeated real root.

root that equals p^*. In other words, *complex roots always occur in complex-conjugate pairs*:

$$p = \sigma + j\omega \qquad \text{and} \qquad p^* = \sigma - j\omega$$

Furthermore, the corresponding constants multiplying the two exponential functions also will be complex conjugates; otherwise, $y_{zi}(t)$ would not be a real function of time.

Under these conditions, we can group the two terms together, writing†

$$Ae^{pt} + A^* e^{p^* t} = (Ae^{pt}) + (Ae^{pt})^* = 2 \operatorname{Re}[Ae^{pt}]$$

thereby emphasizing that the sum is real even though the individual terms are complex. Substituting $p = \sigma + j\omega$ and putting the complex constant A in polar form‡ as $A = |A| e^{j \arg[A]}$ finally yields

$$\begin{aligned} Ae^{pt} + A^* e^{p^* t} &= 2 \operatorname{Re}[|A| e^{j \arg[A]} e^{\sigma t} e^{j\omega t}] \\ &= 2|A| e^{\sigma t} \operatorname{Re}[e^{j(\omega t + \arg[A])}] \\ &= 2|A| e^{\sigma t} \cos(\omega t + \arg[A]) \end{aligned} \tag{2}$$

Therefore, the pair of roots p and p^* leads to a sinusoidal or oscillatory mode whose *envelope* $2|A|e^{\sigma t}$ can be constant, growing, or decaying

†Readers who find this manipulation unfamiliar should review the subject of complex numbers.

‡That is, if $A = A_r + jA_i$ where A_r and A_i are the real and imaginary parts, respectively, then

$$|A| \triangleq \sqrt{A_r^2 + A_i^2} \qquad \arg[A] \triangleq \arctan \frac{A_i}{A_r}$$

(Figure 3.3). We leave the case of a repeated complex root for the reader to contemplate.

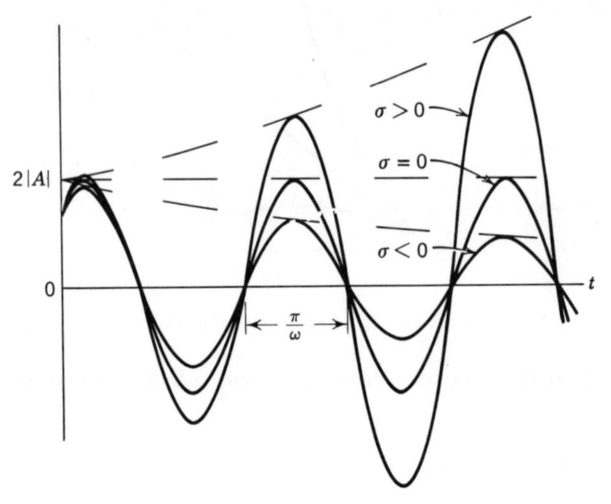

Figure 3.3. Mode functions for a pair of complex-conjugate roots, $2|A|e^{\sigma t}\cos(\omega t + \arg |A|)$.

s-Plane diagrams

The natural-behavior modes of a fixed linear system are uniquely associated with the roots of its characteristic equation. For a given system whose roots are known, this information is conveniently displayed by plotting the roots as points in the *s-plane* – a plane whose axes are the real and imaginary parts of $s = \sigma + j\omega$, with σ and ω treated as independent variables. A specific root such as $p_1 = \sigma_1 + j\omega_1$ is indicated by a $\times$ at the point having rectangular coordinates σ_1 and ω_1. All real roots obviously will fall on the axis of reals (the horizontal axis), while complex roots result in symmetric points with respect to the real axis since they come in complex-conjugate pairs. Repeated roots are represented by multiple $\times$'s.

To interpret such diagrams we associate a possible system mode with each point in the s-plane, e.g., decaying exponential, growing oscillation, etc. This interpretation is illustrated by Figure 3.4, where the lower half plane ($\omega < 0$) has been omitted, consistent with the complex-conjugate symmetry. Thus, bearing in mind the modifications for repeated roots, one can immediately infer a great deal about $y_{zi}(t)$ – and the system itself – by inspecting the s-plane diagram.

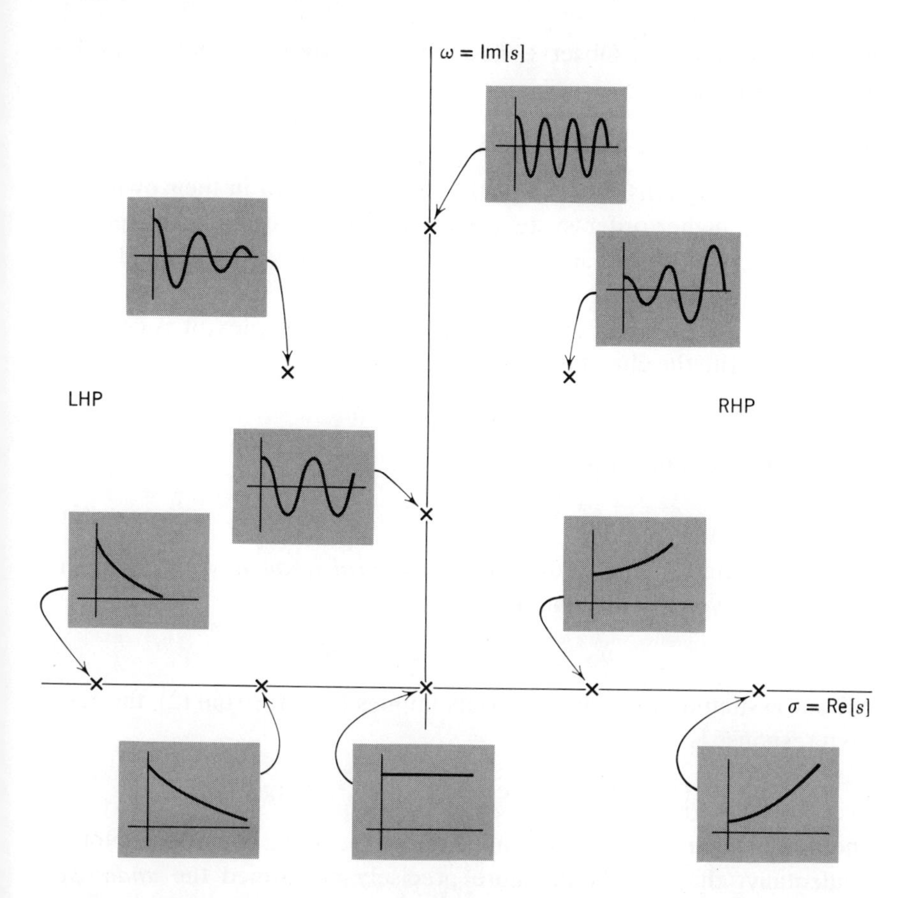

Figure 3.4. Time functions associated with points in the s-plane.

Specifically, points in the right half plane (RHP) correspond to modes involving $e^{\sigma t}$ with $\sigma > 0$; such functions (or their envelopes) increase indefinitely as $t \to \infty$. This diverging behavior also applies to repeated roots on the imaginary axis, including the origin, since the modes are of the form $|A|t^r \cos(\omega t + \arg[A])$ if $s = \pm j\omega$ or At^r if $s = 0$. Simple (non-repeated) roots on the imaginary axis with $\omega \neq 0$ correspond to constant-amplitude oscillations, while a single root at $s = 0$ represents a constant. Finally, all points entirely within the left half plane (LHP) represent modes involving $e^{\sigma t}$ with $\sigma < 0$ which, even if multiplied by a sinusoid or a power of t, will be the dominant factor when t is large. Hence, these modes will go to zero as $t \to \infty$, regardless of whether the roots are simple, repeated, or complex. Observe carefully that only LHP roots yield this decaying behavior.

s-Plane plots for second-order systems

Second-order systems deserve further attention, both in their own right and because higher-order system models often have dominant second-order characteristics or can be treated as coupled first- and second-order systems.

When the roots of a second-order system are complex, it is common practice to write the characteristic equation as

$$s^2 + 2\zeta\omega_n s + \omega_n^2 = 0 \tag{3a}$$

where, in terms of the original coefficients a_1 and a_0,

$$\omega_n \triangleq \sqrt{a_0} \qquad \zeta \triangleq a_1/2\sqrt{a_0} \tag{3b}$$

these parameters being known as the *natural frequency* and *damping ratio*, respectively. In this notation, the roots are

$$p_1, p_2 = -\zeta\omega_n \pm j\omega_n\sqrt{1-\zeta^2} \tag{4}$$

and our assumption of complex roots implies $\zeta^2 < 1$. From (2), the zero-input response is

$$y_{zi}(t) = 2|A|e^{-\zeta\omega_n t} \cos\,(\omega_n\sqrt{1-\zeta^2}\,t + \arg\,[A]) \tag{5}$$

where $|A|$ and $\arg\,[A]$ are determined from the initial conditions. Notice, incidentally, that ω_n should more precisely be termed the *undamped* natural frequency since the oscillation frequency of $y_{zi}(t)$ equals ω_n only when $\zeta = 0$.

The s-plane location of the roots is shown in Figure 3.5*a*, where the parameters ζ and ω_n take on graphical meaning: ω_n is the *radial distance* from the origin to either root, while ζ determines the *angle* θ relative to the negative real axis, i.e.,

$$\theta = \arccos\,\zeta$$

Figure 3.5*b* plots the paths traced out in the s-plane by the roots when the damping ratio is varied over the range $1 > \zeta > -1$ with ω_n held fixed. We see that the roots go into the RHP when $\zeta < 0$, which means that the natural behavior will grow exponentially if the damping is negative.

So far we have assumed $|\zeta| < 1$. If $\zeta^2 = 1$ then the roots are real and

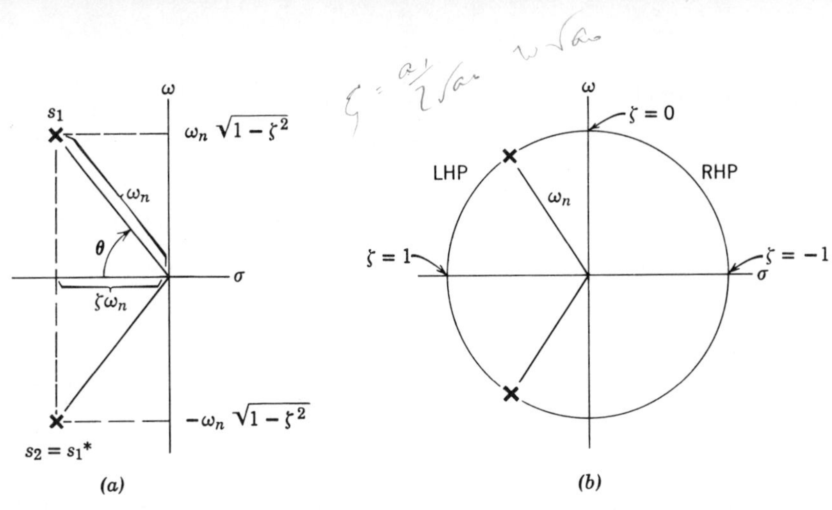

Figure 3.5. s-Plane diagram for a second-order system. (*a*) Geometric interpretation of ω_n and ζ. (*b*) Root locations with fixed ω_n, $1 \geq \zeta \geq -1$.

equal, i.e., repeated roots on the real axis; when $\zeta = +1$ this is referred to *critical damping*. For $\zeta^2 > 1$ the roots are real and distinct, and the designation of natural frequency is no longer meaningful since the modes are no longer oscillatory.

Stability of the zero-input response

Let us now introduce the relationship of s-plane diagrams to the concept of *stability*. Roughly speaking, a system is termed unstable if a small change of its state at some instant results in a disproportionately large effect at a later time; if this does not occur, then the system is said to be stable. Of course, for analytic work one must define a "small change" and a "large effect." Furthermore, stability is properly associated with a *response* of the system, rather than the system alone, so one must specify the response in question.

Here our interest centers on the zero-input response $\mathbf{y}_{zi}(t) = \mathbf{S}[\mathbf{q}(t_0); \mathbf{0}]$. For simplicity, it will be assumed that the system is linear, has only one output, and obeys the equilibrium requirement of Eq. (13), Sect. 2.3, namely, $S[\mathbf{0}; 0] = 0$ for all $t \geq t_0$. Under these conditions,† we take $\mathbf{q}(t_0) \neq \mathbf{0}$ and classify the stability as follows:

Unstable. There is at least one finite initial state such that $y_{zi}(t)$ grows

†La Salle and Lefschetz (1961, Chap. 2) may be consulted for less restricted classifications.

thereafter without bound, i.e.,

$$S[\mathbf{q}(t_0); 0] \to \infty \qquad \text{as} \qquad t \to \infty \tag{6}$$

for some $\mathbf{q}(t_0)$.

Asymptotically stable. For all possible initial states, $y_{zi}(t)$ eventually decays to zero, i.e.,

$$S[\mathbf{q}(t_0); 0] \to 0 \qquad \text{as} \qquad t \to \infty \tag{7}$$

for all $\mathbf{q}(t_0)$.

Marginally stable. For all initial states, $y_{zi}(t)$ remains thereafter within finite bounds, i.e.,

$$|S[\mathbf{q}(t_0); 0]| \leq C \qquad t \geq t_0 \tag{8}$$

where C is a finite positive constant, *but* there is at least one state such that (7) does not hold.

We now apply these definitions to the class of systems at hand and show that the stability can be predicted directly from the s-plane diagram.

A system having one or more RHP roots is obviously *unstable* – that is, its zero-input response is unstable – since the corresponding mode is proportional to $e^{\sigma t}$ with $\sigma > 0$ and, hence, $y_{zi}(t)$ increases exponentially. Similarly, one or more repeated roots on the imaginary axis means that the system has a sinusoidal mode whose envelope grows as a power of t, so $y_{zi}(t)$ is unstable. At the other extreme, for a system to be *asymptotically stable* all roots must lie entirely within the LHP so that all modes decay exponentially. In this light, the imaginary axis plays the role of the dividing line between unstable and asymptotically stable response. A *marginally stable* system would have one or more nonrepeated roots on the imaginary axis and no RHP roots; then, for large enough t, the dominant mode or modes will be constant or constant-amplitude oscillations.

As implied by the preceding paragraph, fixed linear systems have the unique property that the stability of the zero-input response is independent of the specific initial state. In Chapter 4 a parallel conclusion will be demonstrated for the stability of the zero-state response.

3.3 ZERO-STATE RESPONSE AND COMPLETE RESPONSE

We have spent a great deal of time on the zero-input response because it tells us much about the natural behavior of the system and, more relevant here, it is also helpful when finding the zero-state or forced response. This intriguing fact was implied in the examples of Section 2.2. Now we intend to demonstrate it explicitly by calculating the *complete* response $y(t)$ from the differential equation

$$\sum_{i=0}^{n} a_i y^{(i)} = \sum_{k=0}^{m} b_k x^{(k)} \qquad t \geq t_0$$

where the right-hand side is the forcing function $F_x(t)$. We assume that the input $x(t)$ is zero for $t < t_0$ and that the initial conditions $y_0, \dot{y}_0, \ldots y_0^{(n-1)}$ at $t = t_0$ are known.

To avoid undue complications at this point, three additional restrictions will be imposed on the problem statement, as follows:

1. The roots of the characteristic equation are assumed to be *distinct*, i.e.,

$$p_i \neq p_j \qquad \text{for all } i \neq j$$

which eliminates the notationally messy case of repeated roots.

2. The order of the highest derivative of the input is taken to be less than the order of the highest derivative of the output, i.e.,

$$m \leq n-1$$

Most systems fall in this category; they are termed *strictly proper* systems.

3. The initial value of the input and its first $m-1$ derivatives are assumed to be zero, i.e.,

$$x(t_0) = \dot{x}(t_0) = \cdots = x^{(m-1)}(t_0) = 0$$

With respect to this last restriction, if it does not hold then $F_x(t)$ will be undefined at $t = t_0$ because it contains derivatives of discontinuous functions. In the next chapter the tools necessary to deal with such functions are developed, and those tools are applied to initial-value problems in Chapter 8. For the moment, however, it is sufficient to add that one can simply replace t_0 by some earlier time, say $t_0 - \epsilon$, at which $x(t)$ and its derivatives are zero, thereby circumventing the restriction.

As was done when studying the zero-input response, we begin here with a second-order system.

Complete response of a second-order system

Consider solving the second-order inhomogeneous equation

$$\ddot{y} + a_1 \dot{y} + a_0 y = F_x \tag{1}$$

The approach we shall use to tackle the problem is the *variation-of-parameters* method introduced in Section 2.2. For the homogeneous case ($F_x = 0$) the solution has been found to be of the form

$$y(t) = A_1 e^{p_1 t} + A_2 e^{p_2 t}$$

where p_1 and p_2 are roots of the characteristic equation, i.e.,

$$p_1^2 + a_1 p_1 + a_0 = p_2^2 + a_1 p_2 + a_0 = 0 \tag{2}$$

For the case at hand, with $F_x \neq 0$, one starts by assuming that

$$y(t) = \alpha_1(t) e^{p_1 t} + \alpha_2(t) e^{p_2 t} \tag{3}$$

where, instead of the *constants* A_1 and A_2, the unknown *time functions* $\alpha_1(t)$ and $\alpha_2(t)$ are to be determined.

There are numerous ways to proceed from this point. Perhaps the easiest is to replace (1) by two simultaneous first-order equations. Thus, reminiscent of the state-variable viewpoint, we introduce a new variable $q(t) \triangleq dy(t)/dt$, giving the pair of equations

$$\dot{y} = q \tag{4a}$$

$$\dot{q} + a_1 q + a_0 y = F_x \tag{4b}$$

Then, since (3) is the trial solution for y, what we need now is a trial solution for q. Relative to that question, we again refer to the *homogeneous* case for which $q(t) = p_1 A_1 e^{p_1 t} + p_2 A_2 e^{p_2 t}$ since $q = \dot{y}$. Thus, in the *inhomogeneous* case we assume

$$q(t) = p_1 \alpha_1(t) e^{p_1 t} + p_2 \alpha_2(t) e^{p_2 t} \tag{5}$$

Substituting (3) and (5) into (4) one obtains the pair of equations

$$\begin{bmatrix} 1 & 1 \\ p_1 & p_2 \end{bmatrix} \begin{bmatrix} \dot{\alpha}_1 e^{p_1 t} \\ \dot{\alpha}_2 e^{p_2 t} \end{bmatrix} = \begin{bmatrix} 0 \\ F_x \end{bmatrix} = F_x \begin{bmatrix} 0 \\ 1 \end{bmatrix} \tag{6}$$

in which the terms not involving $\dot{\alpha}_1$, $\dot{\alpha}_2$, or F_x have been canceled out or eliminated via (2). Examining (6) then reveals that finding $\alpha_1(t)$ and $\alpha_2(t)$ requires two steps: simultaneous solution for $\dot{\alpha}_1$ and $\dot{\alpha}_2$, followed by integration.

Introducing the new determinants

$$D_{01} = \begin{vmatrix} 0 & 1 \\ 1 & p_2 \end{vmatrix} = -1 \qquad D_{02} = \begin{vmatrix} 1 & 0 \\ p_1 & 1 \end{vmatrix} = 1$$

the solution of (6) is

$$\dot{\alpha}_1 e^{p_1 t} = F_x \frac{D_{01}}{D} \qquad \dot{\alpha}_2 e^{p_2 t} = F_x \frac{D_{02}}{D}$$

where $D = p_2 - p_1$ as defined in Eq. (11a), Sect. 3.1. Therefore,

$$\frac{d\alpha_1(t)}{dt} = F_x(t) \frac{D_{01}}{D} e^{-p_1 t}$$

and integration yields

$$\alpha_1(t) = \int_{t_0}^{t} F_x(\lambda) \frac{D_{01}}{D} e^{-p_1 \lambda} d\lambda + \alpha_1(t_0) \tag{7a}$$

where λ is a dummy variable and $\alpha_1(t_0)$ the constant of integration. By a like procedure,

$$\alpha_2(t) = \int_{t_0}^{t} F_x(\lambda) \frac{D_{02}}{D} e^{-p_2 \lambda} d\lambda + \alpha_2(t_0) \tag{7b}$$

Finally, we insert (7) into (3) and evaluate $\alpha_1(t_0)$ and $\alpha_2(t_0)$ using the initial conditions $y(t_0) = y_0$ and $\dot{y}(t_0) = \dot{y}_0$. When this is done, the complete response can be written as the sum of two terms, one term being identical to the previously found *zero-input* response, the other involving F_x but independent of the initial conditions so it is properly interpreted as the *zero-state* response. To be specific, we have

$$y(t) = y_{zs}(t) + y_{zi}(t) \tag{8}$$

where $y_{zi}(t)$ is given by Eq. (13), Sect. 3.1 and

$$y_{zs}(t) = \int_{t_0}^{t} F_x(\lambda) G(t-\lambda)\, d\lambda \tag{9a}$$

in which we have introduced

$$\begin{aligned} G(t-\lambda) &\triangleq \frac{D_{01}}{D} e^{p_1(t-\lambda)} + \frac{D_{02}}{D} e^{p_2(t-\lambda)} \\ &= \frac{1}{p_1 - p_2} [e^{p_1(t-\lambda)} - e^{p_2(t-\lambda)}] \end{aligned} \tag{9b}$$

Since Section 3.1 was devoted exclusively to the zero-input response, let us now focus attention on the zero-state or forced response.

Zero-state response of a second-order system

When the zero-state solution to a differential equation takes the form of (9) – that is, an integration of the forcing function $F_x(\lambda)$ times something which is a function of $t-\lambda$ – that "something," $G(t-\lambda)$, is *Green's function* for the system in question. Although such expressions for $y_{zs}(t)$ are perfectly correct, they have the liability of involving both the input $x(t)$ and its derivatives, as contained in the forcing function; e.g., for the case at hand

$$F_x(\lambda) = b_1\frac{dx(\lambda)}{d\lambda} + b_0 x(\lambda).$$

But, recalling Example 2.5, it should be possible to recast our results directly in terms of $x(\lambda)$ rather than $F_x(\lambda)$. And, for purposes of systems analysis it is preferable to do just that, even though further manipulation of (9) will be required.

To begin the manipulation, (9a) must be written out in full as

$$\begin{aligned} y_{zs}(t) &= \int_{t_0}^{t}\left[b_1\frac{dx(\lambda)}{d\lambda} + b_0 x(\lambda)\right]\frac{e^{p_1(t-\lambda)} - e^{p_2(t-\lambda)}}{p_1 - p_2}\,d\lambda \\ &= b_0\int_{t_0}^{t} x(\lambda)\,\frac{e^{p_1(t-\lambda)} - e^{p_2(t-\lambda)}}{p_1 - p_2}\,d\lambda \\ &\quad + \frac{b_1}{p_1 - p_2}\left[\mathcal{I}_{p_1}(t) - \mathcal{I}_{p_2}(t)\right] \end{aligned} \tag{10}$$

where, for $p = p_1$ or p_2,

$$\mathcal{I}_p(t) \triangleq \int_{t_0}^{t}\frac{dx(\lambda)}{d\lambda}\,e^{p(t-\lambda)}\,d\lambda$$

Since the first part of (10) does not involve the derivative of $x(\lambda)$, we turn immediately to the second part and consider the integral $\mathcal{I}_p(t)$. Bringing e^{pt} out of the integrand and integrating by parts yields

$$\begin{aligned} \mathcal{I}_p(t) &= e^{pt}\left[x(\lambda)e^{-p\lambda}\Big|_{t_0}^{t} - \int_{t_0}^{t} x(\lambda)\,d(e^{-p\lambda})\right] \\ &= x(t) + \int_{t_0}^{t} x(\lambda)\,p e^{p(t-\lambda)}\,d\lambda \end{aligned}$$

where we have drawn on the stipulation that $x(t_0) = 0$. Then, setting $p = p_1$ and p_2 shows that

$$\mathcal{I}_{p_1}(t) - \mathcal{I}_{p_2}(t) = \int_{t_0}^{t} x(\lambda)[p_1 e^{p_1(t-\lambda)} - p_2 e^{p_2(t-\lambda)}]\, d\lambda$$

in which $x(t)$ has cancelled out in the subtraction.

Inserting the above into (10) and "turning the crank" gives at last

$$y_{zs}(t) = \int_{t_0}^{t} x(\lambda) \left[\frac{b_0 + b_1 p_1}{p_1 - p_2} e^{p_1(t-\lambda)} - \frac{b_0 + b_1 p_2}{p_1 - p_2} e^{p_2(t-\lambda)}\right] d\lambda \qquad (11a)$$

which is the desired expression for the zero-state response directly in terms of the input and the system parameters – remember that p_1 and p_2 involve the coefficients a_0 and a_1. Alternately, using summation notation and the determinants D, D_{01}, and D_{02}, $y_{zs}(t)$ is

$$y_{zs}(t) = \int_{t_0}^{t} x(\lambda) \left[\sum_{i=1}^{2} \left(\sum_{k=0}^{1} b_k p_i^{\,k}\right) \frac{D_{0i}}{D} e^{p_i(t-\lambda)}\right] d\lambda \qquad (11b)$$

a form that more readily generalizes to higher-order systems.

Example 3.2 Step response of a second-order system

By way of illustration, let us find the zero-state response given a constant input applied just after $t = 0$, i.e.,

$$x(t) = \begin{cases} 1 & t > 0 \\ 0 & t \leq 0 \end{cases}$$

In the context of the next chapter, the resulting $y_{zs}(t)$ will be called the system's *step response*.

Setting $t_0 = 0$ and $x(\lambda) = 1$ in (11a) gives, after some routine rearrangements,

$$\begin{aligned} y_{zs}(t) &= \frac{b_0 + b_1 p_1}{p_1 - p_2} e^{p_1 t} \int_0^t e^{-p_1 \lambda} d\lambda - \frac{b_0 + b_1 p_2}{p_1 - p_2} e^{p_2 t} \int_0^t e^{-p_2 \lambda} d\lambda \\ &= \frac{b_0 + b_1 p_1}{p_1(p_2 - p_1)}(1 - e^{p_1 t}) - \frac{b_0 + b_1 p_2}{p_2(p_2 - p_1)}(1 - e^{p_2 t}) \qquad t \geq 0 \end{aligned}$$

which, taking the case of complex roots, is plotted versus $\omega_n t$ in Figure 3.6 for various values of the ratio b_0/b_1 with $\zeta = a_1/2\sqrt{a_0} = 0.3$.

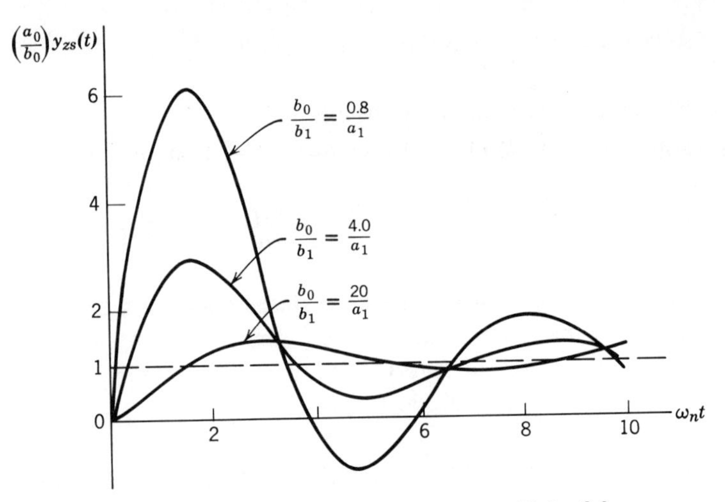

Figure 3.6. Step response of a second-order system with $\zeta = 0.3$.

Complete response of an *n*th-order system

For the complete response of an nth-order system we follow the method used in the second-order case, starting with a set of n first-order equations generated by introducing the new variables

$$\begin{aligned} q_1 &= \dot{y} \\ q_2 &= \dot{q}_1 = \ddot{y} \\ &\vdots \\ q_{n-1} &= \dot{q}_{n-2} = y^{(n-1)} \end{aligned} \tag{12}$$

$$\dot{q}_{n-1} + a_{n-1}q_{n-1} + \cdots + a_1 q_1 + a_0 y = F_x$$

which parallels (4). Assuming distinct roots, the trial solutions are generalized from (3) and (5) as

$$y(t) = \sum_{i=1}^{n} \alpha_i(t) e^{p_i t}$$

$$q_1(t) = \sum_{i=1}^{n} p_i \alpha_i(t) e^{p_i t}$$

$$\vdots$$

$$q_{n-1}(t) = \sum_{i=1}^{n} p_i^{n-1} \alpha_i(t) e^{p_i t}$$

so solving for α_i gives

$$\alpha_i(t) = \int_{t_0}^{t} F_x(\lambda) \frac{D_{0i}}{D} e^{-p_i\lambda} d\lambda + \alpha_i(t_0)$$

Here, D is the nth-order determinant from Eq. (19a), Sect. 3.1, while D_{0i} – the generalization on D_{01} and D_{02} – is given by

$$D_{0i} \triangleq \begin{vmatrix} 1 & \cdots & 0 & \cdots & 1 \\ p_1 & \cdots & 0 & \cdots & p_n \\ \vdots & & \vdots & & \vdots \\ p_1^{n-2} & \cdots & 0 & \cdots & p_n^{n-2} \\ p_1^{n-1} & \cdots & 1 & \cdots & p_n^{n-1} \end{vmatrix} \quad (i\text{th column at the } 0, \ldots, 1 \text{ column}) \tag{13}$$

Note that D_{0i} corresponds to D_i as in Eq. (19b), Sec. 3.1, when

$$\begin{aligned} y_0 = \dot{y}_0 = \cdots = y_0^{(n-2)} &= 0 \\ y_0^{(n-1)} &= 1 \end{aligned} \tag{14}$$

the so-called *homogeneous initial conditions*.

Not unexpectedly, the complete response has the form

$$y(t) = y_{zs}(t) + y_{zi}(t)$$

where $y_{zi}(t)$ is given by Eq. (20), Sec. 3.1, and

$$y_{zs}(t) = \int_{t_0}^{t} F_x(\lambda) \underbrace{\left[\sum_{i=1}^{n} \frac{D_{0i}}{D} e^{p_i(t-\lambda)}\right]}_{G(t-\lambda)} d\lambda \tag{15}$$

so, as indicated, the quantity in brackets is Green's function for an nth-order system with distinct roots. Finally, writing out $F_x(\lambda)$ and repeatedly integrating by parts, the zero-state response comes out paralleling (11b) as

$$y_{zs}(t) = \int_{t_0}^{t} x(\lambda) \left[\sum_{i=1}^{n} \left(\sum_{k=0}^{m} b_k p_i^k\right) \frac{D_{0i}}{D} e^{p_i(t-\lambda)}\right] d\lambda \tag{16}$$

which requires the assumptions $m \leq n-1$ and

$$x(t_0) = \dot{x}(t_0) = \cdots = x^{(m-1)}(t_0) = 0$$

The intervening steps, which are more tedious than illuminating, have been omitted.

One should not be discouraged by this rather formidable appearing expression; the important point to notice is its relation to the *zero-input response* and the *natural-behavior modes*. To develop this relationship, we insert the homogeneous initial conditions of (14) into Eq. (20), Sect. 3.1. Then

$$y_{zi}(t) = \sum_{i=1}^{n} \eta_i(t-t_0)$$

where

$$\eta_i(t) \triangleq \frac{D_{0i}}{D} e^{p_i t} \tag{17}$$

so each η_i is proportional to one mode of the system. But, looking back at (16) shows that, in terms of η_i,

$$y_{zs}(t) = \int_{t_0}^{t} x(\lambda)\Big[\sum_{i=1}^{n} C_i \eta_i(t-\lambda)\Big]\,d\lambda \tag{18a}$$

where

$$C_i \triangleq \sum_{k=0}^{m} b_k p_i^k \tag{18b}$$

Therefore, the *forced* response of a fixed linear system is an integration involving the input and a certain linear combination of the mode functions. A physical interpretation of this special combination of mode functions will be derived in the next chapter.

A second point to be made here is that any system described by the model of Eq. (1), Sect. 3.1, is indeed a *fixed linear* system. We have assumed this to be true all along but have not proven it using the definitions of Section 2.4. In this section it has been shown that the complete response is decomposable into the sum of the zero-state and zero-input responses, and the linearity and time-invariance of $y_{zs}(t)$ and $y_{zi}(t)$ are readily confirmed from (16) and Eq. (20), Sect. 3.1.

As the final point of this section, let us consider another way of breaking down the complete response $y(t)$, namely in terms of the transient and steady-state components.

Transient and steady-state response

Using variation of parameters to solve linear differential equations, we have identified and separated $y_{zs}(t)$ and $y_{zi}(t)$. However, the reader is perhaps more familiar with the classical method which starts by writing

$$y(t) = y_P(t) + y_H(t) \tag{19}$$

where $y_P(t)$ is any *particular* solution of the inhomogeneous equation while $y_H(t)$ is the complementary solution. These terms must be carefully distinguished from the zero-state and zero-input responses, especially since both $y_H(t)$ and $y_{zi}(t)$ satisfy the homogeneous equation.

Distinction is based on how the initial conditions enter the method. In the classical approach the constants in $y_H(t)$ are evaluated after the complete solution has been formed; therefore, they depend in part on $y_P(t)$ which, in turn, depends on the input $x(t)$. On the other hand, $y_{zi}(t)$ is independent of $x(t)$, by definition, and its constants are determined exclusively by the initial conditions. Thus, with one important exception, the latter is preferred for systems analysis, the exception being when we are interested in the transient and steady-state behavior.

Consider an asymptotically stable system which is excited by an input – say a constant or a sinusoid – applied at some instant and continuing indefinitely thereafter. The system's asymptotic stability means that the natural-behavior modes will eventually die away, leaving only an output term whose form is dictated directly by the forcing input. The latter is known as the *steady-state* response, while the initial period of adjustment or "settling down" includes both the steady-state and *transient* responses. Under the conditions we have described, the steady-state and transient components correspond one-to-one with $y_P(t)$ and $y_H(t)$ in (19).

By way of illustration, suppose the RC circuit in Example 2.4 has an initial output y_0 at $t = 0$, at which time $x(t) = At$ is applied. From Eq. (8), Sect. 2.2, the resulting complete response is easily found as being

$$y(t) = \overbrace{\underbrace{A(t - RC)}_{y_P(t)} + ARCe^{-t/RC}}^{y_{zs}(t)} + \overbrace{y_0 e^{-t/RC}}^{y_{zi}(t)} \tag{20}$$

where $y_H(t) = ARCe^{-t/RC} + y_0 e^{-t/RC}$.

Note that the first term, which is $y_P(t)$, is the steady-state response since the remaining terms, marked as $y_H(t)$, go to zero for large t. Also note that $y_{zs}(t)$ contains part of the transient behavior along with the steady-state component. Thus, if one is concerned only with the steady-state situation, the smart thing to do would be to calculate $y_P(t)$ rather than $y_{zs}(t)$. On the other hand, if the system is in the zero state and one is concerned with the complete response, $y_P(t)$ does not give the whole story but $y_{zs}(t)$ does. We shall return again to the matter of transients and steady state in Chapter 8, where the Laplace transform is developed as the most effective weapon.

3.4 SIMULATION DIAGRAMS

Although mathematical methods for system analysis are important – and the main concern of this text – there are times when it is more convenient or even absolutely necessary to resort to experimental techniques, particularly for high-order systems. Rather than working with the actual system, which may not be practical (e.g., the system may not yet be built), many experimental investigations use computers to *simulate* its behavior. Although simulation can be carried out on analog, digital, or hybrid computers, according to the needs of the problem at hand, the material we cover here is most directly applicable to analog simulation.

Analog simulation allows one to model and study in the laboratory virtually any system described by a differential equation like Eq. (1), Sec. 3.1, plus a variety of other types. Furthermore, the block diagrams associated with analog simulation are useful in themselves for analytic purposes, regardless of whether or not simulation is actually going to be done, and the basic ideas carry over to digital and hybrid simulation.

The three principal components available in analog simulation are: *summing junctions*, *scalors* (i.e., amplifiers, attenuators, and inverters), and *integrators*, shown diagramatically in Figure 3.7. In view of the fact

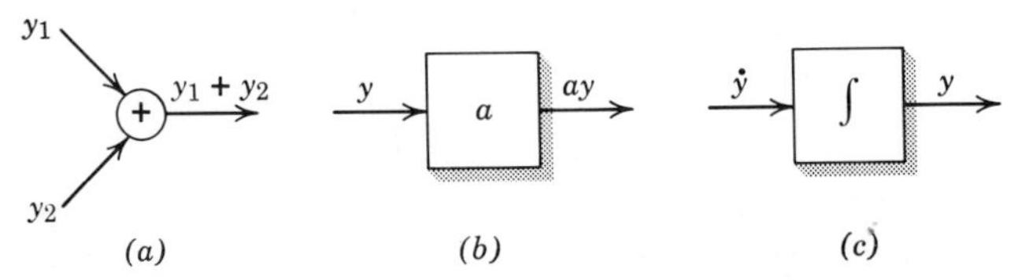

Figure 3.7. Simulation building blocks. (*a*) Summing junction. (*b*) Scalor. (*c*) Integrator.

that we have dealt primarily with *differential* equations it may seem peculiar that *differentiators* are not used. As it happens – and we shall demonstrate the fact immediately below – differentiation essentially is accomplished by using an *integrator* in a *feedback* configuration. Given the choice between a differentiator and an integrator, practical considerations such as drift, noise, etc., favor the latter. Moreover, simulation diagrams constructed in this fashion prove to have greater conceptual merit.

Consider the first-order homogeneous equation $\dot{y} + a_0 y = 0$ or $\dot{y} = -a_0 y$. Assuming we have obtained $\dot{y}$ as part of the simulation, integrating it will yield y. And if y is multiplied by $-a_0$ the product equals $\dot{y}$, implying that the two signals are the result of a circular or "boot-strap" operation. Hence, using an integrator and an amplifier with gain $-a_0$ in the configuration of Figure 3.8, the equation in question can be simulated without requiring a differentiator.

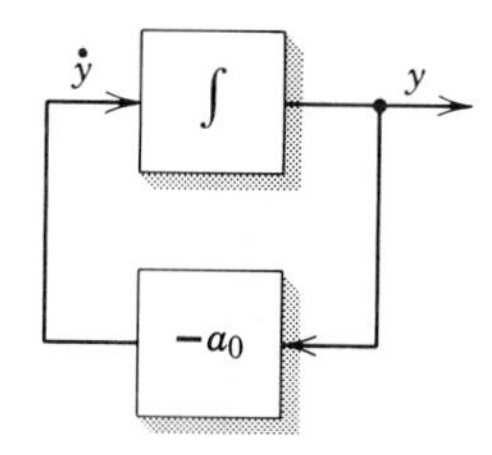

Figure 3.8. Simulation of $\dot{y} + a_0 y = 0$.

By the same reasoning, Figure 3.9 is a simulation of $\ddot{y} + a_1\dot{y} + a_0 y = x$ since the input to the first integrator is $\ddot{y} = x - a_1\dot{y} - a_0 y$. Note carefully the trick here: one simply writes out the terms being summed to form the input to the first integrator, which should equal the highest derivative of

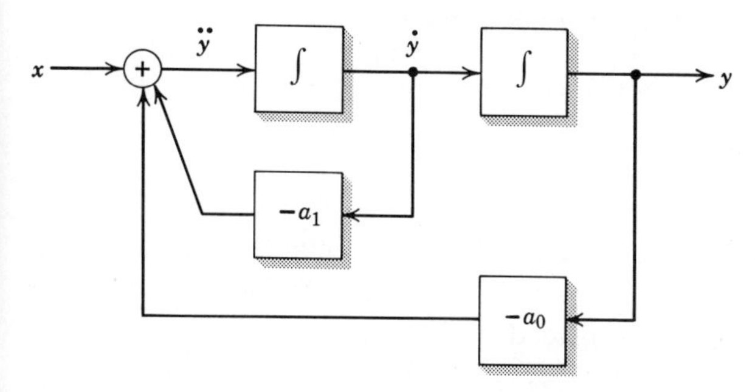

Figure 3.9. Simulation of $\ddot{y} + a_1\dot{y} + a_0 y = x$.

the variable being simulated. This procedure immediately extends to nonhomogeneous equations of the form

$$y^{(n)} + a_{n-1}y^{(n-1)} + \cdots + a_1\dot{y} + a_0 y = x$$

Again, we solve for the highest derivative of the unknown to get

$$y^{(n)} = x - [a_{n-1}y^{(n-1)} + \cdots + a_1\dot{y} + a_0 y] \tag{1}$$

so n integrators and scalors are needed, plus an external generator for $x(t)$. Figure 3.10 is the resulting block diagram.

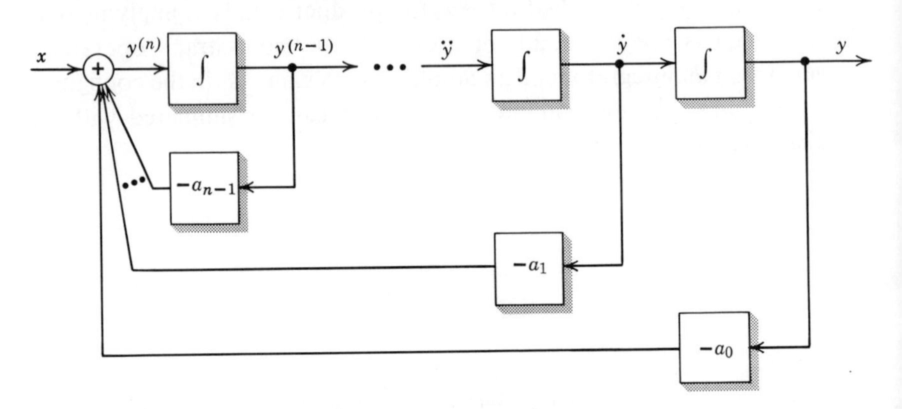

Figure 3.10. Simulation of $y^{(n)} + a_{n-1}y^{(n-1)} + \cdots + a_1\dot{y} + a_0 y = x$.

But how do we handle the more general case when the forcing function $F_x(t)$ contains derivatives of x? To answer this question consider first the equation

$$\ddot{y} + a_1\dot{y} + a_0 y = b_1\dot{x} + b_0 x \tag{2}$$

One method for simulating (2) is based on introducing an auxiliary variable $q(t)$ such that

$$\ddot{q} + a_1\dot{q} + a_0 q = x \tag{3}$$

so q is readily simulated by the previous technique. This auxiliary variable has obviously been chosen with malice of forethought for, if $q(t)$ obeys (3), then (2) can be replaced by

$$y = b_1\dot{q} + b_0 q \tag{4}$$

for simulation purposes. To prove this assertion, substitute (4) into (2):

$$\overbrace{(b_1\dddot{q} + b_0\ddot{q})}^{\ddot{y}} + a_1\overbrace{(b_1\ddot{q} + b_0\dot{q})}^{\dot{y}} + a_0\overbrace{(b_1\dot{q} + b_0 q)}^{y} = b_1\dot{x} + b_0 x$$

or, regrouping terms,

$$b_1\underbrace{(\dddot{q}+a_1\ddot{q}+a_0\dot{q})}_{\dot{x}}+b_0\underbrace{(\ddot{q}+a_1\dot{q}+a_0q)}_{x}=b_1\dot{x}+b_0x$$

Now, as indicated, if (3) and its first derivative are used to eliminate q, we obtain an identity – thereby confirming that (3) and (4) are equivalent to (2). It immediately follows that we can simulate y by simulating q and $\dot{q}$ and forming the linear combination $y=b_1\dot{q}+b_0q$. The corresponding block diagram is given in Figure 3.11.

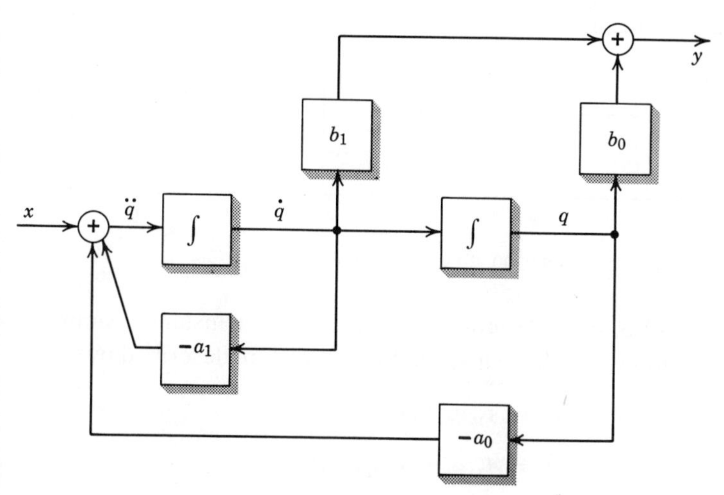

Figure 3.11. Simulation of $\ddot{y}+a_1\dot{y}+a_0y=b_1\dot{x}+b_0x$.

In like manner we can simulate the equation

$$\sum_{i=0}^{n} a_i y^{(i)} = \sum_{k=0}^{m} b_k x^{(k)} \qquad a_n = 1 \tag{5}$$

providing $m \le n$. Figure 3.12 is the complete diagram when $m=n-1$; for $m<n-1$ one simply sets $b_{m+1}=b_{m+2}=\cdots=b_{n-1}=0$. Other equivalent simulations are found in the literature, but Figure 3.12 has the advantage of explicitly displaying the coefficients a_i and b_k from the differential equation. We shall refer to this diagram as the *canonic* form.

Simulation diagrams are particularly easy to construct when the system model is couched in terms of state variables, that is, as a set of first-order

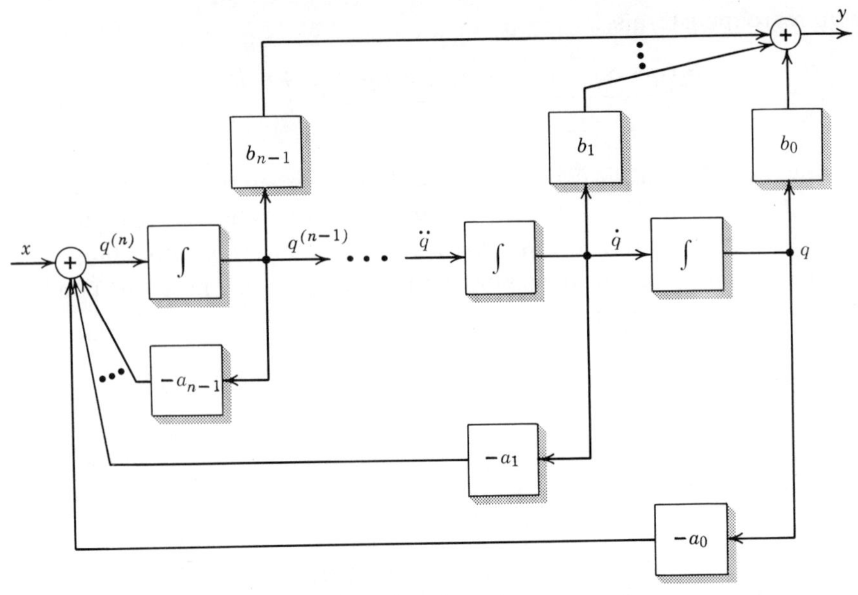

Figure 3.12. Canonic simulation of an nth-order system, $m = n - 1$.

equations instead of a single nth-order equation. For instance, similar to Example 2.6, suppose we have a second-order system described by

$$
\begin{aligned}
\dot{q}_1 &= A_{11}q_1 + A_{12}q_2 + B_1 x \\
\dot{q}_2 &= A_{21}q_1 + A_{22}q_2 + B_2 x \qquad (6) \\
y &= C_1 q_1 + C_2 q_2
\end{aligned}
$$

where the A's, B's, and C's are constants. A direct simulation of this system is given in Figure 3.13, which has the integrators in *parallel* rather than in series. The diagram nicely brings out how the two first-order subsystems are coupled to each other via the scalors A_{12} and A_{21}.

Two more items remain to be mentioned in this brief discussion. First, there is the question of initial conditions and the zero-input response. Although not included in Figure 3.7, analog-computer integrators have a special input terminal for establishing the initial output value at the time that simulation is begun, i.e., if we start the computer at time t_0, and $\dot{y}(t)$ is an integrator's input, then the corresponding output is

$$y(t) = \int_{t_0}^{t} \dot{y}(\lambda)\,d\lambda + y_0 \qquad t \geq t_0$$

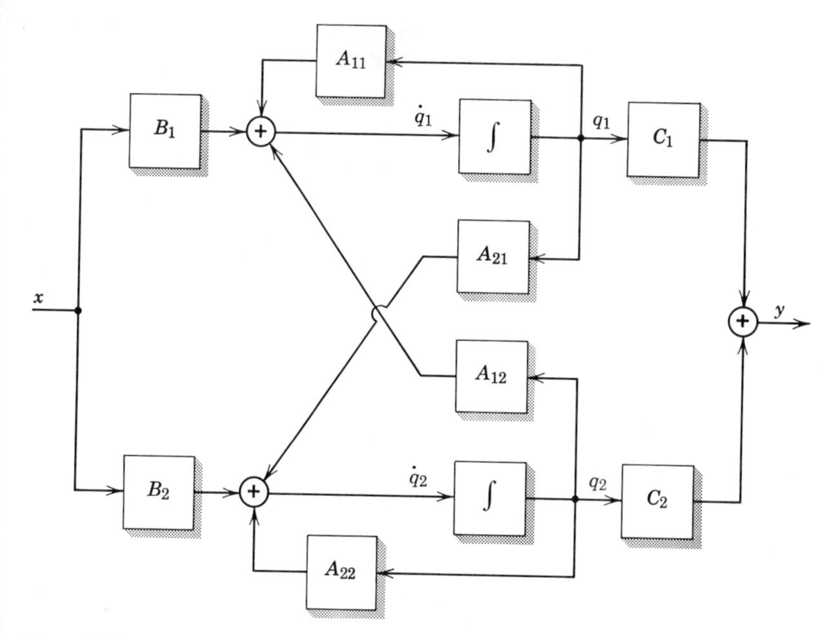

Figure 3.13.

Therefore, one must set the appropriate initial conditions on each and every variable that appears at the output of an integrator.

Second, simulation diagrams such as Figures 3.12 or 3.13 cannot be transferred directly to the analog machine. For one reason computer components have a limited range of linearity so the variables have to be *scaled in magnitude* to keep them within the linear range; for another, the rate of change of the variables may be too fast for the machine's response time or too slow for convenient observation, in which case *time scaling* is necessary. Additionally, most computer amplifiers and integrators produce a sign inversion, and amplifiers have fixed gains so that adjustable gain requires an input potentiometer. Thus, the resulting "patching" diagram looks somewhat different from the simulation diagram. Blum (1969) gives an excellent introduction to the techniques of analog computation, and is suitable for self-study.

3.5 STATE EQUATIONS

Thus far in this chapter we have dealt exclusively with single-input-output systems, and they will continue to be the major subject of later chapters. But some of the most challenging and fascinating problems of contemporary engineering are those of multi-input-output systems of rather high order. In view of the fact that space-age systems are being called on to perform many functions simultaneously, with greater accuracy and in less time, the move toward ever-increasing system complexity is an understandable and inevitable consequence. By the same token, it puts a severe burden on the engineer who designs these complicated systems with very stringent tolerances and little opportunity to experimentally check out his design as a whole.

Because of such demands, the *state-space approach* based on the concept of state variables offers a more convenient means of analysis and design than the straight input-output viewpoint. Here, we shall expand upon our earlier discussion of the state of a system, and show how state equations are related to input-output equations. As a preliminary, we recall that the state vector of an nth-order system is any signal vector $\mathbf{q}(t)$ with n components such that

$$\mathbf{q}(t) = \mathbf{S}_q[\mathbf{q}(t_0); \mathbf{x}(t)] \qquad t \geq t_0 \tag{1a}$$

and

$$\mathbf{y}(t) = \mathbf{S}[\mathbf{q}(t_0); \mathbf{x}(t)] \qquad t \geq t_0 \tag{1b}$$

which are the state and output equations, respectively.

Single-input-output systems

Consider an nth-order linear system (not necessarily fixed) with one input $x(t)$ and one output $y(t)$, whose differential equation is

$$\sum_{i=0}^{n} a_i(t) y^{(i)}(t) = \sum_{k=0}^{m} b_k(t) x^{(k)}(t) \qquad m \leq n-1 \tag{2}$$

We are including the time-varying case by showing the coefficients as functions of time; if the system in question is time-invariant then the coefficients are constants.

Introducing the state variables $q_1(t), q_2(t), \ldots q_n(t)$, the system can be described by n first-order state equations of the general form

$$\begin{aligned} \dot{q}_1 &= A_{11}q_1 + A_{12}q_2 + \cdots + A_{1n}q_n + B_1 x \\ \dot{q}_2 &= A_{21}q_1 + A_{22}q_2 + \cdots + A_{2n}q_n + B_2 x \\ &\vdots \\ \dot{q}_n &= A_{n1}q_1 + A_{n2}q_2 + \cdots + A_{nn}q_n + B_n x \end{aligned} \tag{3a}$$

plus an output equation†

$$y = C_1 q_1 + C_2 q_2 + \cdots + C_n q_n \tag{3b}$$

where the new coefficients are related to the system parameters and may be time-dependent if the system is not fixed.

It is not obvious that the state variables in (3) do, in truth, have the properties required by (1). Verifying this fact would entail solving the equations to obtain the explicit forms $\mathbf{q}(t) = \mathbf{S}_q[\mathbf{q}(t_0); x(t)]$ and $y(t) = S[\mathbf{q}(t_0); x(t)]$. These solutions are not attempted here since they go beyond our intended scope. Suffice it to say that equations like (3) almost always are valid state equations.

To write (3) more compactly we use the column vector

$$\mathbf{q}(t) = [q_1(t) \quad q_2(t) \cdots q_n(t)]^T$$

and define the $n \times n$ *matrix* consisting of the coefficients $A_{ij}(t)$, namely

$$\mathbf{A}(t) \triangleq \begin{bmatrix} A_{11}(t) & A_{12}(t) & \cdots & A_{1n}(t) \\ A_{21}(t) & A_{22}(t) & \cdots & A_{2n}(t) \\ \vdots & \vdots & & \vdots \\ A_{n1}(t) & A_{n2}(t) & \cdots & A_{nn}(t) \end{bmatrix} \tag{4}$$

We also need two $n \times 1$ matrices for the B and C coefficients, say

$$\begin{aligned} \mathbf{b}(t) &\triangleq [B_1(t) \quad B_2(t) \quad \cdots \quad B_n(t)]^T \\ \mathbf{c}(t) &\triangleq [C_1(t) \quad C_2(t) \quad \cdots \quad C_n(t)]^T \end{aligned} \tag{5}$$

With these definitions, the set of state equations becomes the matrix equation

$$\dot{\mathbf{q}}(t) = \mathbf{A}(t)\mathbf{q}(t) + \mathbf{b}(t)x(t) \tag{6a}$$

while the output equation is

$$y(t) = \mathbf{c}^T(t)\mathbf{q}(t) \tag{6b}$$

where $\mathbf{c}^T(t)$, being the transpose of an $n \times 1$ column matrix, is a $1 \times n$ row matrix. If the system is fixed, then **A**, **b**, and **c** are simply constant matrices. But fixed or time-varying, (6) succinctly constitutes a state-space model of the system.

†For $m > n-1$ or some selections of the state variables, the output equation also must include $x(t)$.

Having thus disposed of notational liabilities, one may still ask if the state-space description is worth the effort. Indeed, from a quick comparison of (2) and (6), it might appear that the latter has little to offer in the way of advantages, and has several seeming disadvantages: Where, for instance, do the state variables come from? And why bother with n simultaneous equations plus an output equation when the same information is contained in one nth-order equation?

Taking these questions one at a time, we have already shown that state variables often arise quite naturally in the course of modeling a system from its physical description. In fact, it usually takes additional labor to get a single higher-order equation in terms of just the input and output. But, as previously mentioned, the state vector is not unique; any $\mathbf{q}(t)$ satisfying (1) qualifies as a state vector. This means that we have considerable latitude in choosing the state variables, and that they need not even correspond to physical variables in the system – although that is perhaps the most appealing choice.

Emphasizing this point, suppose all we have is the differential equation (2) instead of a physical description of the system. Several systematic procedures exist for obtaining state equations therefrom, one being based on the canonic simulation, Figure 3.12; that diagram is repeated here as Figure 3.14. Since the input to each integrator is the first derivative of its output, we can immediately write down a set of state equations by taking the state variables as those *integrator outputs*, i.e.,

$$\dot{q}_1 = q_2 \quad \dot{q}_2 = q_3 \quad \cdots \quad \dot{q}_{n-1} = q_n$$

$$\dot{q}_n = -a_0 q_1 - a_1 q_2 - \cdots - a_{n-1} q_n + x$$

Also by inspection of the diagram, the output equation is

$$y = b_0 q_1 + b_1 q_2 + \cdots + b_{n-1} q_n.$$

Thus, for this choice of state variables, the coefficient matrices directly display the coefficients of the differential equation:

$$\mathbf{A} = \begin{bmatrix} 0 & 1 & 0 & \cdots & 0 \\ 0 & 0 & 1 & \cdots & 0 \\ \vdots & \vdots & \vdots & & \vdots \\ -a_0 & -a_1 & -a_2 & \cdots & -a_{n-1} \end{bmatrix} \tag{7a}$$

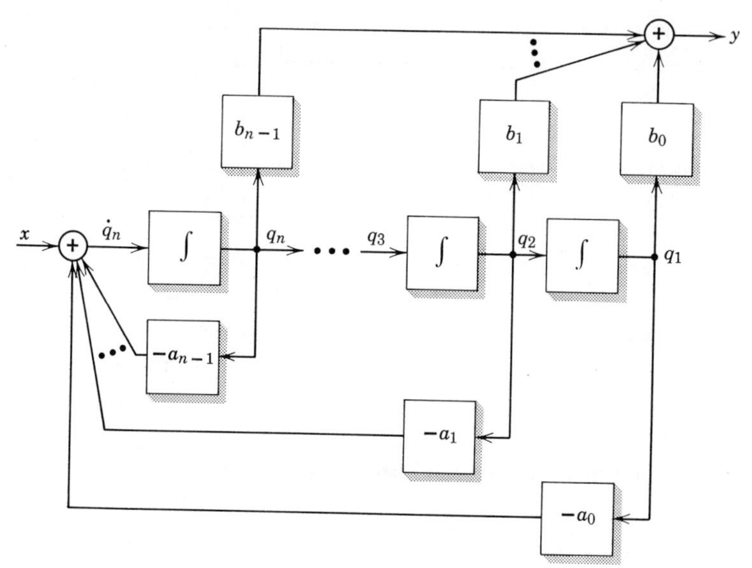

Figure 3.14.

$$\mathbf{b} = [0 \quad 0 \cdots 0 \quad 1]^T \tag{7b}$$

$$\mathbf{c} = [b_0 \quad b_1 \cdots b_{n-1}]^T \tag{7c}$$

which, when inserted in (6), constitutes the *canonic* state-variable model of an SIO system.

Granted that the state equations may be easier to formulate than the corresponding nth-order equation, this in itself is only a minor benefit. A major benefit of the state-space approach is the conceptual clarity and notational convenience when dealing with MIO systems, a subject we shall turn to momentarily. Another important benefit is the knowledge one can gain about what is going on *inside* the system. This aspect is illustrated by the following example.

Example 3.3

Suppose a physical system corresponds to the simulation diagrammed in Figure 3.15. With the state variables as indicated

$$\dot{q}_1 = q_1 + x \qquad \dot{q}_2 = q_1 - 3q_2$$

and

$$y = -\tfrac{1}{4}q_1 + q_2$$

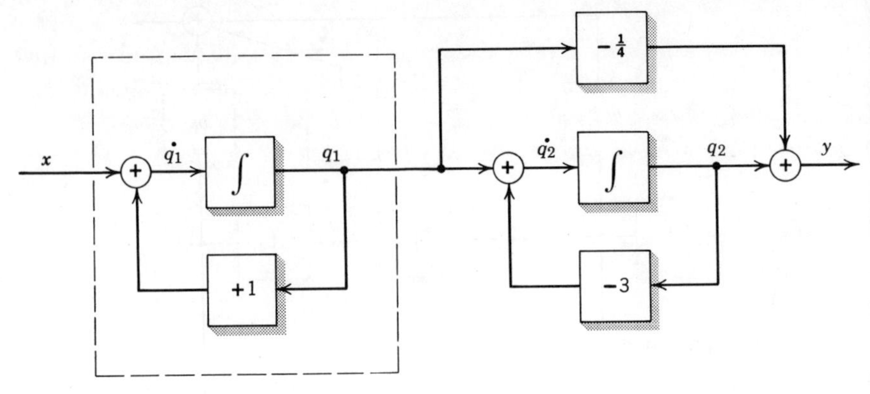

Figure 3.15.

Thus, upon eliminating q_1 and q_2, the input-output relation is

$$\dot{y}+3y=-\tfrac{1}{4}x$$

Judging only from the latter, one would conclude that the system is *first-order* and *asymptotically stable*—as follows from the characteristic equation $s+3=0$. But the simulation has two integrators, implying a *second-order* system. Moreover, the first state equation can be recast as $\dot{q}_1-q_1=x$, which corresponds to an RHP root, namely at $s=+1$; thus, the q_1 mode grows as e^{+t} and the subsystem inside the dashed lines is, by itself, *unstable*.

These seemingly contradicting conclusions are perfectly correct. The contradiction is resolved with further study of the diagram, for such a study will reveal that the unstable behavior is cancelled at the output summing junction and does not appear in the output.† (The reader should verify this for himself by taking $q_1=e^t$ and solving for y.) Consequently, insofar as input-output is concerned, the total system acts as if it were first-order and stable.

Nonetheless, the internal variable q_1 may very well increase without bound while x and y remain finite. Furthermore, a slight change in any of the parameters voids the cancellation effect and the total system becomes unstable. Clearly, the engineer ought to be aware of these potentially grave circumstances. Equally clear, the necessary information may be missing from the input-output equation and contained only in the state equations.

†We therefore say that q_1 is an *unobservable* state. See Zadeh and Desoer (1963, Chap. 11).

Multi-input-output systems

Consider an nth-order linear system with p input signals and r output signals. To formulate the state equations we observe that, paralleling (3a), every element of $\dot{\mathbf{q}}(t)$ will be a linear combination of all the state variables plus the p input signals, i.e.,

$$\dot{q}_i = A_{i1}q_1 + \cdots + A_{in}q_n + B_{i1}x_1 + \cdots + B_{ip}x_p$$

Similarly, there are r output equations of the form

$$y_i = C_{i1}q_1 + \cdots + C_{in}q_n$$

Expressing these in matrix notation we define an $n \times p$ matrix for the B coefficients

$$\mathbf{B}(t) \triangleq \begin{bmatrix} B_{11}(t) & B_{12}(t) & \cdots & B_{1p}(t) \\ \vdots & \vdots & & \vdots \\ B_{n1}(t) & B_{n2}(t) & \cdots & B_{np}(t) \end{bmatrix} \tag{8a}$$

and an $r \times n$ matrix for the C coefficients

$$\mathbf{C}(t) \triangleq \begin{bmatrix} C_{11}(t) & C_{12}(t) & \cdots & C_{1n}(t) \\ \vdots & \vdots & & \vdots \\ C_{r1}(t) & C_{r2}(t) & \cdots & C_{rn}(t) \end{bmatrix} \tag{8b}$$

which correspond to $\mathbf{b}(t)$ and $\mathbf{c}^T(t)$ in the SIO case. Then, with $\mathbf{x}(t)$ and $\mathbf{y}(t)$ as the input and output signal vectors, we have the state-space matrix equations

$$\dot{\mathbf{q}}(t) = \mathbf{A}(t)\mathbf{q}(t) + \mathbf{B}(t)\mathbf{x}(t) \tag{9a}$$

$$\mathbf{y}(t) = \mathbf{C}(t)\mathbf{q}(t) \tag{9b}$$

where $\mathbf{q}(t)$ is the state vector and $\mathbf{A}(t)$ is as previously defined by (4). Corresponding to these equations, one can draw a *matrix block diagram* of the system, Figure 3.16, where double flow lines mean vector signals and boldface symbols are matrix operations.

Probably the most striking feature of (9) or Figure 3.16 is that it is not appreciably more complicated than a *first-order SIO* system. The addition of multiple inputs or outputs changes the *degree* of the problem but not the basic solution methods, all other factors being equal. In witness of this assertion, let us outline how one could tackle the zero-input response of a linear, time-invariant MIO system. The details and the complete response will be discussed in Chapter 8.

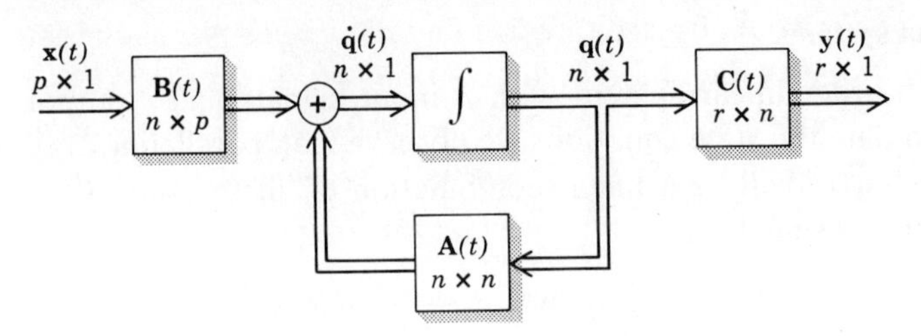

Figure 3.16. Matrix block diagram for an MIO system.

Setting $\mathbf{x}(t) = \mathbf{0}$, $\mathbf{A}(t) = \mathbf{A}$ and $\mathbf{C}(t) = \mathbf{C}$ in (9) gives

$$\mathbf{y}_{zi}(t) = \mathbf{C}\mathbf{q}(t) \tag{10a}$$

where $\mathbf{q}(t)$ must satisfy the homogeneous state equation

$$\dot{\mathbf{q}}(t) - \mathbf{A}\mathbf{q}(t) = \mathbf{0} \tag{10b}$$

subject to the initial condition $\mathbf{q}(t_0)$. We begin, as before, by assuming an exponential trial solution for (10b), say

$$\mathbf{q}(t) = [G_1 e^{st} \quad G_2 e^{st} \cdots G_n e^{st}]^T$$

in which the G's are constants. It then follows that $\dot{\mathbf{q}}(t) = s\mathbf{q}(t)$ and, hence, (10b) becomes

$$(s\mathbf{I} - \mathbf{A})\mathbf{q}(t) = \mathbf{0} \tag{11}$$

where $\mathbf{I}$, the $n \times n$ *identity matrix*, has been introduced to permit factoring.

Discarding the trivial case $\mathbf{q}(t) = \mathbf{0}$, matrix theory says that (11) requires the *determinant* of $(s\mathbf{I} - \mathbf{A})$ to vanish, i.e.,

$$|s\mathbf{I} - \mathbf{A}| = 0 \tag{12}$$

and it can be shown that $|s\mathbf{I} - \mathbf{A}|$ is an nth-order polynomial in s having n roots, $s = p_1, p_2, \ldots p_n$. Accordingly, if the roots are distinct, our trial solution must be modified such that each element of $\mathbf{q}(t)$ takes the form

$$q_i(t) = G_{i1} e^{p_1 t} + G_{i2} e^{p_2 t} + \cdots + G_{in} e^{p_n t}$$

Then, in principle, the G_{ij} can be evaluated from the initial conditions and $\mathbf{q}(t)$ inserted into (10a) to yield $\mathbf{y}_{zi}(t)$. However, that is not the real point of our investigation, since a better method for proceeding from (12) will be given in Chapter 8.

The important point to notice here is that (12) plays the role of the *characteristic equation*, i.e., from it one can find the values of s such that the trial solution satisfies the differential equation. Thus, we shall *define* $|s\mathbf{I}-\mathbf{A}|$ as the *characteristic polynomial* in the MIO case. Having done that, this valuable concept and its interpretations are generalized to MIO systems, and SIO systems become a subclass thereof.

Example 3.4

As a simple check on this last point, consider a third-order SIO system described by $\dddot{y}+a_2\ddot{y}+a_1\dot{y}+a_0y=F_x$. Earlier we said its characteristic polynomial is $s^3+a_2s^2+a_1s+a_0$, which we now compare with the generalized definition $|s\mathbf{I}-\mathbf{A}|$.

Taking the canonic state-variable model, the **A** matrix is given by (7a) as

$$\mathbf{A}=\begin{bmatrix}0 & 1 & 0\\ 0 & 0 & 1\\ -a_0 & -a_1 & -a_2\end{bmatrix}$$

Hence,

$$|s\mathbf{I}-\mathbf{A}|=\left|\begin{bmatrix}s & 0 & 0\\ 0 & s & 0\\ 0 & 0 & s\end{bmatrix}-\begin{bmatrix}0 & 1 & 0\\ 0 & 0 & 1\\ -a_0 & -a_1 & -a_2\end{bmatrix}\right|$$

$$=\begin{vmatrix}s & -1 & 0\\ 0 & s & -1\\ a_0 & a_1 & s+a_2\end{vmatrix}=s^3+a_2s^2+a_1s+a_0$$

so we have perfect agreement. Of course, the choice of **A** is not sacred since the state vector is not unique; the interesting implication is that, for a given system, $|s\mathbf{I}-\mathbf{A}|$ will be the same with *any* valid **A**.

Advantages of the state-space approach

To summarize, the state-space approach has at least three major advantages compared to straight input-output analysis: conceptual clarity, greater information about the system itself, and computational convenience.

From the conceptual viewpoint, the state equations provide a mathematical model of great generality that is readily extended to include time-varying, nonlinear, and MIO systems. Similarly, the matrix notation is a compact vehicle for analytic manipulation, particularly when the powerful techniques of linear algebra are brought to bear on the problem. As a matter of fact, without these features the main results of modern systems theory would have been quite difficult to obtain.

When the state variables are chosen as directly related to internal variables of the system, it is clear that we will get more information about the system itself than we do just from input-output considerations, and sometimes that additional information is critical. But regardless of how the state variables are selected, the state equations further enhance our intuitive grasp of the problem.

Finally, in the case of very complex systems – e.g., large n, nonlinearities, many inputs or outputs, etc. – computer-aided solution is almost always a necessity. For this purpose, the state equations are far better suited to analog simulation or digital computation than an nth-order differential equation, whether or not one is also interested in the state variables themselves.

Having thus expounded upon the merits of state-space analysis, the reader may find it puzzling that the method is not fully detailed in later chapters. There are two reasons for our stopping short on the subject. First, solving the state equations requires a knowledge of matrix theory above the level assumed in this book. Second, the real payoff of the state-space approach comes when one is faced with highly complex systems – whereas we shall have our hands full just investigating rather simple systems with a reasonable degree of thoroughness. However, from time to time we shall invoke the state concept as an aid to interpretation and understanding.

Problems

3.1 Verify that Eq. (13), Sect. 3.1, gives the specified values for y and $\dot{y}$ at $t = t_0$.

3.2 Show that Eq. (15), Sect. 3.1, is a solution of Eq. (7) when the roots are repeated, and confirm that Eq. (16) yields the specified initial values.

3.3 Suppose the roots of a second-order system are nearly, but not exactly equal, so $p_1 = p + \epsilon$ and $p_2 = p - \epsilon$ where $|\epsilon|^2 << |p|^2$. Show that

$$y(t) \approx (B_1 + B_2 \epsilon t) e^{pt}$$

3.4 The characteristic equation of a certain system is $(s+1)(s^2+9)=0$. Solve for $y_{zi}(t)$, $t \geq 0$, if $y_0 = 10$ and $\dot{y}_0 = \ddot{y}_0 = 0$ at $t=0$. *Answer*: $9e^{-t} + \cos 3t + 3 \sin 3t$.

3.5 Consider a third-order system having $s^3 + s^2 + \frac{5}{4}s = 0$. Find and sketch the zero-input response when $\dot{y}_0 = 2$ and $y_0 = \ddot{y}_0 = 0$.

3.6 Make a sketch similar to Figure 3.3 for the mode functions corresponding to a repeated complex root.

3.7 Consider a fourth-order system having $(s^2+s+2)^2 = 0$. Plot the roots in the s-plane and sketch a typical zero-input response.

3.8 Express $|A|$ and arg $[A]$ in Eq. (5), Sect. 3.2, in terms of the initial conditions $y(t_0)$ and $\dot{y}(t_0)$.

3.9 Classify the stability of the zero-input response for each of the following characteristic polynomials: (a) s^2+3s; (b) $(s^2+3s)^2$; (c) $s^2+7s+10$; (d) s^2+2s+5; (e) s^2-2s+5.

3.10 For a second-order system with repeated roots, $p_1 = p_2 = p$, assume the trial solution $y(t) = \alpha_1(t)e^{pt} + \alpha_2(t)te^{pt}$ to derive the complete response

$$y(t) = \int_{t_0}^{t} F_x(\lambda)(t-\lambda)e^{p(t-\lambda)}d\lambda + [\alpha_1(t_0)e^{pt} + \alpha_2(t_0)te^{pt}]$$

3.11 Rewrite $y_{zs}(t)$ in Example 3.2 in terms of σ and ω when $p_1, p_2 = \sigma \pm j\omega$. Simplify your result so that it contains no imaginary quantities.

3.12 Using Eq. (16), Sect. 3.3, obtain an integral expression for $y_{zs}(t)$ given that $\dddot{y} + 5\ddot{y} + 11\dot{y} + 15y = \ddot{x} + 15x$. *Hint*: one of the roots is -3.

3.13 Suppose the RC circuit in Example 2.4 has $y(t) = y_0$ at $t=0$ and

$$x(t) = \begin{cases} -y_0 & 0 \leq t \leq T \\ 0 & \text{otherwise} \end{cases}$$

where $T >> RC$. Find and sketch $y(t)$, identifying the transient and steady-state components.

3.14 Draw a simulation diagram for the system in Problem 3.12.

3.15 Draw a simulation diagram for Eqs. (17) and (18), Sect. 2.3.

3.16 Find the differential equation relating x and y from the simulation diagram of Figure P3.1. *Answer*: $\dddot{y} + 3\ddot{y} + y = 2\ddot{x} + \dot{x} + 5x$.

3.17 Devise a simulation diagram for the circuit in Figure P3.2 such that each circuit element is represented by one and only one scalor. (These are called *isolated-parameter* simulations, and have obvious advantages for experi-

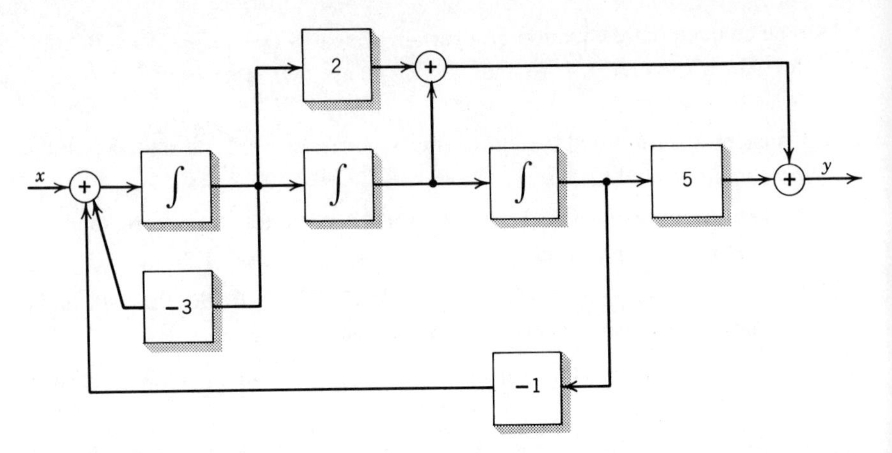

Figure P3.1.

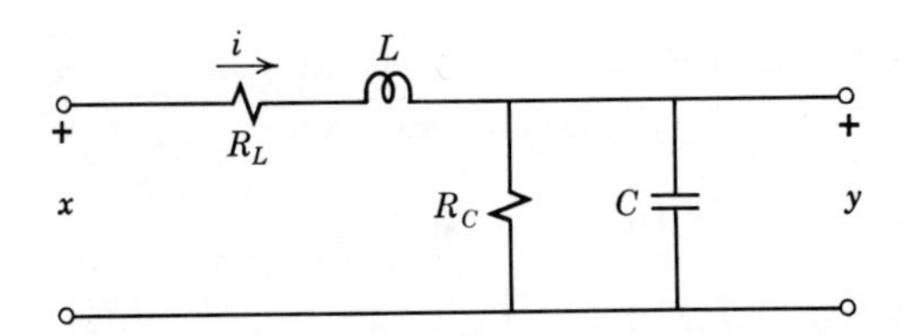

Figure P3.2.

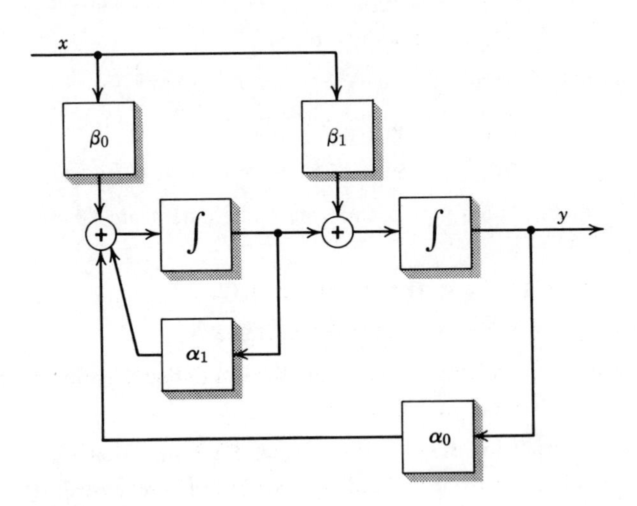

Figure P3.3.

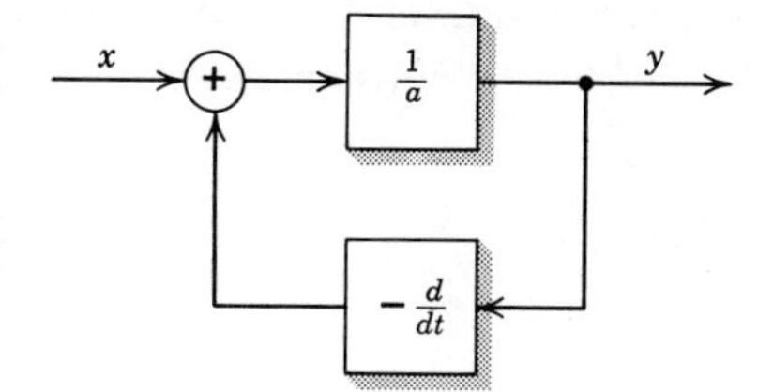

Figure P3.4.

mental design work.) *Hint*: try a configuration like Figure 3.13, with $y(t)$ and $i(t)$ as the integrator outputs.

3.18 By relating the α's and β's to the a's and b's, show that Figure P3.3 simulates $\ddot{y}+a_1\dot{y}+a_0y=b_1\dot{x}+b_0x$. *Answer*: $\alpha_0=-a_0$, $\alpha_1=-a_1$, $\beta_1=b_1$, $\beta_0=b_0-a_1b_1$.

3.19 Suppose *differentiators* were used for analog simulation. (a) Show that Figure P3.4 simulates $\dot{y}+ay=x$. (b) Construct a similar diagram, without integrators, for $\ddot{y}+a_1\dot{y}+a_0y=b_1\dot{x}+b_0x$.

3.20 Taking the state variables as the integrator outputs, find the matrices **A**, **b**, and **c** for the system simulated in Figure P3.1. *Answer*:

$$\mathbf{A}=\begin{bmatrix}0 & 1 & 0\\ 0 & 0 & 1\\ -1 & 0 & -3\end{bmatrix} \qquad \mathbf{b}=[0 \quad 0 \quad 1]^T \qquad \mathbf{c}=[5 \quad 1 \quad 2]^T$$

3.21 Given $\ddot{y}+a_1\dot{y}+a_0y=b_2\ddot{x}+b_1\dot{x}+b_0x$, set up the canonical state equations and show that the output equation must include $x(t)$.

3.22 Select appropriate state variables for the circuit in Figure P3.5, and obtain the state and output equations in the form of Eq. (3), Sect. 3.5. Note that $C_2(t)$ is time-varying, so that $i=d(Ce)/dt$.

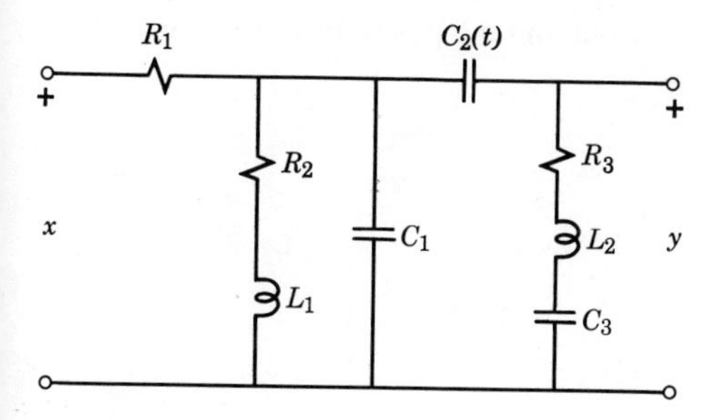

Figure P3.5.

3.23 Taking the state variables as indicated, find **A**, **B**, and **C** for the MIO system simulation of Figure P3.6. Also find the characteristic polynomial. *Answer*: $|s\mathbf{I}-\mathbf{A}| = s^4+5s^3+s^2+5s$.

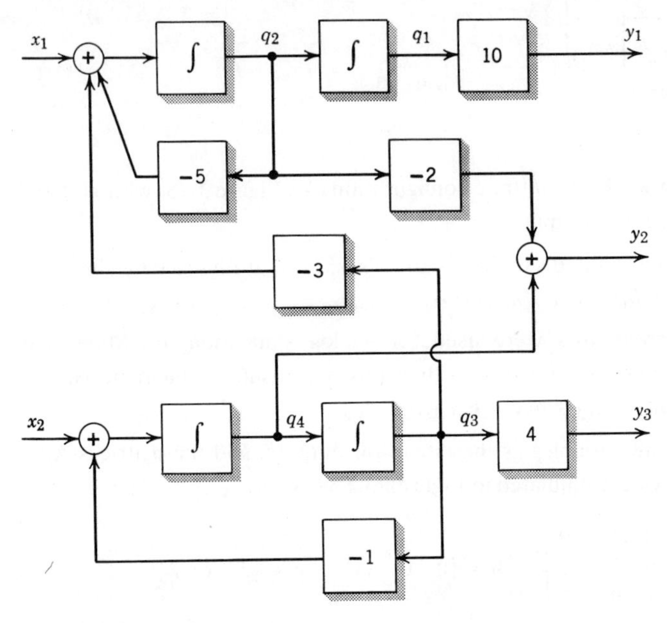

Figure P3.6.

3.24 A train consists of a locomotive, with mass M and bearing friction B, plus N identical cars, each having mass m and bearing friction b. Identical shock-absorbing couplings are used throughout, with the force relationship $F_c = KL + C\dot{L}$, where L is the incremental spacing between units and K and C are constants. Set up a state-variable model for this system taking the locomotive driving force F as the input and the positions of each unit of the train as the outputs. Neglect air drag.

4

Time-domain analysis: Convolution

There are three different but related ways of finding the zero-state response of a fixed linear system: *1*. brute-force solution of the mathematical model; *2*. time-domain analysis in terms of the system's impulse response; and *3*. frequency-domain or transform analysis. The first of these methods is not particularly attractive for complicated systems, as attested to by our studies in Chapter 3. Indeed, transform methods came into the picture around 1900 as a shortcut for handling linear differential equations with constant coefficients. And, by 1950, engineers were using transforms and the frequency-domain viewpoint almost exclusively for systems analysis and design.

Not to belittle the value of transforms, it has been recognized that time-domain analysis is an equally valuable approach that deserves our attention – spurred, in part, by the availability of digital computers that make time-domain calculations more feasible. By working in the time domain the engineer often obtains better insight to the physical characteristics of the system, characteristics that may be obscured under transformation. Furthermore, to properly understand and interpret transforms, we must ultimately relate them to the time domain; after all, that is the domain in which the system "exists."

This chapter presents the elements of time-domain systems analysis. Unless specifically stated to the contrary, it is assumed throughout that the systems are fixed and linear with one input and one output, and that the zero-input response is not of immediate concern. The principal subject, therefore, will be finding the zero-state response by time-domain methods.

As preparation, the first two sections define and discuss some important mathematical concepts: convolution and singularity functions. We then get into the details of systems analysis in terms of the impulse response, including its relationship to the material of the previous chapter. Finally, linking time-domain concepts to transform methods, the chapter closes with an introduction to the system transfer function.

4.1 CONVOLUTION

For the class of systems under consideration in Section 3.3, the zero-state response was found to be

$$S[\mathbf{0}; x(t)] = \int_{t_0}^{t} x(\lambda)\left[\sum_{i=1}^{n} C_i \eta_i(t-\lambda)\right] d\lambda \tag{1}$$

when $x(t) = 0$ for $t < t_0$. Of particular interest here is the fact that the function inside the brackets in the integrand is a property of the system only, and in no way involves the input signal. More generally, it can be shown for *any fixed linear system* that

$$S[\mathbf{0}; x(t)] = \int_{-\infty}^{\infty} x(\lambda) h(t-\lambda)\, d\lambda \tag{2}$$

where the function $h(t)$ is called the system's *impulse response*. The definition and discussion of $h(t)$ is postponed temporarily; here we are concerned with the integral operation in (1) and (2), which is known as *convolution*.

Convolution plays an important role in systems analysis not just because the zero-state response function is a convolution operation, but also because of its applications in transform methods. Indeed, convolution ranks among the most powerful tools available to the systems engineer. However, intelligent usage of this tool requires some grounding in its mathematical basics, which is the reason for this section.

The convolution integral

Given two functions $v(t)$ and $w(t)$, their convolution is defined to be

$$v * w(t) \triangleq \int_{-\infty}^{\infty} v(\lambda) w(t-\lambda)\, d\lambda \tag{3}$$

where $v * w(t)$ is just a symbol† standing for the integral on the right-hand

†Note that the asterisk (*) used here has nothing to do with complex conjugation.

side. It must be stressed and clearly understood that both functions being convolved have the same independent variable – t in this case – and that their convolution produces a new function of the same independent variable. The integration is always carried out with respect to a *dummy variable*, say λ. When there is no danger of confusion about the independent variable we shall write $v * w$ as a shorthand notation for (3) while on other occasions the longer form $[v(t)] * [w(t)]$ will be necessary.

To help understand the operations implied by $v * w$, let us consider a graphical interpretation of convolution. Parts (*a*) and (*b*) of Figure 4.1

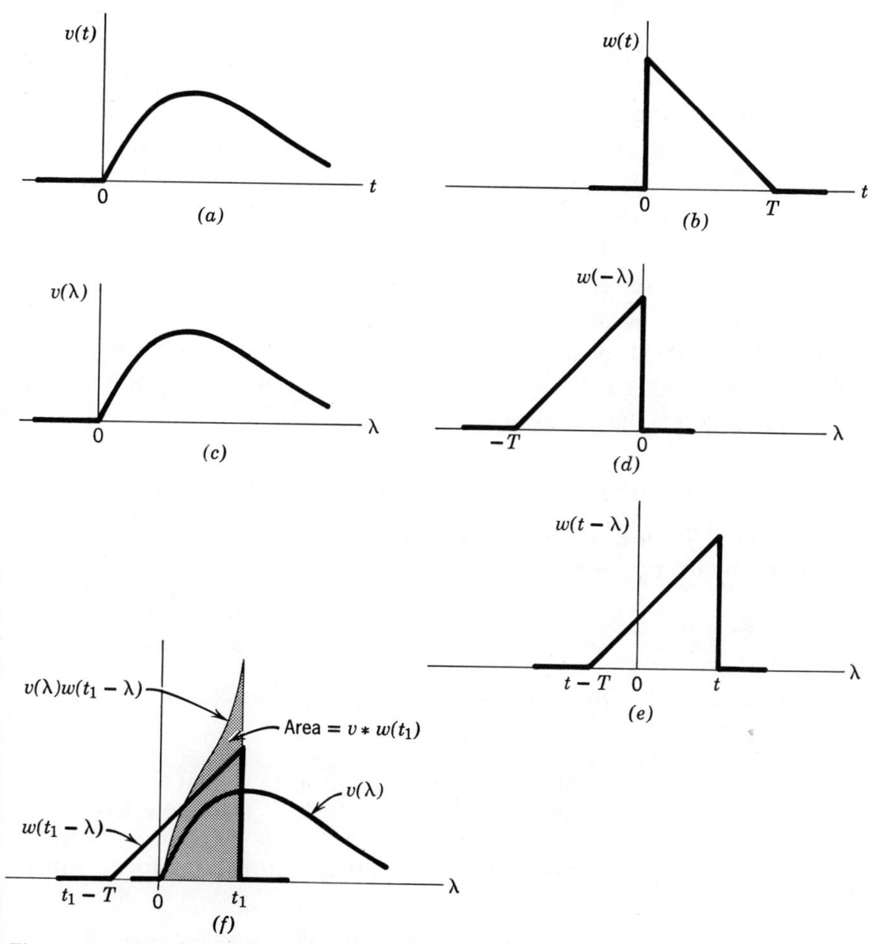

Figure 4.1. The graphical interpretation of $v * w(t)$.

show the assumed functions $v(t)$ and $w(t)$. Since the integration in (3) is performed with respect to λ, the function in the integrand must be plotted with λ as the independent variable. Thus, $v(\lambda)$ is merely $v(t)$ with t replaced by λ, as in part (*c*). The term $w(t-\lambda)$ may be obtained in two steps: first $w(-\lambda)$ is found by taking $w(\lambda)$ and interchanging the positive and negative portions of the λ axis, i.e., folding† over $w(\lambda)$ about $\lambda = 0$, yielding part (*d*); then t is reintroduced by sliding the plot of $w(-\lambda)$ to the right by t units, per part (*e*). As a check on this manipulation, the discontinuity in the function w occurs when its argument is zero, regardless of what variable or combination of variables is used as the argument. Therefore, in the plot of $w(t-\lambda)$ versus λ, the discontinuity must fall at the point $\lambda = t$.

To evaluate $v * w$ at some particular time $t = t_1$, $v(\lambda)$ is multiplied by $w(t_1 - \lambda)$ and their product is integrated over $-\infty < \lambda < \infty$; in other words, $v * w(t_1)$ is the shaded *area* indicated in Figure 4.1*f*. To obtain $v * w$ at some other value of t, the plot of $w(t-\lambda)$ must be shifted accordingly and the multiplication and integration process repeated. Hence, as $v * w(t)$ is computed for increasing t, $w(t-\lambda)$ slides from left to right relative to $v(\lambda)$.

The graphical interpretation is helpful, but not essential, when both functions being convolved are continuous for all t. However, more often than not we must deal with discontinuous or piecewise-continuous functions, in which case the graphical picture becomes indispensable. The following example illustrates this point.

Example 4.1 An illustrative convolution problem

It is desired to find $v * w(t)$ when, corresponding to Figure 4.1, the functions in question are

$$v(t) = \begin{cases} te^{-at} & t \geq 0 \\ 0 & t < 0 \end{cases} \qquad w(t) = \begin{cases} bt & 0 \leq t \leq T \\ 0 & \text{otherwise} \end{cases}$$

Therefore, in the basic definition $v * w(t) = \int_{-\infty}^{\infty} v(\lambda) w(t-\lambda) d\lambda$ we must insert

$$v(\lambda) = \begin{cases} \lambda e^{-a\lambda} & \lambda \geq 0 \\ 0 & \lambda < 0 \end{cases} \qquad w(t-\lambda) = \begin{cases} b(t-\lambda) & 0 \leq t-\lambda \leq T \\ 0 & \text{otherwise} \end{cases}$$

However, direct substitution into the integral turns out to be rather

†The name *convolution* comes from the German word *Faltung*, meaning *folding*.

awkward for this problem because $v(\lambda)$ and $w(t-\lambda)$ each have two possible expressions, depending on the value of λ. A better approach, therefore, is to break up the problem into three cases, corresponding to $t < 0$, $0 \le t \le T$, and $t > T$. The simple sketches in Figure 4.2 reveal these cases immediately.

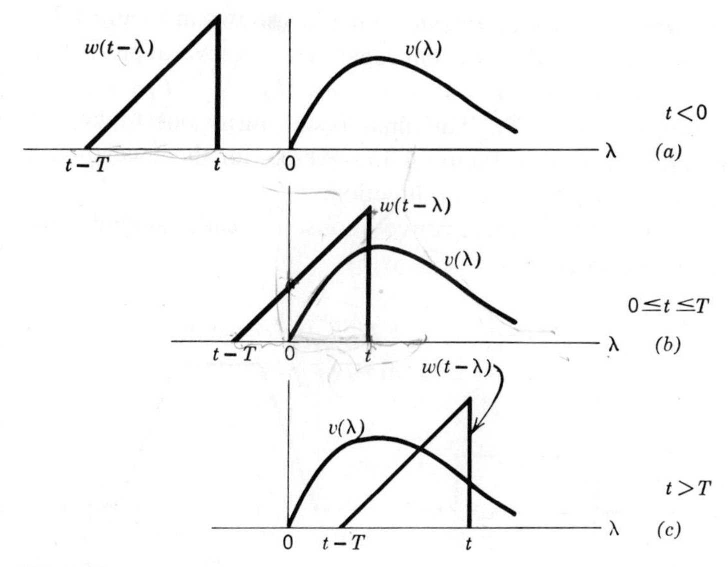

Figure 4.2.

Case (a): $t < 0$. Here $v(\lambda) = 0$ for $\lambda \le 0$ whereas $w(t-\lambda) = 0$ for $\lambda \ge 0$. Thus, the two functions in the integrand do not "overlap," their product is zero, and so

$$v * w(t) = 0 \qquad t < 0$$

Case (b): $0 \le t \le T$. Now the functions are partially overlapping, but only over the range $0 \le \lambda \le t$; thus

$$v * w(t) = \int_0^t \lambda e^{-a\lambda} b(T-t+\lambda)\,d\lambda \qquad 0 \le t \le T$$

Case (c): $t > T$. Finally, the functions fully overlap, but $w(t-\lambda)$ is zero outside of $t-T \le \lambda \le t$; hence

$$v * w(t) = \int_{t-T}^t \lambda e^{-a\lambda} b(T-t+\lambda)\,d\lambda \qquad t > T$$

Note in the last two cases that t appears in the *limits of integration*, as will generally happen for convolutions of this type. Having gotten this far in setting up the problem, the rest of the work is routine and will be skipped.

Example 4.2 Convolution of rectangular functions

Consider convolving the two *rectangular* functions shown in Figure 4.3*a*. Like the previous example, the problem separates into several parts, five to be exact: $t < 0$, $0 < t < T_1$, $T_1 < t < T_2$, $T_2 < t < T_1 + T_2$, and $t > T_1 + T_2$, assuming $T_2 \geq T_1$. The final result turns out to be the *trapezoidal* function plotted in Figure 4.3*b* – which, for the special case of $T_1 = T_2$, degenerates to a *triangular* function.

To acquire some facility in doing convolutions, the reader should work out this example completely.

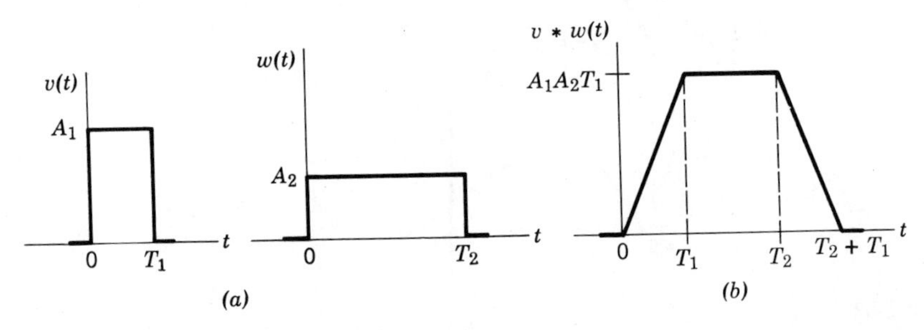

Figure 4.3. Convolution of rectangular functions.

Basic convolution relations

Convolution has a number of mathematical properties that will be of considerable importance and help throughout later work. In particular, like ordinary algebraic operations, convolution is *commutative*, *associative*, and *distributive*, i.e.,

$$v * w = w * v \tag{4}$$

$$v * (w * z) = (v * w) * z \tag{5}$$

$$(\alpha v + \beta w) * z = \alpha(v * z) + \beta(w * z) \tag{6}$$

where v, w, and z are functions of the same variable and α and β are constants.

The distributive property, (6), should be evident from the definition of convolution, while commutativity, (4), is readily justified from the graphical interpretation – i.e., in Figure 4.1 it does not matter whether we fold w and slide it past v or vice versa, since the area of the product will be the same. However, proving the associative property, (5), requires a judicious change of integration variables, as follows. Using μ instead of the dummy variable λ, $w * z(t) = \int_{-\infty}^{\infty} w(\mu)z(t-\mu)\,d\mu$, and

$$[v(t)] * [w * z(t)] = \int_{-\infty}^{\infty} v(\lambda)\overbrace{\left[\int_{-\infty}^{\infty} w(\mu)z(t-\lambda-\mu)\,d\mu\right]}^{w\,*\,z(t-\lambda)}d\lambda$$

But λ is a constant insofar as the inner integral is concerned, so we can let $\gamma = \mu + \lambda$ and $d\gamma = d\mu$, and interchange the order of integrations, thus:

$$\begin{aligned}[v(t)] * [w * z(t)] &= \int_{-\infty}^{\infty} v(\lambda)\left[\int_{-\infty}^{\infty} w(\gamma-\lambda)z(t-\gamma)\,d\gamma\right]d\lambda \\ &= \int_{-\infty}^{\infty}\underbrace{\left[\int_{-\infty}^{\infty} v(\lambda)w(\gamma-\lambda)\,d\lambda\right]}_{v\,*\,w(\gamma)}z(t-\gamma)\,d\gamma \\ &= [v * w(t)] * [z(t)]\end{aligned}$$

which completes the proof. In manipulations such as this, the reader should bear in mind that a dummy variable is a *dummy* variable and may be assigned any symbol not otherwise used.

Another useful relation concerns the derivative of $v * w(t)$:

$$\begin{aligned}\frac{d}{dt}[v * w(t)] &= \frac{d}{dt}\int_{-\infty}^{\infty} v(\lambda)w(t-\lambda)\,d\lambda \\ &= \int_{-\infty}^{\infty} v(\lambda)\,\frac{dw(t-\lambda)}{dt}\,d\lambda \\ &= [v(t)] * \left[\frac{dw(t)}{dt}\right]\end{aligned}$$

Similarly, $d[w * v(t)]/dt = [w(t)] * [dv(t)/dt]$; therefore, in view of the fact that $v * w = w * v$, we combine these relationships as

$$\frac{d}{dt}(v * w) = v * \frac{dw}{dt} = \frac{dv}{dt} * w \tag{7}$$

There are analogous expressions for higher derivatives of $v * w$ and also for integration.

These basic relations will be illustrated by their use in the remainder of this chapter.

4.2 SINGULARITY FUNCTIONS: IMPULSES AND STEPS

We previously asserted, and will prove in the next section, that the zero-state response of any fixed linear system takes the form

$$S[\mathbf{0}; x(t)] = \int_{-\infty}^{\infty} x(\lambda)h(t-\lambda)d\lambda = x * h(t)$$

Thus, knowing the system's impulse response, we can calculate the response to any input signal $x(t)$ – at least in theory. But first one must find $h(t)$; and, as suggested by the work in Chapter 3, that could prove to be a chore. As an alternate tactic, let us search for an experimental means of finding $h(t)$.

To be specific, suppose a system in the zero initial state is excited by the input signal sketched in Figure 4.4, a rectangular pulse having

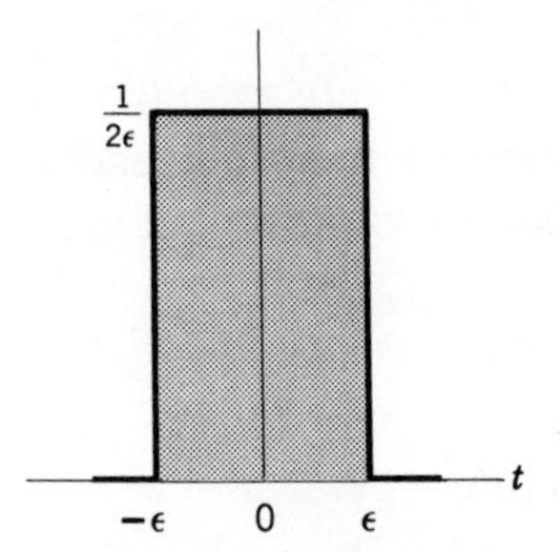

Figure 4.4.

amplitude $1/2\epsilon$ and time duration 2ϵ so that its area is unity. Since $x(t) = 0$ for $|t| > \epsilon$, the corresponding output will be

$$y(t) = \int_{-\infty}^{\infty} x(\lambda)h(t-\lambda)d\lambda = \int_{-\epsilon}^{\epsilon} \frac{1}{2\epsilon} h(t-\lambda)d\lambda$$

Now if ϵ is sufficiently small that the integrand may be approximated as a constant equal to its midpoint value at $\lambda = 0$, then

$$y(t) \approx \int_{-\epsilon}^{\epsilon} \frac{h(t)}{2\epsilon} d\lambda = \frac{h(t)}{2\epsilon} \int_{-\epsilon}^{\epsilon} d\lambda = h(t)$$

Thus, if the input pulse is "short enough," the zero-state response very nearly equals the impulse response, and by this means one can experimentally measure $h(t)$.

It probably comes as no surprise that we are leading up to the concept of the *unit impulse* $\delta(t)$, also known as the Dirac delta or the delta function. In the present context, $\delta(t)$ has the property that if $x(t) = \delta(t)$ then the zero-state response *exactly* equals the impulse response by definition, i.e.,

$$h(t) \triangleq S[\mathbf{0}; \delta(t)] = \delta * h(t)$$

or, written out fully,

$$h(t) = \int_{-\infty}^{\infty} \delta(\lambda) h(t-\lambda) d\lambda \tag{1}$$

for any $h(t)$.

Historical precedent notwithstanding, we avoid calling $\delta(t)$ a *function* for the reason that (1) is insufficient to define a function. Actually, the unit impulse belongs to a family of *singularity functions* commonly used in engineering, many of which do not even qualify as functions in the strict sense of the word. But because scientists and engineers find these functions to be convenient abstractions, mathematicians have sought to put them on a more acceptable and rigorous basis, resulting in the new theory of *generalized functions* or *distributions*. Although we cannot get into that theory here, we shall discuss singularity functions in a fashion consistent with it.

The impulse and its properties

The unit impulse is defined in terms of a *process* that assigns to any ordinary function $v(t)$ the number $v(0)$ according to the rule

$$\int_{-\infty}^{\infty} v(t)\delta(t) dt = v(0) \tag{2}$$

providing $v(t)$ is continuous at $t=0$ so that $v(0)$ has a unique value. It must be emphasized that (2) is an assignment rule, not an equation in the usual sense – that is to say, $\delta(t)$ is defined only in terms of the above integral and one should resist the temptation to give it an independent meaning. Since (2) is our only definition for $\delta(t)$, we shall use it exclusively to develop other relationships, and in this application $v(t)$ will be called the *test function*; correspondingly, we generally assume $v(t)$ is continuous everywhere.

Probably the two most important impulse relations involve the notion of a delayed or displaced impulse $\delta(t-t_d)$, where t_d is a constant. One of these, known as the *sampling* or *sifting* property, is

$$\int_{-\infty}^{\infty} v(t)\delta(t-t_d)dt = v(t_d) \tag{3}$$

which says that integrating the product $v(t)\delta(t-t_d)$ over $-\infty < t < \infty$ "sifts out" or samples the value of $v(t)$ at $t = t_d$. Clearly, (3) is a simple extension of the basic definition obtained by changing the variable of integration.

Related to the sampling integral is *convolution* with impulses. In particular, consider $v(t)$ convolved with the delayed impulse $\delta(t-t_d)$. Using the fact that $v * w = w * v$ we have

$$\begin{aligned}[v(t)] * [\delta(t-t_d)] &= [\delta(t-t_d)] * [v(t)] \\ &= \int_{-\infty}^{\infty} \delta(\lambda-t_d)v(t-\lambda)d\lambda\end{aligned}$$

Then, letting $\lambda = \mu + t_d$ yields

$$\begin{aligned}[v(t)] * [\delta(t-t_d)] &= \int_{-\infty}^{\infty} v(t-t_d-\mu)\delta(\mu)d\mu \\ &= v(t-t_d)\end{aligned} \tag{4}$$

Hence, convolving an ordinary function with $\delta(t-t_d)$ merely reproduces the *entire* function delayed by t_d. If it happens that $t_d = 0$ then (4) reduces to

$$v * \delta(t) = \delta * v(t) = v(t) \tag{5}$$

a peculiar-looking but very significant identity. In fact, (5) is precisely the property of $\delta(t)$ originally specified in (1) so that $\delta * h(t) = h(t)$.

When deriving (3) to (5) it was assumed that $v(t)$ is continuous everywhere. Reexamination of (3) reveals, however, that this stipulation may be relaxed: the sampling integral picks out just the one specific value $v(t_d)$, so we need only require continuity at $t = t_d$, i.e., at the point where the argument of $\delta(t-t_d)$ is zero. On the other hand, convolution with $\delta(t-t_d)$ reproduces the entire function $v(t-t_d)$, and hence continuity is required everywhere. Alternately, one can say that (4) and (5) are valid only for those points in time at which $v(t-t_d)$ or $v(t)$ is continuous, and are *undefined* at points of discontinuity.

From time to time one encounters impulses under integral signs with finite or semifinite limits, say $a \leq t \leq b$. Such cases are handled by introducing a modified test function

$$v_{ab}(t) \triangleq \begin{cases} v(t) & a < t < b \\ 0 & t > b \quad \text{or} \quad t < a \end{cases}$$

so $\int_a^b v(t)\delta(t)\,dt = \int_{-\infty}^{\infty} v_{ab}(t)\delta(t)\,dt$, from which it easily follows that

$$\int_a^b v(t)\delta(t)\,dt = \begin{cases} v(0) & a < 0 < b \\ 0 & b > a > 0 \quad \text{or} \quad b < a < 0 \end{cases} \tag{6}$$

If either $a = 0$ or $b = 0$, i.e., one endpoint of the integration's range falls exactly at $t = 0$, then the integral is undefined because $v_{ab}(t)$ has a discontinuity at that point. When this happens, the integration must be examined in the light of the corresponding physical problem, since the formal mathematics provides no answer.

Interpreting (6) shows that the value of the integral is zero (or undefined) unless the range of integration spans $t = 0$, in which case the value is $v(0)$. This suggests that $\delta(t)$ has *unit area* concentrated at the discrete point $t = 0$ (or, more generally, wherever its argument is zero) and no *net* area elsewhere. To confirm that fact let $v(t) = v(0) = 1$ in (2), giving the familiar relation $\int_{-\infty}^{\infty} \delta(t)\,dt = 1$. But now take (6) with $v(t) = 1$ and an arbitrarily small range of integration centered at $t = 0$, say $-\epsilon < t < \epsilon$, and hence $\int_{-\epsilon}^{\epsilon} \delta(t)\,dt = 1$. Combining these two yields

$$\int_{-\infty}^{\infty} \delta(t)\,dt = \int_{-\epsilon}^{\epsilon} \delta(t)\,dt = 1 \tag{7a}$$

which is often interpreted as

$$\delta(t) = 0 \qquad t \neq 0 \tag{7b}$$

The latter, not being an integral expression, must be used with great care; it is presented here primarily for its merit in visualizing impulses. Distribution theory, strictly interpreted, does not specify values of $\delta(t)$ itself.

Equation (7) leads to the common graphical picture of the unit impulse as shown in Figure 4.5*a*, where the number 1 next to the arrowhead stands for unit area or *weight*. More generally, Figure 4.5*b* represents $A\delta(t - t_d)$, an impulse of weight A located at $t = t_d$.

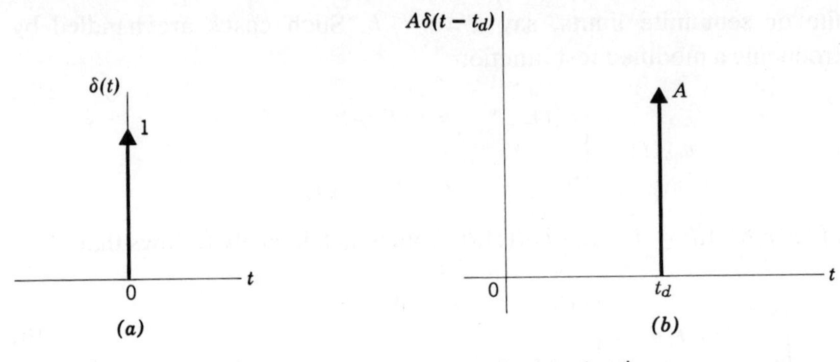

Figure 4.5. (*a*) The unit impulse. (*b*) An impulse of weight A at time t_d.

By definition, the impulse is meaningless – mathematically or physically – unless it appears under integration. Having repeated that statement several times, we now go ahead and list a few additional impulse properties which, like (7b), are not integral relationships. Justification for this is based on the following line of thought: since any expression involving impulses must eventually be integrated, certain simplifications that come about *after integration* might just as well be made *before*. Therefore, bearing that context in mind, the impulse can be manipulated to a limited extent like an ordinary function. But when in doubt, we must return to the integral form.

Elaborating on the above by illustration, consider the product $v(t)\delta(t-t_d)$; in view of (7b) it appears that

$$v(t)\delta(t-t_d) = v(t_d)\delta(t-t_d) \tag{8}$$

since $\delta(t-t_d) = 0$ for $t \neq t_d$. Equation (8) is, in fact, correct, and is proved by integrating each side over $-\infty < t < \infty$. Specifically, from (3), the integral of the left-hand side is

$$\int_{-\infty}^{\infty} v(t)\delta(t-t_d)\,dt = v(t_d)$$

while, using (7) for the right side,

$$\int_{-\infty}^{\infty} v(t_d)\delta(t-t_d)\,dt = v(t_d)\int_{-\infty}^{\infty} \delta(t-t_d)\,dt = v(t_d)$$

and since both sides are equal after integration they may be equated before integration, giving (8).

Similar arguments can be used to prove the scale-change relationship

$$\delta(ct) = \frac{1}{|c|}\delta(t) \qquad c \neq 0 \tag{9a}$$

which means that, relative to the independent variable t, $\delta(ct)$ is an impulse of weight $1/|c|$. The special case $c = -1$ gives

$$\delta(-t) = \delta(t) \tag{9b}$$

so the impulse has the property of *even symmetry.*

Representations for the unit impulse

Consistent with the above properties, there are several important representations for the unit impulse expressed as limiting forms of conventional functions. Specifically, if $\delta_\epsilon(t)$ is a function of t having the parameter ϵ such that

$$\lim_{\epsilon\to 0}\int_{-\infty}^{\infty} v(t)\delta_\epsilon(t)\,dt = v(0) \tag{10a}$$

for any test function $v(t)$, then we say that

$$\lim_{\epsilon\to 0}\delta_\epsilon(t) = \delta(t) \tag{10b}$$

For example, let us try $\delta_\epsilon(t)$ as the rectangular pulse of Figure 4.4, i.e.,

$$\delta_\epsilon(t) = \begin{cases} \dfrac{1}{2\epsilon} & |t| \leq \epsilon \\ 0 & |t| > \epsilon \end{cases} \tag{11}$$

Inserting (11) in (10a) gives

$$\lim_{\epsilon\to 0}\int_{-\infty}^{\infty} v(t)\delta_\epsilon(t)\,dt = \lim_{\epsilon\to 0}\frac{1}{2\epsilon}\int_{-\epsilon}^{\epsilon} v(t)\,dt$$

Since we eventually will take $\epsilon \to 0$, the test function can be expanded in a Taylor series about $t = 0$ as

$$v(t) = v(0) + \frac{dv(0)}{dt}t + \frac{1}{2!}\frac{d^2v(0)}{dt^2}t^2 + \cdots$$

and integrate term by term. By symmetry, all terms involving odd powers of t vanish, leaving

$$\frac{1}{2\epsilon}\int_{-\epsilon}^{\epsilon} v(t)\,dt = \frac{1}{2\epsilon}\left[v(0)\,(2\epsilon) + \frac{1}{2!}\frac{d^2v(0)}{dt^2}\left(\frac{2\epsilon^3}{3}\right) + \cdots\right]$$

$$= v(0) + \frac{1}{3!}\frac{d^2v(0)}{dt^2}\epsilon^2 + \cdots$$

Hence, letting $\epsilon \to 0$,

$$\lim_{\epsilon\to 0}\int_{-\infty}^{\infty} v(t)\delta_\epsilon(t)\,dt = v(0)$$

so the function in question has the necessary properties to represent an impulse in the limit. Similarly one can show this property for all the piecewise-linear functions of Figure 4.6. Note that some of these do not possess

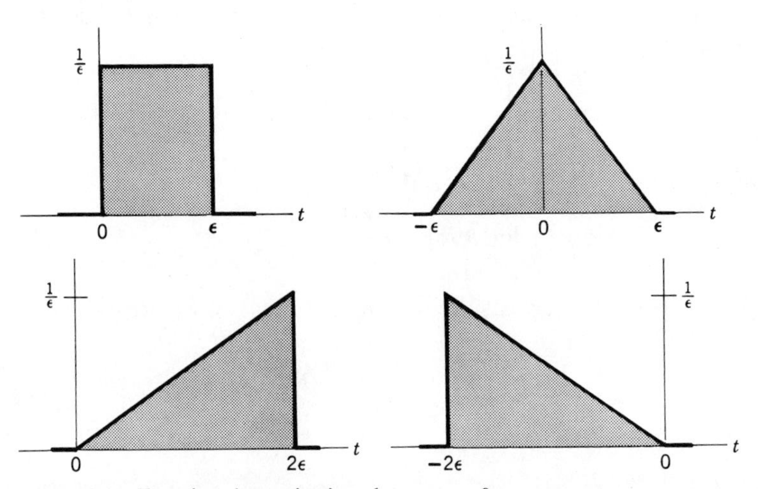

Figure 4.6. Functions becoming impulses as $\epsilon \to 0$.

even symmetry, in apparent contradiction to our previous conclusion that $\delta(-t) = \delta(t)$; the contradiction is resolved as $\epsilon \to 0$.

The practical implication of (10) and Figure 4.6 is that, although an impulse cannot be generated physically, it can be *approximated* by a pulse of finite height and nonzero duration. Thus, to experimentally determine the impulse response of a system, we can apply a very brief pulse of more-or-less arbitrary shape and keep reducing its duration, holding the area constant, until there is no discernible change in the output – which then is a good approximation to $h(t)$.

One of the most unusual functions satisfying (10) is

$$\delta_\epsilon(t) = \frac{1}{\pi t} \sin \frac{\pi t}{\epsilon} \tag{12}$$

plotted in Figure 4.7 for two values of ϵ. Proof that (12) obeys (10a) will be skipped, since it is rather involved. However, the unit-area property can be established from a table of definite integrals, and Figure 4.7 further

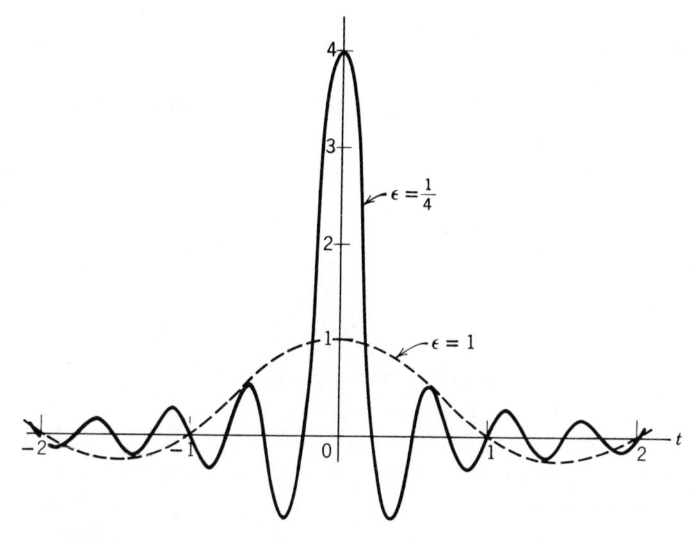

Figure 4.7. $\delta_\epsilon(t) = \dfrac{1}{\pi t} \sin \dfrac{\pi t}{\epsilon}$.

enhances the plausibility – as does a discussion given in Section 7.3. The fact that this particular $\delta_\epsilon(t)$ is not zero for all $t \neq 0$ even when $\epsilon \to 0$ does not violate our earlier assertion that $\delta(t) = 0,\ t \neq 0$; for recall that that property must be interpreted in the *integral* sense and, as $\epsilon \to 0$, (12) does have *zero net area* over any interval that excludes the point $t = 0$. Later, we shall find that (12) plays a pivotal role in Fourier theory.

The unit step and other singularity functions

Other singularity functions can be defined which are related to the unit impulse by integration or differentiation. Specifically, we define† the *unit*

†An alternate definition is given in Problem 4.12.

step $u(t)$ as being the integral of the unit impulse in the form

$$u(t) \triangleq \int_{-\infty}^{t} \delta(\lambda)\,d\lambda = \begin{cases} 0 & t < 0 \\ 1 & t > 0 \\ \text{undefined} & t = 0 \end{cases} \tag{13}$$

where we have used (6) to evaluate the integral. In this sense, and only in this sense,

$$\delta(t) = \frac{du(t)}{dt} \tag{14}$$

which is not a definition of $\delta(t)$ but rather a consequence of (13). Figure 4.8 visually depicts the unit step.

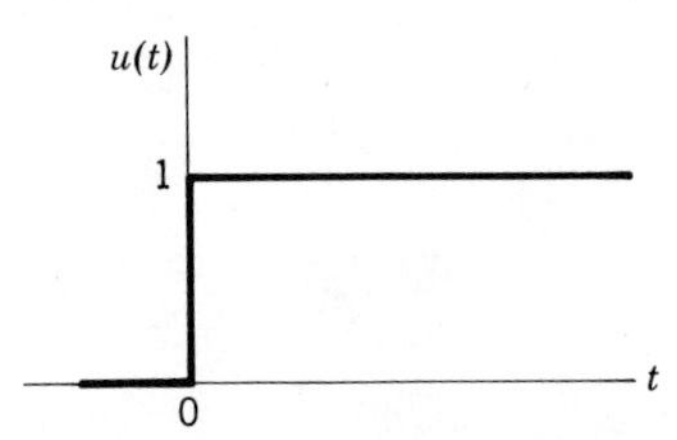

Figure 4.8. The unit step $u(t)$.

It may seem disconcerting at first that $u(t)$ is *undefined* for $t = 0$. But this is consistent with Figure 4.6 in that integrating these various functions gives several possible values for $u(0)$. More practically, a physical variable cannot make a perfect stepwise transition; so if one approximates some signal using $u(t)$, the exact value of $u(0)$ is immaterial. Hence, we are free to take $u(0)$ as befits whatever application is at hand, the choice usually being $0, \frac{1}{2}$, or 1.

A convenient feature of $u(t)$ is its compactness when used to write piecewise-continuous functions. In particular, since a delayed step is

$$u(t - t_d) = \begin{cases} 0 & t < t_d \\ 1 & t > t_d \end{cases}$$

then the rectangular function in Figure 4.9 can be written as

$$A[u(t) - u(t - T)] = \begin{cases} 0 & t < 0 \quad \text{or} \quad t > T \\ A & 0 < t < T \end{cases}$$

which is the sum of a positive step and a delayed negative step.

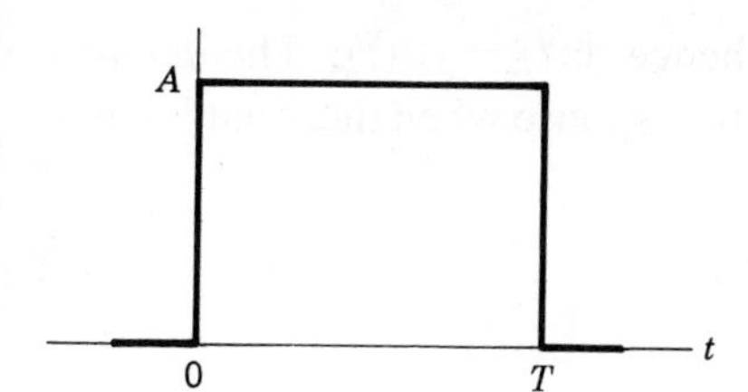

Figure 4.9. The rectangular function $A[u(t)-u(t-T)]$.

Integrating $u(t)$ gives the *unit ramp*

$$\int_{-\infty}^{t} u(\lambda)\,d\lambda = \begin{cases} 0 & t<0 \\ t & t>0 \end{cases}$$
$$= tu(t) \qquad (15)$$

and one might define an entire set of functions based on repeated integration. The kth member of this set is $u^{(-k)}(t) \triangleq (t^k/k!)u(t)$, where $u^{(-k)}(t)$ stands for the kth integral of $u(t)$.

Going in the opposite direction, consider the *derivative* of the unit impulse $\delta'(t) \triangleq d\delta(t)/dt$, known as the *doublet*; it then follows that $\int_{-\infty}^{t} \delta'(\lambda)\,d\lambda = \delta(t)$. If the impulse seems a little farfetched as an abstraction of some physical variable, then a little further thought should convince the reader that the utility of the doublet for describing real-world phenomena is probably nil. Thus, the unit step and impulse are the only singularity functions we shall really need.

4.3 IMPULSE RESPONSE

With the mathematical spadework disposed of, let us get down to the theory of time-domain analysis. We observed in Chapter 2 that if the response of a fixed linear system is known for certain elementary functions then, by virtue of superposition and time-invariance, we can find the response to any excitation that is expressable in terms of the elementary functions. This section pursues that approach, taking the impulse and step as the elementary functions; this leads, in turn, to the previously mentioned convolution, now called the superposition integral.

Impulse response and the superposition integral

Consider an arbitrary fixed linear system whose zero-state response is, in general,

$$y(t) = S[x(t)]$$

where $\mathbf{q}(t_0) = \mathbf{0}$ is understood and hence $y(t) = y_{zs}(t)$. The *impulse response* $h(t)$ is defined as the zero-state response when the input is a unit impulse, i.e.,

$$h(t) \triangleq S[\delta(t)] \tag{1}$$

Similarly, the *step response* † $y_u(t)$ is

$$y_u(t) \triangleq S[u(t)] \tag{2}$$

Now recall that $\delta(t) = du(t)/dt$ and that differentiating (or integrating) the system's input yields the derivative (or integral) of the original output. Therefore, $S[\delta(t)] = S[du(t)/dt] = dS[u(t)]/dt$, so

$$h(t) = \frac{dy_u(t)}{dt} \tag{3}$$

and, conversely,

$$y_u(t) = \int_{-\infty}^{t} h(\lambda)\,d\lambda \tag{4}$$

In short, given either the system's step or impulse response, we can always find the other. And, by extension, we can find the response when the input is any linear combination of singularity functions. For instance, if

$$x(t) = \alpha_1 u(t-t_1) + \alpha_2 u(t-t_2) + \cdots \tag{5a}$$

then

$$y(t) = \alpha_1 y_u(t-t_1) + \alpha_2 y_u(t-t_2) + \cdots \tag{5b}$$

But what about the response to a more-or-less arbitrary input signal beginning, say, at $t = 0$, as in Figure 4.10*a*? Admittedly, this excitation is not a simple combination of singularity functions. Nevertheless, it can be *approximated* in the stepwise fashion indicated by Figure 4.10*b*, where uniform step spacing Δ is used for simplicity. Taking $u(0) = 1$ in this application and noting that the step at $t = k\Delta$ has height $[x(k\Delta) - x(k\Delta - \Delta)]$, we write the approximating function $\hat{x}(t)$ as

$$\begin{aligned}\hat{x}(t) &= x(0)u(t) + [x(\Delta) - x(0)]u(t-\Delta) + [x(2\Delta) - x(\Delta)]u(t-2\Delta) + \cdots \\ &= x(0)u(t) + \sum_{k=1}^{\infty} [x(k\Delta) - x(k\Delta - \Delta)]u(t-k\Delta) \end{aligned} \tag{6}$$

†Unlike $h(t)$, there is no generally accepted notational symbol for the step response; $y_u(t)$ seems as good as any.

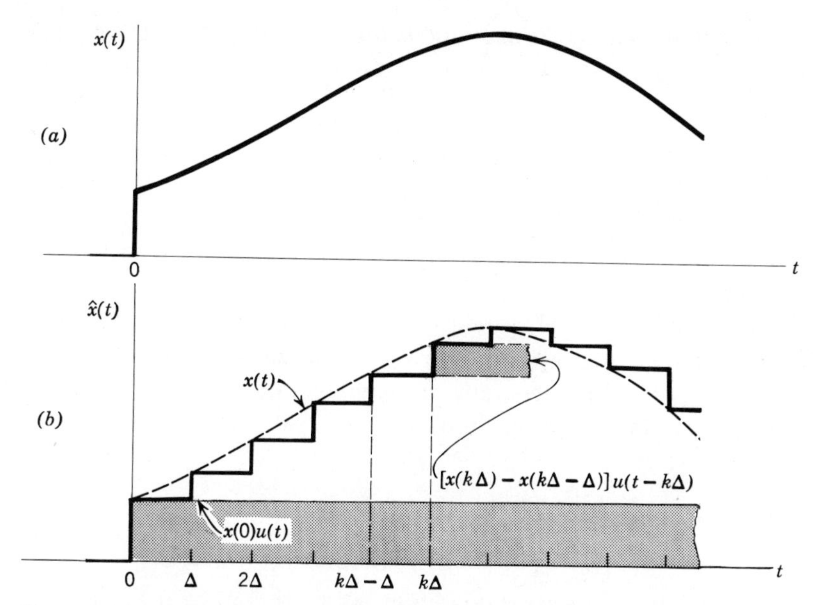

Figure 4.10. A signal and its stepwise approximation.

which exactly equals $x(t)$ at every $t = k\Delta$, $k = 0, 1, 2, \ldots$ Heuristically, it is reasonable to expect that the approximation errors will decrease as the spacing between steps is reduced and, ultimately, $\hat{x}(t) \rightarrow x(t)$ as $\Delta \rightarrow 0$. This certainly will be true wherever $x(t)$ is continuous.

Since $x(0)$, $x(\Delta)$, etc., in (6) are constants for any fixed value of Δ, the corresponding approximation for the output $y(t)$ follows directly from (5) as

$$\hat{y}(t) = x(0)y_u(t) + \sum_{k=1}^{\infty} [x(k\Delta) - x(k\Delta - \Delta)]\, y_u(t - k\Delta)$$

$$= x(0)y_u(t) + \sum_{k=1}^{\infty} \left[\frac{x(k\Delta) - x(k\Delta - \Delta)}{\Delta}\right] y_u(t - k\Delta)\Delta \qquad (7)$$

where we have multiplied and divided by Δ inside the sum. Taking the limiting case where the step spacing becomes vanishingly small, say $\Delta \rightarrow d\lambda$, $k\Delta$ becomes the continuous variable λ, the summation becomes integration over λ, and

$$\frac{x(k\Delta) - x(k\Delta - \Delta)}{\Delta} \rightarrow \frac{x(\lambda) - x(\lambda - d\lambda)}{d\lambda} = \frac{dx(\lambda)}{d\lambda}$$

Hence, assuming that $x(t)$ is continuous for all $t > 0$ so that $\hat{y}(t) \to y(t)$, we arrive at

$$y(t) = x(0)y_u(t) + \int_{0+}^{\infty} \frac{dx(\lambda)}{d\lambda} y_u(t-\lambda)\,d\lambda \tag{8}$$

where the lower limit of integration must be 0+, rather than 0, and $x(0)y_u(t)$ is included as a separate term to account for the fact that $x(t)$ may be discontinuous at $t = 0$.

This equation is a valuable intermediate result, for it gives $y(t)$ as an explicit function of the step response $y_u(t)$ and the input $x(t)$, providing $x(t) = 0$ for $t < 0$. More generally, by repeating the approximation procedure for $x(t)$ starting at any time $t_0 > -\infty$, one obtains the compact expression

$$y(t) = \int_{-\infty}^{\infty} \frac{dx(\lambda)}{d\lambda} y_u(t-\lambda)\,d\lambda \tag{9}$$

which assumes the reasonable condition $x(t_0)y_u(t-t_0) \to 0$ as $t_0 \to -\infty$.

Equation (9) is known as *Duhamel's integral* and, on closer inspection, it should be recognized as a *convolution*, namely $y(t) = [dx(t)/dt] * [y_u(t)]$. Therefore, invoking Eq. (7), Sect. 4.1, we have *three* equivalent expressions for the zero-state response:

$$y(t) = \frac{d}{dt}\{[x(t)] * [y_u(t)]\} \tag{10a}$$

$$= \left[\frac{dx(t)}{dt}\right] * [y_u(t)] \tag{10b}$$

$$= [x(t)] * \left[\frac{dy_u(t)}{dt}\right] \tag{10c}$$

Of these, the last usually proves to be the most convenient since we can replace the derivative of the step response by the *impulse response*, thereby getting rid of the differentiation. Specifically, inserting $h(t) = dy_u(t)/dt$ and writing out (10c) in full,

$$y(t) = \int_{-\infty}^{\infty} x(\lambda)h(t-\lambda)\,d\lambda \tag{11a}$$

as asserted at the very start of this chapter. Alternately, because convolution is commutative, i.e., $x * h = h * x$,

$$y(t) = \int_{-\infty}^{\infty} h(\mu)x(t-\mu)\,d\mu \tag{11b}$$

in which we have used a different dummy variable μ for clarity.

Equation (11) – in either form – is the *superposition integral*; it constitutes the theoretical basis for time-domain analysis of fixed linear systems in terms of the impulse response. Relative to its generality, we should point out that (11) has been derived without referring to any particular system model, assuming only that the zero-state response is a time-invariant linear function of the input. Therefore, when this assumption holds and one is concerned only with the zero-state response, the superposition integral provides an explicit input-output relationship – whether or not the system model is a differential equation like those studied in the last chapter, and whether or not the system is linear in the stronger sense of Eqs. (3b) and (3c), Sect. 2.4, which involve the zero-state response.

Example 4.3 Time-domain analysis of an RC circuit

Consider the RC circuit of Figure 4.11a with the capacitor initially uncharged, i.e., zero-state conditions. If we are to find the zero-state response $y(t)$ for some specific $x(t)$, and if this is to be done using (11), then the problem involves two successive steps: first, calculating the impulse response; second, evaluating a convolution.

Because this particular circuit is so simple, its *step* response can be found by inspection to be

$$y_u(t) = \begin{cases} 1 - e^{-t/RC} & t \geq 0 \\ 0 & t < 0 \end{cases}$$

$$= (1 - e^{-t/RC})u(t)$$

as sketched in Figure 4.11b. Differentiating the step response then gives the impulse response. However, because $y_u(t)$ contains $u(t)$ as a multiplicative factor, this differentiation is not exactly trivial. In particular, using the chain rule and the fact that $du(t)/dt = \delta(t)$,

$$h(t) = \frac{d}{dt}[(1 - e^{-t/RC})u(t)]$$

$$= \frac{1}{RC}e^{-t/RC}u(t) + (1 - e^{-t/RC})\delta(t)$$

The second term then drops out by virtue of Eq. (8), Sect. 4.2, since

$$(1-e^{-t/RC})\delta(t) = (1-e^{-t/RC})|_{t=0}\delta(t) = 0$$

Better yet, reference to the sketch of $y_u(t)$ shows that it is continuous at $t=0$ so its derivative cannot include $\delta(t)$. At any rate, we arrive at

$$h(t) = \frac{1}{RC}e^{-t/RC}u(t)$$

which is plotted in Figure 4.11*c*.

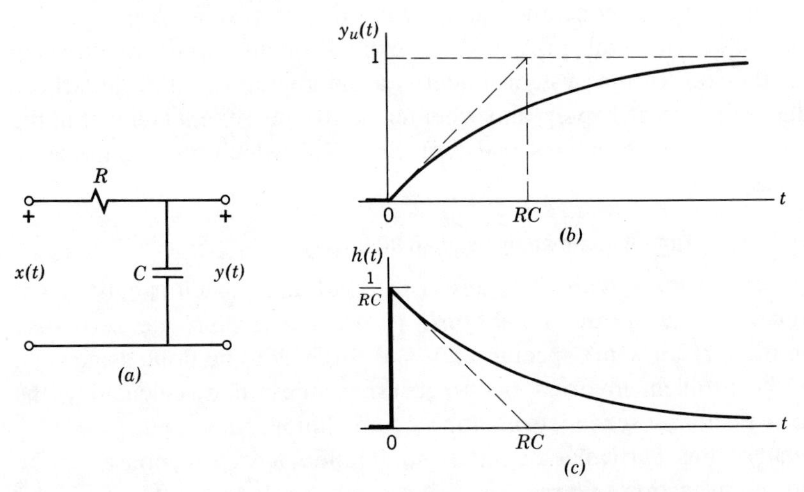

Figure 4.11. (*a*) RC circuit. (b) Step response $y_u(t)$. (c) Impulse response $h(t)$.

Now, going on to the second step, suppose the input signal $x(t)$ is as shown in Figure 4.12,

$$x(t) = \frac{At}{T}[u(t)-u(t-T)]$$

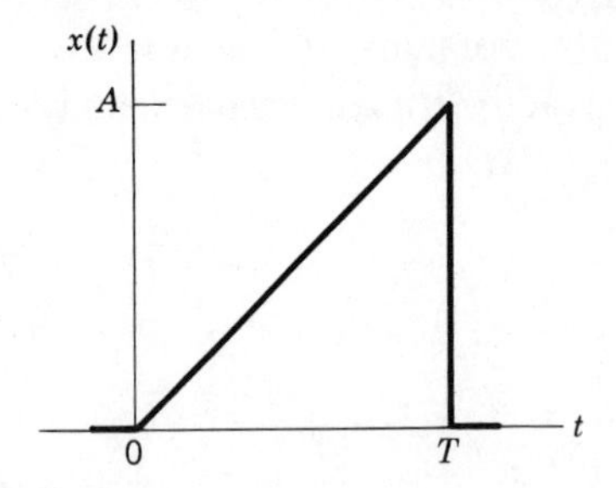

Figure 4.12.

and we want to find $y(t)$ for various values of T relative to the circuit's time constant RC. Taking the superposition integral in the form of (11a),

$$y(t) = \int_{-\infty}^{\infty} \overbrace{\frac{A\lambda}{T}[u(\lambda) - u(\lambda - T)]}^{x(\lambda)} \overbrace{\frac{1}{RC} e^{-(t-\lambda)/RC} u(t-\lambda)}^{h(t-\lambda)} d\lambda$$

and the graphical interpretation of convolution shows that there are three cases to treat, just like Example 4.1. Leaving the details to the reader, one finds that

$$y(t) = \begin{cases} 0 & t < 0 \\ A\left[\dfrac{t}{T} - \dfrac{RC}{T}(1 - e^{-t/RC})\right] & 0 \le t \le T \\ A\left(1 - \dfrac{RC}{T} + \dfrac{RCe^{-T/RC}}{T}\right) e^{-(t-T)/RC} & t > T \end{cases}$$

The output waveform is plotted in Figure 4.13 for two values of the ratio T/RC.

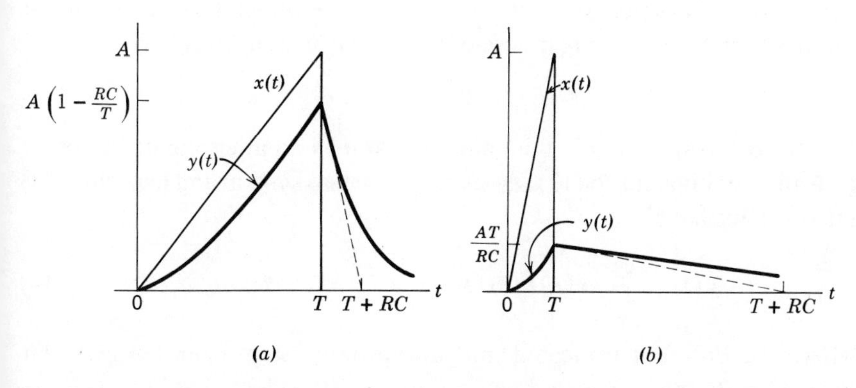

Figure 4.13. Input and output of an RC circuit. (*a*) $T/RC \gg 1$. (*b*) $T/RC \ll 1$.

An interesting feature of this figure is that $y(t)$ approximates $x(t)$ if $T/RC \gg 1$ while it approximates the integral of $x(t)$ if $T/RC \ll 1$. These approximations can be supported analytically, and a frequency-domain interpretation will be given in Chapter 7.

Causal systems with causal inputs

The two general forms of the superposition integral were given in (11). However, when dealing with causal systems excited by causal inputs, these expressions become modified along the lines developed below.

A *causal system* is one whose response at any particular time does not depend on future values of the input. Formally defined, if a system is in the zero state at t_0 and if the input is zero until some later time t_1, i.e.,

$$x(t) = 0 \qquad t < t_1 \tag{12a}$$

then the system is causal if and only if

$$y(t) = S[\mathbf{0}; x(t)] = 0 \qquad t_0 < t < t_1 \tag{12b}$$

This merely says that the output does not precede or *anticipate* the input, or that the system "doesn't laugh before it's tickled." Clearly, the output of a real system cannot appear before the input is applied, and therefore *all real systems are causal.* Noncausal or anticipatory system models necessarily do not represent realizable systems, but sometimes they are very useful idealizations – the ideal lowpass filter, to be discussed in Chapter 7, is one such case.

Relating causality to time-domain analysis, since $h(t)$ is the response to $x(t) = \delta(t)$ and since $\delta(t) = 0$ for $t < 0$, in the integral sense at least, it follows that the impulse response of a causal system must satisfy

$$h(t) = 0 \qquad t < 0 \tag{13}$$

For fixed linear systems, this may be taken as an alternate definition of causality, equivalent to (12). Assuming a causal system and inserting (13) into (11), we have

$$y(t) = \int_{-\infty}^{t} x(\lambda)h(t-\lambda)\,d\lambda = \int_{0}^{\infty} h(\mu)x(t-\mu)\,d\mu \tag{14}$$

where the different truncated integration limits stem from the fact that $h(t-\lambda) = 0$ for $\lambda > t$ while $h(\mu) = 0$ for $\mu < 0$.

Now, suppose we restrict our attention to inputs applied at $t = 0$, i.e.,

$$x(t) = 0 \qquad t < 0$$

which, from the similarity to (13), are called *causal signals* – a somewhat misleading name since it has nothing to do, per se, with the concept of causal systems. But for the case of a *causal system* with a *causal input,* the superposition integral further simplifies to

$$y(t) = \int_0^t x(\lambda)h(t-\lambda)d\lambda = \int_0^t h(\mu)x(t-\mu)d\mu \tag{15}$$

where, in addition to the truncations in (14), we have drawn upon $x(\lambda) = 0$ for $\lambda < 0$ and $x(t-\mu) = 0$ for $\mu > t$. Thus, although the limits are the same in both forms of (15), one should be aware of why this is so.

As a practical matter, all real systems are causal and all real signals have some starting time, which might as well be taken at $t = 0$. Equation (15) would thus appear to be sufficient for all purposes. This fact notwithstanding, the more inclusive form, (11), is necessary when we treat idealized situations. For instance, (15) is not appropriate for the common problem of steady-state analysis wherein a DC or sinusoidal signal is assumed to have been applied long ago, effectively at $t = -\infty$.

Finding the impulse response

Having obtained the superposition integral in terms of the impulse response, we come directly to the question: How does one find $h(t)$ given the mathematical model of a particular system? This sometimes sticky problem is what we intend to tackle now, for without an answer the methods of this chapter would be of no avail in computational work.

Fundamentally, finding $h(t)$ is no more nor less than finding the system's zero-state response to a particular input: the very special input $x(t) = \delta(t)$. After all, the definition of the impulse response is

$$h(t) = S[\delta(t)]$$

Thus, if the system has a simple block-diagram model (i.e., no feedback loops), its impulse response can often be found by setting the input equal to $\delta(t)$ and determining the resulting output by inspection. As a case in point, consider the diagram of Figure 4.14*a*. Clearly, $\dot{y}(t) = Kx(t) - Kx(t-T)$ for any $x(t)$, so

$$y(t) = K\int_{-\infty}^{t} [x(\lambda) - x(\lambda - T)]d\lambda$$

and, with $x(t) = \delta(t)$,

$$h(t) = K\int_{-\infty}^{t} [\delta(\lambda) - \delta(\lambda - T)]d\lambda = K[u(t) - u(t-T)] \tag{16}$$

plotted in Figure 4.14*b*. Incidentally, putting this into the superposition

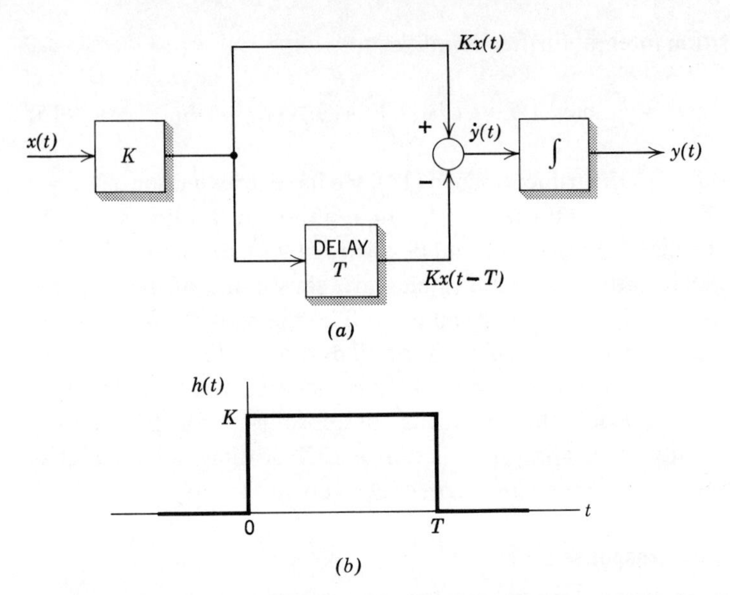

Figure 4.14. The finite-time integrator. (*a*) Block diagram. (*b*) Impulse response.

integral, (11b), yields another, more informative expression for $y(t)$, namely

$$y(t) = K \int_{t-T}^{t} x(\lambda)\,d\lambda \tag{17}$$

which explains why this system is called a *finite-time integrator.*

Alternately, if the *step* response can be deduced by inspection – which is often easier to do – we just differentiate it to get the impulse response, this being the route taken in Example 4.3. But when inspection fails to reveal $h(t)$ or $y_u(t)$, the following systematic approach is called for.

If the system in question is real and therefore *causal*, then we know that $h(t) = 0$ for $t < 0$. Furthermore, since $\delta(t) = 0$ for $t > 0$, we can write $h(t) = S[\mathbf{q}_\delta;\ 0]$, $t > 0$, where $\mathbf{q}_\delta$ is the state at $t = 0+$ as a result of the input $x(t) = \delta(t)$. Thus,

$$h(t) = S[\mathbf{q}_\delta;\ 0]u(t) \tag{18}$$

which says that the impulse response is the same as the zero-*input* response with the special initial state $\mathbf{q}_\delta$. Putting this another way, there is a certain state $\mathbf{q}_\delta$ such that, looking at the output for $t > 0$, we cannot tell whether we are seeing the response *forced* by $x(t) = \delta(t)$ or one particular *natural* response. The factor $u(t)$ is included in (18) to emphasize that this formula does not give the value for $h(0)$, i.e., $\mathbf{q}_\delta$ is calculated at $t = 0+$ to avoid the uncertainties at $t = 0$ where the impulsive input "occurs."

Based on (18), finding the impulse response boils down to three steps.

1. Determine the state $\mathbf{q}_\delta$ at $t = 0+$ caused by $x(t) = \delta(t)$.

2. Write out $h(t)$ for $t > 0$ as the zero-input response with initial state $\mathbf{q}_\delta$.

3. Check for the behavior of $h(t)$ at $t = 0$.

As illustrated below, a simulation diagram proves most helpful for all of these steps.

Example 4.4 Impulse response of a second-order system

Consider the system described by $\ddot{y} + a_1\dot{y} + a_0 y = b_1\dot{x} + b_0 x$. The conditions near $t = 0$ when the input is $\delta(t)$ are indicated on the simulation diagram, Figure 4.15: $q_2 = \int_{-\infty}^{t} \delta(\lambda)\,d\lambda = u(t)$ and $q_1 = \int_{-\infty}^{t} u(\lambda)\,d\lambda = tu(t)$.

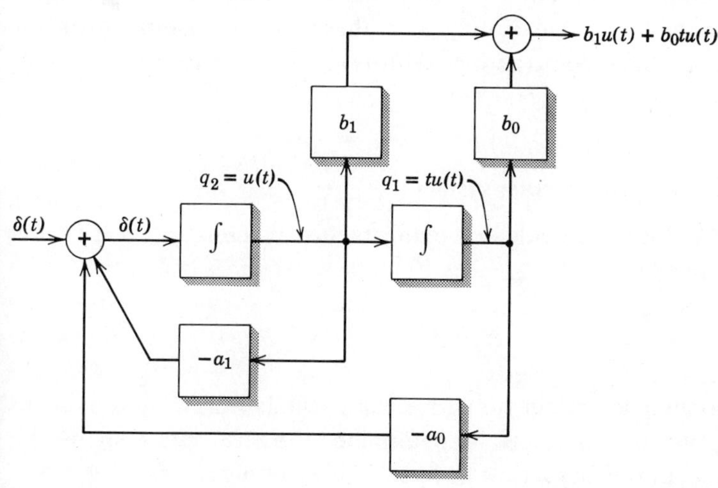

Figure 4.15.

Of course, the feedback loops do add additional terms, but these contributions are negligible at $t = 0+$ compared with the effects of $\delta(t)$. Thus, near $t=0$,

$$h(t)=b_1u(t)+b_0tu(t) \qquad \dot{h}(t)=b_1\delta(t)+b_0u(t)$$

so

$$h(0+)=b_1 \qquad \dot{h}(0+)=b_0$$

Further examination of the diagram shows that the exact value of $h(0)$ is unimportant in this case since $h(t)$ has a finite discontinuity at $t=0$, which disposes of the third step of our method.

As for the second step, assuming distinct roots $p_1 \neq p_2$, the system's modes are e^{p_1t} and e^{p_2t}. Hence, from (18),

$$h(t)=(A_1e^{p_1t}+A_2e^{p_2t})u(t)$$

where A_1 and A_2 are to be evaluated from $h(0+)$ and $\dot{h}(0+)$. The latter task is routine, and we get the final result

$$h(t)=\left(\frac{p_2b_1-b_0}{p_2-p_1}e^{p_1t}-\frac{p_1b_1-b_0}{p_2-p_1}e^{p_2t}\right)u(t) \tag{19}$$

Notice the similarity of (19) to $y_{zi}(t)$ given in Eq. (13), Sect. 3.1.

It might be mentioned that, were we actually simulating this system on an analog computer, no other calculations are needed beyond the initial values of q_1 and q_2. One simply sets these initial conditions on the integrators, sets the input to zero, and turns on the computer; the resulting output waveform will be $h(t)$, $t>0$.

Impulse response of an *n*th-order system

In the same fashion one could derive the impulse response of an nth-order system described by

$$\sum_{i=0}^{n} a_iy^{(i)}=\sum_{k=0}^{m} b_kx^{(k)} \qquad m\leq n-1$$

However, assuming distinct roots, we can profitably draw on the results of the previous chapter where we determined the zero-state response for an arbitrary input starting at time t_0. Specifically, setting $t_0=0$ in Eq. (18a), Sect. (3.3), gives

$$y(t)=\int_0^t x(\lambda)\left[\sum_{i=1}^{n} C_i\eta_i(t-\lambda)\right]d\lambda \tag{20}$$

Now, because these systems are causal and, by taking $t_0 = 0$, the input is a causal signal, (20) is directly comparable with (15), i.e.,

$$y(t) = \int_0^t x(\lambda)h(t-\lambda)d\lambda.$$

It then follows that the bracketed term in (20) must equal $h(t-\lambda)$, and hence

$$h(t) = \Big[\sum_{i=1}^{n} C_i\eta_i(t)\Big]u(t)$$

$$= \Big[\sum_{i=1}^{n}\Big(\sum_{k=0}^{m} b_k p_i^k\Big)\frac{D_{0i}}{D}e^{p_i t}\Big]u(t) \tag{21}$$

in which D and D_{0i} are the determinants defined by Eq. (19a), Sect. 3.1, and Eq. (13), Sect. 3.3.

Formidable though it may appear, (21) is relatively easy to handle with $n \leq 3$. On the other hand, if n is large and there are repeated roots, determining $h(t)$ becomes quite tedious; for such cases, methods other than time-domain analysis would be preferred. But regardless of the system's complexity, (21) clearly indicates that the impulse response is nothing more than a linear combination of the *natural-behavior modes*, again linking the zero-state and zero-input responses. Finally, with $h(t)$ as given by (21), (20) readily generalizes to the form $y(t) = x * h(t)$, thereby tying up the loose ends of Chapter 3 and this section.

4.4 TIME-DOMAIN TECHNIQUES

If the impulse response of a given fixed linear system is known, the zero-state response to almost any input $x(t)$ can be found, at least in theory, from the superposition integral

$$y(t) = \int_{-\infty}^{\infty} x(\lambda)h(t-\lambda)d\lambda \tag{1a}$$

$$= \int_{-\infty}^{\infty} h(\mu)x(t-\mu)d\mu \tag{1b}$$

But because this convolution is often far from trivial to carry out, time-domain analysis may be of limited merit for *quantitative* work unless one can draw on certain simplifications or invoke digital methods.

Some of these special techniques are covered below. Following that, we take up an investigation which shows the value and power of time-domain analysis as a *conceptual* approach.

Response to a rectangular pulse

Consider a system's response to the rectangular input pulse, Figure 4.16,

$$\delta_\Delta(t) \triangleq \frac{1}{\Delta}[u(t)-u(t-\Delta)] \tag{2}$$

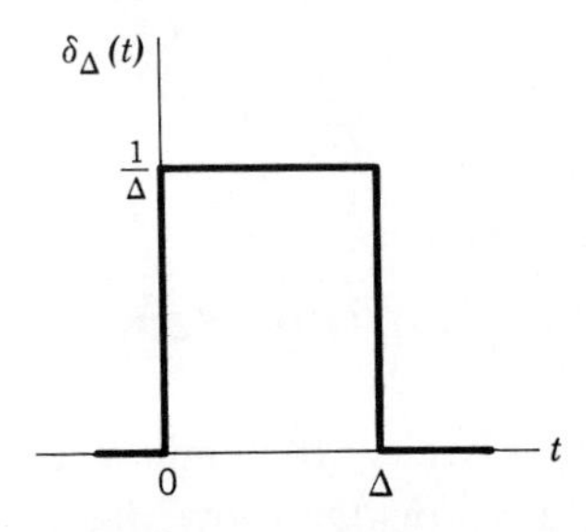

Figure 4.16.

which has duration Δ and unit area. From (1b), the resulting output is

$$\begin{aligned} y_\Delta(t) &= \int_{-\infty}^{\infty} h(\mu)\delta_\Delta(t-\mu)\,d\mu \\ &= \frac{1}{\Delta}\int_{t-\Delta}^{t} h(\mu)\,d\mu \end{aligned} \tag{3}$$

Putting this in words, the rectangular-pulse response at time t is merely the *average* of the impulse response over the preceding Δ units of time.

If a system has the impulse response shown in Figure 4.17, the response at $t = t_1$ to $\delta_\Delta(t)$ can be found graphically merely by locating t_1 on the abscissa and estimating the average value of $h(t)$ between $t_1 - \Delta$ and t_1. Hence, if $h(t)$ has roughly constant slope over that interval, then the average of $h(t_1)$ and $h(t_1 - \Delta)$ will approximately equal the value at the midpoint $t = t_1 - (\Delta/2)$ and, under these conditions, $y_\Delta(t_1) \approx h[t_1 - (\Delta/2)]$. Consequently, any discontinuity in $h(t)$ will be smoothed so as to yield an output that is continuous but with a sharp corner, as seen at $t = \Delta$ in the figure. Notice also, echoing our discussion in Section 4.2, that if Δ is sufficiently small, then $y_\Delta(t) \approx h(t)$, meaning that the output is essentially the impulse response. This follows analytically since, in the sense of Eq. (10), Sect. 4.2, $\delta_\Delta(t) \to \delta(t)$ as $\Delta \to 0$.

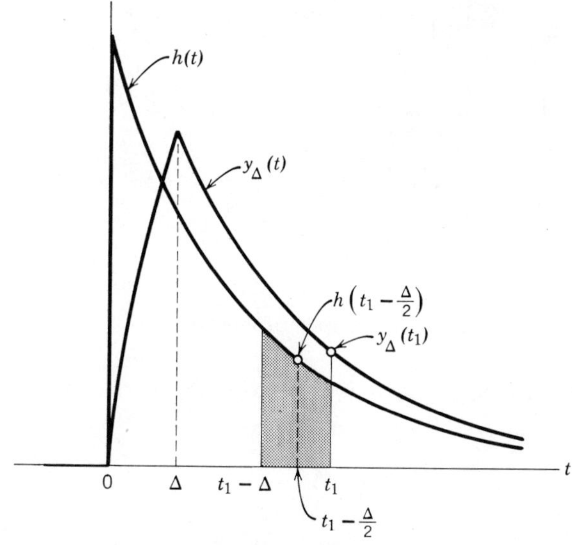

Figure 4.17. An impulse response and the corresponding rectangular-pulse response.

Approximating the response to an arbitrary input

From (3) or Figure 4.17 it is apparent that if $h(t)$ is continuous, $y_\Delta(t)$ may be made an arbitrarily close approximation to the impulse response by taking Δ sufficiently small. Suppose, then, that an arbitrary input $x(t)$ is approximated by a set of narrow rectangular pulses. If the width of these pulses is such that the impulse response is a good approximation to the pulse response, then the output $y(t)$ due to $x(t)$ may be approximated by a summation of impulse responses weighted by the appropriate value of $x(t)$ and delayed by the appropriate multiple of Δ. This line of reasoning leads to the following approximation method.

Assume that $h(t)$ is continuous and that $y_\Delta(t)$ represents the response to $\delta_\Delta(t)$ with Δ such that

$$y_\Delta(t) \approx h\left(t - \frac{\Delta}{2}\right) \tag{4}$$

Furthermore, assume that $x(t)$ may be approximated by $\hat{x}(t)$, Figure 4.18, which is a sequence of rectangular pulses of width Δ *centered* at integer multiples of Δ, i.e., at $t = k\Delta$, $k = \ldots -2, -1, 0, 1, 2, \ldots$ To obtain a good approximation to $x(t)$, it seems intuitively obvious that the heights of the individual pulses should be taken as $x(k\Delta)$.

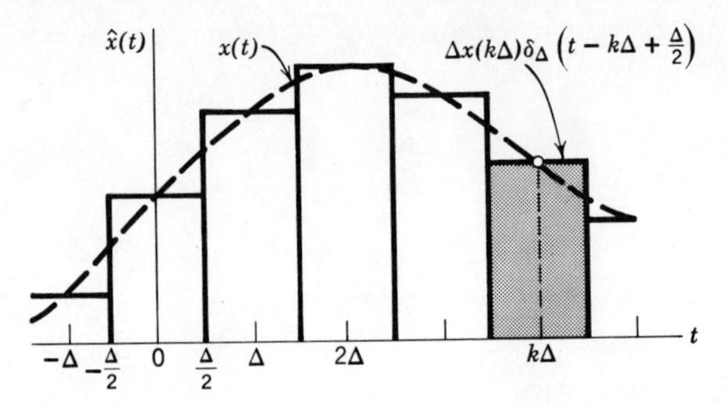

Figure 4.18. Approximating a signal by a sequence of rectangular pulses.

Following this argument, the pulse that approximates $x(t)$ over the interval $k\Delta-(\Delta/2) < t < k\Delta+(\Delta/2)$ is given by

$$\Delta x(k\Delta)\delta_\Delta\left(t-k\Delta+\frac{\Delta}{2}\right)$$

where $\delta_\Delta[t-k\Delta+(\Delta/2)]$ is a unit-area rectangular pulse centered at $t=k\Delta$, and $\Delta x(k\Delta)$ is the *area* of the kth pulse – approximating the area under $x(t)$ in the kth interval. Therefore, we have

$$\hat{x}(t)=\Delta\sum_{k=-\infty}^{\infty}x(k\Delta)\delta_\Delta\left(t-k\Delta+\frac{\Delta}{2}\right) \tag{5}$$

Then, since the response to each pulse $\delta_\Delta[t-k\Delta+(\Delta/2)]$ is $y_\Delta[t-k\Delta+(\Delta/2)]$ – which, by use of (4), approximately equals $h(t-k\Delta)$ – the approximate response $\hat{y}(t)$ becomes

$$\hat{y}(t)=\Delta\sum_{k=-\infty}^{\infty}x(k\Delta)h(t-k\Delta) \tag{6}$$

To restate the significance (6), the input has been broken down into a sequence of narrow pulses whose total area closely approximates the area under the input signal. The response to each individual pulse is then approximated in terms of the impulse response, allowing the approximate output to be computed as the weighted sum of appropriately delayed impulse responses.

Equivalently, if Δ is very small, the narrow pulses in (5) might be replaced by *impulses* occuring at $t=k\Delta$ with weights $\Delta x(k\Delta)$. Then, since $h(t-k\Delta)=S[\delta(t-k\Delta)]$ by definition, the contribution to the response at time t due to the kth impulse is merely its weight $\Delta x(k\Delta)$ times $h(t-k\Delta)$.

In many texts, the arguments used in arriving at (6) are extended to derive the superposition integral

$$y(t) = \int_{-\infty}^{\infty} x(\lambda)h(t-\lambda)d\lambda$$

This follows by letting $\Delta \to d\lambda$, $k\Delta \to \lambda$, etc., as we did to obtain Duhamel's integral from the step-response approximation.

Numerical methods and discrete convolution

Numerical evaluation of the superposition integral is often necessary in quantitative work, and one might draw upon an algorithm such as Simpson's rule, for instance. However, the summation relationship expressed by (6) suggests a related approach which amounts to an approximation of the integration process and requires only minor changes to recast for numerical evaluation. As with any such computational method, there is a tradeoff between accuracy and the number of computations. Heuristically speaking, decreasing Δ by a factor of, say, 2 should certainly reduce the error in $\hat{y}(t)$ as an approximation to $y(t)$; at the same time, the number of numerical operations is doubled. Although no attempt will be made to treat this question in depth, it is instructive to investigate qualitatively how a value for Δ might be chosen by looking at the resulting output approximation.

Assume that Δ has been selected such that $h(t)$ is relatively constant over each interval and let the discrete points at which the output is to be evaluated also be spaced in time by Δ. Then, setting $t = n\Delta$ in (6) gives

$$\hat{y}(n\Delta) = \Delta \sum_{k=-\infty}^{\infty} x(k\Delta)h(n\Delta - k\Delta) \qquad (7)$$

which is *discrete convolution*.† But in order to actually carry out (7) it is necessary to have *finite* limits for the summation; these are established if the system is causal, so $h(t) = 0$ for $t < 0$ and $h(n\Delta - k\Delta) = 0$ for $k > n$, and the input starts at time $t = 0$, so $x(k\Delta) = 0$ for $k < 0$.

†Healy (1969) gives an excellent description of how to do discrete convolutions by hand, along with a discussion of discrete and continuous convolution applied to systems analysis and probability theory.

Under those conditions we have the discrete version of Eq. (15), Sect. 4.3:

$$\hat{y}(n\Delta) = \Delta \sum_{k=0}^{n} x(k\Delta)h(n\Delta - k\Delta) \tag{8a}$$

$$= \Delta \sum_{i=0}^{n} x(n\Delta - i\Delta)h(i\Delta) \tag{8b}$$

where the second expression follows by a simple change of summation index. The two forms of (8) merely reflect that convolution, discrete or continuous, is commutative.

Outlined below is a FORTRAN computer code for carrying out the calculation of (8a). Through the use of $XX(M) \triangleq x[(M-1)\Delta]$, etc., the program accounts for the fact that indices are not permitted to take on zero or negative values in most FORTRAN systems.

```
Input:           READ NMAX, DELTA
                 MAX = NMAX + 1
                 READ (XX(M), M = 1, MAX), (HH(M), M = 1, MAX)

Computations:  ┌→ DØ M = 1, MAX
               │  SUM = 0.
               │┌→ DØ L = 1, M
               │└─ SUM = SUM + XX(L) * HH(M + 1 − L)
               │  YY(M) = DELTA * SUM
               │  TIME = M − 1
               └─ TT(M) = TIME * DELTA

Output:          PRINT ((TT(M), YY(M)), M = 1, MAX)
```

Example 4.5 A numerical solution

Let us use (8) to find the output of an RC circuit (Figure 4.11a) when the input is a ramp function $x(t) = Atu(t)$. Intentionally, this problem is easy enough to be solved directly, the result being

$$y(t) = A[t - RC(1 - e^{-t/RC})]u(t)$$

as the reader can show from Example 4.3. Having the exact answer allows us to test the accuracy of the numerical solution.

Since we know the impulse response is $h(t) = (1/RC)e^{-t/RC}u(t)$, (8a) becomes

$$\hat{y}(n\Delta) = \Delta \sum_{k=0}^{n} \overbrace{Ak\Delta}^{x(k\Delta)} \overbrace{\frac{1}{RC} e^{-(n\Delta - k\Delta)/RC}}^{h(n\Delta - k\Delta)}$$

in which the $u(t)$'s have already been absorbed by the summation limits. Simple rearrangement then gives

$$\hat{y}(n\Delta) = \frac{A\Delta^2}{RC} e^{-n\Delta/RC} \sum_{k=0}^{n} k e^{k\Delta/RC}$$

as compared to

$$y(t)\big|_{t=n\Delta} = A[n\Delta - RC(1 - e^{-n\Delta/RC})]$$

the exact value of the output at $t = n\Delta$. Figure 4.19 shows $\hat{y}(n\Delta)$, along with $y(t)$, for three values of Δ. Clearly, a good approximation requires $\Delta \ll RC$, as might be expected on intuitive grounds.

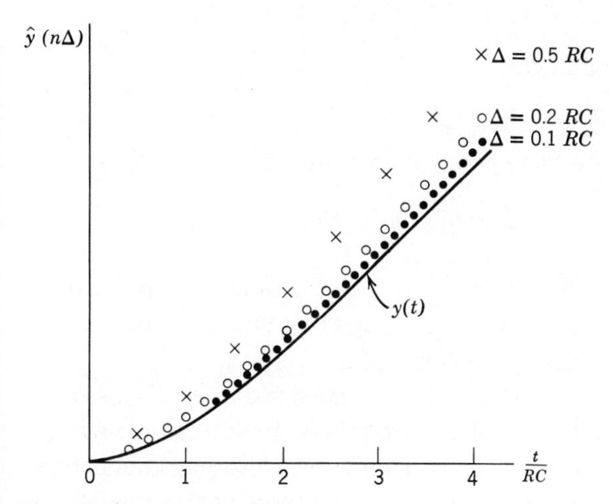

Figure 4.19.

Stability of the zero-state response

Having looked at time-domain analysis in the context of quantitative problem solving, we would be remiss in not demonstrating its value for general analytic studies. This will be done by considering the stability of the zero-state response. Because the concept of stability relates to the nature of a system's response, as distinguished from the system itself, we begin with the so-called *bounded-input-output* (BIO) definition of a stable zero-state response:

> A system is said to be stable with respect to its zero-state response if every bounded (finite) input yields a bounded output, i.e., if
>
> $$|S[\mathbf{0}; x(t)]| < \infty \qquad -\infty < t < \infty \tag{9}$$
>
> with any $x(t)$ such that $|x(t)| < \infty$ for all t.

For a fixed linear system to satisfy this condition, it is necessary and sufficient that

$$\int_{-\infty}^{\infty} |h(t)|dt < \infty \tag{10}$$

which says that the impulse response must be *absolutely integrable.* To prove that (10) guarantees (9) – i.e., sufficiency – we first establish an upper bound on $|S[\mathbf{0}; x(t)]| = |y(t)|$ from (1b), namely

$$|y(t)| = \left| \int_{-\infty}^{\infty} h(\mu)x(t-\mu)d\mu \right| \leq \int_{-\infty}^{\infty} |h(\mu)||x(t-\mu)|d\mu$$

Now take a bounded input, i.e.,

$$|x(t)| < C \qquad -\infty < t < \infty$$

where C is an arbitrary but finite positive constant. With this input

$$|y(t)| \leq C \int_{-\infty}^{\infty} |h(\mu)|d\mu$$

and if (10) holds true, then $|y(t)|$ is finite – so the output is bounded and the system is stable. It can further be shown that (10) is a necessary as well as sufficient condition; see Problem 4.32.

The relationship between BIO stability and our earlier discussion in Section 3.2 is brought out by recalling two things. First, if all the natural-behavior modes of a fixed linear system are either decaying exponentials or sinusoids with decaying envelopes – i.e., all roots within the left half of the s-plane – then $y_{zi}(t) \to 0$ as $t \to \infty$, which we classified as an asymptotically stable zero-input response. Second, as found in Section 4.3, the impulse response is a linear combination of the system's mode functions. Now, one can readily show that (10) holds if and only if all the roots of the characteristic equation are within the LHP, so *BIO stability* of the zero-state (forced) response entails the same conditions as *asymptotic stability* or the zero-input (free) response. However, this conclusion applies, in general, only to fixed linear systems.

4.5 TRANSFER FUNCTIONS

In principle, time-domain analysis is sufficient for dealing with the zero-state response of any fixed linear system. Practically speaking, however, it does have certain drawbacks. For one thing, we must know the impulse response which, as seen in the previous section, can be a rather nasty chore in itself; then, having found $h(t)$, we must solve a convolution integral, another potentially difficult and sometimes impossible job. To get around these problems, systems analysis frequently is tackled not in the time domain but rather through the use of *transform techniques*, techniques based on the concept of the system transfer function introduced here.

Introduction to transfer functions

Consider the superposition integral in the form

$$y(t) = \int_{-\infty}^{\infty} h(\lambda)x(t-\lambda)d\lambda \tag{1}$$

and suppose one could freely choose any input $x(t)$ that expedites the integration. Clearly, it would be advantageous from a computational viewpoint if $x(t)$ had the property $x(t-\lambda) = v(t)w(\lambda)$ so that (1) would become $y(t) = v(t) \int_{-\infty}^{\infty} h(\lambda)w(\lambda)d\lambda$. Then our analysis is greatly simplified providing the product $h(\lambda)w(\lambda)$ lends itself readily to standard integration methods.

One possibility along this line is to take the exponential signal

$$x_P(t) \triangleq e^{st} \qquad -\infty < t < \infty \tag{2}$$

where s is a constant, perhaps complex, and the subscript P denotes the *particular*† nature of the input. Since $x_P(t-\lambda) = e^{s(t-\lambda)} = e^{st}e^{-s\lambda}$, (1) reduces to

$$y_P(t) = e^{st} \int_{-\infty}^{\infty} h(\lambda)e^{-s\lambda}d\lambda \tag{3}$$

in which the integrand and the limits of integration are independent of t. Moreover, the integral in (3) depends only on the system's impulse response and the value of s used to specify $x_P(t)$. Therefore, let us define

†The corresponding output $y_P(t)$ is the *particular solution*, in the same sense as that of the classical method for solving differential equations.

the system *transfer function* (also called simply the system function) as

$$H(s) \triangleq \int_{-\infty}^{\infty} h(\lambda)e^{-s\lambda}d\lambda \tag{4}$$

which we explicitly indicate to be a function of s. Inserting $H(s)$ in (3) yields

$$y_P(t) = e^{st}H(s) \tag{5}$$

So if the input is an exponential for all time and if the transfer function is known, then the output is found merely by *multiplication*, per (5), and we have eliminated convolution.

Obviously (5) represents a much simpler analysis tool than the superposition integral, but *only* if $H(s)$ is known and *only* for the case where $x(t) = e^{st}$. Relative to the latter restriction, one should recall from our discussion of zero-input responses that a linear combination of exponential functions, any of which may be complex, does cover a rich variety of signals. Accordingly, if

$$x(t) = \alpha_1 e^{s_1 t} + \alpha_2 e^{s_2 t} + \cdots \qquad -\infty < t < \infty \tag{6}$$

where the α's and the s's are constants, then

$$\begin{aligned} y(t) &= \alpha_1 e^{s_1 t} \int_{-\infty}^{\infty} h(\lambda)e^{-s_1\lambda}d\lambda + \alpha_2 e^{s_2 t} \int_{-\infty}^{\infty} h(\lambda)e^{-s_2\lambda}d\lambda + \cdots \\ &= \alpha_1 e^{s_1 t}H(s_1) + \alpha_2 e^{s_2 t}H(s_2) + \cdots \qquad -\infty < t < \infty \end{aligned} \tag{7}$$

where $H(s_1) = H(s)|_{s=s_1}$, etc. We see, therefore, that the transfer-function approach can be extended to include the class of excitations expressable as a sum of real and complex exponentials, providing they exist for all time. Going even further in this direction, suppose $x(t)$ can be written as an *integration* with respect to s, such as

$$x(t) = \int_{-\infty}^{\infty} X(s)e^{st}ds \tag{8}$$

which, intuitively, is a limiting case of (6). Then, from (1) and (4), one can show that

$$y(t) = \int_{-\infty}^{\infty} Y(s)e^{st}ds \tag{9a}$$

where

$$Y(s) = X(s)H(s) \tag{9b}$$

It is not apparent at this point whether (9a) represents an easier task than the superposition integral, but at least (9b) is a simple algebraic relationship.

As the reader may have recognized, (6) and (8) are closely related to the *Fourier series* and the *Fourier* and *Laplace transforms*. And the above developments suggest that the mathematical techniques associated with these methods can be brought to bear on the problem of systems analysis once the transfer function is known.

Finding the transfer function

But now to the crucial question: Can we find $H(s)$ by some means other than (4)? For the utility of the transfer-function approach hinges strongly on being able to find $H(s)$ directly, without having to know $h(t)$. The answer to our question is a resounding *yes* for most system models.

The key is whether one can easily determine $y_P(t)$, the response resulting from $x_P(t) = e^{st}$, $-\infty < t < \infty$; if so, then

$$H(s) = y_P(t)e^{-st} \tag{10}$$

which is simply (5) rewritten. Take, for instance, the finite-time integrator diagrammed earlier in Figure 4.14*a*. With $x(t) = e^{st}$, Eq. (17), Sect. 4.3, becomes

$$y_P(t) = K\int_{t-T}^{t} e^{s\lambda}d\lambda = \frac{K}{s}[e^{st} - e^{s(t-T)}]$$

Hence, from (10),

$$H(s) = \frac{K}{s}(1 - e^{-sT}) \tag{11}$$

a result the reader might like to check by inserting $h(t)$, Eq. (16), Sect. 4.3, into the integral definition of $H(s)$.

More generally, consider any system that can be described in differential-equation form by

$$\sum_{i=0}^{n} a_i y^{(i)} = \sum_{k=0}^{m} b_k x^{(k)}$$

If we assume $x(t) = e^{st}$ then, by definition, $y(t) = H(s)e^{st}$; making these substitutions and noting that $x^{(k)} = s^k e^{st}$ while $y^{(i)} = s^i H(s)e^{st}$, $H(s)$ drops out directly as

$$H(s)=\frac{\sum_{k=0}^{m} b_k s^k}{\sum_{i=0}^{n} a_i s^i}=\frac{b_m s^m+b_{m-1}s^{m-1}+\cdots+b_1 s+b_0}{a_n s^n+a_{n-1}s^{n-1}+\cdots+a_1 s+a_0} \tag{12}$$

Three important points are contained in (12). First, the transfer function for systems of this type henceforth can be written down immediately from the differential equation, as contrasted with the effort required to find the impulse response – notice, by the way, that (12) holds even when there are repeated roots and when $m > n$. Second, the denominator of $H(s)$ is precisely the system's *characteristic polynomial.* And this leads us immediately to the third point, namely that $H(s) \to \infty$ whenever s is a *root* of the characteristic equation; so we say that $H(s)$ has *poles* at $s = p_1, p_2, \ldots p_n$.

Essentially the same method as above can be applied when one is dealing with *state equations* rather than a single differential equation. This is particularly advantageous for the case of multi-input-output systems, for which the transfer function is a matrix $\mathbf{H}(s)$ and the denominator of each element of $\mathbf{H}(s)$ will be $|s\mathbf{I}-\mathbf{A}|$. The details are covered in Section 8.4.

In summary, many and wondrous are the simplifications that come about when the system input is assumed to be an exponential signal or a linear combination of exponentials. Notably, the corresponding response is given by an *algebraic* equation involving the transfer function $H(s)$, rather than in terms of an unsolved differential equation or a convolution integral. And finding $H(s)$ is no more work – often less – than finding $h(t)$; indeed, it is a trivial calculation when (12) is applicable. For these reasons, the next four chapters are devoted to expressing arbitrary input signals in exponential form and developing the transform techniques for handling the analysis of continuous systems. Furthermore, in Chapters 12 and 13, these notions are extended to the case of sampled and discrete systems.

Preservation of form

Before leaving this section, we should give some attention to the reason why exponential signals lead to a simplified analysis of fixed linear systems. At the heart of the matter is the property known as *preservation of form*, to wit: If the excitation is proportional to e^{st} for $-\infty < t < \infty$, then the output and all the state variables are proportional to e^{st}. Equation (3) confirms this assertion for $y(t)$. However, a more general and independent proof follows directly from the basic properties of a fixed linear system.

Recall that if $y(t) = S[x(t)]$ is the zero-state response of a fixed linear system then $S[\alpha x(t)] = \alpha y(t)$ and $S[dx/dt] = dy/dt$. With these in mind, take $x(t) = e^{st}$ – so $y(t) = S[e^{st}]$ – and consider

$$S\left[\frac{dx(t)}{dt}\right] = S[se^{st}] = sS[e^{st}] = sy(t)$$

But, regardless of $x(t)$, $S[dx/dt] = dy/dt$; thus, when $x(t) = e^{st}$, $y(t)$ must satisfy

$$\frac{dy(t)}{dt} = sy(t)$$

The solution of this first-order differential equation is known to be

$$y(t) = Ae^{st}$$

where A is independent of time. It does, however, depend on the parameter s, for, referring to (3) shows that $A = H(s)$.

Be that as it may, the important conclusion here is that the output is an exponential time function when the input is an exponential time function. Hence: preservation of form. It can also be shown, although we shall omit the proof, that this property of fixed linear systems is true *only* for exponential signals.

Problems

4.1 Carry out all the details of the convolution of two rectangular functions, Example 4.2.

4.2 Find $v * w(t)$ when

$$v(t) = \begin{cases} at & t \geq 0 \\ 0 & t < 0 \end{cases} \qquad w(t) = \begin{cases} e^{-bt} & t \geq 0 \\ 0 & t < 0 \end{cases}$$

Answer: $(a/b^2)(bt + e^{-bt} - 1)$, $t \geq 0$.

4.3 Find $v * w(t)$ when

$$v(t) = at^2 \qquad -\infty < t < \infty \qquad w(t) = \begin{cases} b & -T/2 \leq t \leq T/2 \\ 0 & \text{otherwise} \end{cases}$$

4.4 Given that both $v(t)$ and $w(t)$ have odd symmetry, show that $v * w(t)$ has even symmetry. See Eq. (19), Sect. 5.1, for the symmetry definitions.

4.5 Prove that, similar to Eq. (7), Sect. 4.1,

$$\int_{-\infty}^{t} [v * w(\lambda)]d\lambda = [v(t)] * \left[\int_{-\infty}^{t} w(\mu)d\mu\right] = \left[\int_{-\infty}^{t} v(\xi)d\xi\right] * [w(t)]$$

4.6 If $y(t) = v * w(t)$, show that $[v(t)] * [w(t-T)] = y(t-T)$.

4.7 Generalize Eq. (6), Sect. 4.2, for $\int_a^b v(t-t_1)\delta(t-t_2)dt$.

4.8 Prove the scale-change relationship Eq. (9a), Sect. 4.2, by showing that $\int_{-\infty}^{\infty} v(t)\delta(ct)dt = v(0)/|c|$, $c \neq 0$. *Hint*: consider the two cases $c > 0$ and $c < 0$ separately.

4.9 Simplify each of the following expressions: (a) $\int_{-\infty}^{\infty} v(t-t_1)\delta(t-t_2)dt$; (b) $\int_{-\infty}^{\infty} v(t)\delta(\lambda)d\lambda$; (c) $[v(t-\tau)] * [\delta(t+\tau)]$; (d) $\int_{-\infty}^{\infty} v(ct)\delta(ct)dt$; (e) $t^n\delta(t)$; (f) $\delta(t-t_1)\delta(t-t_2)$.

4.10 Show that $\delta_\epsilon(t)$ in Figure P4.1 satisfies Eq. (10a), Sect. 4.2.

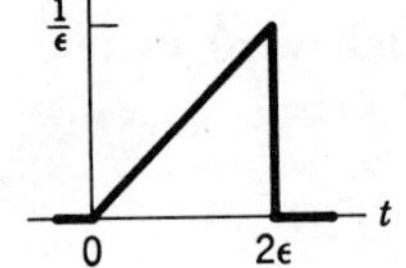

Figure P4.1.

4.11 Use integration by parts to evaluate $\int_{-\infty}^{\infty} u(t)\delta(t)dt$. *Answer*: $\frac{1}{2}$.

4.12 Show that the following expressions, more in the spirit of distribution theory, might be used instead of the definitions given in Section 4.2 for the unit step and doublet, respectively.

$$\int_{-\infty}^{\infty} v(t)u(t)dt = \int_{0}^{\infty} v(t)dt \qquad \int_{-\infty}^{\infty} v(t)\delta'(t)dt = -\dot{v}(0)$$

4.13 Confirm that Eq. (9), Sect. 4.3, reduces to Eq. (8) when $x(t) = 0$ for $t < 0$. *Hint*: write $x(t) = x_c(t)u(t)$ where $x_c(t) = x(t)$ for $t \geq 0$.

4.14 Given that $y_u(t)$ is the step response of a certain system, write an expression for $y(t)$ for each of the inputs shown in Figure P4.2.

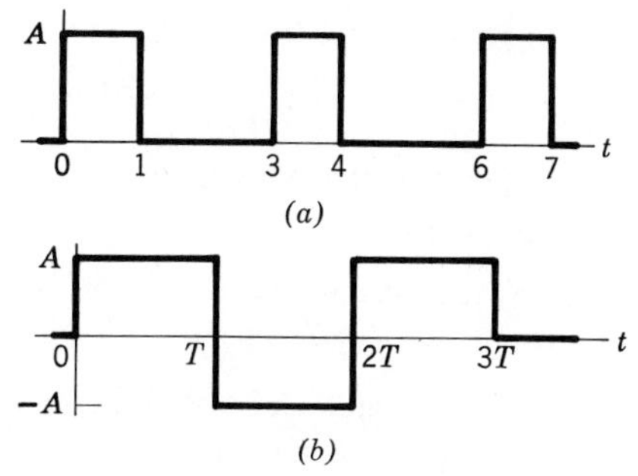

Figure P4.2.

4.15 Carry out the necessary convolution to arrive at $y(t)$ as given in Example 4.3.

4.16 Find the step response of the circuit in Figure P4.3 (by inspection), and differentiate it to find the impulse response. *Answer*: $h(t) = \delta(t) - (R/L)e^{-Rt/L}u(t)$.

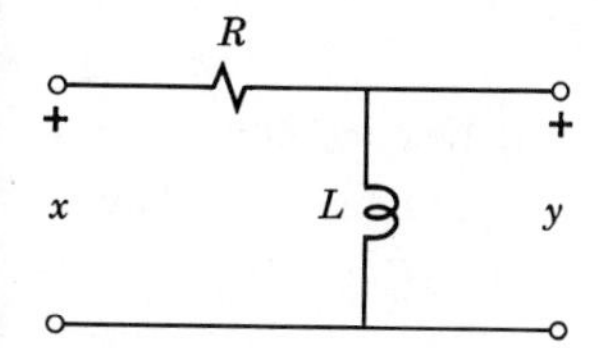

Figure P4.3.

4.17 Find the zero-state response of the circuit in Problem 4.16 when $x(t) = tu(t)$.

4.18 Suppose a system – presumably containing a delay element – has $h(t) = 0$ for $t < T$, where T is a positive constant. Let the input be a signal starting at $t_0 > 0$. Determine the resulting "simplified" forms of the superposition integral, along the lines of Eq. (15), Sect. 4.3, considering the two cases $T < t_0$ and $T > t_0$.

4.19 Find the zero-state response of the RC circuit in Example 4.3 when: (a) $x(t) = A \cos \Omega t$, $-\infty < t < \infty$; (b) $x(t) = A \cos \Omega t\, u(t)$. In the latter case, identify the transient and steady-state terms by comparison with the former.

4.20 Suppose you are given a system whose impulse response is as sketched in Figure P4.4. You also have available a generator that can deliver any waveform $x(t)$, $t \geq 0$, subject only to the condition that $|x(t)| \leq 1$. Based on

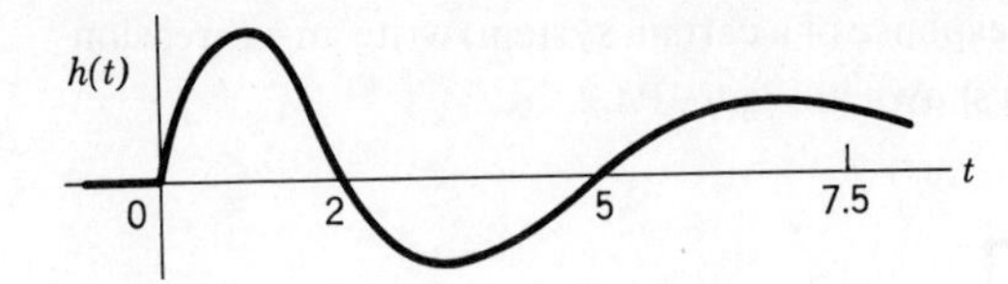

Figure P4.4.

Eq. (15), Sect. 4.3, plot the waveform $x(t)$ you would use to *maximize* $y(t)$ at $t = 7.5$. *Hint*: recall the graphical interpretation.

4.21 The integrator in Figure 4.14*a* is replaced by the RC circuit of Figure 4.11*a*. Find the new $h(t)$ and sketch it assuming $T \ll RC$ and $T \gg RC$.

4.22 Referring to Eq. (16), Sect. 2.2, let $y(t) = \omega(t)$ and $\tau(t) = 0$. Find the impulse response by applying the basic definition $h(t) = S[\mathbf{0}; \delta(t)]$.

4.23 With the aid of a simulation diagram, find the impulse response of the undamped second-order system described by $\ddot{y} + \omega_n^2 y = \omega_n^2 x$.

4.24 Verify that $h(t) = (b_0 - a_0 b_1)e^{-a_0 t}u(t) + b_1 \delta(t)$ is the impulse response of the system described by $\dot{y} + a_0 y = b_1 \dot{x} + b_0 x$.

4.25 Use Eq. (21), Sec. 4.3,to find and sketch $h(t)$ for the system in Problem 3.12.

4.26 Make a plot of $h(t) = te^{-t}u(t)$ and use it to sketch $y_\Delta(t)$ when: (a) $\Delta = 1$; (b) $\Delta = 0.2$.

4.27 Consider any pulse shape $p_\Delta(t)$ that is nonnegative in the interval $0 \le t \le \Delta$, is zero otherwise, and has unit area. If $p_\Delta(t)$ is the input to a system having impulse response $h(t)$, show that the resulting output is bounded by

$$h_{\min}(t, t-\Delta) \le y(t) \le h_{\max}(t, t-\Delta)$$

where $h_{\min}(t, t-\Delta)$ stands for the minimum value of $h(t)$ in the interval $t-\Delta$ to t, etc.

4.28 Given $x(t) = \sin(\pi t/2)\ [u(t) - u(t-2)]$ and $h(t)$ as plotted in Figure P4.5, use Eq. (8), Sect. 4.4, to calculate $\hat{y}(n\Delta)$ at $n\Delta = 1$, 2, and 3, taking $\Delta = 1$. Repeat with $\Delta = \frac{1}{2}$, and compare.

4.29 (Computer solution) For $x(t)$ and $h(t)$ as given in Problem 4.28, calculate all the nonzero values of $\hat{y}(n\Delta)$, taking $\Delta = 0.1$. Plot your result.

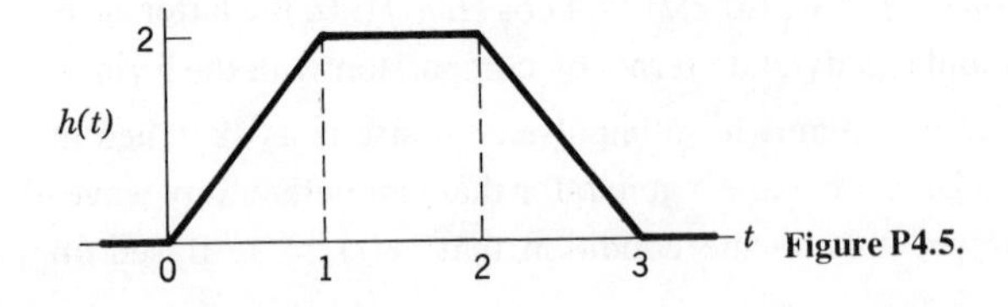

Figure P4.5.

4.30 (Computer solution) Calculate and plot $\hat{y}(n\Delta)$ for $0 \le n\Delta \le 5$, when $x(t)$ is as in Problem 4.28 and $h(t) = (1+t-2t^2)e^{-2t/3}u(t)$. The choice of Δ is left to your discretion. Take $h(0) = 1$.

4.31 Outline a computer program that automatically repeats the calculation of Eq. (8a), Sec. 4.4, with successively smaller values of Δ until a suitably accurate approximation is reached. Note that you must define "suitable accuracy" in some measurable form.

4.32 Show that Eq. (10), Sect. 4.4, is a necessary as well as sufficient condition for BIO stability. *Hint:* note that there is at least one bounded input $x(t)$ such that, for at least one instant of time t_1, $y(t_1) = \int_{-\infty}^{\infty} |h(\mu)| d\mu$.

4.33 If a system has $H(s) = (s+1)^{-1}$, find $y(t)$ when $x(t) = 2 + 4e^{-2t} + 3e^{5t}$, $-\infty < t < \infty$.

4.34 Use the transfer-function approach to find the response of the finite-time integrator, Figure 4.14a, when $K = 1$, $T = \frac{1}{2}$, and $x(t)$ is as in Problem 4.33. Check your result by inserting $x(t)$ into Eq. (17), Sect. 4.3.

4.35 Show that the simple feedback configuration of Figure P4.6 has $H(s) = (s+a)^{-1}$. *Hint:* take $x(t) = e^{st}$ so, by definition, $y(t) = H(s)e^{st}$; then solve the summing-junction equation for $H(s)$.

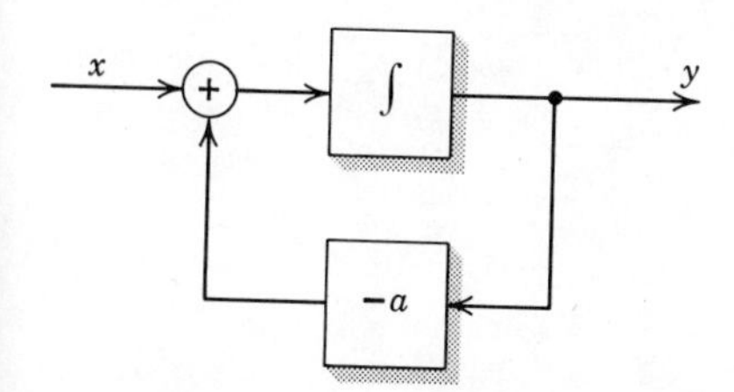

Figure P4.6.

4.36 The zero-state response of a linear *time-varying* system takes the general form $S[x(t)] = \int_{-\infty}^{\infty} x(\lambda)h(t,\lambda)d\lambda$, where the impulse response $h(t,\lambda)$ is a function of both variables, not just a function of $t-\lambda$. If $x(t) = e^{st}$, $-\infty < t < \infty$, show that $y(t) = H(s,t)x(t)$, where

$$H(s,t) \triangleq \int_{-\infty}^{\infty} h(t, t-\mu)e^{-s\mu}d\mu$$

Compare this "transfer function" with Eq. (4), Sect. 4.5, and discuss.

5

Signals and their representation

Just as a mathematical *system model* is necessary in analysis and design work, mathematical *signal representations* are also needed. Like system models, signal representations are abstractions which, although they may not totally conform to physical reality, should convey the essential and pertinent aspects of the signal with a minimum of mathematical complexity. Thus, when we speak of representing a signal, we are usually speaking of an *approximation* or *idealization* of the actual physical variable in question.

Recognizing that our signal representations will be less than exact, it makes good sense to seek out those forms best suited to whatever problem is at hand. In particular, when analyzing fixed linear systems, it would be advantageous to express signals as *linear combinations of exponential functions*, thereby permitting use of the system's transfer function and its collateral simplifications. Accordingly, this chapter deals with the theory underlying signal representations in exponential form.

Following some preliminary classifications, we shall introduce the concept of *signal space*, a space wherein functions of time are viewed as geometric *vectors*. By visualizing signals in this fashion, one can draw upon geometric reasoning and interpretation to give insight to the general problem of signal analysis in much the same manner that vector diagrams are helpful for problems involving complex variables. The signal-space approach is then taken to attack the question of signal approximations and, subsequently, Fourier signal representations, i.e., the Fourier series and transform. Applications of the theory are presented in Chapters 6 and 7.

If, at first glance, the present material seems highly abstract and academic, the reader may take assurance it is probably the easiest way of gaining a real understanding of Fourier analysis, and that practical utilization will come in due course. Just as important is the fact that the ideas introduced here are sufficiently general to encompass several other forms of signal representation, including such topics as the analysis of random signals (noise) and discrete signals. Indeed, a firm grasp of signal-space concepts is essential for many aspects of advanced work in communication and control. Thus, besides immediate applications, this chapter may properly be deemed a long-range investment whose full payoff will be realized as the student progresses in systems engineering.

5.1 CLASSIFICATION OF SIGNAL PROPERTIES

When discussing system modeling, it was noted that a given system may have several different models, depending on the type of signals expected. Consistent with this observation, we begin our coverage of signal analysis by introducing various classifications of signal properties. Throughout this section the reader must remember that when we refer to a signal, say $v(t)$, we really mean a mathematical signal representation.

Energy and power signals

Perhaps the most important classification is based on the common ideas of energy and power. Taking a specific case for illustration, suppose $e(t)$ is the voltage across a resistance R producing a current $i(t)$. The instantaneous power is $p(t) = e(t)i(t) = Ri^2(t)$ and the energy dissipated in a given interval, say $-T \leq t \leq T$, is

$$\text{Energy} = R \int_{-T}^{T} i^2(t)\, dt$$

Since the time interval is $2T$ seconds long, the power averaged over the interval is

$$\text{Power} = \frac{R}{2T} \int_{-T}^{T} i^2(t)\, dt$$

which, like the energy expression, could also be written in terms of the voltage using $i^2(t) = e^2(t)/R^2$. Then, letting $T \to \infty$ so as to encompass all time, we would get the total energy and average power.

In like manner, for any type of system having linear dissipative elements one finds directly analogous expressions for total energy and average power, in the general form

$$\text{Total energy} = \lim_{T\to\infty} K \int_{-T}^{T} v^2(t)\,dt \tag{1a}$$

$$\text{Average power} = \lim_{T\to\infty} \frac{K}{2T} \int_{-T}^{T} v^2(t)\,dt \tag{1b}$$

where $v(t)$ is a physical variable and K an appropriate constant. With minor exceptions, one of two possibilities will then prevail: either the total energy is finite, and hence the average power is zero, or the average power is nonzero, and hence the total energy must be infinite. Thus, we can distinguish between *energy-type* and *power-type* signals, respectively. Clearly, the key factor here is the integral of the squared time function, regardless of system type and the specific proportionality constant.

Now consider some arbitrary signal $v(t)$ which is not necessarily related to actual energy or power and which, for analytic purposes, may be a complex function of time. To accommodate this broader meaning of signals and, at the same time, preserving the notion of energy-type versus power-type, we shall define *energy* E in the abstract as

$$E \triangleq \lim_{T\to\infty} \int_{-T}^{T} |v(t)|^2 dt = \lim_{T\to\infty} \int_{-T}^{T} v(t)v^*(t)\,dt \tag{2}$$

and similarly for *power* P,

$$P \triangleq \lim_{T\to\infty} \frac{1}{2T} \int_{-T}^{T} |v(t)|^2 dt = \lim_{T\to\infty} \frac{1}{2T} \int_{-T}^{T} v(t)v^*(t)\,dt \tag{3}$$

where the magnitude squared is used to insure that E and P will always be *real nonnegative* quantities, even when $v(t)$ is complex.

Based on these abstract definitions one can identify two distinct classes of signals, *energy signals* and *power signals*. We say that $v(t)$ is an energy signal if and only if

$$0 < E < \infty \tag{4}$$

so that $P = 0$. Conversely, $v(t)$ is a *power signal* if and only if

$$0 < P < \infty \tag{5}$$

and hence $E = \infty$. Many signals of practical interest will satisfy either (4) or (5), but not both. For reference purposes Table 5.1 gives the energy, power, and classification for some of the more common time functions; in this table, all $v(t)$ are defined for $-\infty < t < \infty$ unless otherwise indicated by the presence of the unit step $u(t)$.

Table 5.1

Signal $v(t)$	Energy E	Power P	Signal Type
$Ae^{-t/\tau}u(t)$	$A^2\tau/2$	0	Energy
$Ae^{-\lvert t/\tau \rvert}$	$A^2\tau$	0	Energy
$A[u(t)-u(t-\tau)]$	$A^2\tau$	0	Energy
$Ae^{\pm j(\omega t+\theta)}$	∞	A^2	Power
$A \cos (\omega t+\theta)$	∞	$A^2/2$	Power
A	∞	A^2	Power
$Au(t)$	∞	$A^2/2$	Power
$Atu(t)$	∞	∞	Undefined
$Ae^{\pm t/\tau}$	∞	∞	Undefined
$A\delta(t)$	Undefined	Undefined	Undefined

Periodic signals

A signal $v(t)$ is defined to be *periodic* if and only if

$$v(t+T_0) = v(t) \qquad -\infty < t < \infty \tag{6}$$

where the constant T_0 is the period of repetition. By iteration of (6) it follows that

$$v(t \pm nT_0) = v(t) \qquad n = 1, 2, 3, \ldots$$

so if $v(t)$ is periodic in T_0 then it is also periodic in any integer multiple of T_0; Figure 5.1 illustrates this point. To avoid possible confusion, we shall reserve the symbol T_0 for the *smallest* constant satisfying (6), calling it the *fundamental period*, as distinguished from its integer multiples nT_0.

It should be noted from the definition (6) that a periodic signal must *exist for all time*. Of course no actual physical variable will really last forever, but it is often reasonable as well as convenient to make that assumption. After all, that is precisely the assumption upon which AC circuit analysis rests.

An important consequence of (6) and (3) is that, by and large, periodic

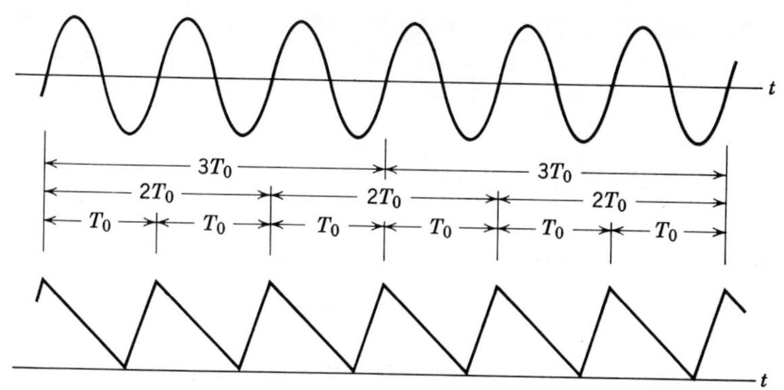

Figure 5.1. Typical periodic signals.

signals are power signals – hence the descriptive phrase *periodic power signals*. To show the conditions under which this statement holds, one must first observe that integrating a periodic signal over any interval exactly one period in duration always yields the same result, regardless of where the integration is taken; that is, if $v(t)$ is periodic in T_0, then

$$\int_{t_1}^{t_1+T_0} v(t)\,dt = \int_{t_2}^{t_2+T_0} v(t)\,dt \tag{7}$$

for any constants t_1 and t_2. (This property should be obvious from Figure 5.1, although analytic proof requires a bit of ingenuity; see Problem 5.2.) In view of (7), and because we shall have frequent need for such integrals, we introduce the shorthand notation

$$\int_{T_0} v(t)\,dt \triangleq \int_{t_1}^{t_1+T_0} v(t)\,dt \tag{8}$$

where t_1 is arbitrary. The practical implication here is that one is free to choose whatever lower or upper integration limit simplifies the calculation, if it can be simplified, as long as the range of integration is T_0.

Applying this conclusion to finding the power of a periodic signal, we let $T = MT_0$ with M being an integer and write (3) as a sum of $2M$ integrals, each covering exactly one period, in this manner:

$$P = \lim_{M\to\infty} \frac{1}{2MT_0}\left[\sum_{m=-M}^{M-1} \int_{mT_0}^{(m+1)T_0} |v(t)|^2 dt\right]$$

Now if $v(t)$ is periodic in T_0 then so is $|v(t)|^2$; thus, from (7), all the integrals in the summation are equal and

$$P = \lim_{M \to \infty} \frac{1}{2MT_0} \left[2M \int_{T_0} |v(t)|^2 dt \right]$$

$$= \frac{1}{T_0} \int_{T_0} |v(t)|^2 dt \tag{9}$$

which means that $v(t)$ is a power signal provided only that

$$0 < \int_{T_0} |v(t)|^2 dt < \infty.$$

A little thought will show that a bounded signal that is at least piecewise continuous will always satisfy this condition. Therefore, almost all of the periodic signals we deal with are power signals, and (9) becomes a very useful formula for computing their power.

Sinusoids and phasors

A particular type of signal having special interest in engineering is the *sinusoid*, written in the general form

$$v(t) = A \cos(\omega t + \theta) \qquad -\infty < t < \infty \tag{10}$$

where A is the *amplitude*, ω the *angular frequency*, and θ the *phase*, all assumed to be real quantities. It is a simple matter to show that sinusoids are periodic power signals, having

$$T_0 = \frac{2\pi}{\omega} \qquad P = \frac{A^2}{2} \tag{11}$$

The fact that P involves neither ω nor θ is a consequence of averaging $|v(t)|^2$.

First cousin to the sinusoid is the *phasor* signal

$$v(t) = Ae^{j(\omega t + \theta)} = Ae^{j\theta}e^{j\omega t} \qquad -\infty < t < \infty \tag{12}$$

which is a periodic power signal with

$$T_0 = \frac{2\pi}{\omega} \qquad P = A^2 \tag{13}$$

and A, ω, and θ have the same names as before. However, unlike (10), (12) represents a complex function of time since, from Euler's law,†

$$Ae^{j(\omega t + \theta)} = A \cos(\omega t + \theta) + jA \sin(\omega t + \theta) \tag{14}$$

†For those who might have forgotten, Euler's law is $e^{\pm j\phi} = \cos\phi \pm j \sin\phi$.

Note, by the way, that a complex signal is periodic if and only if both its real and imaginary parts have the same periodicity.

Because a phasor is a complex quantity, it can be depicted graphically as a rotating vector in the complex plane, Figure 5.2*a*. The amplitude A becomes the vector *length* while θ is the *angle* with respect to the positive real axis at time $t=0$. As time progresses, the vector *rotates* in a counterclockwise sense at a rate of ω radians per second so, at any time t, the vector projection on the axis of reals equals the real part of (14), and similarly for the imaginary part.

Equation (14) also brings out the relationship between sinusoids and phasors, namely

$$A \cos(\omega t+\theta) = \text{Re}\,[Ae^{j(\omega t+\theta)}] \tag{15}$$

which should be old hat to electrical engineering students. There is, however, another equivalent expression,

$$A \cos(\omega t+\theta) = \frac{A}{2}e^{j(\omega t+\theta)} + \frac{A}{2}e^{-j(\omega t+\theta)} \tag{16}$$

which gets rid of the real-part operation in (15) at the expense of involving two terms instead of one. Although the left-hand side of (16) is a real time function, the two phasors on the right are complex – *complex conjugates* to be exact. This observation leads to the phasor diagram of Figure 5.2*b*, in which the two phasors have the same amplitude $A/2$ but rotate in opposite directions such that the phasor sum falls along the real axis for all values of t.

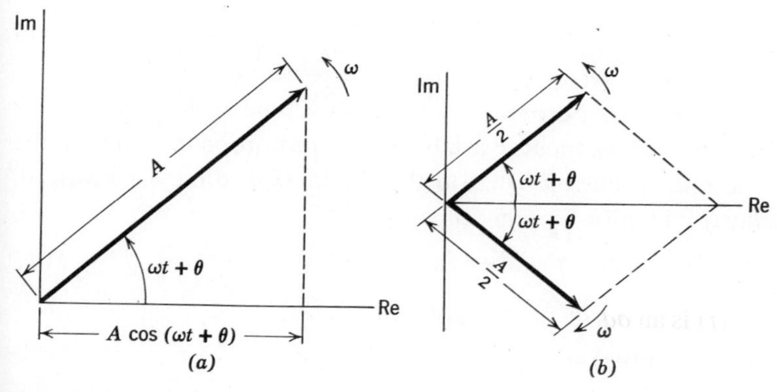

Figure 5.2. (*a*) The phasor $Ae^{j\theta}e^{j\omega t}$. (*b*) Phasor representation of $A\cos(\omega t+\theta)$.

Consider now the linear combination of phasors†

$$v(t) = \sum_m \alpha_m e^{j\omega_m t} \qquad \alpha_m = A_m e^{j\theta_m} \tag{17}$$

a signal that may or may not be periodic, depending on the specific angular frequencies ω_m. If there exists some constant Ω such that all ω_m are integer multiples of it, i.e., if for all m,

$$\omega_m = N_m \Omega \tag{18}$$

where N_m is any positive or negative integer, or zero, then $v(t)$ satisfies (6) and is a periodic signal. Assuming Ω is the *largest* constant satisfying (18), then the fundamental period of $v(t)$ is

$$T_0 = \frac{2\pi}{\Omega}$$

and Ω can justly be designated the *fundamental angular frequency*.

A set of frequencies having the relationship of (18) are said to be *harmonics* of Ω; $\omega_m = 2\Omega$, if in the set, is the second harmonic, 3Ω the third harmonic, and so forth. The first harmonic and the fundamental Ω are one and the same, and need not actually be included in the ω_m's. Another way of describing harmonic frequencies is to say that they are related as *ratios of integers*.

If all ω_m in (17) are not harmonically related, then $v(t)$ is not periodic. It does, however, come close to repeating itself if one waits a sufficiently long time. Such signals which almost, but not quite, have periodic behavior are called *quasi-periodic*, a simple example being $v(t) = e^{jt} + e^{j\pi t}$, whose frequencies have the irrational ratio π.

Symmetric signals

Symmetry, and whether or not a signal possesses it, may be useful at times for classification purposes. A function is said to have symmetry with respect to its independent variable when the portion to the left of the origin is the mirror image of that to the right, either direct or inverted. Analytically, $v(t)$ is an *even* function of t if

$$v(-t) = v(t) \qquad -\infty < t < \infty \tag{19a}$$

Conversely, $v(t)$ is an *odd* function of t if

$$v(-t) = -v(t) \qquad -\infty < t < \infty \tag{19b}$$

†Summations written as $\sum_m$ indicate that the number of terms in the sum is arbitrary.

and, as a consequence of (19b), an odd function must have $v(0) = 0$. Figure 5.3 illustrates these two cases. Note that $\cos \omega t$ and $\sin \omega t$ are even and odd functions, respectively.

For future use we recall two other facts about symmetric functions, both of which follow from (19). First, the product of two functions having

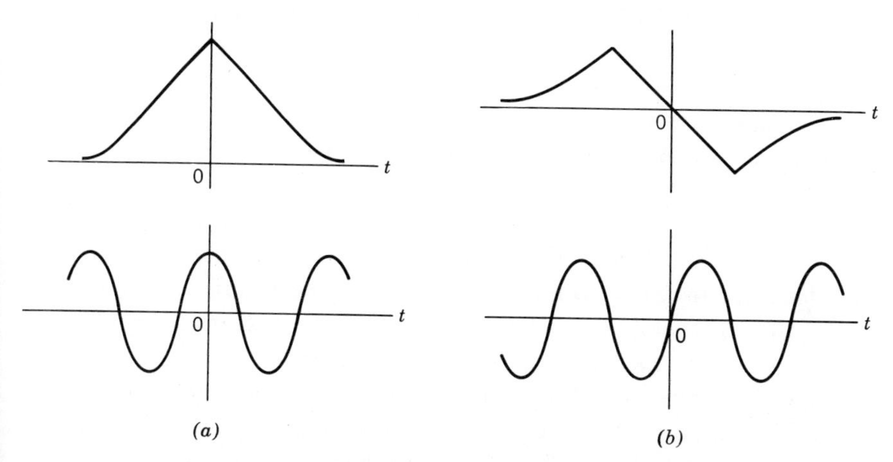

Figure 5.3. Typical symmetric functions. (*a*) Even symmetry. (*b*) Odd symmetry.

like symmetry, both even or both odd, yields an even function, whereas an odd function times an even function always yields an odd function. Second, integrals of the form $\int_{-T}^{T} v(t)\,dt$ have special properties when the integrand is even or odd, namely

$$\int_{-T}^{T} v(t)dt = 2\int_{0}^{T} v(t)dt \qquad v(t) \text{ even} \tag{20a}$$

$$\int_{-T}^{T} v(t)dt = 0 \qquad v(t) \text{ odd} \tag{20b}$$

where T is any constant.

When a signal is nonsymmetric, one can still decompose it into symmetric components by the following procedure. Given $v(t)$, define the even and odd functions

$$\begin{aligned} v_e(t) &\triangleq \tfrac{1}{2}[v(t) + v(-t)] \\ v_o(t) &\triangleq \tfrac{1}{2}[v(t) - v(-t)] \end{aligned} \tag{21}$$

then, clearly,

$$v(t) = v_e(t) + v_o(t)$$

as desired.

Discrete signals and random signals

In the above classifications it was tacitly assumed that the signals are deterministic functions of the continuous variable t. But it is also necessary to consider signals that are nondeterministic, called *random signals*, and signals that are functions of a discrete variable, say t_m, called *discrete-time* or simply *discrete signals*. Further discussion of discrete signals is deferred to the last two chapters where the topic is taken up in some depth.

As for random signals, they differ from the deterministic signals with which we have been dealing in that they defy representation as known or predictable functions of time; instead, they can be described only in terms of certain probabilities and statistical averages. Clearly, the study of random signals requires a knowledge of probability theory and statistics, so it will not be covered in this text. For the interested reader there are several introductory treatments of the subject, e.g., Cooper and McGillem (1967, Chaps. 9 to 12) or Thomas (1969, Chaps. 2 to 4).

5.2 INTRODUCTION TO SIGNAL SPACE

Having completed the necessary classifications, we now enter the world of *signal theory*† by developing the concepts of signal space. Two essential ingredients are involved here: first, a set of signals forming a *linear space*; second, *rules of magnitude and direction measure* for representing signals as vectors.

Consider, for instance, the energy signals $v(t)$ and $w(t)$ and their linear combination $z(t) = \alpha v(t) + \beta w(t)$, where α and β are finite constants, possibly complex. Obviously, $z(t)$ is also an energy signal and, going further, any such combination of two or more energy signals always yields another energy signal. One formalizes this observation by saying that the collection or set of all possible energy signals is *closed under linear combination*, which is the principal characteristic of a *linear space*.

†Those who are interested primarily in the applications of signal analysis, rather than the theoretical background, may defer or omit the rest of this chapter and proceed directly to Chapter 6. Occasional subsequent discussions that draw upon Sections 5.2 to 5.4 are of a supplemental nature and will be indicated by the symbol ⋆ at their headings.

This particular set also has at least one evident rule of measure, namely the energy definition, since every signal in the set can be characterized in part by a simple numerical quantity equal to its energy E. Similarly, we can construct a set consisting of all periodic power signals having the same period; this set is closed under linear combination and the power definition is an appropriate rule of measure.

Suppose, therefore, that we let $\mathbb{S}$ stand for a set of signals forming a linear space; then we can analyze and discuss the general properties of those signals without being pinned down to specific cases. Furthermore, suppose we have the rules of measure needed to treat the signals in $\mathbb{S}$ as vectors; then we can invoke geometric arguments to facilitate our analysis. By so doing, we have established $\mathbb{S}$ as a *signal space*.†

Of course, for the signal-space picture to be of any value, the rules of measure must be such that, given a signal $v(t)$, there are two quantities associated with it that can be interpreted as *vector magnitude* and *direction*, magnitude and direction being the parameters defining an ordinary vector. The equivalent quantities are known as the *norm* and *scalar product* for vectors in $\mathbb{S}$.

Scalar product, norm, and orthogonality

To obtain a suitably general rule of measure, including both energy and power signals, we must allow at least *two* definitions for the scalar product. Therefore, if $v(t)$ and $w(t)$ are both in the same class, their *scalar product*‡ is defined as:

$$\langle v, w\rangle \triangleq \begin{cases} \lim_{T\to\infty} \int_{-T}^{T} v(t)w^*(t)\,dt & \text{Energy signals} \quad (1a) \\ \lim_{T\to\infty} \dfrac{1}{2T}\int_{-T}^{T} v(t)w^*(t)\,dt & \text{Power signals} \quad (1b) \end{cases}$$

In the case of periodic power signals whose periods are equal or harmonically related, (1b) reduces to

$$\langle v, w\rangle = \frac{1}{T_0}\int_{T_0} v(t)w^*(t)\,dt \qquad \text{Periodic signals} \qquad (1c)$$

†But see, for example, Timothy and Bona (1968, Chap. 2) for a rigorous definition.

‡What we have termed the scalar product is also known as the *inner product* or *dot product*, and the notations (v, w) or $\mathbf{v}\cdot\mathbf{w}$ are sometimes used.

where T_0 is the period of the product $v(t)w^*(t)$. A third definition, covering *discrete* signals, will be given in Chapter 12, and there is yet another definition for *random* signals.

But no confusion should arise from the multiple expressions for $\langle v, w \rangle$; the correct formula in a specific application will be dictated by the nature of the signals involved. And regardless of which definition applies, there are certain properties of $\langle v, w \rangle$ that always hold. A few of these, the most important ones, are as follows:

$$\langle w, v \rangle = \langle v, w \rangle^* \tag{2}$$

$$\langle \alpha v, \beta w \rangle = \alpha \beta^* \langle v, w \rangle \tag{3}$$

$$\langle v_1 + w_1, v_2 + w_2 \rangle = \langle v_1, v_2 \rangle + \langle v_1, w_2 \rangle + \langle w_1, v_2 \rangle + \langle w_1, w_2 \rangle \tag{4}$$

where $v(t)$, $w(t)$, etc., are signals from the same set and α and β are constants. Equation (4), the additivity rule, carries no surprises; but (2) and (3) show that the scalar-product operation is neither commutative nor homogeneous when complex quantities are involved – so special care is indicated in such cases.

The other rule of measure in $\mathcal{S}$ is the signal *norm* $\|v\|$, whose square is the scalar product of a signal with itself, i.e.,

$$\|v\| \triangleq \langle v, v \rangle^{1/2} \tag{5}$$

which, like the scalar product, is simply a number and should be distinguished from the function $|v(t)|$. From (5) and (1), together with our prior definitions, it follows that

$$\|v\|^2 = \langle v, v \rangle = \begin{cases} E & \text{Energy signals} \\ P & \text{Power signals} \end{cases} \tag{6}$$

Clearly, then, $\|v\|$ is a *real nonnegative* quantity. Furthermore, $\|v\| = 0$ only when $v(t) = 0$ for all t, i.e., the *trivial signal*. Other properties of the norm are

$$\|\alpha v\| = |\alpha| \; \|v\| \tag{7}$$

$$\|v + w\| \leq \|v\| + \|w\| \tag{8}$$

Equation (8) is known as the *triangle inequality* by analogy with the relationship between the lengths of the hypotenuse and the other two sides of a plane triangle.

Another important relationship, linking $\langle v, w\rangle$ with $\|v\|$ and $\|w\|$, is *Schwarz's inequality*

$$|\langle v, w\rangle| \leq \|v\| \; \|w\| \tag{9a}$$

where equality holds only if the two signals are directly proportional, i.e.,

$$|\langle v, w\rangle| = \|v\| \; \|w\| \qquad \text{if} \qquad w(t) = \alpha v(t) \tag{9b}$$

The proof of Schwarz's inequality affords a nice example of scalar-product manipulations, as well as a significant side result, so we shall do it now.

Consider the linear combination $v(t)+\beta w(t)$ where β is an arbitrary constant and $w(t) \neq 0$. We expand $\|v+\beta w\|^2$ using (6), (4), (3), and (2) successively:

$$\begin{aligned}
\|v+\beta w\|^2 &= \langle v+\beta w, v+\beta w\rangle \\
&= \langle v, v\rangle + \langle v, \beta w\rangle + \langle \beta w, v\rangle + \langle \beta w, \beta w\rangle \\
&= \langle v, v\rangle + \beta^* \langle v, w\rangle + \beta\langle w, v\rangle + \beta\beta^* \langle w, w\rangle \\
&= \|v\|^2 + \beta^* \langle v, w\rangle + \beta\langle v, w\rangle^* + |\beta|^2 \; \|w\|^2
\end{aligned} \tag{10}$$

Now let $\beta = -\langle v, w\rangle/\|w\|^2$ so the last two terms of (10) cancel out and

$$\|v+\beta w\|^2 = \|v\|^2 - \frac{|\langle v, w\rangle|^2}{\|w\|^2}$$

But $\|v+\beta w\|^2 \geq 0$ regardless of β, so rearranging the above gives (9a). Furthermore, adding the fact that $\|v+\beta w\|^2 = 0$ only when $v(t)+\beta w(t) = 0$ for all t immediately yields (9b). Note, by the way, that these arguments were accomplished without resorting to any one particular definition of the scalar product.

Referring back to (10) again, suppose $\beta = 1$ and $v(t)$ and $w(t)$ are such that

$$\langle v, w\rangle = 0 \tag{11}$$

Under this condition

$$\|v+w\|^2 = \|v\|^2 + \|w\|^2 \tag{12}$$

whose form is identical to the *Pythagorean theorem* if one views $\|v+w\|$ as the length of the hypotenuse of a right triangle while $\|v\|$ and $\|w\|$ are the lengths of the other two sides. And because those other two sides are perpendicular or orthogonal, we say that two signals are *orthogonal* if their scalar product is zero, as in (11).

The general physical interpretation of orthogonality comes from (12) and (6). Specifically, since $\|v\|^2$ is proportional to the total energy or average power of the signal $v(t)$, and likewise for $\|w\|^2$ and $\|v+w\|^2$, we see that *superposition of energy or power* applies to the sum of orthogonal signals, and only if they are orthogonal. But this does not say much about their time-domain appearance. For that purpose, it is helpful to write out (11) in detail, say for energy signals, i.e.,

$$\lim_{T\to\infty} \int_{-T}^{T} v(t)w^*(t)\,dt = 0$$

Clearly, the signals are orthogonal if $v(t)w^*(t)$ vanishes for all t; this corresponds to *nonoverlapping* or *disjoint* time functions, one being zero when the other is not, and vice versa. Another condition for orthogonality is when the signals have *opposite symmetry*, say $v(t)$ an even function and $w^*(t)$ odd; for then their product has odd symmetry and the integral is zero even though the integrand is not. The same conclusions apply to orthogonal power signals, and examples are sketched in Figure 5.4.

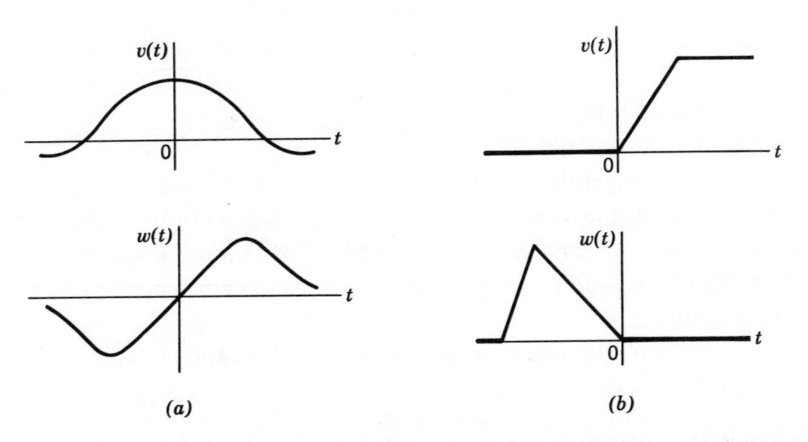

Figure 5.4. Examples of orthogonal signals. (*a*) Opposite symmetry. (*b*) Disjoint in time.

We emphasize, however, that these particular conditions are sufficient but not necessary for orthogonality. When all is said and done, the ultimate and sole orthogonality condition is the basic definition $\langle v, w\rangle = 0$. The significance of orthogonal signals will become more apparent with a discussion of the vector interpretation.

The vector interpretation

Let us now adopt the position that a signal $v(t)$ is a vector in a space $\mathbb{S}$, with v being the vector symbol.† There is no actual difference between v and $v(t)$, since $v = v(t)$; the time dependence has been suppressed merely to indicate the vector viewpoint. This symbolism also underscores the fact that, unlike rotating phasors, a signal vector is stationary and has the same magnitude and direction for all time.

The fundamental characteristic of the space $\mathbb{S}$ is that it is a linear space. Thus, if v and w – representing $v(t)$ and $w(t)$ – are any two vectors in $\mathbb{S}$, then any linear combination

$$z = \alpha v + \beta w \tag{13}$$

must also be in $\mathbb{S}$. An immediate corollary to (13) is that the *trivial* or *zero signal*, equaling zero for all time, must be in $\mathbb{S}$. Or, in other words, a signal space always contains the origin.

The *magnitude* of v is defined to be the signal *norm* $\|v\|$, so we are measuring vector "length" in terms of energy or power, a rather appealing notion. Additionally, if $z = v + w$, then the triangle inequality, (8), says that $\|z\| \leq \|v\| + \|w\|$, in agreement with the familiar vector diagram of Figure 5.5*a*.

Turning to the question of vector *direction*, we measure it in a relative sense via the *scalar product*. Specifically, the angle from v to w is defined by

$$\cos \theta_{vw} = \frac{\langle v, w \rangle}{\|v\| \, \|w\|} \tag{14}$$

as shown in Figure 5.5*b*. Schwarz's inequality, (9), then guarantees that $|\cos \theta_{vw}| \leq 1$; however, unless the signals are real functions, (14) may be a complex quantity.

The meaning of (14) is enhanced by observing that $\cos \theta_{vw}$ serves as a measure of *similarity* between the two signals. Clarifying this point, if $w(t) = \alpha v(t)$ then $|\langle v, w \rangle| = \|v\| \, \|w\|$ and $|\cos \theta_{vw}| = 1$, so *proportional signals* correspond to *colinear vectors*. On the other hand, if $\langle v, w \rangle = 0$ then $\cos \theta_{vw} = 0$, so *orthogonal signals* correspond to *perpendicular vectors*. Between these extremes, $\|v\| \cos \theta_{vw}$ is the component of v along w and tells us, intuitively at least, how much of $v(t)$ is "contained" in

†In contrast to Chapter 2 where the column matrix $\mathbf{v}(t)$ represented m different signals, v stands for just one signal, $v(t)$.

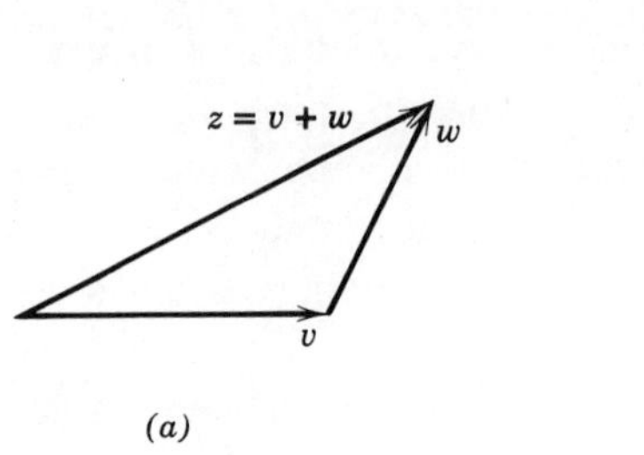

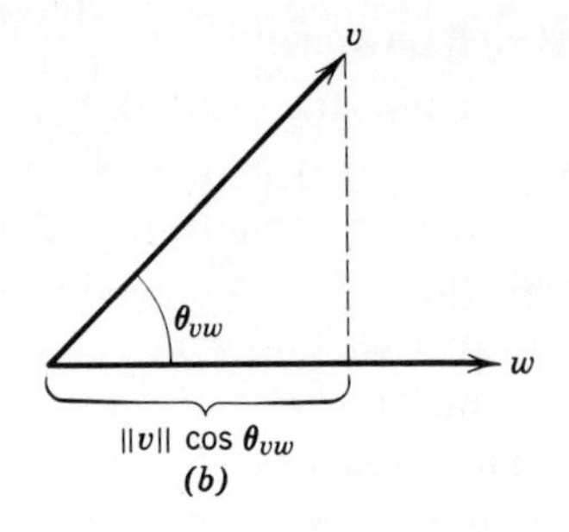

Figure 5.5.

$w(t)$. One should also note from (14), together with $\|v\|^2 = \langle v, v \rangle$, that the scalar product of signal vectors is analogous to the *dot product* of ordinary vectors.

To summarize, we have indicated how functions of time can be interpreted as vectors, although the purpose is perhaps not very obvious at the moment. It remains to be shown that this interpretation is consistent with the common notions of two- and three-dimensional spaces. This will be demonstrated through the introduction of basis functions.

Basis functions

A coordinate system in ordinary space is usually defined by a set of mutually perpendicular unit vectors, such that every vector has a unique set of coordinates. In a signal space, orthonormal basis functions play the role of unit vectors.

Suppose there exists a set of N elementary functions

$$\phi_n(t) \qquad n = 1, 2, \ldots N$$

so chosen that all signals in a given linear space are uniquely expressable as a linear combination of the ϕ_n's. By this we mean that one can write any signal $v(t)$ under consideration in the form

$$v(t) = a_1\phi_1(t) + a_2\phi_2(t) + \cdots + a_N\phi_N(t) = \sum_{n=1}^{N} a_n\phi_n(t)$$

with the coefficients a_n being constants called the coordinates of $v(t)$ with respect to the basis $\phi_n(t)$. We then define the space $\mathcal{S}_N$ as containing all vectors of the form

$$v = \sum_{n=1}^{N} a_n\phi_n \tag{15}$$

If a vector cannot be expressed like this then, by definition, it is not in $\mathcal{S}_N$. The ϕ_n's are *basis functions* for $\mathcal{S}_N$, and N is the *dimensionality* of the space.

Equation (15) generalizes on ordinary space in that: *1*. the basis functions ϕ_n (analogous to unit vectors) are functions of time; *2*. the coefficients a_n (analogous to vector coordinates) may be complex; and *3*. the dimensionality N is unrestricted. But scalar multiplication and vector addition in $\mathcal{S}_N$ obey the usual rules. For instance, if v is given by (15) and w has the same form with coefficients b_n, then

$$\alpha v + \beta w = \alpha\Big(\sum_{n=1}^{N} a_n\phi_n\Big) + \beta\Big(\sum_{n=1}^{N} b_n\phi_n\Big)$$

$$= \sum_{n=1}^{N} (\alpha a_n + \beta b_n)\phi_n$$

so $z = \alpha v + \beta w$ has coefficients $c_n = \alpha a_n + \beta b_n$ and, clearly, z is in $\mathcal{S}_N$.

Although not strictly essential, it is very advantageous to require that the basis functions have unit norm and be mutually orthogonal, that is,

$$\|\phi_n\|^2 = \langle \phi_n, \phi_n \rangle = 1 \qquad n = 1, 2, \ldots N$$

$$\langle \phi_n, \phi_m \rangle = 0 \qquad m \neq n$$

These conditions are compactly lumped together by writing

$$\langle \phi_n, \phi_m \rangle = \delta_{nm} \tag{16}$$

where δ_{nm} is the *Kronecker delta* notation

$$\delta_{nm} \triangleq \begin{cases} 1 & m = n \\ 0 & m \neq n \end{cases} \tag{17}$$

Basis functions satisfying (16) are called *orthonormal.* The *Gram-Schmidt procedure,* found in most linear algebra texts†, may be employed to generate an orthonormal basis for a given space. Obviously, the systems engineer is most interested in orthonormal basis functions that are *exponential,* and these will be considered in Section 5.4. Another type, called the *Walsh* functions, are useful for the study of digital communications; see Harmuth (1969).

†Or see Wozencraft and Jacobs (1965, pp. 266–273) for a treatment specifically in the context of signal theory.

Assuming that (16) holds, let us now find the scalar product of $v(t)$ and $w(t)$ in terms of their coefficients a_n and b_n. By definition,

$$\langle v, w\rangle = \Big\langle \Big(\sum_{n=1}^{N} a_n\phi_n\Big), \Big(\sum_{m=1}^{N} b_m\phi_m\Big)\Big\rangle$$

where different indices are required because the sums are independent of each other and, like dummy integration variables, they must be kept distinct. To carry out this calculation – which is a bit tricky – we interchange the summation and scalar-product operations, by extension of (4), and then use (3) and (16), as follows:

$$\langle v, w\rangle = \sum_{n=1}^{N} \sum_{m=1}^{N} \langle a_n\phi_n, b_m\phi_m\rangle$$

$$= \sum_{n=1}^{N} a_n\Big(\sum_{m=1}^{N} b_m^* \underbrace{\langle \phi_n, \phi_m\rangle}_{\delta_{nm}}\Big)$$

Since $\delta_{nm} = 0$ for $m \neq n$, the inner sum has only one nonzero term as m ranges from 1 to N, namely the term with $m = n$. But that term is $b_n^*\delta_{nn} = b_n^*$, hence

$$\langle v, w\rangle = \sum_{n=1}^{N} a_n b_n^* \tag{18}$$

Hopefully, the reader will recognize in this result a similarity to the dot product of ordinary vectors written out in terms of the vector coordinates, the only difference here being the complex conjugation of the w coefficients. The similarity is more evident when one sets $w = v$ (and hence $b_n = a_n$) to get

$$\langle v, v\rangle = \sum_{n=1}^{N} a_n a_n^*$$

or

$$\|v\|^2 = \sum_{n=1}^{N} |a_n|^2 = |a_1|^2 + |a_2|^2 + \cdots + |a_N|^2 \tag{19}$$

But it must be stressed that (18) and (19) are valid only for an orthonormal basis.

Example 5.1

For illustration purposes, let us define the special class of real energy signals that are zero outside $0 \le t \le 2$, make stepwise transitions at

$t = 0$, 1, and 2, and are otherwise constant. This class forms a linear space, and Figure 5.6*a* shows two members of it. We assert that the functions of Figure 5.6*b* constitute an orthonormal basis for the space containing the signals in question and, therefore, the dimensionality is $N = 2$.

Testing the assertion, ϕ_1 and ϕ_2 are orthogonal since they do not overlap in time, and quick calculation using (6) confirms that $\|\phi_1\|^2 = \|\phi_2\|^2 = 1$. We also see that v and w can be written as the linear combinations

$$v = \phi_1 + 3\phi_2 \qquad w = 2\phi_1 - \tfrac{2}{3}\phi_2$$

corresponding to the vector diagram of Figure 5.6*c*. Further thought should convince the reader that all signals like v and w can be so expressed.

Finally, from (18) and (19), we have

$$\|v\|^2 = 1^2 + 3^2 = 10 \qquad \|w\|^2 = 2^2 + (-\tfrac{2}{3})^2 = 40/9$$

and

$$\langle v, w \rangle = 1 \cdot 2 + 3(-\tfrac{2}{3}) = 0$$

so v and w are orthogonal. These results are easily confirmed by direct evaluation of the integral expressions or by measurements on Figure 5.6*c*.

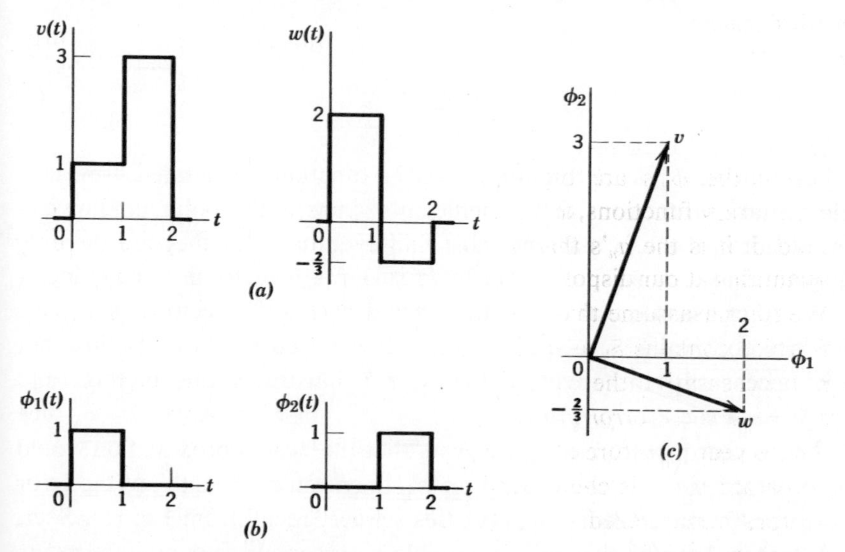

Figure 5.6. A two-dimensional signal space. (*a*) Typical signals. (*b*) Basis functions. (*c*) Vector diagram.

5.3 SIGNAL APPROXIMATIONS

Frequently, a systems engineer desires to represent a signal $v(t)$ in terms of a certain set of elementary functions. If an *exact* representation is impossible, one must settle instead for an *approximation* $\hat{v}(t)$, such that $\hat{v}(t) \approx v(t)$ over the time interval under consideration. And if $\hat{v}(t)$ is an approximation, there will be an *error*

$$\varepsilon(t) = v(t) - \hat{v}(t)$$

so devising a good approximation $\hat{v}(t)$ implies minimizing some measure of the error $\varepsilon(t)$.

Although this problem was easily stated, its solution is far from obvious. First, we need a criterion for the conditions under which a function of time $\varepsilon(t)$ is minimized. Then we must find $\hat{v}(t)$ to achieve this goal. But the main reason behind introducing signal-space concepts was for the geometric insight they lend to just such problems as this. Pursuing the vector analogy leads to the least-square-error approximation.

Error criterion and vector projections

To cast the problem in signal-space terminology, let us write the approximate signal as

$$\hat{v} = \sum_{n=1}^{N} a_n \phi_n \tag{1}$$

where the ϕ_n's are orthonormal basis functions generated from the elementary functions, and the coefficients a_n are, at this point, unknown – indeed, it is the a_n's that one must find eventually for they are the only quantities at our disposal. According to (1), $\hat{v}$ is a vector in some space $\mathbb{S}_N$. We further assume that the original signal $v(t)$ is the vector v in a space $\mathbb{S}$ which contains $\mathbb{S}_N$ as a *subspace*, i.e., every vector in $\mathbb{S}_N$ is also in $\mathbb{S}$, but not necessarily the reverse. Figure 5.7 illustrates this picture, and $\varepsilon = v - \hat{v}$ is the *error vector.*

The vector picture clearly suggests that the best approximation would be obtained if $\hat{v}$ is chosen such that $\|\varepsilon\|$, the norm or "length" of the error vector, is minimized. To support this choice, recall from Eq. (6), Sect. 5.2, that making $\|\varepsilon\|$ as small as possible corresponds to *minimum energy or power* in $\varepsilon(t)$, a very reasonable criterion for minimizing a signal.

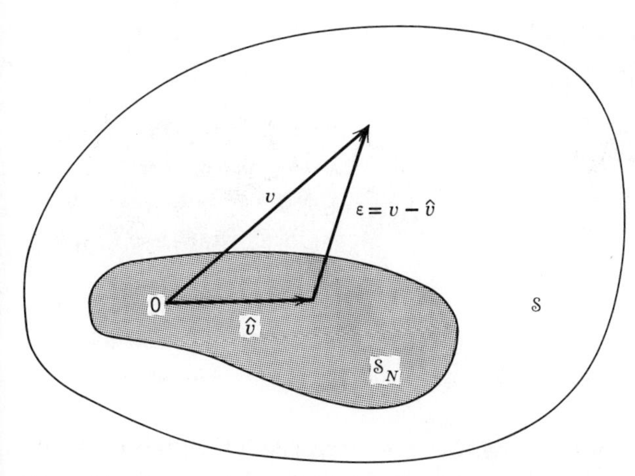

Figure 5.7. Vector picture of the signal-approximation problem.

Since $\|\varepsilon\|$ is nonnegative, our criterion revolves around

$$\|\varepsilon\|^2 = \|v - \hat{v}\|^2 \tag{2}$$

and minimized expressions of this sort occur in a variety of circumstances, both within and without the field of signal theory. They are known as *least-mean-square* or *least-square-error* approximations. While there are other possible criteria, the one we have chosen has distinct computational advantages as well as being intuitively pleasing.

Having decided on the criterion, we need to know something about picking minimum-length vectors. As a step in that direction, consider an ordinary three-dimensional space containing a vector $\mathbf{Z}$ and a plane $\mathcal{P}$, Figure 5.8*a*. The *shortest distance* from the tip of $\mathbf{Z}$ to $\mathcal{P}$ is found by dropping a perpendicular whose intersection with $\mathcal{P}$ defines the projection of $\mathbf{Z}$ on $\mathcal{P}$. The projection $\mathbf{Z}_p$ is unique and has the property that $\mathbf{Z} - \mathbf{Z}_p$ is perpendicular to every vector in $\mathcal{P}$.

If the word "perpendicular" is replaced by "orthogonal" then all of the preceding paragraph applies equally well to signal space and vector projections on N-dimensional subspaces. Specifically, referring to Figure 5.8*b*, let v be in $\mathcal{S}$ and let w be any vector in $\mathcal{S}_N$. The projection v_p of v on $\mathcal{S}_N$ is defined by

$$\langle v - v_p, w \rangle = 0 \tag{3}$$

and

$$\|v - w\| \geq \|v - v_p\| \tag{4}$$

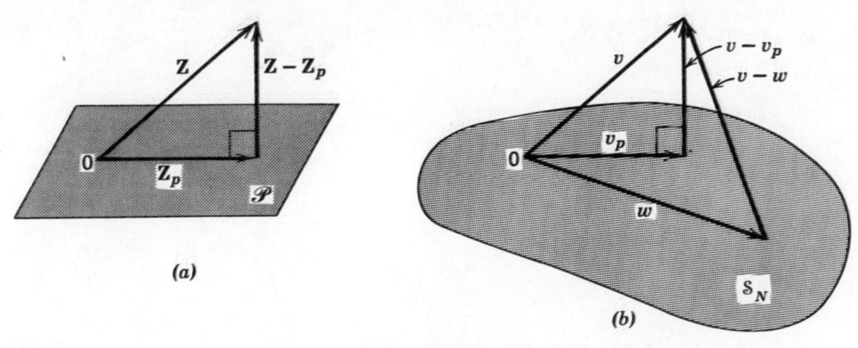

Figure 5.8. Vector projections. (*a*) Three-dimensional space. (*b*) Signal space.

with $\|v-w\|$ minimized if and only if $w = v_p$. Equation (4) is the *projection theorem*, whose proof is left as an exercise for the reader in Problem 5.21.

So much for the preliminaries; now we are ready for the particulars of finding $\hat{v}$.

Least-square-error approximations

Given $\hat{v}$ to be in the form of (1), the best approximation to $v(t)$ – "best" in the sense of minimum $\|\varepsilon\|$ – comes about by taking the a_n's so that $\hat{v}$ is the projection of v on $\mathbb{S}_N$. For this to be true, $\varepsilon = v - \hat{v}$ must be orthogonal to every vector in $\mathbb{S}_N$, including the basis functions, i.e., $\langle v-\hat{v}, \phi_n\rangle = 0$. But $\langle v-\hat{v}, \phi_n\rangle = \langle v, \phi_n\rangle - \langle \hat{v}, \phi_n\rangle$, so the requirement (3) becomes

$$\langle \hat{v}, \phi_n\rangle = \langle v, \phi_n\rangle \qquad n = 1, 2, \ldots N \tag{5}$$

and this relationship can be reworked to obtain a formula for the a_n's in terms of v and ϕ_n.

Operating on the left-hand side of (5), we insert the summation (1) for v, using a different dummy index, say m. Then

$$\begin{aligned}\langle \hat{v}, \phi_n\rangle &= \left\langle \left(\sum_{m=1}^{N} a_m\phi_m\right), \phi_n\right\rangle \\ &= \sum_{m=1}^{N} a_m\langle \phi_m, \phi_n\rangle = a_n\end{aligned}$$

where the sum reduces to just one term by the same argument used in conjunction with Eq. (18), Sect. 5.2. Hence, substitution back in (5) gives the desired expression

$$a_n = \langle v, \phi_n \rangle \qquad n = 1, 2, \ldots N \tag{6}$$

a very simple result.

It should be observed from (6) that any one particular a_n depends only on $v(t)$ and the related basis function $\phi_n(t)$. This is because the basis functions are mutually orthogonal and the projection of v on the ϕ_n "axis" is independent of all other coordinates. (This feature holds only for orthogonal basis functions, underscoring the practical advantages of orthogonality for signal approximations.) Another interesting and significant aspect of our result is that the a_n's are *independent of N*. Therefore, if it is desired to increase the number of terms in $\hat{v}$, one has only to compute the additional coefficients, the others remaining unchanged – again, a consequence of orthogonality. Presumably, increasing N also improves the approximation, but we should check that presumption by examining the measure of error $\|\varepsilon\|^2$.

The least-square-error approximation requires ε to be orthogonal to every vector in $\mathbb{S}_N$, including $\hat{v}$. Thus, with $v = \hat{v} + \varepsilon$ and $\langle \hat{v}, \varepsilon \rangle = 0$, Eq. (12), Sect. 5.2, can be applied to yield $\|v\|^2 = \|\hat{v} + \varepsilon\|^2 = \|\hat{v}\|^2 + \|\varepsilon\|^2$ or

$$\|\varepsilon\|^2 = \|v\|^2 - \|\hat{v}\|^2 \tag{7}$$

where, from (1) and Eq. (19), Sect. 5.2,

$$\|\hat{v}\|^2 = \sum_{n=1}^{N} |a_n|^2 \tag{8}$$

Clearly, $\|\hat{v}\|^2$ increases with N, or at least it cannot decrease. Therefore, the energy or power in the error signal generally does become smaller as we increase the number of terms in the approximation. More will be said on this subject when we consider signal representation, the topic of the next section.

5.4 FOURIER SIGNAL REPRESENTATIONS

The Fourier series and transform provide ways of representing certain classes of signals in terms of exponential functions. And that attribute will be drawn upon in later chapters to expedite systems analysis. Here, we shall concern ourselves with Fourier signal representations per se, in isolation from the applications. This is done not just for completeness' sake but rather because intelligent engineering usage of these mathematical tools demands some understanding of their basic nature and inherent limitations.

Periodic power signals: Fourier series

Let $v(t)$ be a periodic power signal, with period T_0, that we desire to approximate as a linear combination of exponential functions. The spadework having been done in Section 5.3, all we need now is an appropriate set of basis functions – where "appropriate" translates into three requirements. The $\phi_n(t)$ must be: *1. exponential* functions, as dictated by the problem statement; *2. periodic* with period T_0, to be of the same type as the signal in question; and *3. orthonormal*, so we can use the previous theory.

In view of these, we are led to try

$$\phi_n(t) = e^{jn\Omega t} \qquad -\infty < t < \infty \tag{1}$$

where

$$\Omega = \frac{2\pi}{T_0}$$

By inspection, the angular frequency of $\phi_n(t)$ is $n\Omega = 2\pi n/T_0$, so if n is an integer then every $\phi_n(t)$ will be periodic in T_0. It only remains to test orthonormality, for which the scalar-product definition Eq. (1c), Sect. 5.2, applies. Leaving the details to the reader,

$$\langle \phi_n, \phi_m \rangle = \frac{1}{T_0}\int_{T_0} \overbrace{(e^{jn\Omega t})}^{\phi_n(t)}\overbrace{(e^{-jm\Omega t})}^{\phi_m^*(t)}dt = \begin{cases} 1 & m = n \\ 0 & m \neq n \end{cases} \tag{2}$$

and $\langle \phi_n, \phi_m \rangle = \delta_{nm}$ as desired.

But one small catch has been glossed over, namely that the basis functions are *complex*. To approximate real as well as complex signals, the ϕ_n's must be taken in *complex-conjugate pairs*, i.e., $n = 0, \pm 1, \pm 2$, etc. Hence, the equivalent to Eq. (1), Sect. 5.3, becomes

$$\hat{v} = \sum_{n=-N}^{N} c_n e^{jn\Omega t} \tag{3}$$

where, for a least-square-error approximation,

$$c_n = \langle v, e^{jn\Omega t} \rangle = \frac{1}{T_0}\int_{T_0} v(t)e^{-jn\Omega t}dt \tag{4}$$

Note that the exponential has a negative sign in (4), but not in (3), because the scalar product for c_n involves $\phi_n^*(t)$.

If $\hat{v}$ is a "good" approximation with a certain N, then it should get better as N increases. Carried to the logical extreme, we let $N \to \infty$ and write (3) as

$$\hat{v}_\infty(t) = \sum_{n=-\infty}^{\infty} c_n e^{jn\Omega t} \tag{5}$$

in which, from prior consideration, the c_n's are still correctly given by (4). By taking $N \to \infty$ we have arrived at the best possible least-square-error approximation for $v(t)$ using a linear combination of exponentials. That is, all basis functions satisfying our three requirements have been included in $\hat{v}_\infty(t)$ and the error power $\|\varepsilon\|^2 = \|v - \hat{v}_\infty\|^2$ is at an absolute minimum consistent with the given constraints. Granted that $\hat{v}_\infty(t)$ is the best approximation of this type, the crucial question becomes this: Is $\hat{v}_\infty(t)$ a *representation* for $v(t)$, i.e., does it equal $v(t)$? The question is crucial because the right-hand side of (5) happens to be the *exponential Fourier series* expansion of $v(t)$. We now inquire as to whether that series *converges* to $v(t)$.

If the summation in (5) is to converge then, certainly, $v(t)$ must be sufficiently well behaved that the coefficients c_n *exist*, i.e., so the integration (4) can be performed at least in principle. Using a broader interpretation of integration (*Lesbesgue* integration), mathematicians have proved the existence of the c_n's providing only that $\int_{T_0} |v(t)|^2 dt$ exists. But we are dealing with periodic power signals for which

$$\|v\|^2 = \frac{1}{T_0} \int_{T_0} |v(t)|^2 dt = P$$

and hence the existence of the coefficients is ensured if the signal *power* P is well defined. What is more, the stipulation of well-defined power also is sufficient to prove that

$$\|\hat{v}_\infty\|^2 = \|v\|^2 \tag{6a}$$

or, invoking Eq. (7), Sect. 5.3,

$$\|\varepsilon\|^2 = \|v\|^2 - \|\hat{v}_\infty\|^2 = 0 \tag{6b}$$

which says that the power in the error is not only minimized but actually goes to zero as $N \to \infty$. When (6) holds, $\hat{v}_\infty(t)$ is said to *converge in the mean* to $v(t)$ since the average or mean value of $|\varepsilon(t)|^2$ goes to zero.

While convergence in the mean does not absolutely imply that $\hat{v}_\infty(t) = v(t)$ for all t, the distinction is a rather fine one. Specifically, if $\|\varepsilon\|^2 = 0$, then $\hat{v}_\infty(t)$ can differ at most from $v(t)$ only by a waveform $\varepsilon(t)$

that has zero energy or power. Thus, for most engineering purposes, one infers that $\hat{v}_\infty(t)$ is a valid representation for $v(t)$. That inference will be supported subsequently; in the interim, let us turn to the other major signal classification, nonperiodic energy signals.

Energy signals: Fourier transforms

Given a nonperiodic energy signal, we shall proceed as before by seeking suitable basis functions, developing the least-square-error approximation, and passing in the limit from approximation to representation. But right at the start one runs into a snag trying to find an *index* for the approximation, since there is no periodicity nor any other physical property that naturally suggests an integer index. To get around this problem, consider the periodic-signal approximation (3) written in the form

$$\hat{v}(t) = \sum_{n=-N}^{N} \langle v, e^{j2\pi n f_0 t} \rangle e^{j2\pi n f_0 t} \qquad f_0 \triangleq \frac{\Omega}{2\pi}$$

$$= \sum_{n=-N}^{N} c(nf_0)\phi(nf_0, t)$$

which brings out the dependence of $c_n = c(nf_0)$ on nf_0 as a *discrete* variable and shows that $\phi_n(t) = \phi(nf_0, t)$ is a function of two variables, one discrete and the other continuous.

Now, with a small stretch of the imagination, let nf_0 be replaced by the *continuous* variable f and, correspondingly, let the summation over $-Nf_0 \leq nf_0 \leq Nf_0$ become an integration with respect to f over, say $-F \leq f \leq F$, i.e.,

$$\hat{v}(t) = \int_{-F}^{F} V(f)\phi(f, t)\,df \tag{7}$$

where

$$\phi(f, t) = e^{j2\pi ft} \qquad -\infty < t < \infty \tag{8}$$

and, presumably,

$$V(f) = \langle v, e^{j2\pi ft} \rangle$$

Having thereby eliminated the index n, it seems plausible that (7) might be used for *nonperiodic* signals. This, in fact, is the case providing the energy definition of the scalar product is taken for $V(f)$, as we now demonstrate.

If $\hat{v}(t)$ in (7) is a least-square-error approximation, then it is necessary that

$$\langle \hat{v}, e^{j2\pi ft} \rangle = \langle v, e^{j2\pi ft} \rangle \qquad -F \leq f \leq F \tag{9}$$

this being the equivalent of Eq. (5), Sect. 5.3. Therefore, as was done to derive Eq. (6), Sect. 5.3, we shall manipulate the left-hand side of (9) to determine the unknown function $V(f)$. The manipulation begins by inserting (7) and (8), with the dummy variable λ replacing f, into the scalar product in energy form and then interchanging the order of integration, i.e.,

$$\begin{aligned}\langle v, e^{j2\pi ft} \rangle &= \lim_{T\to\infty} \int_{-T}^{T} \overbrace{\left[\int_{-F}^{F} V(\lambda) e^{j2\pi\lambda t} d\lambda\right]}^{v(t)} \overbrace{e^{-j2\pi ft}}^{\phi^*(f,t)} dt \\ &= \int_{-F}^{F} V(\lambda) \left[\lim_{T\to\infty} \int_{-T}^{T} e^{j2\pi(\lambda - f)t} dt\right] d\lambda \end{aligned} \tag{10}$$

So, to get any further, one must investigate the inner integration.

For that purpose, consider the related expression

$$\begin{aligned}\mathcal{I}(\mu) &= \lim_{T\to\infty} \int_{-T}^{T} e^{j2\pi\mu t} dt \\ &= \lim_{T\to\infty} \frac{1}{\pi\mu} \sin 2\pi\mu T \\ &= \lim_{\xi\to 0} \frac{1}{\pi\mu} \sin \frac{\pi\mu}{\xi}\end{aligned}$$

where, in the last line, we have let $T = 1/2\xi$. Now, a similar expression was mentioned in Section 4.2, namely† $\delta_\xi(t) = (\pi t)^{-1} \sin(\pi t/\xi)$, and it was argued that $\delta_\xi(t) \to \delta(t)$ as $\xi \to 0$. Therefore, by comparison, $\mathcal{I}(\mu) = \delta(\mu)$ or, written out fully,

$$\lim_{T\to\infty} \int_{-T}^{T} e^{j2\pi\mu t} dt = \delta(\mu) \tag{11}$$

a strange but important integral relationship for the unit impulse.

Going back to (10), the inner integral is seen to be $\mathcal{I}(\lambda - f) = \delta(\lambda - f)$, and

$$\langle \hat{v}, e^{j2\pi ft} \rangle = \int_{-F}^{F} V(\lambda)\delta(\lambda - f) d\lambda = V(f)$$

†Here, the symbol ϵ is replaced by ξ to avoid confusion with the error signal $\varepsilon(t)$.

where the stipulation $-F \leq f \leq F$ in (9) insures that $\delta(\lambda - f)$ is "located" within the range of integration so the sampling property of impulses can be invoked. With this result put in (9) we finally obtain

$$V(f) = \langle v, e^{j2\pi ft} \rangle = \lim_{T \to \infty} \int_{-T}^{T} v(t) e^{-j2\pi ft} dt \tag{12}$$

which might be recognized as the *Fourier transform* of $v(t)$. Summarizing the situation to this point, we have shown that (7) is a least-square-error approximation for a nonperiodic energy signal $v(t)$ when $V(f)$ is the Fourier transform of $v(t)$, assuming $V(f)$ exists.

Paralleling the periodic case, a sufficient condition for the existence of $V(f)$ is that $v(t)$ have well-defined energy E. If so, then we can go from approximation to representation by letting $F \to \infty$, i.e.,

$$\hat{v}_\infty(t) = \int_{-\infty}^{\infty} V(f) e^{j2\pi ft} df \tag{13}$$

which represents $v(t)$ in the sense that $\|\hat{v}_\infty\|^2 = \|\hat{v}\|^2$ so $\|\varepsilon\|^2 = 0$ and the error signal has zero energy.

Proof of the Fourier representations: Kernel functions

In both the periodic and nonperiodic case we asserted, but did not prove, the validity of the representation $\hat{v}_\infty(t)$. A limited proof that $\hat{v}_\infty(t) = v(t)$ will now be developed by introducing the so-called kernel functions. The proof is limited by the assumption that c_n or $V(f)$ exists, and that $v(t)$ is continuous.

Consider the energy-signal approximation (7). Inserting the integral (12) for $V(f)$ – again with a dummy variable – yields

$$\begin{aligned} \hat{v}(t) &= \int_{-F}^{F} \left[\lim_{T \to \infty} \int_{-T}^{T} v(\lambda) e^{-j2\pi f\lambda} d\lambda \right] e^{j2\pi ft} df \\ &= \lim_{T \to \infty} \int_{-T}^{T} v(\lambda) \left[\int_{-F}^{F} e^{j2\pi f(t-\lambda)} df \right] d\lambda \\ &= \int_{-\infty}^{\infty} v(\lambda) K_F(t - \lambda) d\lambda \end{aligned} \tag{14}$$

in which we have let $T \to \infty$ and introduced the *Fourier kernel*

$$K_F(t) \triangleq \int_{-F}^{F} e^{j2\pi ft} df = \frac{\sin 2\pi Ft}{\pi t} \tag{15}$$

Notice, however, that the kernel appears in (14) as a function of $t-\lambda$ so that $\hat{v}(t)$ is expressed as a *convolution*, namely $\hat{v}(t) = [v(t)] * [K_F(t)]$.

Similar manipulations on the periodic-signal approximation (3) give

$$\hat{v}(t) = \int_{T_0} v(\lambda) D_N(t-\lambda)\, d\lambda \tag{16}$$

where $D_N(t)$ is the *Dirichlet kernel*

$$D_N(t) \triangleq \frac{1}{T_0} \sum_{n=-N}^{N} e^{jn\Omega t} = \frac{\sin[(2N+1)\pi t/T_0]}{T_0 \sin(\pi t/T_0)} \tag{17}$$

in which the summation has been converted to closed form via the rule for geometric progressions. Then, taking the integration range in (16) as $t-(T_0/2) \le \lambda \le t+(T_0/2)$ and defining the *truncated Dirichlet kernel*

$$K_N(t) \triangleq \begin{cases} D_N(t) & |t| \le T_0/2 \\ 0 & |t| > T_0/2 \end{cases} \tag{18}$$

we have

$$\hat{v}(t) = \int_{-\infty}^{\infty} v(\lambda) K_N(t-\lambda)\, d\lambda \tag{19}$$

which, like (14), is a convolution integral.

We pass from approximation to Fourier representation by letting $F \to \infty$ in (14) or $N \to \infty$ in (19); expressed compactly,

$$\begin{aligned} \hat{v}_\infty(t) &= \int_{-\infty}^{\infty} v(\lambda) K_\infty(t-\lambda)\, d\lambda \\ &= [v(t)] * [K_\infty(t)] \end{aligned} \tag{20}$$

where $K_\infty(t)$ stands for the corresponding limit of $K_F(t)$ or $K_N(t)$ depending on whether $v(t)$ is a nonperiodic energy signal or a periodic power signal. If the representation does equal the original signal, that is, if $\hat{v}_\infty(t) = v(t)$, then $[v(t)] * [K_\infty(t)] = v(t)$, which ascribes to $K_\infty(t)$ the same attribute as a *unit impulse* under convolution. Therefore, demonstrating that $K_\infty(t) = \delta(t)$, at least in the limiting sense, constitutes justification of the Fourier representations.

Testing first the Fourier kernel $K_F(t)$, we use the integral form in (15), with the result

$$\lim_{F\to\infty} K_F(t) = \lim_{F\to\infty} \int_{-F}^{F} e^{j2\pi ft}\, df = \delta(t) \tag{21}$$

where (11) has been invoked to prove the hypothesis. For the truncated Dirichlet kernel $K_N(t)$ with large but finite N, two approximations need to be made in the closed-form expression (17): first,

$$\sin\left[(2N+1)\pi t/T_0\right] \approx \sin\left(2\pi Nt/T_0\right)$$

since $N >> 1$; second, most of the net area of $K_N(t)$ is near $t = 0$, where $\sin(\pi t/T_0) \approx \pi t/T_0$. Hence, for use under integration, we have

$$K_N(t) \approx \frac{\sin 2\pi(N/T_0)t}{\pi t} \qquad \begin{array}{l} N >> 1 \\ |t| \leq T_0/2 \end{array}$$

and, by comparison with (15) and (21),

$$\lim_{N\to\infty} K_N(t) = \lim_{N\to\infty} \int_{-N/T_0}^{N/T_0} e^{j2\pi ft}\,df = \delta(t) \tag{22}$$

Finally, in view of (21) and (22), (20) becomes

$$\hat{v}_\infty(t) = \int_{-\infty}^{\infty} v(\lambda)\delta(t-\lambda)\,d\lambda \tag{23}$$

which completes the proof.

Based on our knowledge of impulses, (23) tells us that $\hat{v}_\infty(t)$ equals or converges to $v(t)$ at every point in time where $v(t)$ is *bounded* and *continuous*. And this applies whether we are speaking of periodic power signals or nonperiodic energy signals, i.e., the Fourier series or Fourier transform representation. Bear in mind, of course, that the corresponding coefficients c_n or transform $V(f)$ must exist. As a practical consideration, any real physical signal will have well-defined energy or power, thereby guaranteeing the necessary existence; and certainly any physical signal will be everywhere bounded and continuous, thereby guaranteeing convergence.

Troubles sometimes do arise when we work with *idealized* signals that usually have discontinuities, the rectangular pulse being an example. In such cases, the Fourier representation suffers an unusual behavior called *Gibbs phenomenon*† at each discontinuity point. Nevertheless, as long as the signal energy or power is well defined, it is still true that $\|v - \hat{v}_\infty\|^2 = 0$ and, except at the discontinuities, $\hat{v}_\infty(t)$ is a useful representation for $v(t)$. Therefore, in the next two chapters we shall equate $v(t)$ to its Fourier series or transform representation.

†Papoulis (1962, pp. 29–32) covers the details.

5.1 Classify each of the following signals as to energy-type or power-type by calculating E and P: (a) $A|\cos \omega t|$; (b) $Ate^{-t/\tau}u(t)$; (c) $A\tau/(\tau+jt)$; (d) $A \exp[-(t/\tau)^2]$.

5.2 Prove Eq. (7), Sect. 5.1, by first assuming that $0 < (t_2-t_1) < T_0$ and writing the left-hand side in terms of three integrals, $t_1 \le t \le t_2$, $t_2 \le t \le t_2+T_0$, and $t_1+T_0 \le t \le t_2+T_0$. Then generalize to the case $|t_2-t_1| \ge T_0$.

5.3 Use Eq. (9), Sect. 5.1, to find P for each of the following periodic signals. The definitions given are for just one period.

$$\text{(a)}\ v(t) = \begin{cases} A \sin(2\pi t/T_0) & 0 < t < T_0/2 \\ 0 & T_0/2 < t < T_0 \end{cases}$$

$$\text{(b)}\ v(t) = At/T_0 \qquad -T_0/2 < t < T_0/2$$

$$\text{(c)}\ v(t) = \begin{cases} A & 0 < t < \tau \\ B & \tau < t < T_0 \end{cases}$$

Answer: (a) $A^2/4$; (b) $A^2/12$.

5.4 Consider the signal $v(t) = A \exp[j(\omega t+\theta)]$ where A is *complex*. Find $|v(t)|$, $\arg[v(t)]$, $\text{Re}[v(t)]$, and $\text{Im}[v(t)]$.

5.5 Prove that Eq. (17), Sect. 5.1, satisfies $v(t \pm nT_0) = v(t)$ when Eq. (18) holds and $T_0 = 2\pi/\Omega$.

5.6 Find the fundamental period T_0 when the ω_m's in Eq. (17), Sect. 5.1, are: (a) 0, 3, 6, 12; (b) 10, 12, 14; (c) −5, 0, 10; (d) $\sqrt{2}$, 2, 4.

5.7 Find the even and odd parts of $v(t) = A\cos(\omega t+\theta)$.

5.8 Give simple examples of signals satisfying the following descriptions: (a) A real energy signal with odd symmetry; (b) A complex power signal with even symmetry; (c) A nonsymmetric real signal whose even part is power-type and whose odd part is energy-type.

5.9 Show that the energy-signal definition of $\langle v, w\rangle$ has the properties of Eqs. (2) to (4), Sect. 5.2.

5.10 Taking the appropriate definition, calculate $\langle v, w\rangle$ for each of the following pairs of signals: (a) $e^{-|t|}$, $2e^{-5t}u(t)$; (b) $e^{-(2+j3)t}u(t)$, $e^{-(3+j2)t}u(t)$; (c) $\cos 2\pi t$, $\sin^2 2\pi t$; (d) $\cos 2\pi t$, $10\,u(t)$.

5.11 Prove the triangle inequality in the form $\|v+w\|^2 \le (\|v\|+\|w\|)^2$ by starting with Eq. (10), Sect. 5.2, setting $\beta = 1$, and invoking Schwarz's inequality. *Hint:* note that, for any complex number Z, $Z+Z^* \le 2|Z|$.

5.12 Drawing upon the above hint and Schwarz's inequality, show that

$$\left| \int_{-\infty}^{\infty} [v(t)w^*(t) + v^*(t)w(t)]dt \right|^2 \le 4 \int_{-\infty}^{\infty} |v(t)|^2 dt \int_{-\infty}^{\infty} |w(t)|^2 dt$$

5.13 Give an example of a pair of orthogonal signals that are not time-disjoint and do not have opposite symmetry.

5.14 Evaluate $\|v\|$, $\|w\|$, $\|z\|$, $\langle w, v\rangle$, and $\langle z, v\rangle$ for the waveforms in Figure P5.1. Use these numbers to construct a vector diagram and graphically verify that $z = v + w$. *Answer:* $\|v\| = \sqrt{2}$, $\|w\| = \sqrt{\frac{2}{3}}$, $\|z\| = 2\sqrt{\frac{2}{3}}$, $\langle w, v\rangle = 0$, $\langle z, v\rangle = 2$.

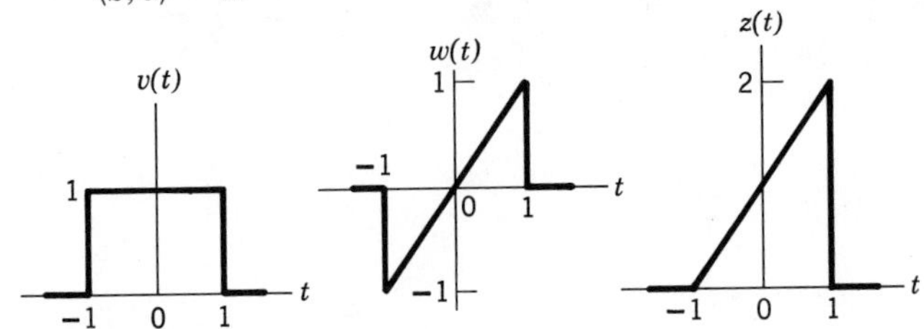

Figure P5.1.

5.15 Repeat Problem 5.14 with $v(t) = te^{-|t|/2}$, $w(t) = (\sin t)/t$, and $z(t) = v(t) + w(t)$.

5.16 Consider the set of all signals that satisfy the following specifications: $v(t) = 0$ for $t \le 0$ or $t \ge 4$; at $t = 1, 2,$ or 3, $v(t)$ is real and finite but otherwise arbitrary; between the aforementioned instants, $v(t)$ consists of straight line segments connecting $v(0)$ to $v(1)$, $v(1)$ to $v(2)$, etc. (a) Verify that this set defines a linear space. (b) Devise and sketch an appropriate set of orthonormal basis functions. *Hint:* the dimensionality of the space is 8. (c) Using your basis functions, determine the corresponding coefficients for the signal having $v(t) = t^2, t = 1, 2, 3$.

5.17 Evaluate $\left\| \sum_{n=1}^{N} \phi_n \right\|$ when the ϕ_n's are orthonormal but otherwise unrestricted. *Answer:* $\sqrt{N}$.

5.18 Verify Schwarz's inequality for $v = \sum_{n=1}^{N} a_n\phi_n$ and $w = \sum_{n=1}^{N} b_n\phi_n$ when the ϕ_n's are orthonormal.

5.19 Expand $\|v\|^2$ when $v(t) = a_1\psi_1(t) + a_2\psi_2(t)$ where ψ_1 and ψ_2 are *not* orthonormal. Compare with Eq. (19), Sect. 5.2.

5.20 Look up the Gram-Schmidt procedure in any convenient reference and use it to generate a set of orthonormal basis functions from the set of signals $v_k(t) = e^{-kt}u(t), k = 1, 2, 3$.

5.21 Prove the projection theorem, Eq. (4), Sect. 5.3, by the following method: referring to Figure 5.8, let $x = v - v_p$, $y = v_p - w$, and $z = v - w$, so $\langle x, y\rangle = 0$ (why?) and $\|z\|^2 = \|x\|^2 + \|y\|^2$.

5.22 Referring to Eqs. (1) and (6), Sect. 5.3, assume that $v(t)$ and $\phi_n(t)$ are real energy signals so the a_n's must be real. Derive $a_n = \langle v, \phi_n\rangle$ by requiring that $\partial\|\varepsilon\|^2/\partial a_n = 0, n = 1, 2, \ldots N$.

5.23 Consider the approximation procedure of Figure 4.18. What should be the amplitude of the kth pulse if it is required that $\hat{x}(t)$ be a least-square-error approximation?

5.24 Obtain a least-square-error approximation over $-10 \le t \le 10$ for $v(t) = 2 - 0.4|t|$ in terms of $\phi_n(t) = 10^{-1/2} \cos(2\pi nt/20)$, $n = 1, 2, 3$. Plot $v(t)$, $\hat{v}(t)$, and $\varepsilon(t)$, and evaluate $\|\varepsilon\|^2/\|v\|^2$.

5.25 Let $v(t)$ and $w(t)$ be real signals of finite duration, say over $0 \le t \le 1$. Obtain a least-square-error approximation for $w(t)$ in the form $\hat{w}(t) = \alpha + \beta v(t)$. This is called *linear regression analysis*. *Hint:* take $\phi_1(t) = A[u(t) - u(t-1)]$ and $\phi_2(t) = B[v(t) - C][u(t) - u(t-1)]$, where A, B, and C are constants such that $\langle\phi_n, \phi_m\rangle = \delta_{nm}$. You will also find it convenient to introduce the notational symbol $\bar{v} = \int_0^1 v(t)dt$, etc. *Answer:* $\beta = (\langle v, w\rangle - \bar{v}\bar{w})/(\|v\|^2 - \bar{v}^2)$, $\alpha = \bar{w} - \beta\bar{v}$.

5.26 Verify Eq. (2), Sect. 5.4.

5.27 Starting from Eq. (6a), Sect. 5.4, derive Parseval's power theorem

$$P = \sum_{n=-\infty}^{\infty} |c_n|^2$$

5.28 The sine-cosine Fourier series for a real periodic signal has the form

$$v(t) = A_0 + \sum_{n=1}^{\infty} (A_n \cos n\Omega t + B_n \sin n\Omega t)$$

Identify the basis functions, check for orthogonality, and obtain formulas for A_0, A_n, and B_n.

5.29 Carry out the necessary manipulations to get Eq. (16) from Eq. (3), Sect. 5.4.

5.30 When $v(t)$ is an energy signal having discontinuities, the Gibbs-phenomenon overshoot can be avoided by taking the approximation

$$\hat{v}_F(t) = \int_{-F}^{F} V(f)\left(1 - \frac{|f|}{F}\right) e^{j2\pi ft}\,df$$

which is not a least-square-error approximation but does yield $\hat{v}_F(t) \to v(t)$ as $F \to \infty$. Find the corresponding kernel function, known as the *Fejér kernel*, and give a *simple* argument supporting the assertion that $\hat{v}_F(t)$ will not have overshoots when $v(t)$ is discontinuous.

6

Periodic steady-state analysis: Fourier series

When applicable, the transfer-function method may greatly simplify the analysis of fixed linear systems, but it hinges upon having the input signal expressed as a linear combination of exponentials. Thus, we are led to seek out those situations where the necessary signal representation is possible. One such case is the special but important class of problems falling under the general heading of *periodic steady-state analysis*, which is the subject of this chapter.

Actually, periodic steady-state analysis is merely a generalization of AC (alternating-current) circuit analysis commonly used in electrical engineering. Therefore, we shall begin with a short treatment of the sinusoidal steady state, and then proceed to the main problem. Generalization is achieved by drawing upon the Fourier series to expand more or less arbitrary periodic signals as sums of exponentials – complex exponentials to be exact. The series expansion, coupled with a form of the system transfer function called the frequency response function, finally gives the periodic steady-state response.

Besides analytic simplifications, the techniques presented here also suggest the concept of the *frequency domain*, a concept opening up another avenue for interpreting signals and the properties of fixed linear systems. It is an essential concept in communications work, as well as having profound significance for systems engineering in general. Thus, a second major theme of this chapter will be the frequency-domain interpretation.

6.1 SINUSOIDAL STEADY-STATE ANALYSIS

No doubt the reader has done AC circuit problems using complex impedance (or admittance) and assuming all time functions are of the form $e^{j\omega t}$. Underlying this approach are several conditions that often are not explicitly mentioned and sometimes overlooked. To set the record straight in this regard, we must first state rather carefully what is meant by the sinusoidal steady state, a definition that later will be extended to cover the periodic steady state.

The sinusoidal steady state

Suppose that a fixed linear system, electrical or otherwise, is excited by a sinusoidal input signal starting at some time t_0 and lasting until $t_0 + T$ such that many cycles of the waveform have elapsed during this span, Figure 6.1*a*. Further suppose that the system is asymptotically stable so its natural-behavior modes decay with time, becoming vanishingly small after $\tau \ll T$ seconds. Under these conditions, and during the interval $t_0 + \tau < t < t_0 + T$, any transients will have died away leaving only the forced response due to the input, and it too will be sinusoidal, Figure 6.1*b*.

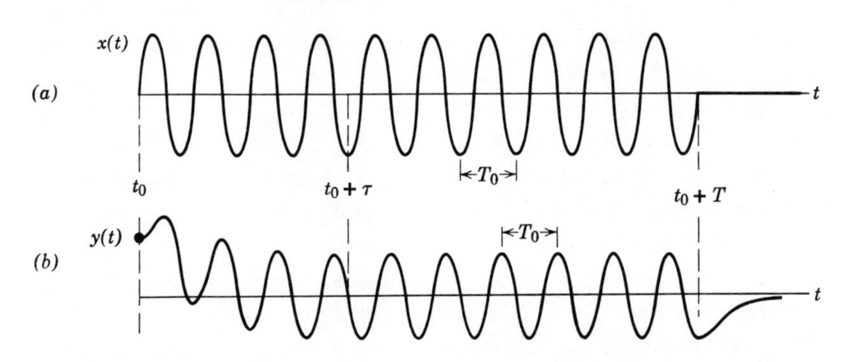

Figure 6.1. Sinusoidal (AC) steady state. (*a*) Input. (*b*) Output.

The fact that the forced response is a sinusoid having the same frequency as the input follows from the preservation-of-form property of fixed linear systems; but we shall shortly solve for $y(t)$ and confirm this fact.

If we now agree to restrict our attention to the time interval where $y(t)$ is sinsuoidal, it does no harm to pretend that $t_0 \to -\infty$ and $t_0 + T \to +\infty$ for analysis purposes. Thus, we write the input and output as

$$x(t) = A_x \cos(\omega t + \theta_x) \qquad -\infty < t < \infty \tag{1a}$$

$$y(t) = A_y \cos(\omega t + \theta_y) \qquad -\infty < t < \infty \tag{1b}$$

which defines the *sinusoidal steady state*. Putting the definition in words, the input is a sinusoid whose duration is very large compared to its period, the system is fixed, linear, and asymptotically stable, and we observe the output only during that time interval when it consists of just the forced sinusoidal response. In view of the several restrictions it is perhaps surprising, but nonetheless true, that the sinusoidal steady state is a reasonable model for many practical problems.

Further inspection of (1) reveals one other characteristic of the sinusoidal steady state: the response $y(t)$ is completely predetermined except for two quantities, its *amplitude* A_y and *phase* θ_y. (Remember that the angular frequency ω must equal that of the input.) Therefore, solving such problems entails finding only these quantities in terms of the corresponding input quantities and the system's characteristics.

Sinusoidal steady-state response

Turning to that task, we rewrite $x(t)$ as a sum of exponentials, namely

$$x(t) = \frac{A_x}{2} e^{j\theta_x} e^{j\omega t} + \frac{A_x}{2} e^{-j\theta_x} e^{-j\omega t} \tag{2}$$

which is the *conjugate phasor* representation of a sinusoid per Eq. (16), Sect. 5.1. Now, recalling the material of Section 4.5, if

$$x(t) = \alpha_1 e^{s_1 t} + \alpha_2 e^{s_2 t} + \cdots \qquad -\infty < t < \infty$$

then

$$y(t) = H(s_1)\alpha_1 e^{s_1 t} + H(s_2)\alpha_2 e^{s_2 t} + \cdots \qquad -\infty < t < \infty$$

where $H(s)$ is the transfer function

$$H(s) \triangleq \int_{-\infty}^{\infty} h(t) e^{-st} dt$$

Applied to the case at hand, it immediately follows that

$$y(t) = H(j\omega) \frac{A_x}{2} e^{j\theta_x} e^{j\omega t} + H(-j\omega) \frac{A_x}{2} e^{-j\theta_x} e^{-j\omega t} \tag{3}$$

wherein $H(j\omega)$ is $H(s)$ with $s = j\omega$, etc.

Equation (3) can be simplified by drawing upon the fact that $H(j\omega)$ and $H(-j\omega)$ are *complex conjugates* of each other, providing only that the impulse response $h(t)$ is a *real* function of time. To demonstrate this property we note that, by definition,

$$H(\pm j\omega) = H(s)|_{s=\pm j\omega} = \int_{-\infty}^{\infty} h(t) e^{\mp j\omega t} dt$$

so, when $h(t)$ is real,

$$H(-j\omega) = H^*(j\omega) \tag{4a}$$

or, in polar terms,

$$|H(-j\omega)| = |H(j\omega)| \qquad \arg[H(-j\omega)] = -\arg[H(j\omega)] \tag{4b}$$

Using (4), our output expression finally reduces to the real sinusoid

$$y(t) = |H(j\omega)| A_x \cos(\omega t + \theta_x + \arg[H(j\omega)])$$

$$= A_y \cos(\omega t + \theta_y) \tag{5a}$$

in which we have introduced

$$A_y = |H(j\omega)| A_x \qquad \theta_y = \theta_x + \arg[H(j\omega)] \tag{5b}$$

This succinct result contains two features of importance. First, as previously asserted, $y(t)$ is a sinusoid having the same frequency as the input. Second, although $H(j\omega)$ is generally complex, it appears here as two real numbers, $|H(j\omega)|$ and $\arg[H(j\omega)]$, the one relating input and output amplitude, the other relating input and output phase. More will be said about this when we give the frequency-domain interpretation.

Because there are notational differences, the inherent equivalence of the above method to conventional AC circuit analysis should be pointed out now. In AC circuit analysis one works with single phasors such as $A_x e^{j\theta_x} e^{j\omega t}$ – so $x(t) = \text{Re}[A_x e^{j\theta_x} e^{j\omega t}]$ – and the time factor $e^{j\omega t}$ is suppressed in intermediate steps; those intermediate steps involve multiplying by the complex impedence and, at the end, one reintroduces $e^{j\omega t}$ and takes the real part to find the actual time function, a step seldom explicitly taken. Here, in contrast, we have carried along $e^{\pm j\omega t}$ in the conjugate phasor representation, which explains why both $H(j\omega)$ and $H(-j\omega)$ crop up, these being analogous to the complex impedence; then, at the end, all terms involving $-j\omega$ drop out because the right-hand side of (3) is a sum of complex conjugates when (4) holds. Performing that sum then yields a real function of time without having to take the real part of an intermediate expression.

Generalization

It might seem that this rather simple problem has been unduly belabored by using the longer approach. And certainly the student henceforth should set down answers such as (5) without any of the additional steps we have gone through here. But by going through those steps at least once, and by using the conjugate phasor representation, we have preserved the generality that allows us to tackle a variety of other problems in addition to simple sinusoidal inputs.

Following up the last remark, suppose $x(t)$ is an arbitrary linear combination of phasors, say

$$x(t) = \sum_m \alpha_m e^{j\omega_m t} \qquad -\infty < t < \infty \tag{6}$$

which is just a sum of terms of the form $\alpha_m e^{j\omega_m t}$ and includes as a special case any linear combination of sinusoids. Then, similar to (3), the steady-state response to this excitation is

$$y(t) = \sum_m H(j\omega_m)\alpha_m e^{j\omega_m t} \qquad -\infty < t < \infty \tag{7}$$

providing the system in question is stable as well as fixed and linear.

Given the fact that almost any periodic signal can be expanded in the form of (6), (7) contains all the essentials of periodic steady-state analysis in a nutshell. The remainder of this chapter therefore is devoted to the detailed development, interpretation, and application of this expression.

6.2 LINE SPECTRA AND FREQUENCY RESPONSE

Although a signal is a function of time and therefore *exists* only in the time domain, it is often possible and profitable to speak of a *frequency-domain representation* – meaning that the signal *consists* of certain well-defined frequency components. The distinction between the two domains is the independent variable: in the time-domain representation the independent variable is *time* t; in the frequency domain it is cyclical or rotational *frequency* f, measured in cycles per second (hertz). Of course, one could take *angular* frequency $\omega = 2\pi f$ (radians per second) as the independent variable of the frequency domain, but cyclical frequency has the advantage of being the direct reciprocal of the period and, in practice, one seldom measures frequency in radians per second.

This section introduces the frequency-domain picture, called a *line spectrum*, for any signal that is expressable as a sum of phasors. Then, as the counterpart of this signal representation, it will be shown how systems can be viewed in the frequency domain via the *frequency response function*, a special form of the transfer function. The results of the previous section allow us to combine and interpret these two new ideas.

Two words of warning at the start. First, because we will be speaking of f as an independent variable, special care must be taken with notation; in particular, any specific frequency, being a constant, is hereafter identified by a subscript to distinguish it from the variables f and $\omega = 2\pi f$. Second, wherever the symbol ω appears, the reader should mentally replace it by $2\pi f$ for which ω is the shorthand version.

Phasors and line spectra

Consider, first, the sinusoidal signal

$$\begin{aligned} v(t) &= A \cos\,(2\pi f_0 t + \theta) \\ &= \frac{A}{2} e^{j\theta} e^{j2\pi f_0 t} + \frac{A}{2} e^{-j\theta} e^{-j2\pi f_0 t} \end{aligned} \tag{1}$$

with Figure 6.2*a* being its phasor diagram. In the phasor description this signal consists of *two* well-defined frequency components – one at $+f_0$ and the other at $-f_0$, corresponding to the two directions of rotation – and each component is completely specified by two additional parameters, the *amplitude* and *phase*. Since these parameters are independent of time, they can just as well be represented in the frequency domain, say as shown in Figure 6.2*b*, where at the frequency $f = f_0$ we associate an amplitude $A/2$ and a phase angle θ, at $f = -f_0$ we associate an amplitude $A/2$ and a phase angle $-\theta$, and zero amplitude and phase elsewhere. Notice that amplitude and phase have thus become *dependent variables*, i.e., functions of the independent variable f.

Figure 6.2*b*, called a line spectrum, completely represents the time function of (1) since forming and summing the indicated phasors yields $v(t)$. This frequency-domain representation has *negative frequencies* simply because there are two rotational directions and, as we have repeatedly observed, it takes a pair of complex-conjugate phasors to give

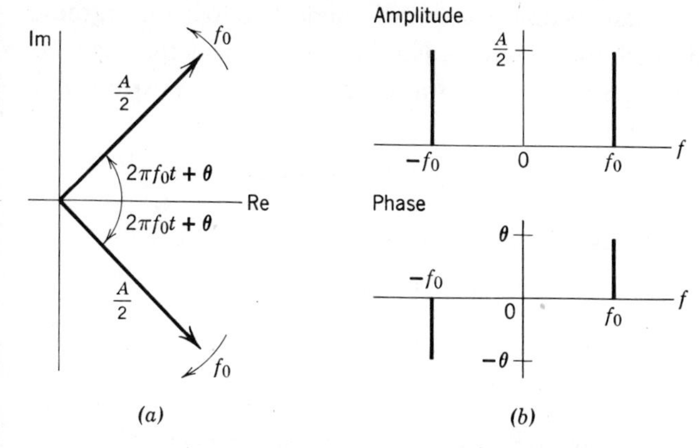

Figure 6.2. $v(t) = A \cos(2\pi f_0 t + \theta)$. (*a*) Phasor diagram. (*b*) Line spectrum.

a real sum. Line spectra having only positive frequencies are sometimes encountered, primarily in modulation theory, but their analytic value is limited.

Three standard conventions about phasors and line spectra need to be stated at this point. First, the phase reference is taken with respect to the positive real axis of the phasor diagram or, equivalently, with respect to *cosine* waves in the time domain, e.g., if $\theta = 0$ in (1) then $v(t) = A \cos 2\pi f_0 t$. Accordingly, sine waves must be converted to cosines by using the relationship

$$\begin{aligned} \sin \omega t &= \cos(\omega t - 90°) \\ &= \tfrac{1}{2} e^{-j90°} e^{j\omega t} + \tfrac{1}{2} e^{j90°} e^{-j\omega t} \end{aligned} \tag{2}$$

Second, amplitude is always a positive quantity, so negative signs are absorbed in the phase via

$$-A \cos \omega t = A \cos(\omega t \pm 180°) \tag{3}$$

Note that it makes no difference whether the phase shift is $+180°$ or $-180°$ since, from the phasor picture, one ends up at the same angle either way; thus, we are at liberty to take whichever sign is more convenient.

Third, phase angles are frequently expressed in *degrees*, rather than radians, even though other angles are expressed in radians – e.g., in (2) and (3), ωt is in radians while $\pm 90°$ and $\pm 180°$ are, obviously, in degrees. This mixed notation should cause no confusion if one bears in mind that angles are implicitly in radians unless explicitly identified to be in degrees.

By direct extension of the above reasoning, any signal that can be written as a sum of phasors can be represented by a line spectrum. Consider, for instance,

$$\begin{aligned} v(t) &= 2-6\cos(2\pi 10t-60^\circ)+\sin 2\pi 30t \\ &= 2+6\cos(2\pi 10t+120^\circ)+\cos(2\pi 30t-90^\circ) \\ &= 2+3e^{j120^\circ}e^{j2\pi 10t}+3e^{-j120^\circ}e^{-j2\pi 10t} \\ &\quad +0.5e^{-j90^\circ}e^{j2\pi 30t}+0.5e^{j90^\circ}e^{-j2\pi 30t} \end{aligned}$$

in which we have used both (2) and (3) to get the standard form. Noting that a constant term is a zero-frequency phasor, the corresponding line spectrum becomes as shown in Figure 6.3. Inspection of the figure reveals that the *amplitude* spectrum has *even symmetry* while the *phase* spectrum

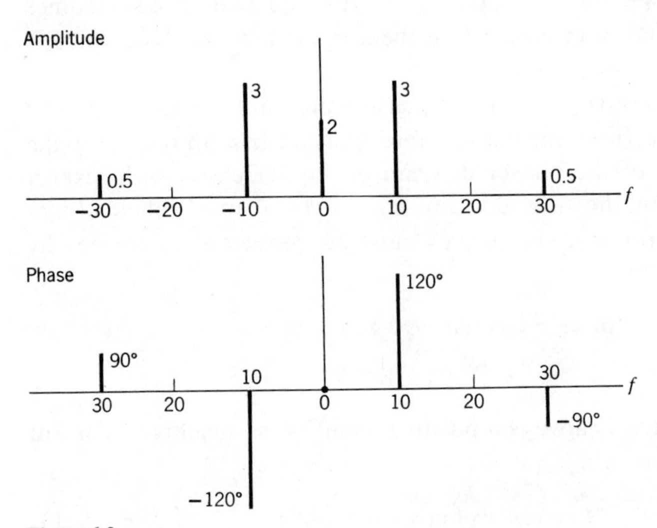

Figure 6.3.

has *odd symmetry*. This situation is symptomatic of the fact that, in this example, the signal is a *real function of time*; conversely, if symmetry is lacking, the signal in question must be complex. Reference to the phasor diagram of Figure 6.2*a* shows why this is so, for the pairs of phasors must have equal amplitudes and opposite angles so that their sum will fall along the real axis.

On the other hand, line spectra are not restricted to real signals. As an important illustration, Figure 6.4 is an asymmetric spectrum representing the complex signal

$$Ae^{j\theta}e^{j2\pi f_0 t} = A\cos(2\pi f_0 t + \theta) + jA\sin(2\pi f_0 t + \theta) \tag{4}$$

From this figure it should be clear that single phasors like (4) are the fundamental building blocks of the frequency-domain concept, rather than sinusoidal signals.

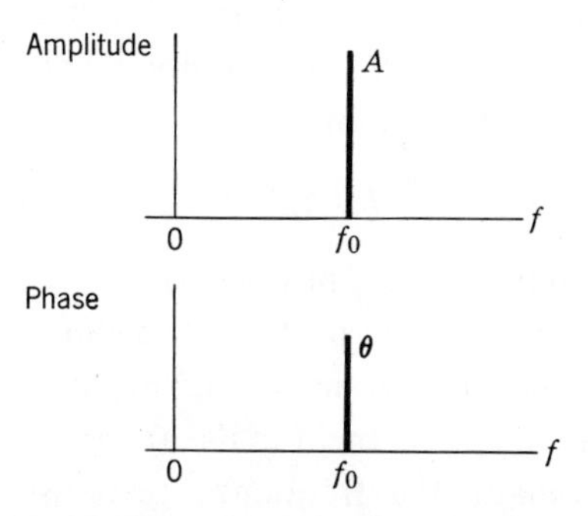

Figure 6.4. Line spectrum for $Ae^{j\theta}e^{j2\pi f_0 t}$.

As a concluding comment we point out that, in at least one respect, the amplitude spectrum has greater significance than the phase spectrum. To be sure, the phase spectrum is required to complete the frequency-domain picture and to correctly return to the time domain. But it is the amplitude spectrum that tells us which frequency components are present in the signal and in what proportion. Putting this more succinctly, the amplitude spectrum shows the signal's *frequency content*. And, in many applications, that is all we need to know.

Transfer functions and frequency response

Having developed line spectra to represent signals that are sums of phasors, we can also speak of a frequency-domain *system* description. For that purpose, take the input to be a single phasor at frequency f_0, i.e.,

$$x(t) = A_x e^{j\theta_x} e^{j2\pi f_0 t} \qquad -\infty < t < \infty \tag{5a}$$

so

$$y(t) = H(j2\pi f_0) A_x e^{j\theta_x} e^{j2\pi f_0 t} \qquad -\infty < t < \infty \tag{5b}$$

providing the system is linear, time-invariant, and asymptotically stable.

Because expressions such as $H(j2\pi f_0)$ are awkward to write repeatedly, it is convenient at this time to introduce a new symbol $H(f)$, defined by

$$H(f) \triangleq H(s)|_{s=j2\pi f} = \int_{-\infty}^{\infty} h(t)e^{-j2\pi ft}dt \tag{6}$$

which we call the *frequency response function* of the system. Actually, (6) is slightly illegal in the strict mathematical sense, in that $H(f)$ is not the same function of f that $H(s)$ is of s; however, the notation is reasonably standard and should cause no confusion. Obviously, $H(f)$ is identical to $H(j\omega)$ with $\omega = 2\pi f$ written out explicitly, the change being simply a matter of convenience and emphasis. Like $H(j\omega)$, $H(f)$ is complex and, when $h(t)$ is real, it has the property $H(-f) = H^*(f)$, or

$$|H(-f)| = |H(f)| \qquad \arg[H(-f)] = -\arg[H(f)] \tag{7}$$

Complex functions obeying (7) are said to have *hermitian symmetry*.

As to interpreting the frequency response function, its definition shows that $H(f)$ is a *continuous function of frequency*, including negative frequency, and is independent of time. Thus, just as the impulse response $h(t)$ describes a system in the time domain, the frequency response function $H(f)$ is the system's frequency-domain description – subject, in either case, to zero-initial-state conditions or steady-state conditions. Justifying this interpretation, we return to (5b) and write it as

$$\begin{aligned} y(t) &= |H(f_0)|e^{j\arg[H(f_0)]}A_x e^{j\theta_x} e^{j2\pi f_0 t} \\ &= \underbrace{|H(f_0)|A_x}_{A_y} \exp[j\underbrace{(\theta_x + \arg[H(f_0)])}_{\theta_y}]\, e^{j2\pi f_0 t} \end{aligned}$$

Hence, the input and output signal parameters are related by

$$\frac{A_y}{A_x} = |H(f_0)| \qquad \theta_y - \theta_x = \arg[H(f_0)] \tag{8}$$

which, in retrospect, is seen to be the same as that for *sinusoidal* signals, Eq. (5b), Sect. 6.1. Notice carefully in (8) that the amplitude relationship is multiplicative while the phase relationship is additive.

Based on (8) we say in general that $|H(f)|$ is the system's *amplitude ratio* (sometimes called the *gain*) and $\arg[H(f)]$ is the *phase shift*, both being functions of frequency. Plotting these two versus f gives a frequency-domain representation of the system in the same fashion that the amplitude and phase spectrum represents a signal. Furthermore, from (7), the amplitude ratio has even symmetry while the phase shift has odd symmetry, the same symmetry in f as the line spectrum of a real signal.

With respect to finding $H(f)$ for a given system, one simply calculates $H(s)$ by any of the methods given in Section 4.5 and then sets $s = j2\pi f$. One of those methods, recall, was based on finding the particular solution $y_P(t)$ when $x(t) = e^{st}$ for all t; translated to the present context it becomes

$$H(f) = \frac{y_P(t)}{x_P(t)} \quad \text{when} \quad x_P(t) = e^{j\omega t} \tag{9}$$

so all the standard electrical engineering tools such as Ohm's law for complex impedence, the voltage-divider relation, etc., are applicable when dealing with electrical systems. And with the aid of analogs and simulation diagrams, (9) may be used for nonelectrical systems too.

Finally, if analytic methods are inappropriate because we do not know what is inside the system – except that it is fixed, linear, and stable – then (8) suggests an experimental procedure, as follows. Apply a sinusoidal input at some frequency f_0 and wait until the output reaches steady-state as another sinusoid; the ratio of output to input amplitude is the system's amplitude ratio at that frequency, i.e., $|H(f_0)|$, and the output phase relative to the input is the phase shift, $\arg[H(f_0)]$. These measurements are then repeated at specific frequencies throughout the range of interest, giving a sampled plot of $H(f)$.

Example 6.1 Frequency response of an RC lowpass filter

The RC circuit of Figure 6.5*a* is, by now, an old friend. In earlier chapters we found its differential equation and impulse response to be

$$RC\dot{y} + y = x \qquad h(t) = \frac{1}{RC} e^{-t/RC} u(t)$$

Here, let us investigate the frequency response function $H(f)$.

Applying the method of (9), we assume $x(t) = e^{j\omega t}$ and the voltage-divider relation for complex impedence gives

$$\frac{y_P(t)}{x_P(t)} = \frac{(1/j\omega C)}{R + (1/j\omega C)} = \frac{1}{1 + j\omega RC}$$

Hence

$$H(f) = \frac{1}{1 + j2\pi f RC} = \frac{1}{1 + j(f/B)} \tag{10}$$

in which we have introduced the system parameter

$$B \triangleq \frac{1}{2\pi RC} \tag{11}$$

Then, converting to polar form,

$$H(f) = \frac{1}{\sqrt{1+(f/B)^2}} \exp\left(-j \arctan \frac{f}{B}\right)$$

so the amplitude ratio and phase shift are

$$|H(f)| = \left[1 + \left(\frac{f}{B}\right)^2\right]^{-1/2} \tag{12a}$$

$$\arg [H(f)] = -\arctan \frac{f}{B} \tag{12b}$$

as plotted in Figure 6.5*b* with arg $[H(f)]$ expressed in degrees.

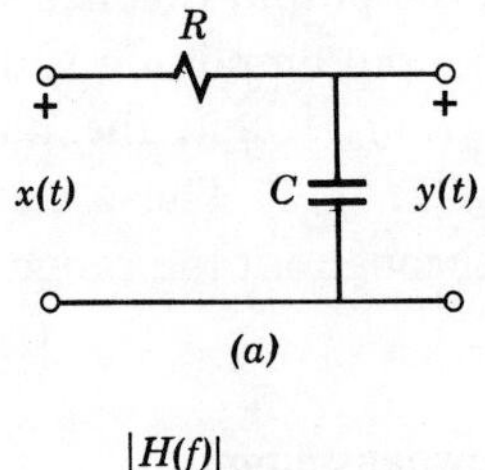

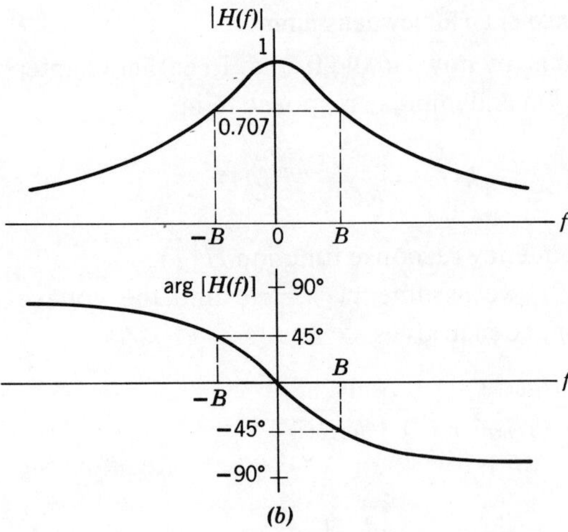

Figure 6.5. The RC lowpass filter. (*a*) Circuit diagram. (*b*) Frequency response $H(f)$.

Filters and bandwidth

Taken together, $|H(f)|$ and arg $[H(f)]$ constitute a system's frequency response in the sense that the steady-state response of a given system to a given sinusoidal input signal depends only on the excitation frequency. The concept of frequency response is especially relevant to those systems that exhibit *frequency-selective* characteristics, meaning that they selectively pass or reject a range or band of frequencies. In other words, frequency-selective systems are *filters*.

The RC circuit we have just considered is said to be a *lowpass filter* because it passes without appreciable change only "low" frequencies, say $|f| < B$, while severely diminishing the output amplitude at "high" frequencies, $|f| \gg B$. Similarly, one can speak of *highpass* and *bandpass filters* and *band-rejection filters*. We use magnitude signs – e.g., $|f| < B$ – and measure passbands only in terms of *positive* frequency in recognition of the symmetry of $H(f)$ plus the fact that negative frequencies do not appear in our final results when the signals in question are real functions of time.

Ideally, a filter should totally reject all frequencies outside its passband, i.e., $|H(f)| = 0$ outside the passband, and therefore an *ideal filter* would have a clearly defined *bandwidth*. But, sad to say, such filters are physically unrealizable, although they may be closely approximated. *Real filters* – the RC lowpass filter for instance – have a more or less gradual transition between passband and rejection, and defining the bandwidth of a real filter becomes somewhat arbitrary.

One convention is to take the bandwidth as that range of positive frequencies over which the amplitude ratio drops no lower than $1/\sqrt{2} = 0.707$ times the maximum value in the passband. For example, from Figure 6.5*b* the bandwidth of the RC filter is $B = 1/2\pi RC$ since $|H(f)|_{\max} = H(0) = 1$ while $|H(B)| = 1/\sqrt{2}$. Further exploring the physical significance of this convention we note that a sinusoidal input with amplitude A and frequency $f = B$ would come out with amplitude $A/\sqrt{2}$; the output power therefore equals $(\frac{1}{2})(A/\sqrt{2})^2 = A^2/4$ as compared with an output power of $A^2/2$ if the input frequency were much lower. For this reason B is called the *half-power bandwidth*.

Because a practical filter does not give perfect rejection, it is sometimes important to accurately evaluate the response at frequencies well beyond the passband. For that purpose, linear frequency-response plots such as Figure 6.5*b* are inadequate, and it becomes necessary to use logarithmic coordinates. Specifically, the *gain* expressed in *decibels* (abbreviated db) is defined to be

$$|H(f)|_{\text{db}} \triangleq 20 \log_{10} |H(f)| = 10 \log_{10} |H(f)|^2 \tag{13}$$

From (13), a positive gain in db corresponds to amplification, i.e., $|H(f)| > 1$, while negative values correspond to attenuation, i.e., $|H(f)| < 1$; the dividing line is $|H(f)|_{\text{db}} = 0$ db, meaning that $|H(f)| = 1$. By plotting $|H(f)|_{\text{db}}$ versus a logarithmic frequency scale, a much wider range can be compactly displayed; the phase shift arg $[H(f)]$ in degrees is also plotted versus $\log f$. Of course, using $\log f$ restricts one to $f > 0$ but this is no limitation in view of the symmetry of $H(f)$.

Figure 6.6 illustrates this display for the RC lowpass filter. An interesting feature of the gain curve is that at very high or very low frequencies it is accurately approximated by *linear asymptotes* shown dashed. These asymptotes follow analytically from (12a) and (13) since

$$|H(f)|_{\text{db}} = 20 \log_{10} [1 + (f/B)^2]^{-1/2}$$
$$\approx \begin{cases} 0 & |f| \ll B \\ -20 \log_{10} (f/B) & |f| \gg B \end{cases}$$

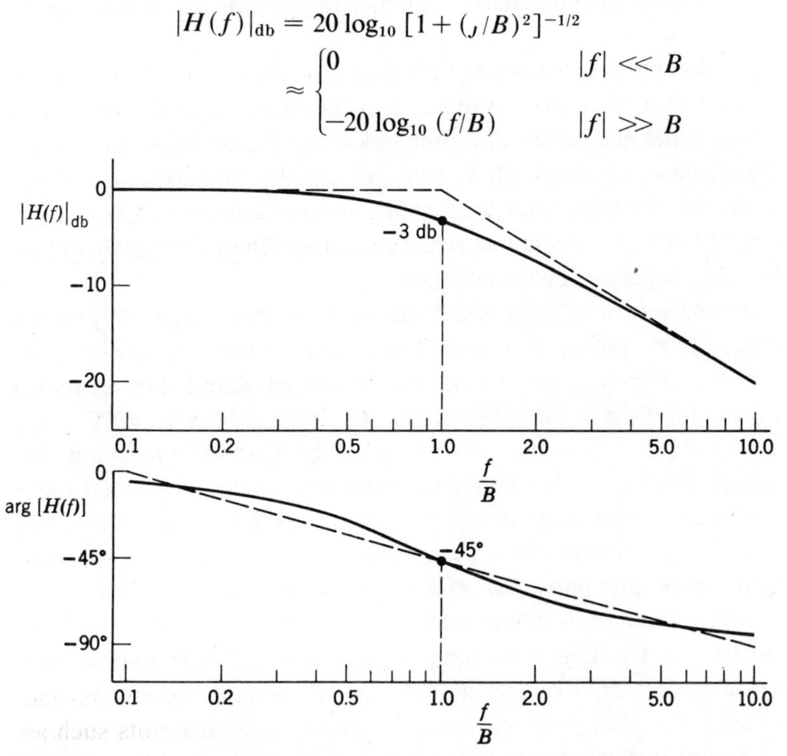

Figure 6.6. Logarithmic plot of the frequency response of an RC lowpass filter.

At low frequencies the gain is essentially constant at 0 db, while at high frequencies it falls off at a rate of -20 db *per decade* (a frequency increase by a factor of 10) or -6 db *per octave* (a frequency increase by a factor of 2). The low- and high-frequency asymptotes intersect at $f = B$, called the corner frequency,† where the approximation has its maximum error of 3 db. Although not so obvious analytically, a reasonable linear approximation for the phase shift of the RC filter may be constructed using a line having slope $-45°$ per decade passing through the exact value $\arg[H(B)] = -45°$ at the corner frequency.

When plotted versus $\log \omega$ rather than $\log f$, such frequency-response curves are called *Bode plots.* They are used extensively to analyze feedback systems.

6.3 FOURIER SERIES

The exponential Fourier series for a periodic signal was developed in Section 5.4 as a linear combination of orthonormal functions having the property of being a least-square-error approximation. Also shown was the fact that when the signal in question has well-defined average power, its Fourier series may be deemed a valid signal representation for most engineering purposes. That premise is adopted here, where we put the theory to work. Specifically, we shall use the Fourier series to express periodic signals as sums of phasors, from which their line spectra drop out immediately. Then, in Section 6.5, the phasor sum is employed to carry out periodic steady-state analysis.

Exponential Fourier series

Let $v(t)$ be a periodic power signal with fundamental period T_0. It can be expanded as a linear combination of phasors via the exponential Fourier series

$$v(t) = \sum_{n=-\infty}^{\infty} c_n e^{jn\Omega t} \tag{1}$$

where Ω is the fundamental angular frequency

$$\Omega = \frac{2\pi}{T_0} \tag{2}$$

†It also happens in this case that the corner frequency numerically equals the half-power bandwidth and, at that point, $|H(B)|_{\text{db}} = -20 \log_{10} (1/\sqrt{2}) = 3.01 \approx -3$ db, which explains why B commonly is referred to as the *3-db bandwidth.*

and the coefficients c_n are given by

$$c_n = \frac{1}{T_0} \int_{T_0} v(t) e^{-jn\Omega t} dt \tag{3}$$

As in Chapter 5, the symbol $\int_{T_0}$ stands for integration over any period $t_1 \leq t \leq t_1 + T_0$, with t_1 being arbitrary.

Perhaps the most striking feature of the series is that only the c_n's depend on the explicit details of the signal's behavior, all else being predetermined once it is known that $v(t)$ is periodic. Putting this another way, the Fourier-series expansions for all power signals having the same period differ only in the coefficients. For this reason we concentrate our discussion on the properties of c_n implied by (3), especially the following four points.

First, the c_n's are generally *complex* quantities, whether or not the signal is complex. To illustrate, if $v(t)$ is in fact real (noncomplex), then the real and imaginary parts of c_n are given by

$$c_n = \underbrace{\left[\frac{1}{T_0} \int_{T_0} v(t) \cos n\Omega t dt\right]}_{\text{Re}\,[c_n]} + j \underbrace{\left[-\frac{1}{T_0} \int_{T_0} v(t) \sin n\Omega t dt\right]}_{\text{Im}\,[c_n]} \tag{4}$$

which comes about from applying Euler's law to $e^{-jn\Omega t}$.

Second, because (3) is a definite integral with t the variable of integration, c_n is *independent of time*. Underscoring this fact, we introduce the change of variable $\psi = \Omega t$ so that (3) becomes

$$c_n = \frac{1}{2\pi} \int_{2\pi} v\left(\frac{\psi}{\Omega}\right) e^{-jn\psi} d\psi \tag{5}$$

which, with its compact form, is sometimes handy when calculating c_n or demonstrating other relationships.

Third, setting $n = 0$ in (3), the zeroth-order coefficient c_0 is

$$c_0 = \frac{1}{T_0} \int_{T_0} v(t) dt \tag{6}$$

which, upon examination, should be recognized as the *time-average value* of $v(t)$. Thus, the constant term in the Fourier series is just the average value of the signal.

Fourth, if the signal in question is a *real* function of time, then the negative and positive coefficients are related simply by *complex conjugation*, namely

$$c_{-n} = c_n^* \tag{7}$$

as follows from (4) by replacing n with $-n$.

Waveform symmetry

Besides these basic properties, there are certain simplifications when $v(t)$ has *symmetry* of one type or another. In particular, if $v(t)$ is an *even* function then the integrand in the first term of (4) has even symmetry while the second has odd symmetry. Therefore, taking the range of integration to be $-T_0/2 \leq t \leq T_0/2$ and invoking Eq. (20), Sect. 5.1,

$$c_n = \frac{2}{T_0}\int_0^{T_0/2} v(t)\cos n\Omega t\, dt \tag{8a}$$

By the same procedure, if $v(t)$ is an *odd* function,

$$c_n = -j\frac{2}{T_0}\int_0^{T_0/2} v(t)\sin n\Omega t\, dt \tag{8b}$$

and it follows that $c_0 = 0$.

Another type of symmetry, called *half-wave* or *rotation* symmetry, is defined by the property

$$v\left(t \pm \frac{T_0}{2}\right) = -v(t) \tag{9}$$

which can hold only for a periodic signal. As illustrated in Figure 6.7, such signals retain odd symmetry even when the time origin is shifted an integer number of half periods in either direction. Under this condition it

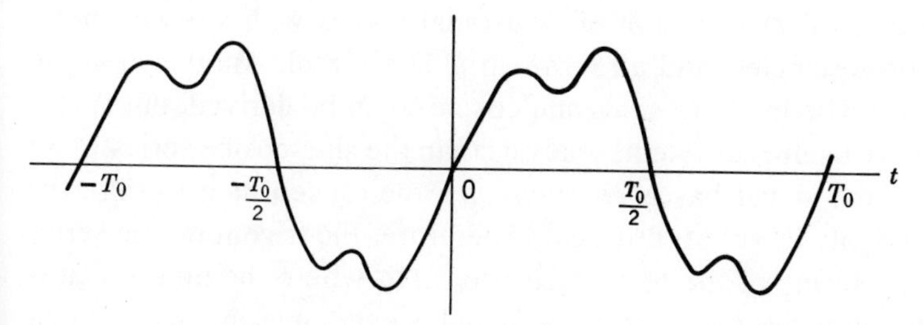

Figure 6.7. An example of half-wave symmetry.

turns out that

$$c_n = 0 \qquad n = 0, \pm 2, \pm 4, \ldots \tag{10}$$

so the Fourier series consists entirely of odd-order terms.

Occasionally one encounters signals with *hidden symmetry*, i.e., actual symmetry that is obscured by the presence of an additive constant term, or potential symmetry that can be realized by shifting the location of the time origin. In the former case the constant can be subtracted out and then added to the zeroth coefficient c_0 after performing the Fourier-series expansion of the symmetric signal. In the latter case, since the time origin is not physically unique we are usually free to redefine it so as to gain simplifications of symmetry. A theorem that covers the effects of time shifting will be presented in Section 6.4

Further understanding of waveform symmetry is gained by examining the *trigonometric* Fourier series.

Trigonometric Fourier series

When $v(t)$ is a *real* signal – and hence $c_{-n} = c_n^*$ – its exponential Fourier series can be converted to a trigonometric form by the following manipulation. Regrouping pairwise all but the zeroth term in (1), so the summation index is always positive, we have

$$v(t) = c_0 + \sum_{n=1}^{\infty} (c_n e^{jn\Omega t} + c_{-n} e^{-jn\Omega t})$$

But $c_n e^{jn\Omega t}$ and $c_{-n} e^{-jn\Omega t}$ now form a complex-conjugate pair so, with c_n in polar form,

$$v(t) = c_0 + \sum_{n=1}^{\infty} |2c_n| \cos(n\Omega t + \arg[c_n]) \tag{11}$$

which expresses $v(t)$ as a sum of sinusoidal waves with various amplitudes and phase angles, and all terms in (11) are real. Another trigonometric form, involving both sines and cosines, can be derived, but (11) is generally more useful in systems analysis than the sine-cosine series.

Now if a real signal has *even symmetry* then, according to (8a), the series coefficients c_n are strictly real. Therefore, the trigonometric series involves only terms of the form $\pm|2c_n| \cos n\Omega t$, where the minus sign is needed when c_n is negative. Since the signal is even, it is only natural that the series should reduce to a sum of *even functions*, $\cos n\Omega t$. Similarly, when $v(t)$ has *odd symmetry* the series terms become $\pm|2c_n| \sin n\Omega t$, so we have a sum of *odd functions*, $\sin n\Omega t$.

These comments, along with our previous observations, are best illustrated by a few examples of calculating the Fourier series of a given signal.

Example 6.2 A rectangular pulse train

The rectangular pulse train of Figure 6.8 is a very important idealized signal. Formally, it is written by specifying its value over one period and citing the periodicity requirement, as below:

$$v(t) = \begin{cases} A & |t| \leq \tau/2 \\ 0 & \tau/2 < |t| \leq T_0/2 \end{cases}$$

$$v(t) = v(t \pm mT_0) \qquad m = 0, \pm 1, \pm 2, \ldots$$

The parameter τ is the pulse *duration* and A the *amplitude*.

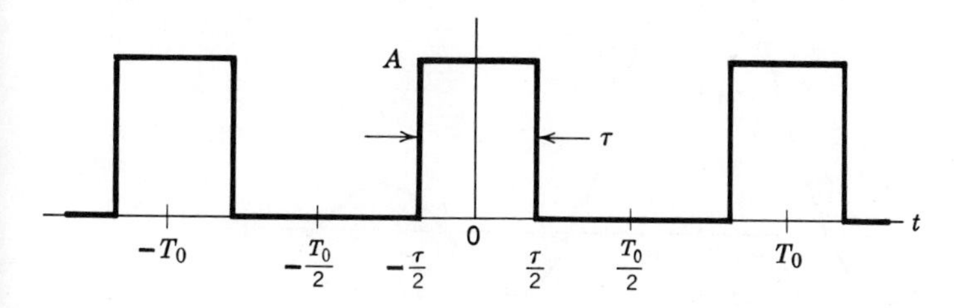

Figure 6.8. Rectangular pulse train.

Since the integration for c_n is straightforward, we shall ignore the fact that $v(t)$ has even symmetry and use the basic expression (3) to find the series coefficients. Taking the integration limits as $-T_0/2$ and $T_0/2$ gives

$$c_n = \frac{1}{T_0} \int_{-T_0/2}^{T_0/2} v(t) e^{-jn\Omega t} dt = \frac{1}{T_0} \int_{-\tau/2}^{\tau/2} A e^{-jn\Omega t} dt$$

$$= \frac{A}{-jn\Omega T_0} (e^{-jn\Omega \tau/2} - e^{jn\Omega \tau/2})$$

$$= \frac{A}{\pi n} \sin \frac{\pi n \tau}{T_0}$$

where we have used $e^{j\phi} - e^{-j\phi} = 2j \sin \phi$ and substituted $\Omega = 2\pi/T_0$.

This expression can be further tidied up by introducing a new function, called the *sinc function*,† defined as

$$\text{sinc } z \triangleq \frac{\sin \pi z}{\pi z} \tag{12}$$

and plotted in Figure 6.9. Being the product of two odd functions, sinc z is an *even* function and, with $\sin \pi z$ in the numerator, it has *zero crossings* at all nonzero integer values of its argument, i.e.,

$$\text{sinc } z = 0 \qquad z = \pm 1, \pm 2, \ldots \tag{13a}$$

while the indeterminate case of $z = 0$ yields

$$\text{sinc } 0 = \lim_{z \to 0} \frac{\sin \pi z}{\pi z} = 1 \tag{13b}$$

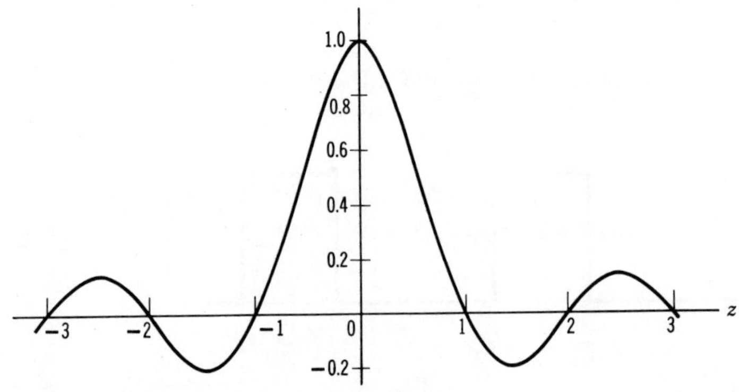

Figure 6.9. The sinc function $\text{sinc } z \triangleq \dfrac{\sin \pi z}{\pi z}$.

Table 6.1

z	sinc z	z	sinc z	z	sinc z
0.0	1.000	1.0	0.000	2.5	0.127
0.2	0.935	1.2	−0.156	3.5	−0.091
0.4	0.757	1.4	−0.216	4.5	0.071
0.6	0.505	1.6	−0.189	5.5	−0.058
0.8	0.234	1.8	−0.104	6.5	0.049

†Some authors define instead $Sa(z) \triangleq (\sin z)/z$, called the *sampling function* because of another context in which it occurs. The two functions differ only by a factor of π in the argument, i.e., $\text{sinc } z = Sa(\pi z)$.

as found by simple limiting. Selected values of sinc z are listed in Table 6.1.

Using the sinc-function notation, the series coefficients become

$$c_n = \frac{A\tau}{T_0} \frac{\sin(\pi n\tau/T_0)}{(\pi n\tau/T_0)}$$

$$= \frac{A\tau}{T_0} \operatorname{sinc} \frac{n\tau}{T_0} \tag{14}$$

which we note is strictly real – because $v(t)$ is real and even – and independent of time. Setting $n = 0$ yields

$$c_0 = \frac{A\tau}{T_0}$$

which clearly is the average value of the signal. Finally, inserting (14) into (1) gives

$$v(t) = \frac{A\tau}{T_0} \sum_{n=-\infty}^{\infty} \operatorname{sinc} \frac{n\tau}{T_0}\, e^{jn\Omega t} \tag{15a}$$

for the exponential Fourier series representation of the rectangular pulse train. Or, since $v(t)$ is real and even, its trigonometric Fourier series can be written as

$$v(t) = \frac{A\tau}{T_0}\left(1 + \sum_{n=1}^{\infty} 2 \operatorname{sinc} \frac{n\tau}{T_0} \cos n\Omega t\right) \tag{15b}$$

Example 6.3 A half-rectified sine wave

Passing a sine wave of angular frequency Ω through a half-wave rectifier produces the signal shown in Figure 6.10. There is no symmetry here,

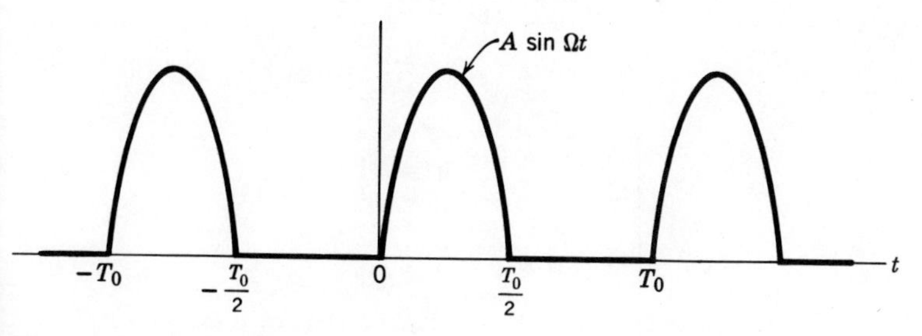

Figure 6.10. Half-rectified sine wave.

but c_n can be found using (5) and a table of integrals; i.e., $T_0 = 2\pi/\Omega$, $v(\psi/\Omega) = A \sin \psi$ for $0 \le \Omega t \le \pi$, and

$$c_n = \frac{1}{2\pi} \int_0^{\pi} A \sin \psi \, e^{-jn\psi} \, d\psi$$

$$= \frac{A}{2\pi(1-n^2)} (1 + e^{-jn\pi}) \qquad n^2 \ne 1$$

But $(1 + e^{-jn\pi})$ equals +2 when n is even and 0 when n is odd, so

$$c_n = \begin{cases} \dfrac{A}{\pi(1-n^2)} & n = 0, \pm 2, \pm 4, \ldots \\ 0 & n = \pm 3, \pm 5, \ldots \end{cases} \tag{16a}$$

and all odd-order terms are zero save for the indeterminate case of $n = \pm 1$. By separate integration or limiting one finds for that case

$$c_{\pm 1} = \mp j \frac{A}{4} \tag{16b}$$

Example 6.4 Sinusoidal waves

Both previous examples have resulted in series representations with an *infinite* number of terms. This is usually the case, but not always, an important counter example being the sinusoidal signal

$$v(t) = A \cos(\Omega t + \theta)$$

Without bothering with integration, the series coefficients can be found directly from the phasor representation

$$v(t) = \underbrace{\left(\frac{A}{2} e^{j\theta}\right)}_{c_1} e^{j\Omega t} + \underbrace{\left(\frac{A}{2} e^{-j\theta}\right)}_{c_{-1}} e^{-j\Omega t}$$

Hence

$$c_n = \begin{cases} \dfrac{A}{2} e^{j\theta} & n = +1 \\ \dfrac{A}{2} e^{-j\theta} & n = -1 \\ 0 & n \ne \pm 1 \end{cases} \tag{17}$$

Perhaps because the results are so simple, students often do not realize that the phasor representation of a sinusoid is also its exponential Fourier series.

Line spectra for periodic signals

Bringing together Fourier analysis and frequency-domain representation, we point out that the exponential Fourier-series expansion of a periodic signal is a *sum of phasors*. Therefore, the line spectrum can be found directly from the series coefficients c_n. For this purpose it is advantageous to think of c_n as a complex function of the *continuous* variable f but defined only for the *discrete* values $f = nf_0$, where $f_0 = 1/T_0 = \Omega/2\pi$.

More formally, we introduce the notation $c(nf_0) = c_n$ and rewrite (3) with $\Omega = 2\pi f_0$, i.e.,

$$c(nf_0) \triangleq \frac{1}{T_0} \int_{T_0} v(t) e^{-j2\pi nf_0 t} dt \tag{18}$$

Then the exponential Fourier series becomes

$$v(t) = \sum_{n=-\infty}^{\infty} |c(nf_0)| e^{j \arg [c(nf_0)]} e^{j2\pi nf_0 t} \tag{19}$$

in which $c(nf_0)$ has been written in polar form. We interpret (19) as saying that a periodic power signal consists of a sum (usually infinite) of phasors at the frequencies $f = 0, \pm f_0, \pm 2f_0, \ldots$, the amplitude and phase of the nth component being $|c(nf_0)|$ and $\arg [c(nf_0)]$, respectively. Therefore, $|c(nf_0)|$ is the *amplitude spectrum* of $v(t)$ while $\arg [c(nf_0)]$ is the *phase spectrum*. From the previously derived properties of the Fourier series coefficients we can make several additional statements about the line spectra of periodic signals, summarized below.

1. All lines in the spectrum are located only at integer multiples of the fundamental frequency $f_0 = 1/T_0$. Hence, a periodic signal consists entirely of frequencies which are *harmonically related to* f_0.

2. The zero-frequency or DC component equals the average value of the signal, since

$$c(0) = c_0 = \frac{1}{T_0} \int_{T_0} v(t) dt$$

3. If $v(t)$ is a *real* function of time then $c(nf_0)$ has hermitian symmetry, i.e.,

$$|c(-nf_0)| = |c(nf_0)| \qquad \arg[c(-nf_0)] = -\arg[c(nf_0)]$$

which means that the amplitude and phase spectra have even and odd symmetry respectively, as observed in conjunction with Figure 6.3.

4. If the signal has *half-wave symmetry*, then

$$c(nf_0) = 0 \qquad n = 0, \pm 2, \pm 4, \ldots$$

so all the even harmonics will be absent from the line spectrum.

5. If a *real* signal has *even* symmetry in time, then $c(nf_0)$ is strictly real and hence

$$\arg[c(nf_0)] = 0 \qquad \text{or} \qquad \pm 180°$$

the latter being required when $c(nf_0)$ is negative. On the other hand, for a real signal with *odd* symmetry, $c(nf_0)$ is strictly imaginary and

$$\arg[c(nf_0)] = \pm 90°$$

since $\pm j = e^{\pm j\pi/2} = e^{\pm j90°}$.

These points, plus the mechanics of constructing line spectra, are illustrated in the following examples.

Example 6.5 Spectrum of a half-rectified sine wave

Using the results of Example 6.3, $|c(nf_0)|$ and $\arg[c(nf_0)]$ have been listed in Table 6.2 for the first few values of n. Negative amplitudes have been converted to phase angles of $+180°$ or $-180°$, the choice being dictated by symmetry considerations since there is no physical difference. Figure 6.11 is the resulting spectrum.

Example 6.6 Spectrum of a rectangular pulse train

To continue Example 6.2, the line spectrum of a rectangular pulse train is given by (14) when rewritten as

$$c(nf_0) = \frac{A\tau}{T_0} \operatorname{sinc} nf_0\tau \tag{20}$$

The amplitude spectrum is then $|c(nf_0)| = (A\tau/T_0)|\operatorname{sinc} nf_0\tau|$, shown in Figure 6.12$a$ for the case of $\tau/T_0 = \frac{1}{5}$ so $f_0 = 1/5\tau$. This plot has been facilitated by regarding the *continuous* function $(A\tau/T_0)|\operatorname{sinc} f\tau|$ as the

Table 6.2

n	c_n	$\lvert c(nf_0)\rvert$	$\arg[c(nf_0)]$
0	$\dfrac{A}{\pi}$	$\dfrac{A}{\pi}$	0
±1	$\mp j\dfrac{A}{4}$	$\dfrac{A}{4}$	$\mp 90°$
±2	$-\dfrac{A}{3\pi}$	$\dfrac{A}{3\pi}$	$\mp 180°$
±3	0	0	0
±4	$-\dfrac{A}{15\pi}$	$\dfrac{A}{15\pi}$	$\mp 180°$

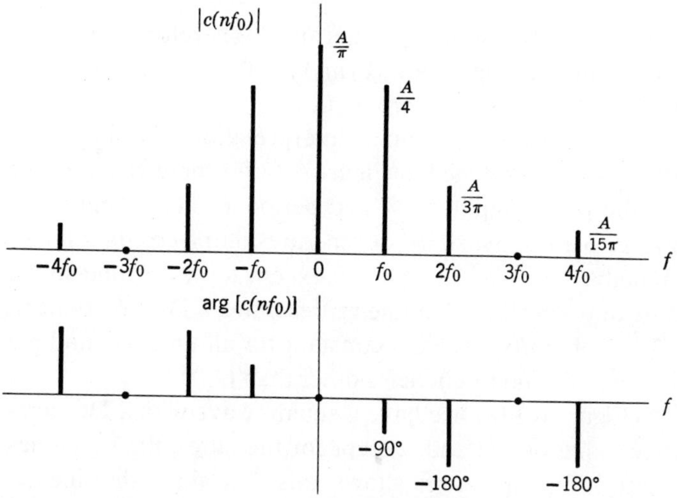

Figure 6.11. Spectrum of a half-rectified sine wave.

envelope of the amplitude lines – the dashed curve in the figure. Features to be noted here are: the uniform line spacing, save where lines are "missing" because they have zero amplitude; the even symmetry, reflecting the fact that the signal is real; and the DC component, $c(0) = A\tau/T_0$.

The phase spectrum, Figure 6.12b, has been constructed by noting

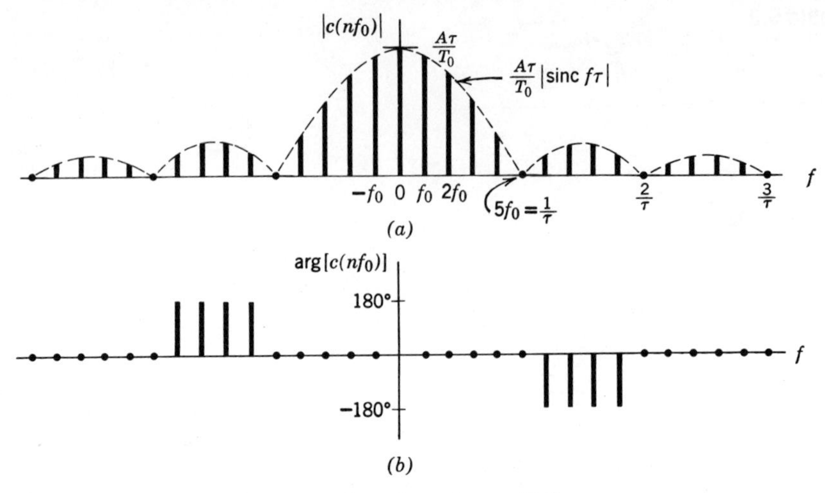

Figure 6.12. Spectrum of a rectangular pulse train, $\tau = T_0/5$.

that $c(nf_0)$ is always real but sometimes negative. Therefore, absorbing negative amplitudes in the phase, arg $[c(nf_0)] = 0$ when sinc $nf_0\tau \geq 0$ while arg $[c(nf_0)] = \pm 180°$ when sinc $nf_0\tau < 0$.

As a further aid to frequency-domain interpretation, the amplitude spectra and waveforms are sketched in Figure 6.13 for three values of the pulse duration τ, the pulse amplitude A and period T_0 being held fixed. When $\tau = T_0$ (Figure 6.13*a*), the signal degenerates into a constant for all time; correspondingly, sinc $nf_0\tau =$ sinc $n = 0$ except for $n = 0$, and so the spectrum contains only one line, that line representing a DC component. This is quite logical, of course, since a constant for all time has no time variation and thus contains no frequencies other than $f = 0$.

When $\tau = T_0/2$ (Figure 6.13*b*), we have a square wave with a DC component or average value of $A/2$ and, except for the latter, the signal has half-wave symmetry. The spectrum shows this fact since the lines at $f = \pm 2f_0, \pm 4f_0, \ldots,$ fall at the zero-crossing of sinc $f\tau$ and thus have zero amplitude.

Going to smaller values of pulse duration (Figure 6.13*c*), the DC component decreases – after all, the area of $v(t)$ is reduced – and the components at higher frequencies become increasingly important. Physically, these higher frequencies are required to represent the more rapid time variation of the short pulses. Thus, as the pulses are contracted in the time domain, the spectrum spreads out in the frequency domain,

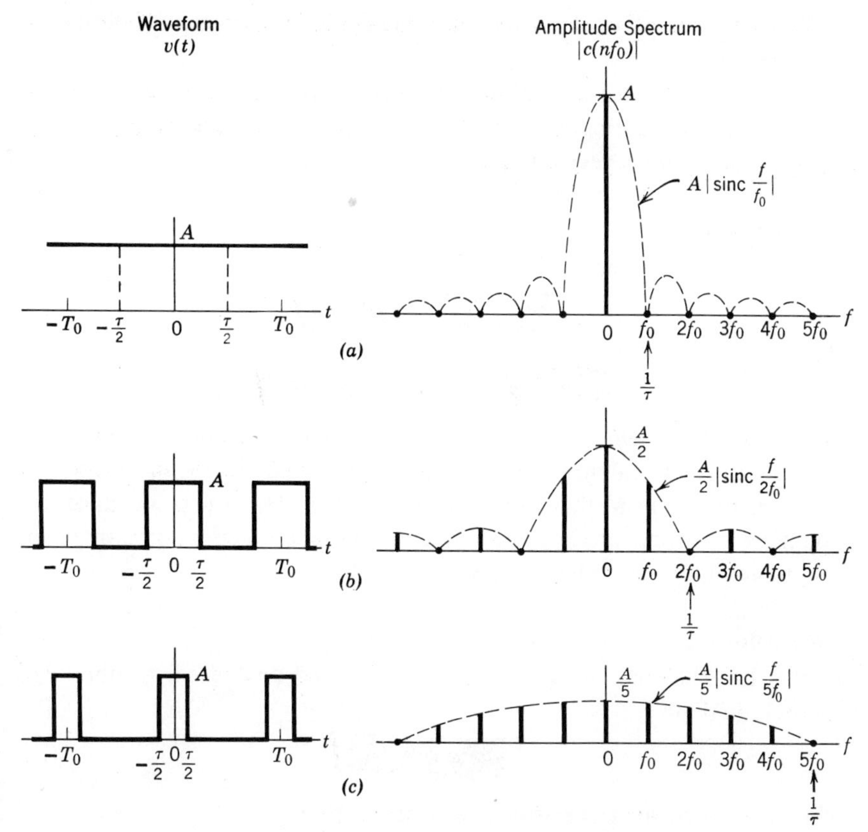

Figure 6.13. Waveform and amplitude spectrum of a rectangular pulse train. (*a*) $\tau = T_0$. (*b*) $\tau = T_0/2$. (*c*) $\tau = T_0/5$.

and vice versa. This phenomenon is known as the *reciprocal spreading* effect; it holds generally for all signals, not just for rectangular pulse trains.

6.4 FOURIER SERIES THEOREMS

There are numerous relations covering the frequency-domain effect of time-domain operations on periodic signals. Some of the more important ones are given here in the form of Fourier series theorems. These theorems are of interest for two reasons: *1.* they aid in interpreting frequency-domain properties from time-domain information, and vice versa; and

2. they are often of value as shortcuts in calculating series coefficients and line spectra.

In stating the theorems below we assume that $v(t)$, $w(t)$, and $z(t)$ are periodic signals having the *same period*, and whose spectra are $c_v(nf_0)$, $c_w(nf_0)$, and $c_z(nf_0)$, respectively.

Superposition

If α and β are constants and

$$z(t) = \alpha v(t) + \beta w(t) \tag{1a}$$

then

$$c_z(nf_0) = \alpha c_v(nf_0) + \beta c_w(nf_0) \tag{1b}$$

Hence, a linear combination in the time domain becomes a linear combination in the frequency domain. This simple and significant theorem is easily proved from the definition of $c(nf_0)$. However, because the coefficients are generally complex, care must be taken when converting (1b) to amplitude and phase spectra.

Time shift

If $z(t)$ has the same shape as $v(t)$ but delayed or shifted in time by t_d seconds so that

$$z(t) = v(t - t_d) \tag{2a}$$

where t_d may be negative as well as positive, then

$$c_z(nf_0) = c_v(nf_0)\, e^{-jn\Omega t_d} \qquad \Omega = 2\pi f_0 \tag{2b}$$

This means that *translating the time origin* affects only the *phase spectrum*, as brought out by writing (2b) in polar form

$$|c_z(nf_0)| = |c_v(nf_0)| \qquad \arg\,[c_z(nf_0)] = \arg\,[c_v(nf_0)] - n\Omega t_d$$

We prove the theorem very simply by replacing t with $t - t_d$ on both sides of Eq. (1), Sect. 6.3, giving

$$\begin{aligned} v(t - t_d) &= \sum_{n=-\infty}^{\infty} c_v(nf_0)\, e^{jn\Omega(t - t_d)} \\ &= \sum_{n=-\infty}^{\infty} \underbrace{c_v(nf_0)\, e^{-jn\Omega t_d}}_{c_z(nf_0)}\, e^{jn\Omega t} \end{aligned}$$

Hence, if $z(t) = v(t - t_d)$ then $c_z(nf_0) = c_v(nf_0)e^{-jn\Omega t_d}$, thereby completing the proof.

Example 6.7 Spectrum of a full-rectified sine wave

Since we know the spectrum of a half-rectified sine wave, the spectrum of a full-rectified sine wave can be found directly using the superposition and time-shift theorems. Specifically, if $v(t)$ is the half-rectified wave shown in Figure 6.10, then the full-rectified wave in Figure 6.14*a* is

$$z(t) = v(t) + v\left(t - \frac{T_0}{2}\right)$$

and therefore

$$\begin{aligned} c_z(nf_0) &= c_v(nf_0) + c_v(nf_0)e^{-jn\Omega T_0/2} \\ &= c_v(nf_0)(1 + e^{-jn\pi}) = \begin{cases} 2c_v(nf_0) & n \text{ even} \\ 0 & n \text{ odd} \end{cases} \end{aligned}$$

Inserting Eq. (16), Sect. 6.3, for $c_v(nf_0)$ we finally obtain

$$c_z(nf_0) = \begin{cases} \dfrac{2A}{\pi(1-n^2)} & n = 0, \pm 2, \pm 4, \ldots \\ 0 & n = \pm 1, \pm 3, \ldots \end{cases} \tag{3}$$

and the amplitude spectrum is plotted in Figure 6.14*b*.

The fact that *all odd harmonics* are missing – including f_0 – agrees with the fact that the fundamental period of the full-rectified wave is actually $T_0/2$ instead of T_0. But perhaps more interesting is the observation that, except for the scale factor of 2, this spectrum differs from Figure 6.11 only by the absence of the first harmonic.

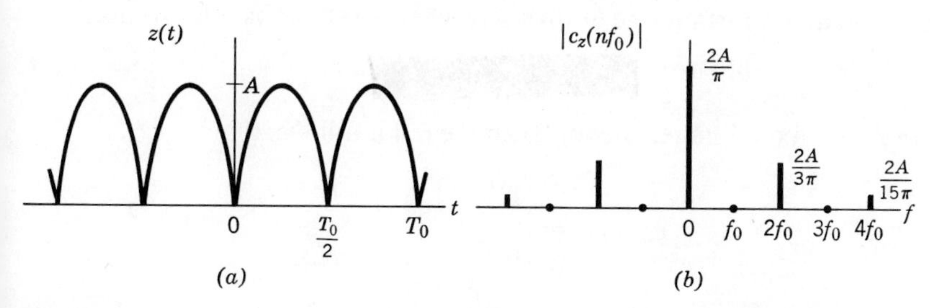

Figure 6.14. Full-rectified sine wave. (*a*) Waveform. (*b*) Amplitude spectrum.

Multiplication and modulation

Suppose that $z(t)$ is the product of two signals, $v(t)$ and $w(t)$. Its Fourier series coefficients, in terms of $c_v(nf_0)$ and $c_w(nf_0)$, are found in the following manner. Given

$$z(t)=v(t)w(t) \tag{4a}$$

we have

$$c_z(nf_0)=\frac{1}{T_0}\int_{T_0} v(t)w(t)e^{-jn\Omega t}dt$$

or, inserting the series representation for $v(t)$,

$$c_z(nf_0)=\frac{1}{T_0}\int_{T_0}\left[\sum_{m=-\infty}^{\infty} c_v(mf_0)e^{jm\Omega t}\right]w(t)e^{-jn\Omega t}dt$$

$$=\sum_{m=-\infty}^{\infty} c_v(mf_0)\underbrace{\left[\frac{1}{T_0}\int_{T_0} w(t)e^{-j(n-m)\Omega t}dt\right]}_{c_w[(n-m)f_0]}$$

where we use the different summation index m for clarity. Because the bracketed integral is just the series coefficient for $w(t)$ at frequency $(n-m)f_0$, the nth coefficient of $z(t)$ is an *infinite summation*

$$c_z(nf_0)=\sum_{m=-\infty}^{\infty} c_v(mf_0)c_w[(n-m)f_0] \tag{4b}$$

which may be recognized as a *discrete convolution*. Thus, *multiplication* in the time domain becomes *convolution* in the frequency domain.

Although (4) has conceptual value, it is not particularly useful for hand calculations unless $v(t)$ or $w(t)$ has only a few spectral lines. One such case – and an important one at that – is when $w(t)=\cos N\Omega t$ so that

$$z(t)=v(t)\cos N\Omega t \tag{5a}$$

where N is a fixed integer. From Example 6.4 it follows that

$$c_w(kf_0)=\begin{cases}\frac{1}{2} & k=\pm N\\ 0 & k\neq \pm N\end{cases}$$

and hence there are only two nonzero terms in (4b), those being for

$m=n\pm N$. Therefore, we get

$$c_z(nf_0)=\tfrac{1}{2}c_v[(n-N)f_0]+\tfrac{1}{2}c_v[(n+N)f_0] \tag{5b}$$

which says that the spectrum of $z(t)$ consists of the spectrum of $v(t)$ *translated up and down in frequency by* Nf_0 and multiplied by $\frac{1}{2}$. Because of the frequency-translation aspect and its relevance to amplitude modulation, (5) is known as the *modulation theorem.*

Average power: Parseval's theorem

The average power of a periodic signal was previously defined as

$$P=\frac{1}{T_0}\int_{T_0}|v(t)|^2dt=\frac{1}{T_0}\int_{T_0}v(t)v^*(t)\,dt \tag{6}$$

Parseval's theorem relates P to the Fourier series coefficients c_n in a very simple manner, namely

$$P=\sum_{n=-\infty}^{\infty}c_nc_n^*=\sum_{n=-\infty}^{\infty}|c_n|^2 \tag{7}$$

whose spectral interpretation is brought out by remembering that $|c_n|=|c(nf_0)|$ is the *amplitude spectrum.* Thus, the signal power is just the sum obtained by squaring and adding the heights of the amplitude lines. The fact that (7) does not involve the phase spectrum, $\arg[c(nf_0)]$, reinforces our earlier remark about the dominant role of the amplitude spectrum in determining a signal's frequency content.

Further interpretation of the theorem is afforded by assuming $v(t)$ to be *real* so c_0 is real, $|c_{-n}|=|c_n|$, and (7) becomes

$$P=c_0^2+\sum_{n=1}^{\infty}\tfrac{1}{2}|2c_n|^2 \tag{8}$$

Now, recalling the trigonometric Fourier series of Eq. (11), Sect. 6.3, we see that each sinusoidal wave of amplitude $|2c_n|$ contributes $|2c_n|^2/2$ to P. But the average power of a sinusoid having amplitude A is $A^2/2$; therefore (8) implies *superposition of average power* in that the total average power of $v(t)$ is the sum of the average powers of its sinusoidal components.

Proving Parseval's theorem is relatively routine, and will be left to the reader (Problem 6.24).

6.5 SYSTEM RESPONSE

Finally we are prepared to investigate the response of a system under *periodic steady-state conditions*. Those conditions are the same as for the sinusoidal steady state except that the input and output are arbitrary periodic power signals, rather than just sinusoids. From our discussion of the exponential Fourier series we know that the input $x(t)$ can be decomposed into a linear combination of harmonically related phasors; therefore, the output $y(t)$ will consist only of those frequency components, their amplitudes and phases being modified by $H(f)$, the frequency response of the system, evaluated at the frequencies in question. That, in words, is the method used below.

Periodic steady-state response

We assume the periodic input $x(t)$ has the representation

$$x(t) = \sum_{n=-\infty}^{\infty} c_x(nf_0)e^{j2\pi nf_0t} \qquad -\infty < t < \infty \tag{1}$$

with $c_x(nf_0)$ given by Eq. (18), Sect. 6.3. Because each input phasor $c_x(nf_0)\ \exp\,(j2\pi nf_0t)$ gives rise to the output term $H(nf_0)c_x(nf_0)$ $\exp\,(j2\pi nf_0t)$, and because the system is linear, the total response to $x(t)$ is

$$y(t) = \sum_{n=-\infty}^{\infty} c_y(nf_0)e^{j2\pi nf_0t} \qquad -\infty < t < \infty \tag{2}$$

where

$$c_y(nf_0) = H(nf_0)c_x(nf_0) \tag{3}$$

Thus, the periodic steady-state response is a periodic signal whose exponential Fourier series coefficients equal the respective coefficients of the input, each multiplied by the frequency response function evaluated at nf_0.

The spectrum of the output signal follows directly from (3) as

$$\begin{aligned} |c_y(nf_0)| &= |H(nf_0)||c_x(nf_0)| \\ \arg\,[c_y(nf_0)] &= \arg\,[c_x(nf_0)] + \arg\,[H(nf_0)] \end{aligned} \tag{4}$$

which says that the output amplitude spectrum is simply the input amplitude spectrum times the amplitude ratio while the phase spectrum equals the input phase plus the system phase shift.

To determine the average power at the output, we apply Parseval's theorem, with the result that

$$P_y = \sum_{n=-\infty}^{\infty} |c_y(nf_0)|^2 = \sum_{n=-\infty}^{\infty} |H(nf_0)|^2 |c_x(nf_0)|^2 \tag{5}$$

Again we have agreement with the frequency-domain interpretation, for the output power is found by squaring and summing the lines in the output amplitude spectrum.

When dealing with real signals, the trigonometric form of (2) may be more convenient. Specifically

$$y(t) = H(0)c_x(0) + \sum_{n=1}^{\infty} 2|H(nf_0)||c_x(nf_0)|$$
$$\times \cos\left(2\pi nf_0 t + \arg\left[c_x(nf_0)\right] + \arg\left[H(nf_0)\right]\right) \tag{6}$$

in which $c_y(nf_0)$ has been written out using (3) and (4).

In principle, (1) through (6) allow us to fully analyze the response under the imposed conditions; but in practice, unless the series has a finite number of terms or there are only a few significant lines in the output spectrum, it is difficult to obtain an accurate picture of the output waveform or a value for the output power. Thus, when this information must be found, numerical evaluation using a digital computer is required to carry out the many calculations. However, there are circumstances for which the analysis is practical to do by hand, primarily when the purpose of the system is to remove all but one or two components from the input signal. Two such problems are illustrated below.

Example 6.8 A DC power supply

Figure 6.15 shows a system whose purpose is converting AC power to DC power. The source, an AC-voltage $A \sin \Omega t$, passes through a rectifier

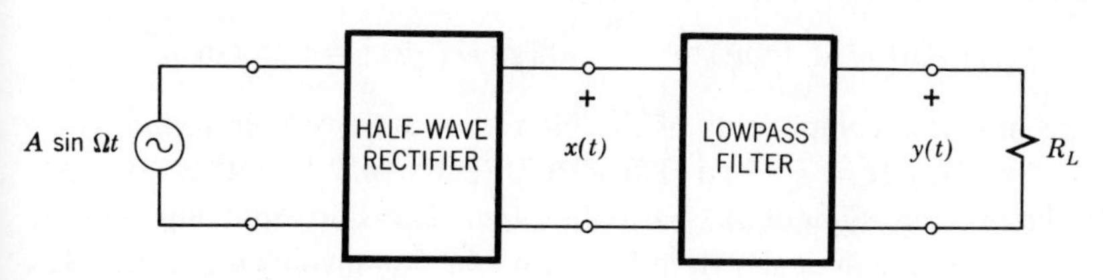

Figure 6.15. A DC power supply.

giving $x(t)$ as a half-rectified sine wave. Our task is to investigate what can be done to obtain DC power from $x(t)$.

Note that, while the source voltage has just one positive-frequency component, namely, $f_0 = \Omega/2\pi$, the rectified signal $x(t)$ contains an infinite number of frequencies, $0, f_0, 2f_0, 4f_0, \ldots$, as shown in its spectrum, Figure 6.11. This production of *new frequency components* is only possible with a *nonlinear* or *time-varying* device, such as a rectifier, since fixed linear systems have the preservation-of-form property – meaning that they can remove existing frequency components but cannot generate new ones.

And removing unwanted frequency components is precisely the next step in the system. Specifically, we want something that will keep the DC component in $x(t)$ and reject all the rest. An *ideal lowpass filter* would do just that, but ideal filters do not exist so we must be content with a real circuit such as the RC lowpass filter. Therefore, let us assume that we have a lowpass filter whose frequency response function takes the form

$$H(f) = \left[1 + j\left(\frac{f}{B}\right)\right]^{-1}$$

which could be an RC filter, but not necessarily. We further assume that this expression for $H(f)$ includes any input and output *loading* effects. For instance, if an RC filter is used, its resistance R must be small compared to the load resistance R_L so that $y(t)$ is essentially the open-circuit voltage; otherwise $H(f)$ as written above will not be valid.

Recalling that the output amplitude spectrum equals the input spectrum multiplied by $|H(f)|$, and comparing Figures 6.5 and 6.11, it follows that we want the filter's bandwidth B to be very small compared to f_0. But physical considerations prevent achieving arbitrarily small bandwidths, regardless of the specific filter circuit employed; i.e., there will always be some minimum feasible bandwidth $B_{\min}$ such that $B \geq B_{\min}$. For the sake of illustration we shall take $B = f_0/3$ so

$$|H(nf_0)| = (1+9n^2)^{-1/2} \qquad \arg\left[H(nf_0)\right] = -\arctan 3n$$

The first few components of the filtered output are then found using $c_y(nf_0) = H(nf_0)c_x(nf_0)$ and Table 6.2. Table 6.3 lists these values, introducing the symbol $A_0 \triangleq c_y(0) = A/\pi$. The corresponding output amplitude spectrum is shown in Figure 6.16*a* from which one concludes that there will be little error in neglecting those frequency components at

Table 6.3

n	$\frac{\lvert c_x(nf_0)\rvert}{A_0}$	$\lvert H(nf_0)\rvert$	$\frac{\lvert c_y(nf_0)\rvert}{A_0}$	arg $[c_x(nf_0)]$	arg $[H(nf_0)]$	arg $[c_y(nf_0)]$
0	1	1	1.000	0	0	0
± 1	$\frac{\pi}{4}$	$\frac{1}{\sqrt{10}}$	0.248	$\mp 90°$	$\mp 72°$	$\mp 162°$
± 2	$\frac{1}{3}$	$\frac{1}{\sqrt{37}}$	0.055	$\mp 180°$	$\mp 81°$	$\mp 261°$
± 4	$\frac{1}{15}$	$\frac{1}{\sqrt{145}}$	0.006	$\mp 180°$	$\mp 85°$	$\mp 265°$

$|nf_0| > 2f_0$. Hence, taking the trigonometric form (6), the output waveform consists predominantly of three terms, i.e.,

$$y(t) \approx A_0 + 0.496A_0 \cos(2\pi f_0 t - 162°) + 0.110A_0 \cos(2\pi 2f_0 t - 261°)$$

This approximation is sketched in Figure 6.16*b*, which reveals the first- and second-harmonic *ripple* on top of the DC component.

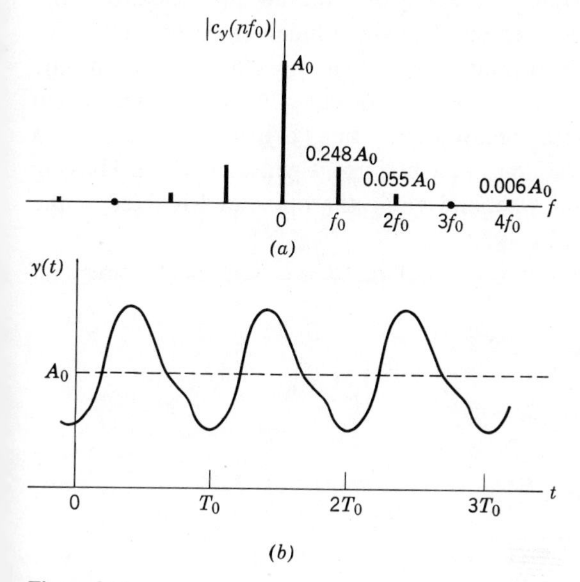

Figure 6.16.

To the extent that the three-term approximation is valid, the total output power is

$$P_y = |c_y(0)|^2 + 2|c_y(f_0)|^2 + 2|c_y(2f_0)|^2 = 1.13A_0^2$$

whereas the DC power is just $|c_y(0)|^2 = A_0^2$. As a measure of the system's performance we can take the ratio of DC to total power at the output:

$$\frac{|c_y(0)|^2}{P_y} = \frac{1}{1.13} = 88.5\%$$

However, because of the abstract nature of our definition of power, it is not meaningful to compare P_y directly with P_x. Nor can we say anything about power dissipated in the filter without specifying the exact circuit.

As a final comment, we point out that the system's efficiency could be substantially improved by using a *full-wave* rectifier since, as follows from Example 6.7, the DC component would be doubled in amplitude and the first-harmonic ripple would be absent.

Example 6.9 A mechanical vibration sensor

A certain mechanical system is shaking or vibrating periodically in the vertical direction. It is suspected that the fundamental frequency of vibration f_0 is approximately equal to f_c, but the exact nature of the vibration is unknown. The problem is to design a simple device which senses whether f_0 is above or below f_c. For this purpose a *vibrating-reed tachometer* can be used, the heart of which is a slender metal reed having a small mass or bob on the free end while the other end is fixed to the vibrating body, Figure 6.17.

Relative to the rest positions, let $x(t)$ be the vertical displacement of

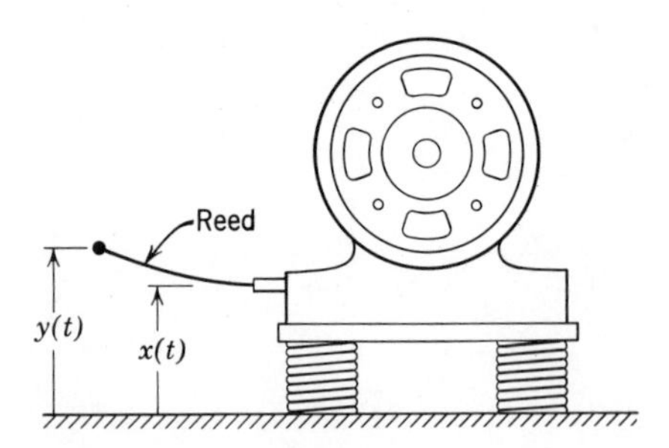

Figure 6.17. Reed attached to vibrating body.

the body and $y(t)$ the displacement of the bob. The differential equation relating the displacements takes the form

$$\ddot{y}+\frac{\omega_r}{Q}\dot{y}+\omega_r^2 y=\omega_r^2 x \qquad Q \gg 1$$

where ω_r^2 and Q are constant parameters, the former proportional to the reed's length and the latter a dimensionless quantity. Using Eq. (12), Sect. 4.5, to find $H(s)$, the displacement frequency response function is

$$H(f)=\frac{\omega_r^2}{(j\omega)^2+(\omega_r/Q)(j\omega)+\omega_r^2}$$

$$=\frac{-jQ(f_r/f)}{1+jQ\left(\dfrac{f}{f_r}-\dfrac{f_r}{f}\right)} \qquad f_r=\frac{\omega_r}{2\pi} \tag{7}$$

The corresponding amplitude ratio is plotted in Figure 6.18 for positive f, taking $Q=10$. As the figure shows, the reed exhibits *mechanical*

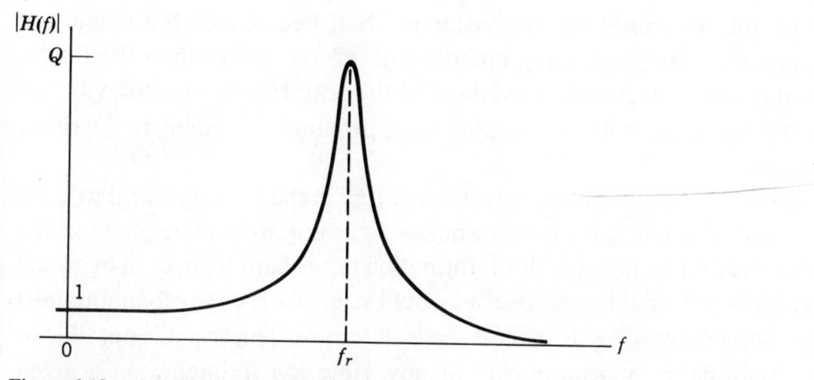

Figure 6.18. Amplitude ratio for vibrating reed.

resonance in that $|H(f)|$ has a pronounced peak at $f=f_r$, the resonant frequency; specifically

$$|H(f_r)|=Q \gg 1$$

whereas

$$H(f)\approx\begin{cases}1 & |f| \ll f_r\\ 0 & |f| \gg f_r\end{cases}$$

In this problem we do not have an exact expression for $x(t)$, but we know that it is a real periodic function of time, and hence $y(t)$ may be written in the trigonometric form

$$y(t) = \sum_{n=1}^{\infty} A_y(nf_0) \cos [2\pi n f_0 t + \theta_y(nf_0)]$$

where

$$A_y(nf_0) = 2|H(nf_0)||c_x(nf_0)|$$

$$\theta_y(nf_0) = \arg [c_x(nf_0)] + \arg [H(nf_0)]$$

and any constant term $A_y(0)$ is omitted since $y(t)$ is an incremental variable. We also know that $f_0 \approx f_c$ so if the length of the reed is such that $f_r \approx f_c$, then

$$A_y(f_0) \approx 2|H(f_r)||c_x(f_0)| = 2Q|c_x(f_0)|$$

and it is reasonable to assume that

$$A_y(nf_0) << A_y(f_0) \qquad n \neq 1$$

Putting this in words we are assuming that, because of the resonance phenomenon, any frequency components of $x(t)$ other than the fundamental produce negligible movement of the bob. Hence, the bob vibrates sinusoidally at f_0 with an amplitude approximately equal to Q times $2|c_x(f_0)|$.

Clearly, a single reed can detect when f_0 is in the vicinity of f_c but does not reveal whether f_0 is above or below f_c, as required. (The phase of the bob's vibration would give this information but is hard to discern by visual observation.) For this purpose we could use *two* reeds resonanting just above and below f_c. With several reeds of various lengths, it is possible to make reasonable measurements of any vibration frequency in a given range; such a device is known as *Frahm's tachometer*.

Problems

6.1 Suppose that $H(-j\omega) \neq H^*(j\omega)$ in Eq. (3), Sect. 6.1. Show, by finding the real and imaginary parts, that $y(t)$ will be complex.

6.2 Given $x(t)$ per Eq. (1a), Sect. 6.1, an electrical engineer usually solves for the steady-state response by writing $x(t) = \text{Re}[Xe^{j\omega t}]$, where $X = A_x e^{j\theta x}$, and finding the output phasor $Y = H(j\omega)X$. Show that $\text{Re}[Ye^{j\omega t}]$ equals $y(t)$ as given in Eq. (5).

6.3 Explain why Eq. (5), Sect. 6.1, is not applicable when: (a) the system is nonlinear or time-varying; (b) the system is not asymptotically stable.

6.4 Plot the line spectrum of

$$v(t) = -5 + 4\cos 2\pi 10t + 6\sin 2\pi 10t - 2\cos(2\pi 15t + 30°) + 3e^{j2\pi 20t}$$

6.5 Expand each of the following signals as a sum of sinusoids and plot the line spectrum: (a) $\cos^3 2\pi t$; (b) $(1 - \sin 2\pi t)\cos 2\pi 10t \cos 2\pi 15t$.

6.6 Use Eq. (12), Sect. 4.5, to find $|H(f)|$ and $\arg[H(f)]$ for the system described by $\ddot{y} + 8\pi\dot{y} + 80\pi^2 y = 40\pi\dot{x} + 80\pi^2 x$. *Answer:*

$$|H(f)| = [(1+f^2)/(1 - 0.06f^2 + 0.0025f^4)]^{1/2}$$

$$\arg[H(f)] = \arctan f - \arctan[0.2f/(1 - 0.05f^2)]$$

6.7 Write expressions for $|H(f)|$ and $\arg[H(f)]$ for $\ddot{y} + 2\zeta\omega_n\dot{y} + \omega_n^2 y = \omega_n^2 x$. Taking $\omega_n = 2\pi f_n$, plot $|H(f)|$, $0 \le f \le 2f_n$, when $\zeta = \frac{1}{4}$, $\frac{1}{2}$, and $\frac{3}{4}$. Put all plots on the same set of axes, and discuss your results.

6.8 Consider the system having $\dot{y} + a_0 y = b_1\dot{x} + b_0 x$. Show, by writing $|H(f)|^2$, that this can be either a highpass or lowpass filter, depending on the specific parameter values.

6.9 Plot $|H(f)|_{\text{db}}$ and $\arg[H(f)]$ vs. $\log f$, for: (a) $H(f) = [1 + j(f/B)]^{-2}$; (b) $H(f) = [1 + j(f/B)]^{-1}[1 + j(f/10B)]^{-1}$. Indicate any linear asymptotes.

6.10 Use Eq. (9), Sect. 6.2, to find $H(f)$ for the circuit in Figure P6.1, expressing your answer in terms of the parameter $B = 1/(2\pi\sqrt{LC})$. Plot $|H(f)|_{\text{db}}$ and $\arg[H(f)]$ vs. $\log f$, $0.1 \le f/B \le 10$, and compare with Figure 6.6. *Answer:* $H(f) = [1 - (f/B)^2 + j(f/B)]^{-1}$.

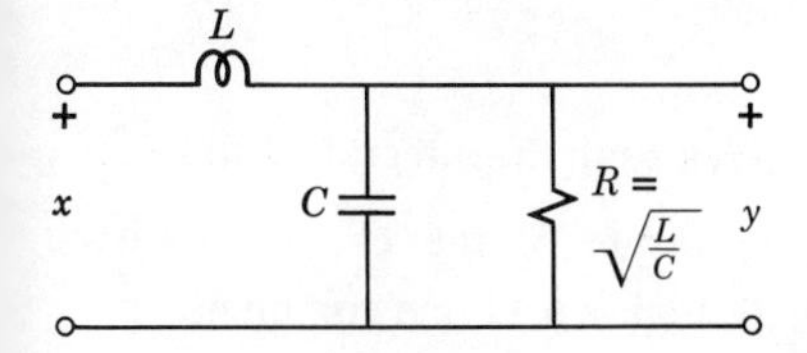

Figure P6.1.

6.11 Prove that $c_n = 0$ for all even n when $v(t)$ has half-wave symmetry, as in Eq. (9), Sect. 6.3.

6.12 In Example 6.2 we assumed that $\tau \leq T_0$. Sketch the waveform that results if $T_0 < \tau < 2T_0$ and prove, by direct calculation, that Eq. (14), Sect. 6.3., still holds.

6.13 Find c_n for the *square wave* shown in Figure P6.2, and write the trigonometric Fourier series. Show that Eq. (15b), Sect. 6.3, reduces to your result when $\tau = T_0/2$ and the constant term $c_0 = A\tau/T_0$ is removed. *Answer:* $c_n = (A/2)\ \text{sinc}\ (n/2), n$ odd.

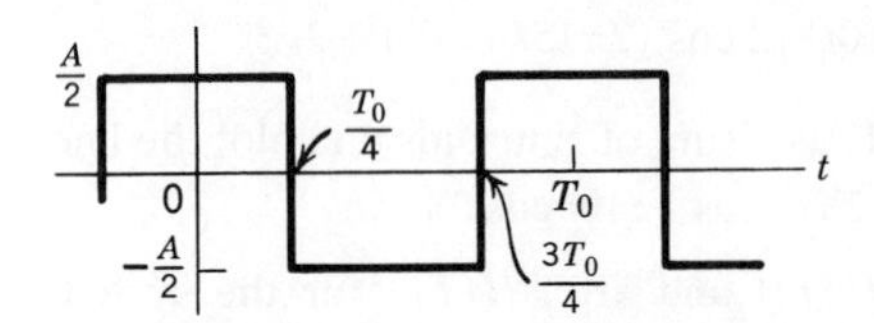

Figure P6.2.

6.14 Calculate c_n for a half-rectified *cosine* wave and compare with Example 6.3.

6.15 Consider the *clipped sinusoid* of Figure P6.3, where

$$v(t) = \begin{cases} \dfrac{A}{1-\alpha}[(\cos \Omega t) - \alpha] & |t| \leq \tau/2 \\ 0 & \tau/2 < |t| < T_0/2 \end{cases}$$

with $\alpha = \cos\ (\Omega\tau/2)$. (a) Find c_n expressed in terms of the sinc function. (b) Show that $|c_n|$ reduces to that of the half-rectified sine wave when $\tau = T_0/2$. *Answer:* $c_n = [A\tau/2(1-\alpha)T_0][\text{sinc}\ (1-n)\tau/T_0 + \text{sinc}\ (1+n)\tau/T_0 - 2\alpha\ \text{sinc}\ n\tau/T_0]$.

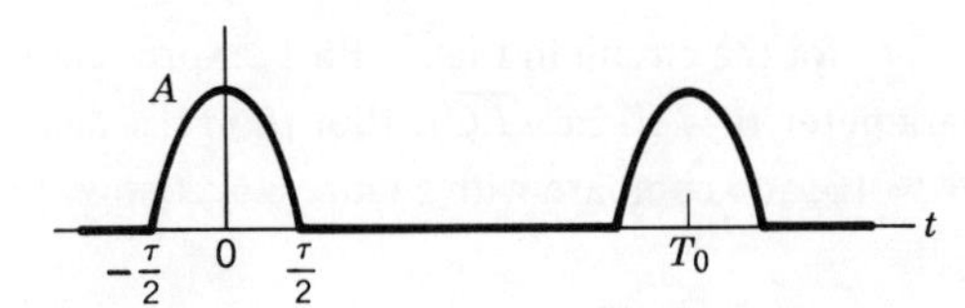

Figure P6.3.

6.16 Confirm the results of Example 6.4 by carrying out the integration for c_n.

6.17 Taking $\tau = T_0/4$, plot the amplitude spectrum of the clipped sinusoid, Problem 6.15, for $|f| \leq 8f_0$. Compare with Figure 6.11 and comment.

6.18 Plot the waveform and amplitude spectrum of a rectangular pulse train having $\tau = 6T_0/5$; see Problem 6.12. Compare with Figure 6.12*a*.

6.19 Obtain c_n for the waveform in Figure P6.2 by applying the superposition and time-shift theorems to the results of Example 6.2.

6.20 Find and plot $c_z(nf_0)$ when $z(t) = v(t) + jv(t - 0.25T_0)$, where $v(t)$ is as in Problem 6.13.

6.21 Generalize the modulation theorem, Eq. (5), Sect. 6.4, to the case of $z(t) = v(t) \cos(N\Omega t + \theta)$. In particular, show that if $z(t) = v(t) \sin N\Omega t$ then $c_z(nf_0) = -(j/2)c_v[(n-N)f_0] + (j/2)c_v[(n+N)f_0]$.

6.22 Mathematically, a full-rectified sine wave can be generated by multiplying $A \sin \Omega t$ times an appropriate square wave, similar to that in Problem 6.13. Take this approach to derive Eq. (3), Sect. 6.4.

6.23 The periodic waveform in Figure P6.4 is called a sinusoidal pulse burst train. It can be viewed as the product of $A \cos 2\pi N f_0 t$ with a rectangular pulse train. (a) Use the modulation theorem to find $c_z(nf_0)$. (b) Sketch the amplitude spectrum for $0 \leq f \leq 2Nf_0$ taking $N = 20$ and $\tau = T_0/4$.

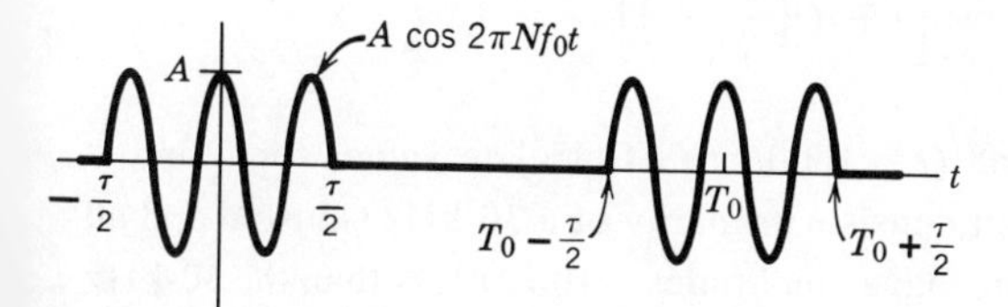

Figure P6.4.

6.24 Prove Parseval's theorem, Eq. (7), Sect. 6.4, by substituting the exponential Fourier series for $v(t)$ in $P = (1/T_0) \int_{T_0} v(t)v^*(t)dt$, leaving $v^*(t)$ as is; then interchange the order of integration and summation.

6.25 Use Eq. (6), Sect. 6.4, to calculate the average power in a square wave (Problem 6.13) and a half-rectified sine wave. Then insert values for c_n into Eq. (8), Sect. 6.4, and evaluate the sums with enough terms to convince yourself of the validity of Parseval's theorem.

6.26 In the DC power supply of Example 6.8, let the source frequency be 60 Hz and let the filter be an RC circuit with $R = 10$ ohms $\ll R_L$. (a) Find the value of C such that the amplitude of the first-harmonic ripple at the output is no greater than 10% of the DC component. (b) What is the amplitude of the second harmonic under this condition? (c) Repeat part (a) for the second-harmonic ripple assuming a full-wave rectifier is used.

6.27 Carry out the analysis of Example 6.8 assuming the filter is an RC circuit and the effect of the load resistance R_L is not negligible. Take $R = R_L/2$ and $2\pi RC = 4/f_0$ for numerical calculations.

6.28 Professor Blank needs a 1-kHz sinusoidal waveform for some tests, but he has misplaced the oscillator. As a substitute he takes a 1-kHz triangular-wave generator, having 5 ohms source resistance, and puts a capacitor C across the output terminals. It is known for triangular waves that $|c_n|$ is proportional to n^{-2}, n odd, and is zero for even n. (a) Give a brief explanation of the professor's strategy, illustrated with frequency-domain pictures. (b) Find the minimum value of C such that the power of the 1-kHz output component is 500 times as large as any other output term.

6.29 A certain frequency multiplier consists of a saturating amplifier followed by a parallel-resonant LC circuit. The amplifier is fed with a 10-kHz sinusoid, and converts this input to a square wave (Problem 6.13) at the same fundamental frequency. The resonant circuit acts as a bandpass filter, and has

$$H(f) = \left[1 + jQ\left(\frac{f}{f_r} - \frac{f_r}{f}\right)\right]^{-1}$$

where $f_r = 1/(2\pi\sqrt{LC})$ and $Q = 1000/\pi f_r L$. Calculate values for L and C such that the filtered output consists primarily of a 30-kHz sinusoid and all other frequency components have amplitudes 1/100 or less than the 30-kHz term. *Answer:* $L = 96.5\ \mu h$, $C = 0.286\ \mu f$.

6.30 In the previous problem the frequency-multiplication factor is 3. However, a much greater factor is sometimes necessary, and for that purpose a class-C amplifier, which converts sinusoids to clipped sinusoids (Problem 6.15), may be employed. Explain the advantages of the latter technique.

6.31 An ideal *quadrature phase shifter* has

$$H(f) = \begin{cases} e^{-j90^\circ} & f > 0 \\ e^{+j90^\circ} & f < 0 \end{cases}$$

If the input is a square wave (Problem 6.13), sketch an approximation for the output waveform using the first three nonzero terms of its trigonometric Fourier series.

6.32 The step response of a given system is $y_u(t) = e^{-at}u(t)$. If the input is a rectangular pulse train with $\tau = 1/a$ and $T_0 = 5\tau$, write the trigonometric Fourier series for the resulting output and plot an approximation using the first four terms. Compare this plot with the exact waveform obtained by direct calculation.

6.33 A certain system – known to be fixed, linear, and stable – is inside a black box. It is driven by a fixed-frequency square wave (Problem 6.13), and the resulting output waveform is recorded. Discuss in detail the information about the system that can be determined from this measurement, stating any intermediate calculations required and the relevant formulas.

7

Frequency-domain analysis: Fourier transforms

The frequency-domain approach for periodic steady-state analysis can be generalized to the broader problem of finding the zero-state response of a stable fixed linear system when the input signal is not restricted to being periodic. It is this problem that we shall examine here; not surprisingly, the mathematical tool involved is the Fourier transform.

The chapter begins by defining the Fourier transform for energy signals, which in turn leads to the concept of continuous spectra as distinguished from line spectra. After examining the properties of continuous spectra and some useful transform theorems, we extend the transform definition to include certain signals which are not energy signals. By doing so, one obtains a spectral representation encompassing a very large class of time functions.

We then reconsider the frequency response function $H(f)$ and show that it is the Fourier transform of the impulse response $h(t)$. This fact, together with the transform of the superposition integral, means that many systems analysis problems can be carried out directly in the frequency domain. Illustrative applications of the method are given, drawn from the area of signal transmission.

7.1 FOURIER TRANSFORMS AND CONTINUOUS SPECTRA

It was stated in the last chapter that a *periodic* time function having well-defined average power could be expanded as the exponential Fourier series

$$v(t) = \sum_{n=-\infty}^{\infty} \overbrace{\left[\frac{1}{T_0}\int_{T_0} v(t)e^{-j2\pi nf_0t}\,dt\right]}^{c_v(nf_0)} e^{j2\pi nf_0t} \tag{1}$$

where $f_0 = 1/T_0$ is the fundamental frequency. A nonperiodic signal, not having an identifiable fundamental frequency, cannot be expressed in quite the same way. There is, however, a representation similar to (1) for the nonperiodic case given by the *Fourier integral theorem*

$$v(t) = \int_{-\infty}^{\infty} \left[\int_{-\infty}^{\infty} v(t) e^{-j2\pi ft} dt \right] e^{j2\pi ft} df \tag{2}$$

This representation is valid for any time function having well-defined energy E, i.e., the class we designate as *nonperiodic energy signals*.

Comparing (1) and (2) shows they are of essentially the same form, one being a discrete summation over the index nf_0, the other an integration or "continuous summation" over the variable f. Previously, we interpreted the Fourier series as decomposing the signal into a sum of harmonically related components at the discrete frequencies nf_0. This, in turn, led to the frequency-domain picture of a line spectrum described by $c_v(nf_0)$. By parallel reasoning, the Fourier integral theorem suggests a continuous spectrum in the frequency domain, that spectrum found from the Fourier transform of the signal.

Fourier transforms

The bracketed quantity in (2) is the *Fourier transform* of $v(t)$, denoted by

$$V(f) = \mathcal{F}[v(t)] \triangleq \int_{-\infty}^{\infty} v(t) e^{-j2\pi ft} dt \tag{3}$$

an integration over all time. The theorem (2) then states that the *inverse Fourier transform* yields the original signal by an integration over all frequency, i.e.,

$$v(t) = \mathcal{F}^{-1}[V(f)] \triangleq \int_{-\infty}^{\infty} V(f) e^{j2\pi ft} df \tag{4}$$

providing the signal energy is well defined. Taken together, these two expressions are referred to as the Fourier integrals or the *Fourier transform pair.*†

Interpreting (4), the Fourier integral theorem analyzes a nonperiodic energy signal as an integration over a continuous distribution of frequency components. The frequency distribution function or weighting function is

†Slightly different definitions result when $\omega = 2\pi f$ is taken as the independent variable of the frequency domain, in which case factors of $1/2\pi$ or $1/\sqrt{2\pi}$ will appear in (3) and (4).

$V(f)$ which, we note from (3), has a definition highly reminiscent of $c_v(nf_0)$. Indeed, $V(f)$ plays the same role in (4) that $c_v(nf_0)$ does in the exponential Fourier series. The difference between the two cases, at once both small and yet significant, is that a periodic signal is deemed to be constructed by *summing* discrete frequency components, while a nonperiodic signal entails *integrating* a continuous frequency function.† Therefore, the Fourier transform of a nonperiodic energy signal can be taken as its frequency-domain representation. But, whereas $c_v(nf_0)$ is a discrete function, $V(f)$ generally is continuous; hence, the resulting frequency-domain picture is a *continuous spectrum*.

The fact that $V(f) = \mathcal{F}[v(t)]$ and $\mathcal{F}^{-1}[V(f)] = v(t)$ form a complete circle of transformations may seem puzzling at first. To clarify that matter, in a given problem we will know either the signal or its spectrum and desire to find the other by transformation. Thus, when finding $V(f)$ from $v(t)$, one goes from time domain to the frequency domain via (3), eliminating the time variable by integrating with respect to t; and vice versa when finding $v(t)$ from $V(f)$. With this in mind there should be no question about the correct variable of integration, t or f.

Properties of continuous spectra

Just as there is a basic physical distinction between periodic power signals and nonperiodic energy signals, there is a basic distinction between their spectra: one is discrete and the other is continuous. Nonetheless, the expressions for $V(f)$ and $c_v(nf_0)$ are nearly identical operations. Having made this observation, a number of important properties of continuous spectra drop out almost immediately based on our prior study of line spectra.

To begin with, $V(f)$ in general is a *complex* function of frequency so that, writing

$$V(f) = |V(f)| e^{j \arg [V(f)]}$$

we get the *amplitude spectrum* $|V(f)|$ and the *phase spectrum* $\arg [V(f)]$. If the signal in question is *real*, then $V(f)$ has the hermitian symmetry

$$V(-f) = V^*(f) \tag{5}$$

†In this section we shall use the term "continuous function" to denote a function of a continuous variable, as distinguished from a function of a discrete variable.

as easily shown from (3). Thus, as before, the amplitude spectrum of a real signal has even symmetry and the phase spectrum has odd symmetry.

When the *signal* has symmetry in the time domain, it follows that

$$V(f) = 2\int_0^{\infty} v(t)\cos\omega t\,dt \qquad v(t) \text{ even} \tag{6a}$$

$$V(f) = -j2\int_0^{\infty} v(t)\sin\omega t\,dt \qquad v(t) \text{ odd} \tag{6b}$$

where

$$\omega = 2\pi f$$

Incidentally, hereafter we shall often write ω instead of $2\pi f$ for compactness. This practice should cause no confusion providing the reader remembers that $\omega = 2\pi f$, particularly when doing integrations with respect to f.

As the final basic property, the value of the spectrum at zero frequency is simply the *total area* under the time function, i.e., setting $f = 0$ in (3) gives

$$V(0) = \int_{-\infty}^{\infty} v(t)\,dt \tag{7}$$

This property compares with the DC component of a line spectrum, $c_v(0)$, which equals the *time average* of $v(t)$.

Example 7.1 Spectrum of a rectangular pulse

In Chapter 6 we obtained the line spectrum for a periodic train of rectangular pulses. Thus, it is interesting to consider here the continuous spectrum of a single rectangular pulse. As a notational simplification we shall introduce the *unit rectangle function* $\Pi(t)$, Figure 7.1*a*, defined as

$$\Pi(t) \triangleq u(t+\tfrac{1}{2}) - u(t-\tfrac{1}{2}) = \begin{cases} 1 & |t| < \tfrac{1}{2} \\ 0 & |t| > \tfrac{1}{2} \end{cases} \tag{8}$$

with $\Pi(t)$ at $|t| = \frac{1}{2}$ being unspecified, similar to our convention about the unit step. In terms of $\Pi(t)$ a pulse with amplitude A and duration τ is written $v(t) = A\Pi(t/\tau)$, Figure 7.1*b*.

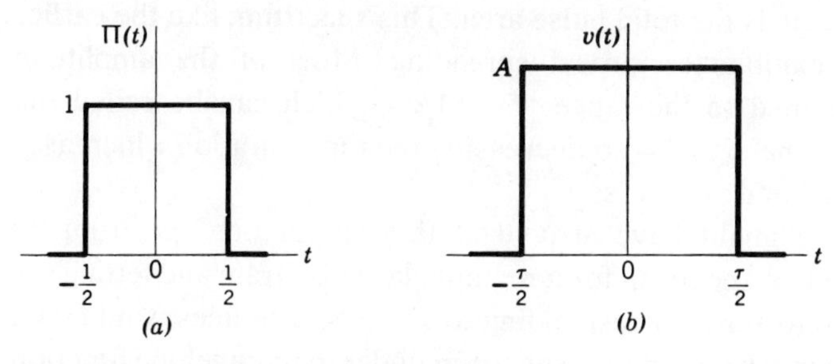

Figure 7.1. (*a*) The unit rectangle function $\Pi(t)$. (*b*) A rectangular pulse $v(t) = A\Pi(t/\tau)$.

From the even symmetry of $v(t)$, its Fourier transform is

$$V(f) = 2\int_0^{\infty} v(t)\cos\omega t\,dt$$

$$= 2\int_0^{\tau/2} A\cos\omega t\,dt = \frac{2A}{\omega}\sin\frac{\omega\tau}{2}$$

$$= A\tau\,\mathrm{sinc}\,f\tau \tag{9}$$

where we have used the sinc function $\mathrm{sinc}\,z \triangleq (\sin\pi z)/\pi z$.

The amplitude and phase spectra based on (9) are plotted in Figure 7.2, clearly showing the expected frequency-domain symmetry and the fact

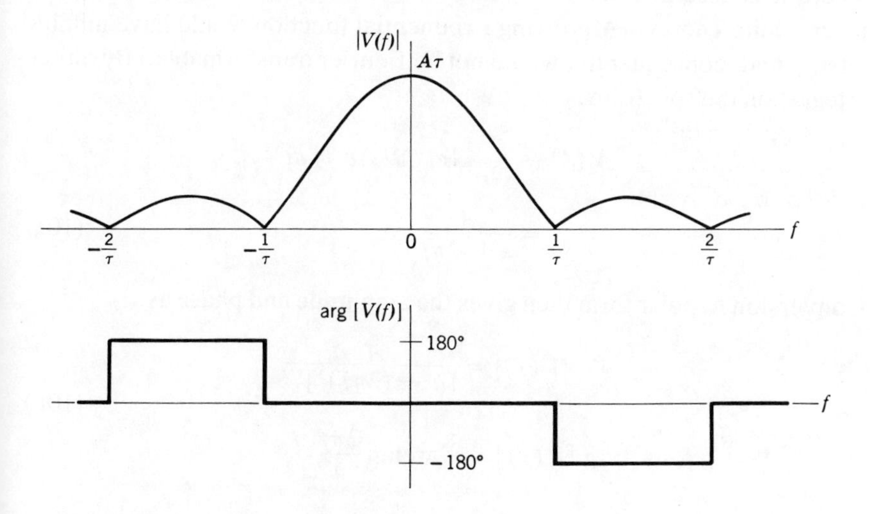

Figure 7.2. Spectrum of a rectangular pulse.

that $V(0) = A\tau$ equals the total pulse area. This spectrum, like the earlier line spectrum, exhibits reciprocal spreading. Most of the amplitude spectrum is confined to the range $|f| < 1/\tau$ – which can be called the *spectral width* of the signal – so decreasing the time duration τ increases the spectral width, and vice versa.

Intuitively, one might have arrived at this continuous spectrum by starting with the line spectrum for a rectangular pulse train and letting the line spacing f_0 go to zero, corresponding to $T_0 \to \infty$. The lines would then appear to merge into a continuous spectrum under their envelope function sinc $f\tau$. In fact, the reader has probably seen the Fourier transform "derived" from the Fourier series by just such a limiting procedure. We have avoided that approach because it tends to obscure the fundamental difference between the two types of spectra. Moreover, as discussed later on, it is preferable to view periodic signals as a special subclass of non-periodic signals, rather than the reverse.

Example 7.2 Spectrum of a decaying exponential

Figure 7.3*a* shows the decaying exponential time function

$$v(t) = Ae^{-at}u(t) = \begin{cases} Ae^{-at} & t > 0 \\ 0 & t < 0 \end{cases}$$

where it is assumed that $a > 0$ and we set $v(t) = 0$ for negative time to insure finite energy. (A growing exponential function would have infinite energy and, consequently, would not be Fourier transformable.) By direct integration the spectrum is

$$\begin{aligned} V(f) &= \int_{-\infty}^{\infty} Ae^{-at}u(t)e^{-j\omega t}\,dt \\ &= \frac{A}{a + j2\pi f} \end{aligned} \tag{10a}$$

Conversion to polar form then gives the amplitude and phase as

$$\begin{aligned} |V(f)| &= \frac{A}{[a^2 + (2\pi f)^2]^{1/2}} \\ \arg[V(f)] &= -\arctan\frac{2\pi f}{a} \end{aligned} \tag{10b}$$

which are sketched in Figure 7.3*b*.

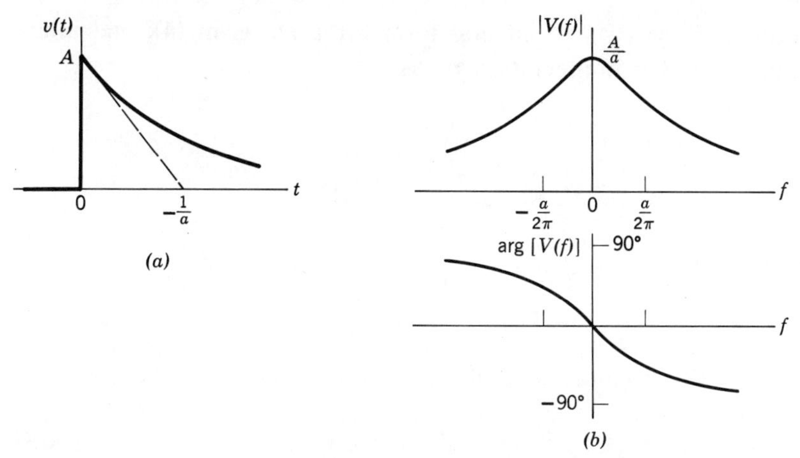

Figure 7.3. A decaying exponential. (*a*) Waveform. (*b*) Spectrum.

Hopefully, the reader has detected a similarity between this example and the RC lowpass filter studied in Chapter 6. To bring out the similarity let $A = a = 1/RC$ so that $v(t) = (1/RC)e^{-t/RC}u(t)$, which should be recognized as the RC filter's impulse response $h(t)$. Furthermore, under these conditions the spectrum becomes $V(f) = [1+j(f/B)]^{-1}$ with $B = 1/2\pi RC$, identical to the filter's frequency response function $H(f)$. Now, by definition, $V(f) = \mathfrak{F}[v(t)]$ and apparently $H(f) = \mathfrak{F}[h(t)]$ for the RC filter. The fact that the frequency response function *always* equals the Fourier transform of the impulse response is confirmed by returning to our definition of $H(f)$, namely

$$H(f) \triangleq \int_{-\infty}^{\infty} h(t)e^{-j2\pi ft}dt$$

so comparison with (3) shows that $H(f) = \mathfrak{F}[h(t)]$ in general. This important result forms the basis of frequency-domain system analysis, as developed in Section 7.4.

Energy spectra and Rayleigh's theorem

We can further enhance the physical meaning of the Fourier transform by relating it to the total signal energy †

$$E = \int_{-\infty}^{\infty} |v(t)|^2 dt = \int_{-\infty}^{\infty} v(t)v^*(t)\,dt \tag{11}$$

†For convenience, the limiting notation used in the original definition, Eq. (2), Sect. 5.1, has been replaced here by infinite limits.

Substituting $\mathcal{F}^{-1}[V(f)]$ in integral form for $v(t)$, as in (4), and interchanging the order of integration yields

$$E = \int_{-\infty}^{\infty} \overbrace{\left[\int_{-\infty}^{\infty} V(f) e^{j\omega t} df\right]}^{v(t)} v^*(t)\, dt$$

$$= \int_{-\infty}^{\infty} V(f) \underbrace{\left[\int_{-\infty}^{\infty} v^*(t) e^{j\omega t} dt\right]}_{V^*(f)} df$$

where, from (3), the inner integral is recognized to be $V^*(f)$. Hence

$$E = \int_{-\infty}^{\infty} V(f) V^*(f)\, df = \int_{-\infty}^{\infty} |V(f)|^2 df \tag{12}$$

which says that the signal energy can be found by integrating $|V(f)|^2$ over all frequency. Alternately, combining (11) and (12) gives

$$E = \int_{-\infty}^{\infty} |v(t)|^2 dt = \int_{-\infty}^{\infty} |V(f)|^2 df \tag{13}$$

This relationship, sometimes called *Rayleigh's energy theorem*, is obviously the analog to *Parseval's power theorem*.

But what is the interpretation of $|V(f)|^2$ itself in the frequency domain? Since $V(f)$ is the spectrum of $v(t)$ and since the total area under $|V(f)|^2$ equals the signal energy E, it would appear that $|V(f)|^2$ gives the energy distribution in frequency, and can thus be termed the *energy spectral density* or the energy spectrum. In support of this notion, suppose we construct a new spectrum $V_\Delta(f)$ by taking a narrow band of frequency components from $V(f)$ as in Figure 7.4. The corresponding time function is

$$v_\Delta(t) = \mathcal{F}^{-1}[V_\Delta(f)] = \int_{f_c-\Delta/2}^{f_c+\Delta/2} V(f) e^{j\omega t} df$$

with total energy

$$E_\Delta = \int_{-\infty}^{\infty} |V_\Delta(f)|^2 df = \int_{f_c-\Delta/2}^{f_c+\Delta/2} |V(f)|^2 df$$

But with Δ being sufficiently small, such that $V(f)$ is essentially constant and equal to $V(f_c)$ over the band, then $E_\Delta \approx |V(f_c)|^2 \Delta$ and hence

$$|V(f_c)|^2 = \frac{E_\Delta}{\Delta}$$

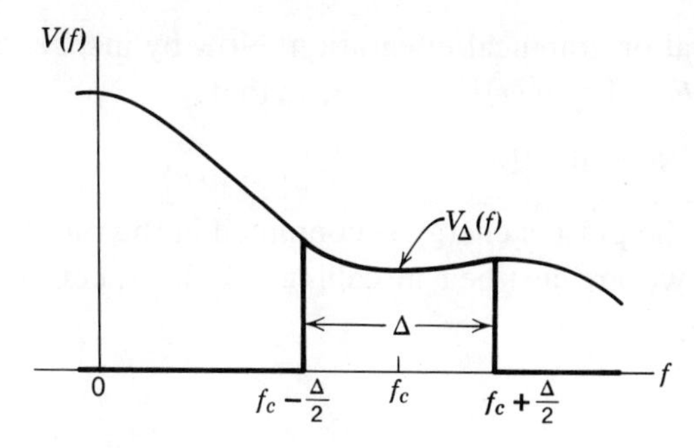

Figure 7.4.

whose dimensions are energy per unit frequency. Because one can perform this analysis for any f_c and with arbitrarily small Δ, we see that $|V(f)|^2$ can be interpreted as giving the energy density of $v(t)$ in the frequency domain.

Example 7.3 Energy spectrum of a rectangular pulse

Drawing upon (9), the energy spectral density of a rectangular pulse is

$$|V(f)|^2 = (A\tau)^2 \operatorname{sinc}^2 f\tau$$

as sketched in Figure 7.5. With this information we can quantitatively test our earlier statement that $1/\tau$ is the spectral width of this signal, meaning that the band $|f| \leq 1/\tau$ contains most of the signal's spectrum.

For this purpose let E_τ denote the energy within the frequency range $|f| \leq 1/\tau$, i.e.,

$$E_\tau = \int_{-1/\tau}^{1/\tau} (A\tau)^2 \operatorname{sinc}^2 f\tau \, df = 0.92A^2\tau$$

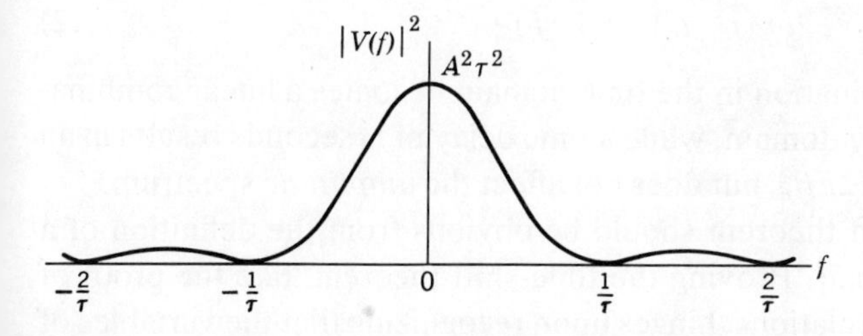

Figure 7.5. Energy spectrum of a rectangular pulse.

a result that requires numerical or graphical integration. Now by inspection of $v(t)$ the *total* energy is $E = \int_{-\infty}^{\infty} |v(t)|^2 dt = A^2\tau$, so that

$$E_\tau = 0.92E$$

Therefore, more than 90% of the pulse's energy is contained in the range $|f| \leq 1/\tau$ and, in this sense, we are justified in calling $1/\tau$ the spectral width.

7.2 TRANSFORM THEOREMS

Covered in this section are several theorems pertaining to Fourier transforms, many of which are direct parallels to those in Section 6.4. Like the Fourier series theorems, these theorems have importance as an aid to spectral analysis and interpretation of the frequency domain. Additionally, they often can be invoked to find new transforms without resorting to brute-force integration.

As a succinct notation we shall denote a Fourier transform pair by

$$v(t) \leftrightarrow V(f)$$

meaning that $V(f) = \mathcal{F}[v(t)]$ and $v(t) = \mathcal{F}^{-1}[V(f)]$, and similarly for $w(t) \leftrightarrow W(f)$, etc. It is also assumed below that all time functions are Fourier transformable.

Superposition and time shift

Completely analogous to the superposition and time-shift theorems of Fourier series, in the case of Fourier transforms we have

$$\alpha v(t) + \beta w(t) \leftrightarrow \alpha V(f) + \beta W(f) \tag{1}$$

$$v(t - t_d) \leftrightarrow V(f) e^{-j\omega t_d} \tag{2}$$

Thus, a linear combination in the time domain becomes a linear combination in the frequency domain, while a time delay of t_d seconds results in an added *phase* term, $-2\pi f t_d$, but does not affect the *amplitude* spectrum.

The superposition theorem should be obvious from the definition of a Fourier transform pair. Proving the time-shift theorem, like the proof of so many transform relations, hinges upon recognizing that the variables of integration in the transform integrals are dummy variables. Thus,

$$\mathfrak{F}[v(t-t_d)] = \int_{-\infty}^{\infty} v(t-t_d)e^{-j\omega t}dt$$

so making the change of variable $\lambda = t - t_d$ gives

$$\mathfrak{F}[v(t-t_d)] = \int_{-\infty}^{\infty} v(\lambda)e^{-j\omega(\lambda+t_d)}d\lambda$$

$$= \underbrace{\left[\int_{-\infty}^{\infty} v(\lambda)e^{-j\omega\lambda}d\lambda\right]}_{V(f)}e^{-j\omega t_d}$$

and (2) thereby follows.

Example 7.4

Consider the pair of rectangular pulses shifted $\pm t_d$ from the origin, Figure 7.6. With the unit-rectangle notation, this signal is

$$z(t) = A\Pi\left(\frac{t-t_d}{\tau}\right) + A\Pi\left(\frac{t+t_d}{\tau}\right)$$

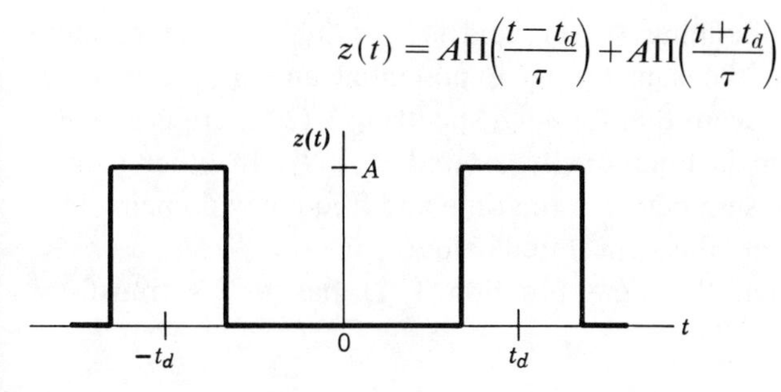

Figure 7.6.

Rather than directly transforming $z(t)$ we can take advantage of the known transform pair from Example 7.1,

$$A\Pi\left(\frac{t}{\tau}\right) \leftrightarrow A\tau \operatorname{sinc} f\tau$$

Then, combining the superposition and time-shift theorems one immediately gets

$$Z(f) = A\tau \operatorname{sinc} f\tau\, e^{-j\omega t_d} + A\tau \operatorname{sinc} f\tau\, e^{+j\omega t_d}$$

$$= 2A\tau \operatorname{sinc} f\tau \cos 2\pi f t_d$$

Plotting this spectrum is left for the reader.

Time scaling

Time shifting is a translation of the time origin. Time *scaling* – e.g., $v(ct)$ with c being a constant – is a *compression* or *expansion* of the shape of $v(t)$, depending on whether $|c| > 1$ or $|c| < 1$. If c is negative, then $v(ct)$ is the waveform $v(t)$ *reversed* in time as well as being expanded or compressed. The time scaling theorem is

$$v(ct) \leftrightarrow \frac{1}{|c|} V\left(\frac{f}{c}\right) \qquad c \neq 0 \tag{3}$$

which formalizes the reciprocal-spreading effect, i.e., if a signal is compressed in time by a factor of c its spectrum is expanded in frequency by $1/c$, and vice versa. Proving the theorem is another exercise in change-of-variables.

Duality

Because the integrals defining $\mathcal{F}[v(t)]$ and $\mathcal{F}^{-1}[V(f)]$ are so much alike – they differ only in the sign of the exponential and the variable of integration – it would seem that for each spectrum $V(f)$ there is a *signal* $V(t)$ whose transform is functionally related to $v(t)$. In other words, there is a dual relationship between the time and frequency domain. This relationship is the duality theorem stated below.

If $v(t) \leftrightarrow V(f)$ then the time function $V(t)$ has as its transform $v(-f)$, that is

$$V(t) \leftrightarrow v(-f) \tag{4}$$

Observe in particular that when $v(t)$ is both real and even, then $V(f)$ is real and even and hence $V(t) \leftrightarrow v(f)$.

The duality theorem is proved by letting $V(t)$ be a signal, so its transform is

$$\mathcal{F}[V(t)] = \int_{-\infty}^{\infty} V(t) e^{-j\omega t} dt = \int_{-\infty}^{\infty} V(\lambda) e^{j2\pi(-f)\lambda} d\lambda$$

But if $v(t)$ is a signal with transform $V(f)$, then

$$v(t) = \mathcal{F}^{-1}[V(f)] = \int_{-\infty}^{\infty} V(f) e^{j\omega t} df = \int_{-\infty}^{\infty} V(\lambda) e^{j2\pi t\lambda} d\lambda$$

Comparing the two integrals shows that $\mathcal{F}[V(t)]$ equals $v(t)$ with t replaced by $-f$, and (4) thereby results.

Example 7.5 The sinc pulse

To illustrate the duality theorem and show how it can be used to generate new transform pairs, consider the known pair

$$A\Pi\left(\frac{t}{\tau}\right) \leftrightarrow A\tau \operatorname{sinc} f\tau$$

Because both the rectangle and sinc functions are even functions, (4) gives us $A\tau \operatorname{sinc} t\tau \leftrightarrow A\Pi(f/\tau)$ or, replacing τ by $2W$ to clarify notation and dividing both sides by $2W$,

$$A \operatorname{sinc} 2Wt \leftrightarrow \frac{A}{2W}\Pi\left(\frac{f}{2W}\right) \tag{5}$$

which are sketched in Figure 7.7.

Although the notion of a "sinc pulse" in the time domain seems perhaps a bit contrived, it actually plays a major role in the theory of sampled signals and digital data transmission. Specifically, we point out that the spectrum is *identically zero for* $|f| \geq W$ and, for the first time, we have run into a spectrum that covers only a finite portion of the frequency domain. Signals having spectra like Figure 7.7*b* are said to be *band-limited* in W and, without qualification, W is the signal's *bandwidth*.

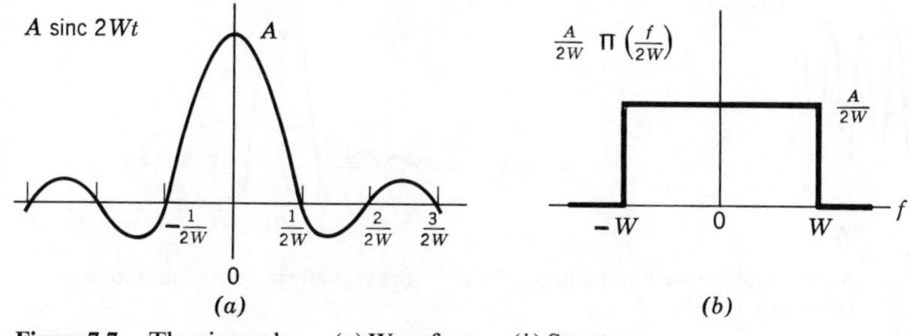

Figure 7.7. The sinc pulse. (*a*) Waveform. (*b*) Spectrum.

Frequency translation and modulation

Theorems, like transform pairs, are subject to duality. Thus, as the dual of time shifting, there is the *frequency-shift* or *frequency-translation* theorem

$$v(t)e^{j\omega_c t} \leftrightarrow V(f-f_c) \tag{6}$$

where $\omega_c = 2\pi f_c$. In words, multiplying a signal by the phasor $e^{j\omega_c t}$ results in a translation of its spectrum by f_c hertz.

A corollary to (6), known as the *modulation theorem*, is

$$v(t)\cos\omega_c t \leftrightarrow \tfrac{1}{2}[V(f-f_c)+V(f+f_c)] \tag{7}$$

which says that multiplying a signal by a sinusoid causes its spectrum to be split into two identical and equal parts, these parts being translated up and down in the frequency domain by f_c hertz. Equation (7) can be simply derived from (6) by writing $\cos\omega_c t = \frac{1}{2}(e^{j\omega_c t}+e^{-j\omega_c t})$ and using superposition. Proof of the basic frequency-shift theorem is left as an exercise.

Example 7.6 A sinusoidal pulse burst

Figure 7.8*a* shows a finite-duration sinusoid called a *sinusoidal pulse burst*; it is written

$$w(t) = A\Pi(t/\tau)\cos\omega_c t \tag{8a}$$

Applying the modulation theorem with $v(t) = A\Pi(t/\tau)$ yields

$$W(f) = \frac{A\tau}{2}[\text{sinc}\,(f-f_c)\tau + \text{sinc}\,(f+f_c)\tau] \tag{8b}$$

whose amplitude spectrum is sketched in Figure 7.8*b* (omitting negative f) for the case of $f_c >> 1/\tau$.

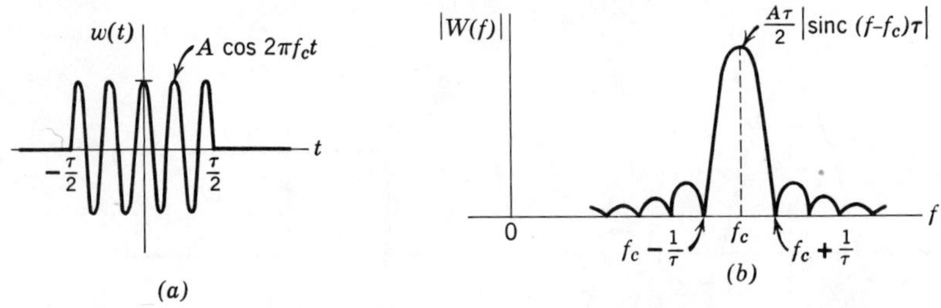

Figure 7.8. A sinusoidal pulse burst with $f_c >> 1/\tau$. (*a*) Waveform. (*b*) Amplitude spectrum, $f \geq 0$.

This example affords a good opportunity for understanding one difference between line spectra and continuous spectra. Had we been dealing with a sinusoid existing for all time, say $v(t) = A\cos\omega_c t$ for $-\infty < t < \infty$, then the frequency-domain representation would have a single amplitude line of height $A/2$ at $f = f_c$ (and, of course, its negative-frequency counterpart). For the finite-duration sinusoid there is a continuous spread of frequency components centered at $f = f_c$, where the maximum value is $|W(f_c)| = A\tau/2$, and the spectral width is $2/\tau$ measured in positive frequency. In this case, as contrasted with the eternal sinusoid, the signal is zero for $|t| > \tau/2$; therefore, other frequency components besides $f = f_c$ are necessary to provide total cancellation outside the time interval $|t| < \tau/2$.

Differentiation and integration

When an energy signal $v(t)$ is differentiated or integrated with respect to time and the result is also an energy signal, then the spectrum of the resultant is related to $V(f)$ by the theorems below. One word of warning, however: the theorems do not guarantee that the resulting signal is Fourier transformable; they merely give the spectrum when it is valid. Thus, erroneous conclusions sometimes result if the theorems are used indiscriminately without checking the nature of the integrated or differentiated signal.

Subject to the above restriction, the Fourier transform of $dv(t)/dt$ equals $(j2\pi f)V(f)$, and by iteration for the nth derivative,

$$v^{(n)}(t) \leftrightarrow (j2\pi f)^n V(f) \tag{9}$$

which is the general form of the differentiation theorem. Going in the opposite direction, let $v^{(-n)}(t)$ stand for the nth integral of $v(t)$, i.e., $v^{(-1)}(t) \triangleq \int_{-\infty}^{t} v(\lambda)\,d\lambda$, etc. The integration theorem then states that

$$v^{(-n)}(t) \leftrightarrow (j2\pi f)^{-n} V(f) \tag{10}$$

providing $V(0) = 0$, which is necessary (but not always sufficient) to insure that the integration yields an energy signal.

To interpret these theorems, recall that differentiating a signal tends to accentuate its time variations while integrating tends to smooth them. Correspondingly, in the frequency domain we see that high-frequency components (and hence rapid time variations) are enhanced by differentiation and diminished by integration, because the original spectrum $V(f)$ is multiplied or divided by f^n, respectively.

To prove (9) we write $v(t)$ in terms of $\mathcal{F}^{-1}[V(f)]$ and differentiate with respect to time:

$$\begin{aligned} \frac{d^n v(t)}{dt^n} &= \frac{d^n}{dt^n}\left[\int_{-\infty}^{\infty} V(f)e^{j\omega t}\,df\right] \\ &= \int_{-\infty}^{\infty} V(f)\frac{d^n}{dt^n}(e^{j\omega t})\,df \\ &= \int_{-\infty}^{\infty} [V(f)(j\omega)^n]e^{j\omega t}\,df \end{aligned}$$

Hence, from the definition of the inverse transform, we identify $V(f)(j\omega)^n$ as the transform of $V^{(n)}(t)$. Proof of (10) follows similar lines.

Convolution and multiplication

Second only to the Fourier integral itself, the convolution theorem is at the heart of frequency-domain system analysis. It has two facets which, stated together, are

$$v * w(t) \leftrightarrow V(f)W(f) \tag{11}$$

$$v(t)w(t) \leftrightarrow V * W(f) \tag{12}$$

In other words, *convolution in the time domain becomes multiplication in the frequency domain,* while *multiplication in the time domain becomes convolution in the frequency domain.* We shall prove (12) and appeal to duality for the justification of (11).

Transforming the product $v(t)w(t)$ and replacing $v(t)$ by its inversion integral gives

$$\mathcal{F}[v(t)w(t)] = \int_{-\infty}^{\infty} \overbrace{\left[\int_{-\infty}^{\infty} V(\lambda)e^{j2\pi\lambda t}\,d\lambda\right]}^{v(t)} w(t)e^{-j2\pi ft}\,dt$$

$$= \int_{-\infty}^{\infty} V(\lambda) \underbrace{\left[\int_{-\infty}^{\infty} w(t)e^{-j2\pi(f-\lambda)t}\,dt\right]}_{W(f-\lambda)} d\lambda$$

$$= \int_{-\infty}^{\infty} V(\lambda)W(f-\lambda)\,d\lambda = V * W(f)$$

which completes the proof.

Example 7.7 Sinc-squared and triangular pulses

An obvious consequence of (12) is that $\mathcal{F}[v^2(t)] = V * V(f)$. Thus, consider the sinc pulse of Example 7.5, whose spectrum was found to be a rectangular frequency function. Now, as mentioned in Section 4.1, a rectangular function convolved with itself yields a triangular function of twice the original width, so the spectrum of a sinc-squared pulse is a triangular frequency function. Leaving the details to the reader, we have the new transform pair

$$A \operatorname{sinc}^2 2Wt \leftrightarrow \frac{A}{2W}\Lambda\left(\frac{f}{2W}\right) \tag{13}$$

where $\Lambda(f/2W)$ is the *triangle function*

$$\Lambda\left(\frac{f}{2W}\right) \triangleq \begin{cases} 1 - \dfrac{|f|}{2W} & |f| \leq 2W \\ 0 & |f| > 2W \end{cases} \tag{14}$$

On the other hand, convolving a rectangular time function with itself yields a *triangular pulse*; so, using (11),

$$A\Lambda\left(\frac{t}{\tau}\right) \leftrightarrow A\tau \operatorname{sinc}^2 f\tau \tag{15}$$

Notice that (15) also could have been obtained by applying duality to (13).

7.3 IMPULSES IN TIME AND FREQUENCY

Classical Fourier transform theory considers only energy-type signals, that is, time functions for which $\int_{-\infty}^{\infty} |v(t)|^2 dt$ is finite and well defined. However, because certain signals of interest and importance in systems analysis do not fall in this category, it becomes desirable to extend the theory beyond energy signals.

Specifically, we are thinking of the functions $v(t) = A$, a constant for all time, and $v(t) = \delta(t)$, the unit impulse. Although neither of these correspond exactly to physical signals, they are certainly useful idealizations. And both do have frequency-domain representations as "continuous" spectra, at least in a limiting sense – i.e., in the sense of distribution theory. However, when dealing with the transforms of such nonenergy signals, the interpretation of $|V(f)|^2$ as an energy spectral density necessarily must be abandoned.

Impulses in the frequency domain

To find the spectrum of $v(t) = A$, which is not an energy signal, we shall start with an energy signal which, under a limiting operation, becomes a constant A for all time. In particular, referring to Figure 7.7, take the function A sinc $2Wt$ and let W go to zero so that

$$\lim_{W\to 0} A \operatorname{sinc} 2Wt = A \qquad -\infty < t < \infty$$

Now, since $\mathcal{F}[A \operatorname{sinc} 2Wt] = (A/2W)\Pi(f/2W)$, it seems plausible that

$$\mathcal{F}[A] = \lim_{W\to 0} \frac{A}{2W} \Pi\left(\frac{f}{2W}\right)$$

But, as covered in Section 4.2 for time-domain impulses, if

$$\delta_W(f) \triangleq \frac{1}{2W}\Pi\left(\frac{f}{2W}\right) = \begin{cases} \dfrac{1}{2W} & |f| < W \\ 0 & |f| > W \end{cases}$$

then $\delta_W(f)$ has the properties required to write $\lim_{W\to 0} \delta_W(f) = \delta(f)$, where $\delta(f)$ is the unit impulse in the frequency domain. Therefore, for the case at hand, we tentatively conclude that $\mathcal{F}[A] = \lim_{W\to 0} A\delta_W(f) = A\delta(f)$, or

$$A \leftrightarrow A\delta(f) \tag{1}$$

so the spectrum of a *constant* in the time domain is a *frequency-domain impulse* at $f = 0$.

Does this result agree with physical reasoning? Upon reflection the answer is affirmative, for a constant signal has no time variation and hence its spectral content must be entirely at $f = 0$. The reason why we have an impulse is simply that returning to the time domain via the inverse transform entails *integration*, and the impulse is required to concentrate nonzero area at a single frequency. Checking this reasoning by actually taking the inverse transform, we have

$$\begin{aligned} \mathcal{F}^{-1}[A\delta(f)] &= \int_{-\infty}^{\infty} A\delta(f)e^{j2\pi ft}\,df \\ &= Ae^{j2\pi ft}\Big|_{f=0} = A \end{aligned}$$

which, for our purposes, is sufficient justification of (1). The fact that we integrate over the impulse to find a physical quantity (the time function) is consistent with the theory of distributions and our treatment of impulses in Chapter 4.

Applying the frequency-translation and modulation theorems to (1) yields two more transform pairs:

$$Ae^{j\omega_c t} \leftrightarrow A\delta(f - f_c) \tag{2}$$

$$A\cos\omega_c t \leftrightarrow \frac{A}{2}[\delta(f - f_c) + \delta(f + f_c)] \tag{3}$$

The corresponding spectra are given in Figure 7.9, where we note that the single phasor $Ae^{j\omega_c t}$ has its spectral content concentrated at $f = f_c$ whereas $A\cos\omega_c t$ requires two components, at $+f_c$ and $-f_c$, since it is the

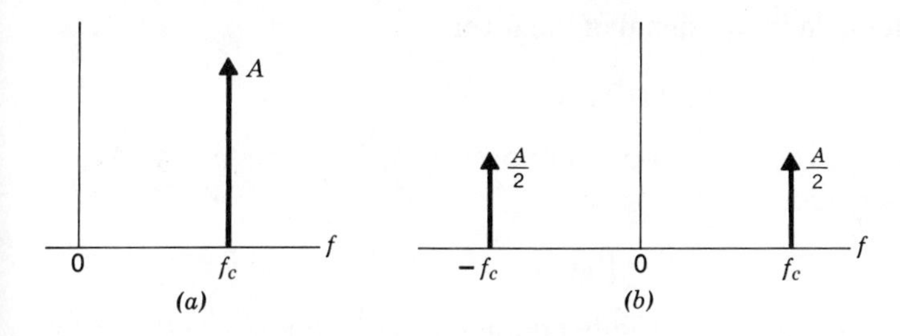

Figure 7.9. (*a*) Spectrum of the phasor $Ae^{j\omega_c t}$. (*b*) Spectrum of the sinusoid $A \cos \omega_c t$.

sum of *two* phasors. The reader can verify these results by inverse transformation.

In retrospect, (1) through (3) are seen to be Fourier transforms of *power* signals, signals we have previously represented by line spectra in the frequency domain. Thus, with the aid of the frequency-domain impulse, both periodic power signals and nonperiodic energy signals can be described by continuous spectra. In this sense, then, periodic power signals become a subclass of nonperiodic signals. (However, the reverse argument does not apply since nonperiodic signals defy representation as line spectra.)

Pursuing this line of thought, let us derive the Fourier transform of an arbitrary periodic signal having the exponential Fourier series expansion

$$v(t) = \sum_{n=-\infty}^{\infty} c_v(nf_0) e^{j2\pi n f_0 t}$$

From superposition, the summation can be transformed term by term and, since the coefficients $c_v(nf_0)$ are constants, (2) gives

$$V(f) = \sum_{n=-\infty}^{\infty} c_v(nf_0) \delta(f - nf_0) \tag{4}$$

Hence, the continuous spectrum of a periodic power signal consists of impulses at all the harmonics, the impulse weights being the *complex* numbers $c_v(nf_0)$. Based on this result, we have the following general recipe for converting any line spectrum into a continuous spectrum: replace all lines in the line spectrum by impulses whose weights are complex numbers representing the amplitude and phase of the respective frequency components. This rule thus includes quasi-periodic signals as well as periodic signals.

Now consider a "mixed" signal of the form

$$v(t) = v_E(t) + v_P(t)$$

where $v_E(t)$ and $v_P(t)$ are energy and power signals, respectively. Again invoking superposition, the transform is

$$V(f) = V_E(f) + V_P(f)$$

so the spectrum consists of a strictly continuous component $V_E(f)$ and an impulsive component $V_P(f)$. The example below illustrates an important case in point.

Example 7.8 Spectrum of an amplitude-modulated wave

When an energy signal $v_\mu(t)$ amplitude-modulates a carrier wave $A_c \cos \omega_c t$, the resulting modulated wave is

$$\begin{aligned} v(t) &= [1 + v_\mu(t)]A_c \cos \omega_c t \\ &= A_c \cos \omega_c t + A_c v_\mu(t) \cos \omega_c t \end{aligned} \tag{5}$$

which is a mixed signal. Transforming $v(t)$ term by term gives

$$V(f) = \frac{A_c}{2}[\delta(f-f_c) + \delta(f+f_c)] + \frac{A_c}{2}[V_\mu(f-f_c) + V_\mu(f+f_c)] \tag{6}$$

where we have used (3) and the modulation theorem.

Taking $V_\mu(f)$ as the bandlimited spectrum shown in Figure 7.10*a*, the spectrum of the AM wave is the mixed spectrum of Figure 7.10*b*, where the negative-frequency portion has been omitted. An interesting and significant feature here is that the bandwidth of the modulated signal

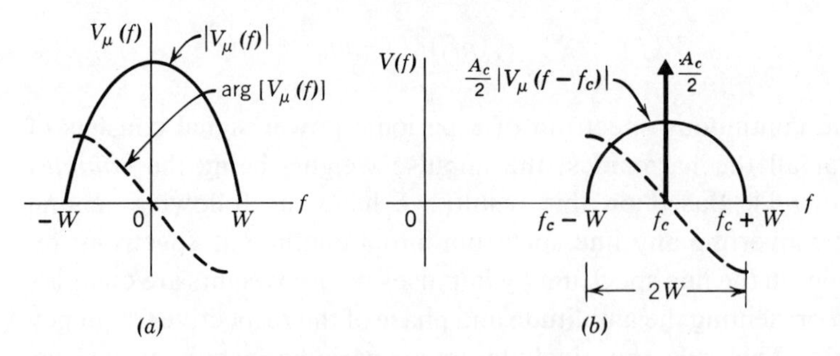

Figure 7.10. Amplitude modulation. (*a*) Spectrum of bandlimited modulating signal. (*b*) Spectrum of modulated wave, $f \geq 0$.

$v(t)$ is $2W$ or *twice* the bandwidth of $v_\mu(t)$. This results because the negative-frequency portion of $V_\mu(f)$, which did not count in measuring its bandwidth, now appears in the positive-frequency range $f_c - W \le f \le f_c$ of $V(f)$. The components residing above and below the carrier frequency f_c in the modulated spectrum are known as the *upper* and *lower sidebands.*

Impulses in the time domain

Let us now focus on the impulse in the time domain by considering the Fourier transform of $A\delta(t)$. In view of the duality theorem and (1), it immediately follows that

$$A\delta(t) \leftrightarrow A \tag{7}$$

Therefore, the amplitude spectrum of an impulse is a constant for $-\infty < f < \infty$, meaning that *all frequency components are present in equal proportion.* Another way of looking at (7), together with (1), is that the transform pairs $A \leftrightarrow A\delta(f)$ and $A\delta(t) \leftrightarrow A$ are the extremes of reciprocal spreading: on one hand, a constant signal of infinite time duration has "zero" spectral width; on the other, an impulsive signal of "zero" time duration has infinite spectral width.

Generalizing (7) to the case of a time-shifted impulse gives

$$A\delta(t-t_d) \leftrightarrow Ae^{-j\omega t_d} \tag{8}$$

Because this transform pair is so simple, it sometimes proves useful for calculating spectra when coupled with the differentiation theorem. Problem 7.20 outlines this technique.

Impulses and sinc functions

To derive $A \leftrightarrow A\delta(f)$ at the beginning of this section we started with a *sinc* pulse and took an appropriate limit. Alternately, suppose we take the *rectangular* pulse $A\Pi(t/\tau)$ and let the duration τ go to infinity, which also yields a constant for all time. Then, since $A\Pi(t/\tau) \leftrightarrow A\tau \operatorname{sinc} f\tau$, agreement with the previous result implies that

$$\lim_{\tau\to\infty} \tau \operatorname{sinc} f\tau = \delta(f) \tag{9}$$

so a sinc function with amplitude τ and zero-crossings spaced by $1/\tau$ becomes a unit impulse in the limit as $\tau \to \infty$.

Changing this over to the time domain and renaming the parameters, let

$$\delta_\epsilon(t) = \frac{1}{\epsilon}\operatorname{sinc}\frac{t}{\epsilon} = \frac{1}{\pi t}\sin\frac{\pi t}{\epsilon} \tag{10a}$$

so, by comparison with (9),

$$\lim_{\epsilon\to 0}\delta_\epsilon(t) = \delta(t) \tag{10b}$$

Equation (10) should be recognized as the impulse representation asserted but not justified in Section 4.2; it is also the same relationship invoked to derive Eq. (11), Sect. 5.4, and used to prove the Fourier integral theorem. Here, while we have not presented a rigorous argument, we have added strong evidence in support of this crucial relationship. Papoulis (1962, pp. 277–281) may be consulted for a more complete treatment.

7.4 SYSTEMS ANALYSIS

Having developed the appropriate tools we shall now treat the problem of systems analysis using frequency-domain concepts. For this approach to be valid, the following conditions must be satisfied:

1. The system initially is in the zero state, $\mathbf{q}(t_0) = \mathbf{0}$, so the complete response equals the zero-state response. Therefore, the initial time can be taken as $t_0 = -\infty$ even if the input $x(t)$ "starts" at some later instant.

2. The system in question is fixed, linear, and asymptotically stable.

3. Either the system's impulse response $h(t)$ or frequency response function $H(f)$ is known.

4. The input signal $x(t)$ is Fourier transformable, i.e., $X(f)$ exists at least in a limiting sense.

Input-output relations

Under the above conditions our starting point is the superposition integral, repeated here as

$$y(t) = x * h(t) = h * x(t) \tag{1}$$

Combining (1) with the convolution theorem, the Fourier transform of the output signal equals the Fourier transform of the input times the transform of the impulse response, that is

$$\mathcal{F}[y(t)] = \mathcal{F}[x * h(t)] = \mathcal{F}[x(t)] \quad \mathcal{F}[h(t)]$$

But by definition

$$\mathcal{F}[h(t)] = \int_{-\infty}^{\infty} h(t)e^{-j2\pi ft}\,dt$$

and this integral has been defined to be the frequency response function $H(f)$, referred to as the *transfer function* for short in the remainder of this chapter. Therefore

$$h(t) \leftrightarrow H(f) \tag{2}$$

and

$$Y(f) = H(f)X(f) \tag{3}$$

where $Y(f) = \mathcal{F}[y(t)]$.

Equation (3) is an elegantly simple relationship, saying that the output spectrum equals the input spectrum multiplied by the transfer function. Clearly, here we have the key to frequency-domain systems analysis assuming the conditions stated at the outset. The distinction between time-domain and frequency-domain analysis then is this: in the former we work with *time functions* (signals and impulse response) linked by *convolution*, while in the latter we work with *frequency functions* (spectra and frequency response) linked by *multiplication*. Figure 7.11 illustrates this point schematically.

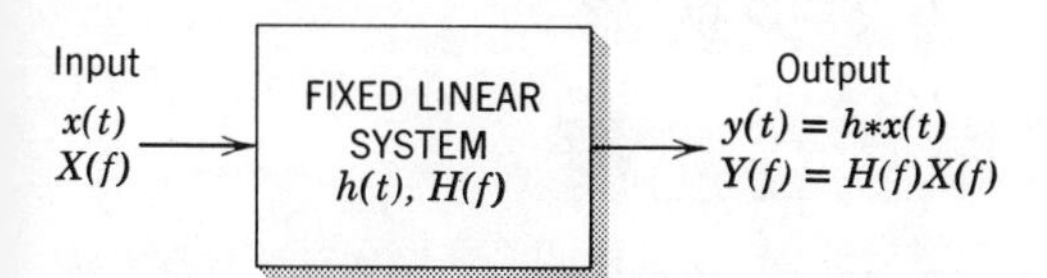

Figure 7.11. Time- and frequency-domain input-output relations.

For further interpretation of (3), suppose that the input signal is a unit impulse; then, since $\mathcal{F}[\delta(t)] = 1$, the output spectrum becomes

$$Y(f) = H(f) \qquad x(t) = \delta(t) \tag{4}$$

This, of course, is so because $H(f) = \mathcal{F}[h(t)]$ and the impulse response, by definition, is the output signal resulting from a unit impulse at the input. Two less obvious points to be gained from (4) follow.

First, although $H(f)$ describes the system in the frequency domain and is not a spectrum in the usual sense, it does appear as an identifiable spectrum when the system excitation is $\delta(t)$. Second, by applying a reasonable approximation of $\delta(t)$ to the input, the system's transfer function – or at least the amplitude ratio – can be observed directly as the output spectrum, this in agreement with the fact that $\delta(t)$ contains all frequencies in equal proportion so the shape of the output spectrum depends only on $H(f)$.

With an arbitrary input signal, the amplitude and phase spectra at the output are given by (3) expressed in polar form as

$$|Y(f)| = |H(f)||X(f)| \tag{5a}$$

$$\arg[Y(f)] = \arg[X(f)] + \arg[H(f)] \tag{5b}$$

where $|H(f)|$ and $\arg[H(f)]$ are the system's amplitude ratio and phase shift, respectively. Equation (5) generalizes on the periodic steady-state case where

$$|c_y(nf_0)| = |H(nf_0)||c_x(nf_0)|$$

$$\arg[c_y(nf_0)] = \arg[c_x(nf_0)] + \arg[H(nf_0)]$$

It is left for the reader to show that these expressions can be derived from (5) when $X(f)$ is the impulsive spectrum of Eq. (4), Sect. 7.3.

A rather obvious consequence of (5a), together with Rayleigh's energy theorem, is that the output signal's energy can be found entirely from frequency-domain terms. Specifically, the output energy spectrum is

$$|Y(f)|^2 = |H(f)|^2|X(f)|^2$$

so

$$E_y = \int_{-\infty}^{\infty} |H(f)|^2|X(f)|^2 df \tag{6}$$

providing, of course, that $x(t)$ is an energy signal. In view of (6), $|H(f)|^2$ sometimes is termed the energy transfer function.

Finally, we come to the question of completing the analysis by returning to the time domain. To get the output time function $y(t)$ from the output spectrum one must perform the inverse Fourier transformation

$$y(t) = \int_{-\infty}^{\infty} \overbrace{H(f)X(f)}^{Y(f)} e^{j\omega t} df \tag{7}$$

Frequently, it is a moot point whether the inversion (7) or the convolution (1) is the easier, for both can be rather nasty chores. In fact, when the exact output time function is required, the frequency-domain approach may actually involve more work than straight time-domain analysis.

But, in such instances, one should ask if finding the waveform $y(t)$ is really a necessity. Perhaps its spectrum $Y(f)$ provides sufficient information; if so, then spectral analysis – stopping short of inverse transformation – will be a true labor-saving device. This is one reason why we have given considerable attention to the interpretation of various spectra. The example below demonstrates just how far one can go using frequency-domain concepts without explicitly solving for the time function.

Example 7.9 A frequency-domain analysis

Suppose an energy signal is processed by an RC lowpass filter having

$$H(f) = \frac{1}{1 + j(f/B)} \qquad B = \frac{1}{2\pi RC}$$

and we wish to find the resulting output signal $y(t)$ without resorting to (7) if at all possible. To discuss this situation we note the approximations

$$H(f) \approx \begin{cases} 1 & |f| \ll B \qquad \text{(8a)} \\ \dfrac{B}{jf} & |f| \gg B \qquad \text{(8b)} \end{cases}$$

and consider the three cases shown in Figure 7.12, where the negative-frequency portions have been omitted.

In the first case, Figure 7.12a, the input signal has negligible content at frequencies large compared to the filter bandwidth B, i.e.,

$$|X(f)| \approx 0 \qquad |f| \geq W, \; W \ll B$$

Using this fact and (8a), the output spectrum becomes

$$Y(f) = H(f)X(f) \approx X(f)$$

Thus, since $Y(f)$ is essentially the same as $X(f)$ in both amplitude and phase, we immediately deduce that

$$y(t) \approx x(t)$$

omitting the formality of inverse transformation.

In the second case, Figure 7.12b, the input signal has very little low-frequency content, such that

$$|X(f)| \approx 0 \qquad |f| \leq W, \; W \gg B$$

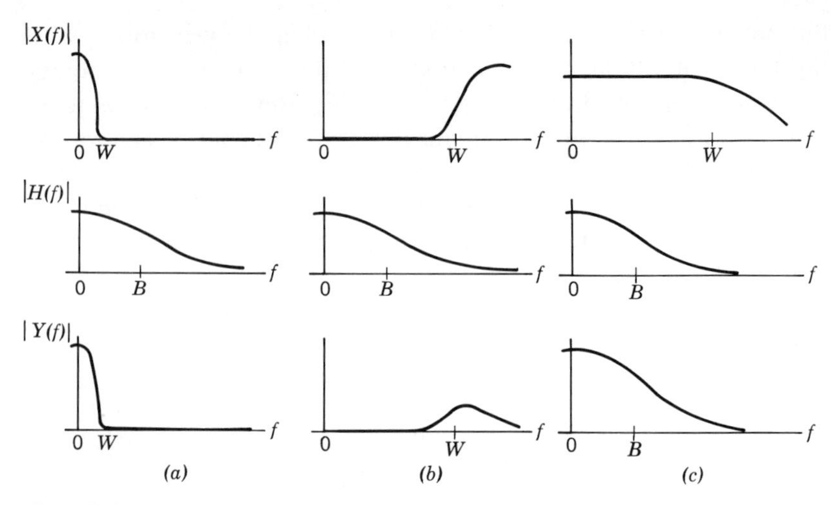

Figure 7.12.

Hence, from (8b),

$$Y(f) \approx \frac{B}{jf}X(f) = \frac{1}{RC}(j2\pi f)^{-1}X(f)$$

so the integration theorem tells us that

$$y(t) \approx \frac{1}{RC}\int_{-\infty}^{t} x(\lambda)\,d\lambda$$

Under these conditions, the filter is acting like an *integrator*. Recall that similar conclusions were reached in Example 4.3 from time-domain calculations of a specific output signal; here we support those conclusions on more general grounds.

Finally, suppose the input spectrum is essentially constant or "flat" over the filter bandwidth, Figure 7.12c, i.e.,

$$|X(f)| \approx K \qquad |f| \leq W,\ W >> B$$

Then, since $|H(f)| << 1$ for $|f| > W$,

$$Y(f) \approx KH(f)$$

and thus

$$y(t) \approx Kh(t)$$

Here, the input signal has such a short time duration (as implied by the large spectral width) that, insofar as the system is concerned, $x(t)$ is virtually an impulse.

Parallel and cascade systems

Often — almost always, in fact — a system consists of a number of different units or subsystems connected in some fashion. If the individual subsystems are linear and time-invariant and the total system is also, then one can meaningfully speak of the overall impulse response and transfer function. Here, we shall relate these system functions to those of the constituent subsystems for two common arrangements, the parallel and cascade connections.

Consider the system consisting of two units in *parallel*, Figure 7.13, such that both have the same input $x(t)$ and their individual outputs are

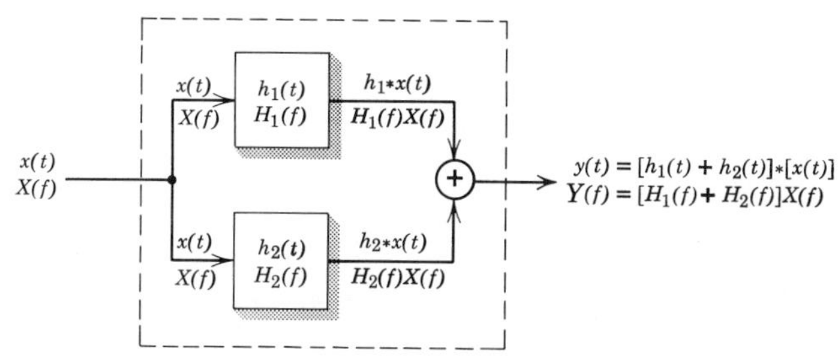

Figure 7.13. Two subsystems in parallel.

summed to form the system output $y(t)$. Using the distributive property of convolution we have

$$y(t) = [h_1 * x(t)] + [h_2 * x(t)] = [h_1(t) + h_2(t)] * [x(t)]$$

so the overall impulse response is $h(t) = h_1(t) + h_2(t)$ and it follows that $H(f) = H_1(f) + H_2(f)$ is the overall transfer function.

In the case of N parallel-connected subsystems the above results immediately generalize to

$$h(t) = h_1(t) + h_2(t) + \cdots + h_N(t) \tag{9a}$$

$$H(f) = H_1(f) + H_2(f) + \cdots + H_N(f) \tag{9b}$$

and hence

$$y(t) = (h_1 + h_2 + \cdots + h_N) * x(t) \tag{10a}$$

$$Y(f) = [H_1(f) + H_2(f) + \cdots + H_N(f)]X(f) \tag{10b}$$

We therefore conclude that a parallel configuration yields *additivity* in both the time and frequency domains.

Although (9) and (10) are quite simple and obvious, one not-so-obvious assumption has been implicit in their derivation, namely that any *interaction* or loading effects that may be present as a result of the interconnection of subsystems must be accounted for in the individual subsystem functions, $h_1(t)$, $H_1(f)$, etc. This means, for instance, that $h_1(t)$ is the response actually delivered by the first subsystem when $x(t) = \delta(t)$ and all other subsystems are connected; it does not necessarily equal the impulse response of the first subsystem when it is disconnected from the others.

In contrast with the parallel connection, Figure 7.14 shows two subsystems in *cascade*. Again, interaction effects have been assumed to be

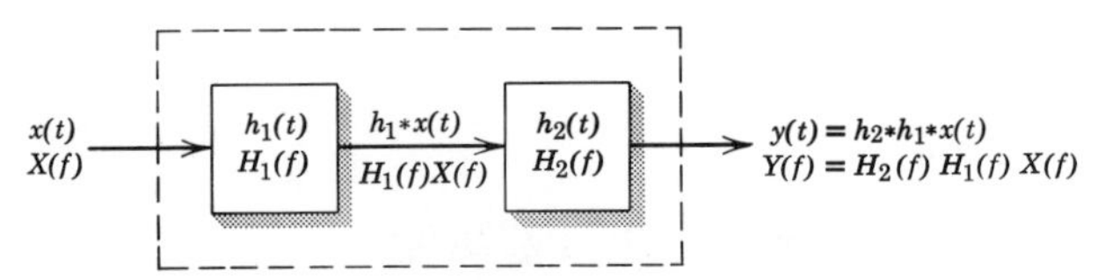

Figure 7.14. Two subsystems in cascade.

absorbed in the subsystem functions. Repeated application of the superposition integral gives the output signal as

$$y(t) = [h_2(t)] * \{[h_1(t)] * [x(t)]\} = (h_1 * h_2) * x(t)$$

where the associative property of convolution has been used to regroup terms. Thus, the impulse response of the cascade configuration is $h(t) = h_1 * h_2(t)$ whereas $H(f) = H_1(f)H_2(f)$, as follows from either the convolution theorem or direct inspection of the figure.

Extrapolating to the case of N systems in cascade then yields

$$h(t) = h_1 * h_2 * \cdots * h_N(t) \tag{11a}$$

$$H(f) = H_1(f)H_2(f) \cdots H_N(f) \tag{11b}$$

and

$$y(t) = (h_1 * h_2 * \cdots * h_N) * x(t) \tag{12a}$$

$$Y(f) = [H_1(f)H_2(f) \cdots H_N(f)]X(f) \tag{12b}$$

so we see that the cascade configuration results in *convolution* in the time domain and *multiplication* in the frequency domain, quite different from the parallel case of additivity in both domains.

An interesting implication of (12b) is that, as far as the final output is concerned, the specific order of the cascaded blocks does not matter, providing the loading is always the same. Thus, at least for analysis purposes, they can be shifted around at will – a property we shall use to advantage in the study of feedback loops. Notice, however, that this is valid only for fixed linear systems, and if nonlinear or time-varying components are present the order of operation generally cannot be altered without substantially changing the output. A case in point is the DC power supply of Example 6.8 where, for instance, putting the RC filter before the rectifier will seriously alter performance.

Returning to parallel versus cascade systems, the difference between (10) and (12) has particular relevance to the problem of finding the output time function $y(t)$. With a parallel-connected system, the inversion of (10b) generally will be as difficult as the single convolution in (10a), if not more so; thus, as previously observed, frequency-domain analysis may have no advantage over time-domain analysis. But with the cascade system, time-domain analysis as in (12a) entails N successive convolutions whereas the inversion of (12b) is just a single integration,

$$y(t) = \int_{-\infty}^{\infty} [H_1(f)H_2(f) \cdots H_N(f)]X(f)e^{j\omega t}\,df \tag{13}$$

so the frequency-domain approach (complete with inverse transformation) usually is the easier method. Moreover, because cascades occur in practice far more often than parallel connections, (13) represents one of the major payoffs of transform analysis as a labor-saving tool.

Example 7.10 Frequency response of the finite-time integrator

As an exercise involving the parallel and cascade formulas, let us calculate the overall transfer function for the finite-time integrator diagrammed in Figure 7.15*a*. This system was discussed briefly in Chapter 4, but here we shall start again from scratch.

The first step is to set down the transfer functions of the individual blocks, as follows: the ideal amplifier with gain K has $H_1(f) = K$; the upper branch of the parallel group is a direct path, equivalent to $H_2(f) = 1$; the lower branch consists of pure delay by T seconds so, using the time-shift theorem, $H_3(f) = -e^{-j\omega T}$ where the negative sign at the summing junction has been absorbed as a gain of -1; finally, for the ideal integrator, reference to the integration theorem gives $H_4(f) = (j2\pi f)^{-1} = 1/j\omega$. Thus, in terms of these transfer functions, Figure 7.15*b* is the new block diagram.

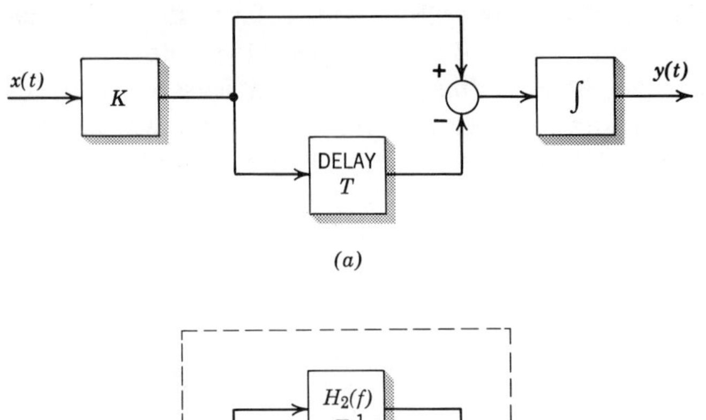

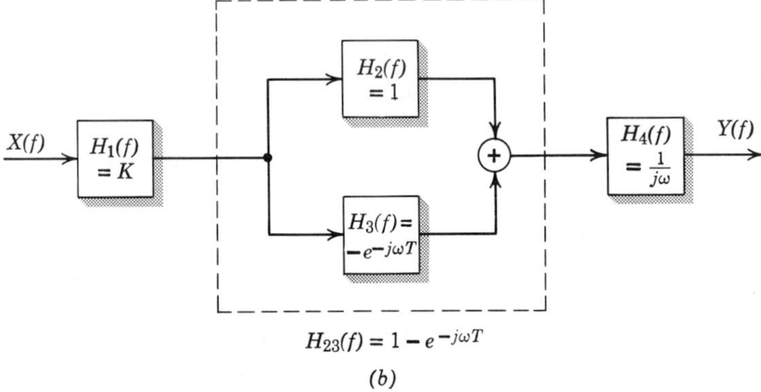

Figure 7.15. Finite-time integrator. (*a*) Simulation diagram. (*b*) Frequency-domain block diagram.

Invoking (9b), the two blocks in parallel are then reduced to a single unit having $H_{23}(f) = H_2(f) + H_3(f) = 1 - e^{-j\omega T}$ and, since $H_{23}(f)$ is in cascade with $H_1(f)$ and $H_4(f)$, the overall transfer function becomes

$$H(f) = H_1(f)H_{23}(f)H_4(f) = K(1 - e^{-j\omega T})\left(\frac{1}{j\omega}\right)$$

To further simplify expressions of this type we can factor out $\exp(-j\omega T/2)$ so that

$$H(f) = KT\frac{e^{j\omega T/2} - e^{-j\omega T/2}}{j\omega T}e^{-j\omega T/2} = KT \operatorname{sinc} fTe^{-j\pi fT} \tag{14}$$

Hence, the system's amplitude ratio is a *sinc function in frequency*, a result the reader may wish to check against the already known impulse response. However, because perfect integrators do not exist, this transfer function can only be approximated in practice.

Distortionless transmission and ideal filters

As brought out by Example 7.9, a system may be a *signal processor*, e.g., an integrator, such that the output waveform is generally quite different from the input. Or the system may have very little effect upon the waveform, so that the output is a reproduction of the input. Indeed, the usual function of a *communication* system is *signal reproduction* rather than signal processing, which leads us to the concept of *distortionless transmission*.

A system is said to give distortionless transmission if its output has the same form as the input, to within a constant scale factor and a finite time delay. Analytically, an undistorted version of the signal $x(t)$ would be

$$y(t) = Cx(t - t_d)$$

where C and t_d are real constants. Since the corresponding spectrum is

$$Y(f) = CX(f)e^{-j\omega t_d} = Ce^{-j\omega t_d}X(f)$$

and since $Y(f) = H(f)X(f)$ for a fixed linear system, distortionless transmission requires a system having

$$H(f) = Ce^{-j\omega t_d} \tag{15}$$

for all frequencies where $X(f) \neq 0$. Thus, the amplitude ratio must be *constant*,

$$|H(f)| = |C|$$

representing ideal amplification ($|C| > 1$) or attentuation ($|C| < 1$), while the phase shift must be a negative linear function

$$\arg[H(f)] = -2\pi t_d f + \arg[C]$$

where the second term equals $0°$ or $\pm 180°$, depending on whether C is positive or negative. However, as an exception to these requirements, $H(f)$ can take on *any* values over those frequencies where the input spectrum is zero.

Drawing upon (15) we now define an *ideal filter* as a system that passes without distortion all frequencies in a certain band, say $f_1 < |f| < f_2$, and completely rejects all other frequencies. Thus, the transfer function of such a filter is

$$H(f) = \begin{cases} Ce^{-j\omega t_d} & f_1 < |f| < f_2 \\ 0 & \text{otherwise} \end{cases} \tag{16}$$

and the width of the positive-frequency passband is clearly the *bandwidth* $B = f_2 - f_1$. Ideal filters are classified as being lowpass, bandpass, or highpass, according to the values of f_1 and f_2 – e.g., an ideal lowpass filter has $f_1 = 0$ and $B = f_2 < \infty$, as shown in Figure 7.16. Notice that the output of such a filter will be a bandlimited signal, regardless of the input, since $Y(f)$ must be zero for $|f| \geq B$.

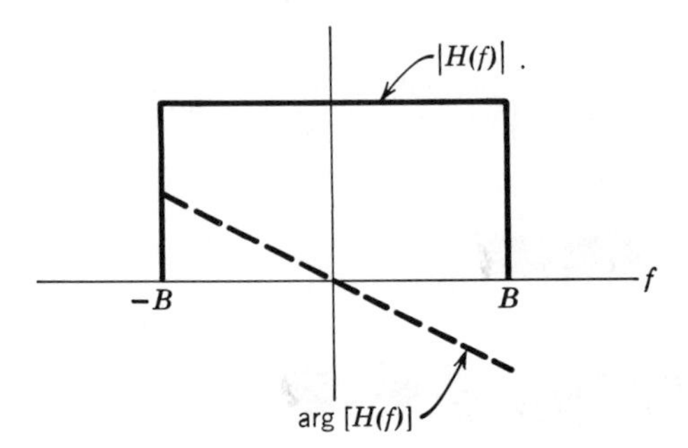

Figure 7.16. Frequency response of an ideal lowpass filter.

Not surprisingly, ideal filters are *nonrealizable*, meaning they cannot be constructed as an actual network with a finite number of lumped elements. However, since their characteristics can be *approximated* by real networks to any reasonable degree of accuracy, the concept of ideal filtering remains a valid and useful tool in system design. The principal use of such filters is for signal selection and separation, a task occurring in both communication and control. This function will be illustrated later under the heading of frequency-division multiplexing. For the moment we wish to show, by example, the nonrealizability of ideal filters.

Consider the ideal lowpass-filter characteristic of Figure 7.16 taking $C = 1$ (unit gain) and $t_d = 0$ (zero time delay) for convenience. Since $f_1 = 0$ and $B = f_2$, (16) can be rewritten as

$$H(f) = \Pi\left(\frac{f}{2B}\right) \tag{17a}$$

Inverse transformation then gives the impulse response as

$$h(t) = 2B \operatorname{sinc} 2Bt \tag{17b}$$

which has nonzero values for $t < 0$, Figure 7.17*a*. But $h(t)$ is the response to $x(t) = \delta(t)$ which is zero for $t < 0$; therefore, *the output appears before the input is applied*, an obvious physical impossibility. Stated another way, the impulse response is *not causal* and hence the system must be nonrealizable. Similar arguments can be made for ideal bandpass and highpass filters.

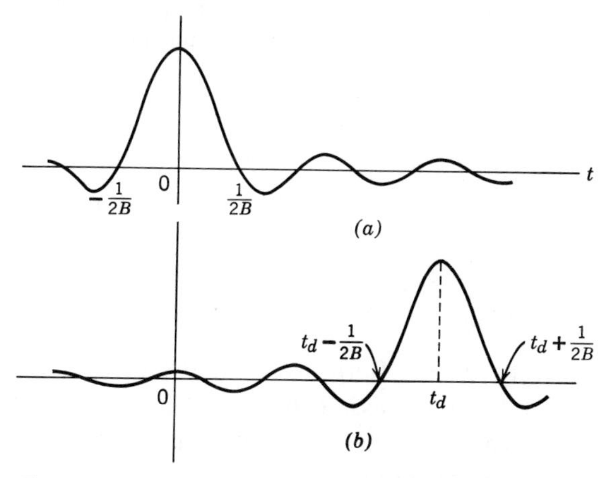

Figure 7.17. Impulse response of an ideal lowpass filter. (*a*) $t_d = 0$. (*b*) $t_d >> 1/B$.

Suppose, however, we add a time delay t_d to the lowpass filter, such that $H(f) = \Pi(f/2B)e^{-j\omega t_d}$ and, as plotted in Figure 7.17*b*,

$$h(t) = 2B \operatorname{sinc} 2B(t - t_d)$$

If t_d is large enough, say $t_d >> 1/B$, then $h(t)$ will be negligibly small for $t < 0$ and, presumably, one could approximate this impulse response with a real filter – the better the approximation, the larger t_d must be. Thus, practical filters with sharp-cutoff characteristics will inevitably introduce *time delay*, the delay increasing with increasing selectivity.

Example 7.11 Butterworth lowpass filters

As an illustration of just how close one can come to ideal filters with actual networks, consider the Nth-order Butterworth lowpass filter whose amplitude ratio is

$$|H(f)| = \left[1 + \left(\frac{f}{B}\right)^{2N}\right]^{-1/2} \tag{18}$$

with B being the half-power bandwidth and N the number of reactive circuit elements. In retrospect, one sees that the RC lowpass filter is a first-order Butterworth filter, i.e., $N = 1$. Circuit diagrams of higher-order Butterworths are found in any text on network synthesis.

An informative graphical comparison of (18) with the ideal case is afforded by plotting the gain in decibels versus $\log f$, that is, a Bode plot. For an ideal lowpass filter with unit amplification the gain definition gives

$$|H(f)|_{\text{db}} = \begin{cases} 0 & |f| < B \\ -\infty & |f| \geq B \end{cases}$$

while the linear asymptotes of a Butterworth are found from the fact that

$$|H(f)|_{\text{db}} \approx \begin{cases} 0 & |f| \ll B \\ -20\,N \log_{10}(f/B) & |f| \gg B \end{cases}$$

Figure 7.18 shows the gain curves for Butterworth filters of various orders superimposed along with the ideal case. Thus, if N is large enough, the difference between a Butterworth filter and an ideal filter is negligible, at least insofar as *amplitude ratio* is concerned.

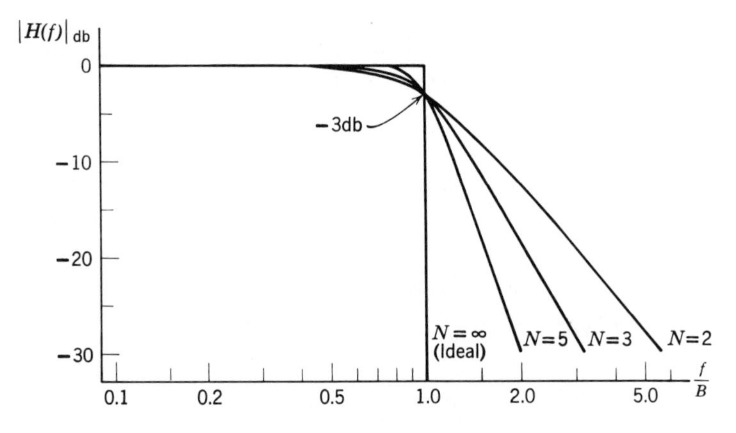

Figure 7.18. Amplitude response of Butterworth lowpass filters.

But, to be completely honest, we should also mention that the phase shift of a Butterworth filter is not very linear, and hence its impulse response differs markedly from a sinc function even if N is very large. (In this regard one must always bear in mind that amplitude and phase are equally important when determining the exact shape of a time function from its frequency-domain representation.) For those applications where linear phase is critical – phase- or frequency-modulation systems for instance – alternate designs such as the Chebyshev filter may be preferable. Christian and Eisenmann (1966) is a handy reference on these various filter designs.

7.5 SIGNAL TRANSMISSION SYSTEMS

We now have covered all the essentials of systems analysis using frequency-domain concepts, the sum and substance being just the input-output relationship

$$Y(f) = H(f)X(f)$$

Like other powerful tools, however, it takes some pondering and a few examples to gain a firm grasp of the meaning. Therefore, this section closes the chapter with three illustrative applications of frequency-domain analysis drawn from the general area of signal transmission systems.

Linear distortion and equalizers

The purpose of a signal transmission system is to present at the output a faithful reproduction of the input waveform, differing at most by an amplitude scale factor and some reasonable amount of time delay. Standing in the way of this goal generally are two kinds of problems, *distortion* and *contaminating signals.*

Contaminating signals are those signals independent of the input which, for one reason or another, also appear at the output and tend to mask or even obliterate the desired signal. Included under this category are *interference*, contaminating signals coming from sources external to the system, and *noise*, undesired random signals frequently arising from sources within the system itself. In principle, man-made interference can be eliminated completely by turning off its source; when this is not possible, techniques such as radio shielding, directive antennas, and appropriate channel allocations are required to minimize the interference. On the other hand, noise from internal sources can never be "turned off" or otherwise eliminated. Thus it is that sophisticated signal-processing techniques – frequency modulation, coding, the matched filter, etc. – have evolved as means for combating noise in communication systems. The interested reader will find numerous texts devoted to these subjects.

Distortion, in contrast to contamination, is a "warping" of the input signal itself as it passes through the system. For instance, if there are pronounced nonlinearities in the system so that the output is a nonlinear function of the input then, clearly, the result will be a distorted waveform at the output – this being termed *nonlinear* distortion. But even when the system has a fixed linear model, distortion will take place if the conditions for distortionless transmission are not satisfied. This case is called *linear distortion* and, unlike nonlinear distortion, it has a perfect remedy – in theory, at least – through the use of *equalizers*.

Suppose that a signal transmission system is fixed and linear, with transfer function $H_s(f)$, and suffers from linear distortion. To compensate for the distortion one merely adds an additional subsystem in cascade, as shown in Figure 7.19, giving the net effect of distortionless

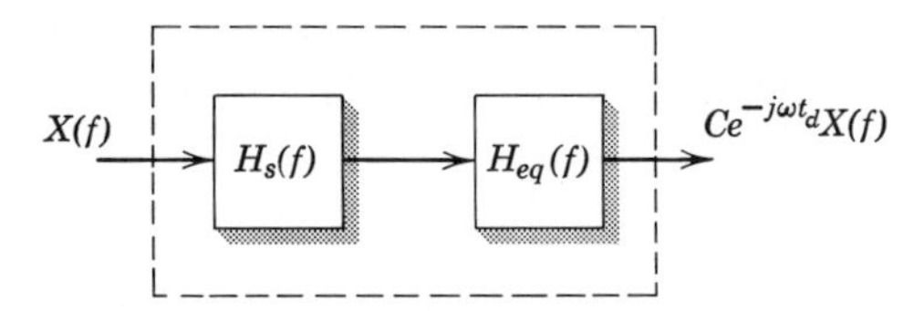

Figure 7.19. Signal transmission system with equalization.

transmission. The compensating subsystem is called an *equalizer* and its transfer function $H_{eq}(f)$ must be such that the overall system frequency response $H(f) = H_s(f)H_{eq}(f)$ obeys Eq. (15), Sect. 7.4, i.e.,

$$H_s(f)H_{eq}(f) = Ce^{-j\omega t_d}$$

where C and t_d are constants. Thus, we require that

$$H_{eq}(f) = \frac{Ce^{-j\omega t_d}}{H_s(f)} \tag{1}$$

at least over the frequency band covered by the input signal spectrum.

Usually, to the designer's regret, perfect compensation is impossible because the required equalizer characteristic turns out to be physically unrealizable. However, thanks to such ingenious devices as the transversal filter and the all-pass network, much can be done to *minimize* linear distortion with relatively simple and realizable networks. The following example illustrates this point, while Problem 7.32 suggests a more general design technique.

Example 7.12 Equalization of multipath distortion

A common phenomenon in radio communication is reflections or "echoes" resulting from the existence of two (or more) possible propagation paths from the transmitter to receiver, Figure 7.20*a*. Since both paths involve attentuation and time delay, with the reflection path having greater attenuation and delay, the total received signal takes the form

$$y(t) = C_p x(t - t_p) + aC_p x[t - (t_p + \Delta)]$$

where $|a| < 1$. A block-diagram model of this situation is shown in Figure 7.20*b*.

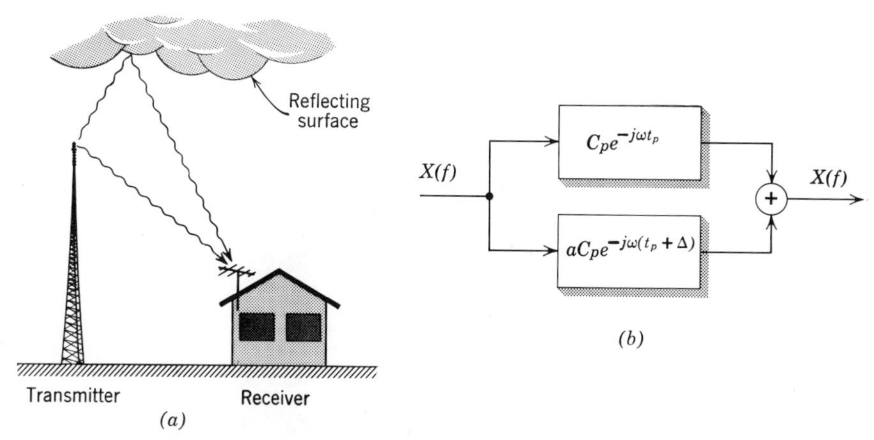

Figure 7.20. Radio propagation system with a reflection path. (*a*) Physical diagram. (*b*) Block-diagram model.

In view of the parallel configuration, it follows that the transfer function is

$$H_s(f) = C_p e^{-j\omega t_p} + aC_p e^{-j\omega(t_p+\Delta)}$$

$$= C_p e^{-j\omega t_p}(1 + ae^{-j\omega\Delta}) \tag{2}$$

And, since this is not in the form for distortionless transmission, there is said to be *multipath distortion* in the transmission system. Figure 7.21 plots the corresponding amplitude ratio and phase shift using the reasonable assumption of a weak reflection, i.e., $a^2 \ll 1$, such that

$$|H_s(f)| \approx C_p(1 + a\cos 2\pi f\Delta)$$
$$\arg[H_s(f)] \approx -2\pi f t_p - a \sin 2\pi f\Delta \tag{3}$$

Note that the presence of a weak reflection path results in frequency-domain *ripples*.

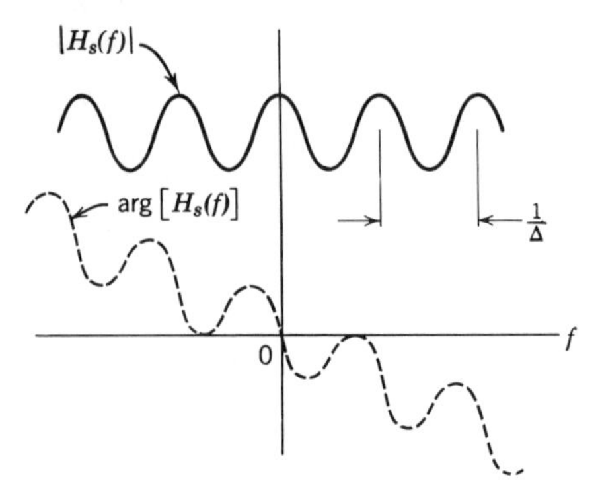

Figure 7.21. Frequency response with weak multipath distortion.

According to (1), the equalizer needed to compensate this distortion must have

$$H_{eq}(f) = \frac{Ce^{-j\omega(t_d - t_p)}}{C_p(1 + ae^{-j\omega\Delta})}$$

or, taking $C = C_p$ and $t_d = t_p$ for simplicity,†

$$\begin{aligned} H_{eq}(f) &= (1 + ae^{-j\omega\Delta})^{-1} \\ &= 1 - ae^{-j\omega\Delta} + a^2 e^{-j2\omega\Delta} - a^3 e^{-j3\omega\Delta} + \cdots \end{aligned}$$

where, drawing upon the fact that $|ae^{-j\omega\Delta}| < 1$, the binomial series expansion has been used. Obviously, perfect equalization entails an infinite number of delayors and scalors in a parallel configuration; equally obvious, this is a physical impossibility. But we said before that $a^2 \ll 1$, so the higher-order terms may be neglected with no great loss. Thus, for instance, dropping all but the first four terms yields the approximation

$$H_{eq}(f) \approx 1 - ae^{-j\omega\Delta} + a^2 e^{-j2\omega\Delta} - a^3 e^{-j3\omega\Delta} \tag{4}$$

A convenient way of building this equalizer comes about if we use a delay line having unit gain and total time delay 3Δ, with uniformly spaced taps; Figure 7.22 shows the resulting *tapped-delay-line equalizer*, also called a *transversal filter*.

†Recall from our discussion of distortionless transmission that the constants C and t_d are arbitrary within reason.

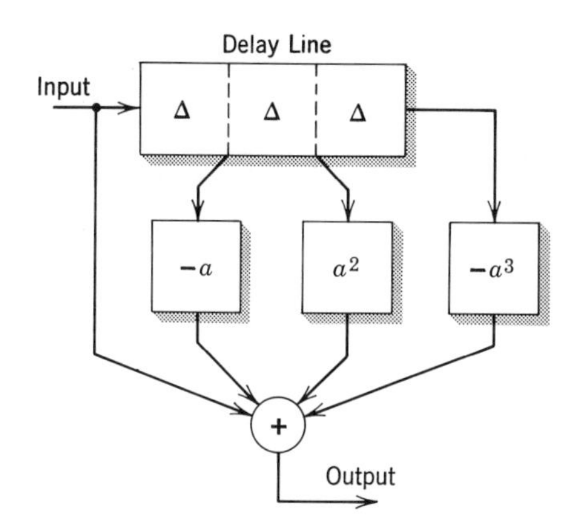

Figure 7.22. Tapped delay line equalizer.

Frequency-division multiplexing

Multiplexing is the transmission of several different signals over one facility such that they can be sorted out and individually recovered at the receiving end. One technique for doing this, known as *frequency-division multiplexing* (FDM), is based on translating each input spectrum to a separate frequency "slot" and summing the translated signals to form the multiplexed signal that will be transmitted. Thus, although the original signals are all jumbled together in the time domain, they are disjoint (nonoverlapping) in the frequency domain and can be separated at the receiver by bandpass filtering, after which their spectra must be translated back to the original positions on the frequency axis.

Figure 7.23*a* diagrams a very simple two-channel FDM system, the two input signals $x_1(t)$ and $x_2(t)$ being bandlimited in W_1 and W_2, respectively. Each input is first multiplied by a sinusoid, giving the multiplexed signal

$$x(t) = x_1(t) \cos \omega_1 t + x_2(t) \cos \omega_2 t \tag{5}$$

Drawing upon the modulation theorem, the spectrum $X(f)$ will be as shown in Figure 7.23*b* (omitting negative f), where it is assumed that f_1 and f_2 are such that the translated spectra do not overlap. Appropriate bandpass filtering at the receiver will then pick out the two signals

$$y_1(t) = x_1(t) \cos \omega_1 t \qquad y_2(t) = x_2(t) \cos \omega_2 t$$

from which $x_1(t)$ and $x_2(t)$ can be recovered.

But how is the recovery implemented? Curiously, in the same way that

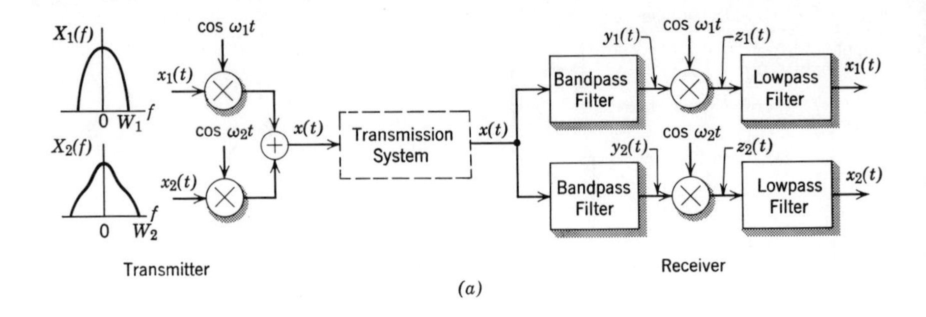

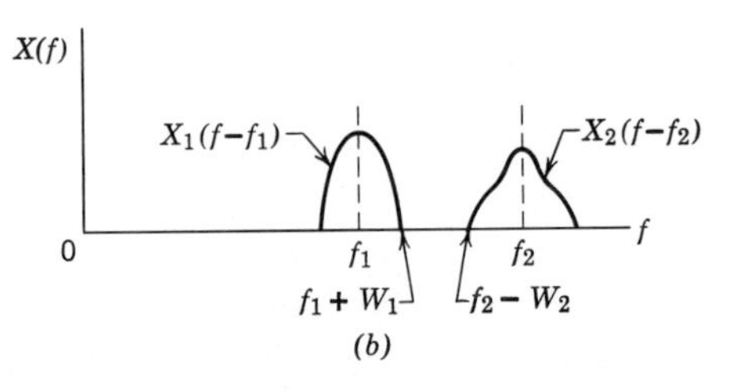

Figure 7.23. A two-channel FDM system. (*a*) Block diagram. (*b*) Multiplexed spectrum, $f \geq 0$.

the multiplexing process was begun, i.e., multiplying by a sinusoid. To show that this works, consider the product

$$\begin{aligned} z_1(t) &= \overbrace{[x_1(t)\cos \omega_1 t]}^{y_1(t)} 2\cos(\omega t + \theta) \\ &= x_1(t)[\cos(\omega_1 t - \omega t - \theta) + \cos(\omega_1 t + \omega t + \theta)] \end{aligned}$$

where we have used the trigonometric identity for the product of two cosines. Now if $\omega = \omega_1$ and $\theta = 0$ then

$$z_1(t) = x_1(t) + x_1(t)\cos 2\omega_1 t$$

whose spectrum is sketched in Figure 7.24. Clearly, lowpass filtering will yield just $X_1(f)$ in the frequency domain or $x_1(t)$ in the time domain. Hence, complete recovery has been achieved. The other signal $x_2(t)$ is recovered in a similar fashion.

A number of interesting points have arisen in this illustration. First, both the multiplexing and recovery – or the *modulation* and *demodulation* – are frequency-translation processes, one being up-translation and the other down-translation. We recall, however, from the modulation

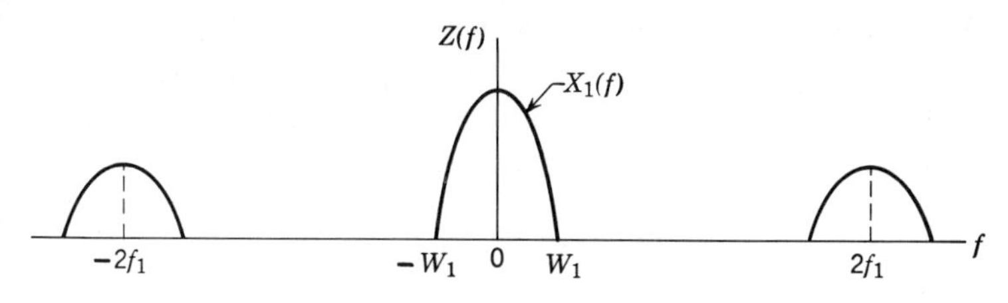

Figure 7.24.

theorem that multiplying by a sinusoid produces translation in *both* directions, which explains why this process works for both modulation and demodulation. Second, exact recovery requires the multiplying sinusoid to have the same phase and frequency as the corresponding one at the transmitter; therefore, *synchronization* between the two may be necessary, and hence the designation *synchronous demodulation*. Third, because multipliers are not fixed linear devices, the FDM system in Figure 7.23*a* must include either nonlinear or time-varying components, even though it acts *overall* as a fixed linear system. Finally, referring to Figure 7.23*b*, we see that bandwidth might be conserved if only the positive-frequency portion of the input spectra were translated, i.e., just the upper sideband; one way to achieve this is to put an appropriate filter right after the multiplier at the transmitter, in which case one gets *single-sideband modulation* (SSB). Further details on all of these aspects can be found in any text on modulation theory, e.g., Carlson (1968).

Orthogonality and multiplexing*

It was specified that the various components of an FDM spectrum, such as Figure 7.23*b*, must be *disjoint in frequency*. Another type of multiplexing, time-division multiplexing, will be described in Chapter 12 where it is shown that the various components of the multiplexed waveform must be *disjoint in time*. Underlying these two multiplexing techniques is the concept of *orthogonality* introduced in Section 5.2.

The theoretical basis for this latter statement comes from the fact that

$$\int_{-\infty}^{\infty} v(t)w^*(t)\,dt = \int_{-\infty}^{\infty} V(f)W^*(f)\,df \tag{6}$$

where $v(t)$ and $w(t)$ are energy signals with transforms $V(f)$ and $W(f)$, respectively. Equation (6), a handy relationship in its own right, is proved in a manner nearly identical to our derivation of Rayleigh's energy theorem. Here, however, we are interested in its signal-space interpretation. For that purpose we note that both integrals are scalar products in the energy sense; thus, in the notation of Section 5.2,

$$\langle v(t), w(t) \rangle = \langle V(f), W(f) \rangle$$

so the time-domain and frequency-domain scalar products are equal. Then, if either scalar product is zero, the other is also zero; hence, orthogonality in the time domain implies orthogonality in the frequency domain, and vice versa.

Bandpass transmission systems

Several times in the past we have encountered signals of the form

$$\begin{aligned} v_{BP}(t) &= A_v(t) \cos [\omega_c t + \phi_v(t)] \\ &= \text{Re} [A_v(t) e^{j\phi_v(t)} e^{j\omega_c t}] \end{aligned} \tag{7}$$

This is the general expression for a *modulated sinusoid*, with *carrier frequency* f_c, *amplitude modulation* $A_v(t)$, and *phase modulation* $\phi_v(t)$. Such signals occur not only in communication systems but also in some types of control systems, especially those having AC servomechanisms. In view of the modulation theorem, one would expect $V_{BP}(f)$, the spectrum of $v_{BP}(t)$, to be concentrated in the vicinity of $f = \pm f_c$ with negligible content for $|f| << f_c$. A representative waveform and spectrum of this sort are illustrated in Figure 7.25.

Because of the nature of $V_{BP}(f)$, we call $v_{BP}(t)$ a *bandpass signal*. And, as a rule, bandpass signals are processed by *bandpass systems*, i.e., a system whose transfer function $H_{BP}(f)$ has a bandpass-filter characteristic, passing only frequencies in a band around $f = \pm f_c$. If an input signal $x_{BP}(t)$ is a bandpass signal then, obviously, the output $y_{BP}(t)$ is also bandpass and can be found from $Y_{BP}(f) = H_{BP}(f) X_{BP}(f)$. But it is frequently more convenient in analysis or design to work instead with an *equivalent lowpass system*. The transformation making that possible is the subject pursued here.

We begin by defining the lowpass equivalent for $V_{BP}(f)$ as

$$V_{LP}(f) \triangleq V_{BP}(f+f_c) u(f+f_c) = \begin{cases} V_{BP}(f+f_c) & f > -f_c \\ 0 & f \leq -f_c \end{cases} \tag{8}$$

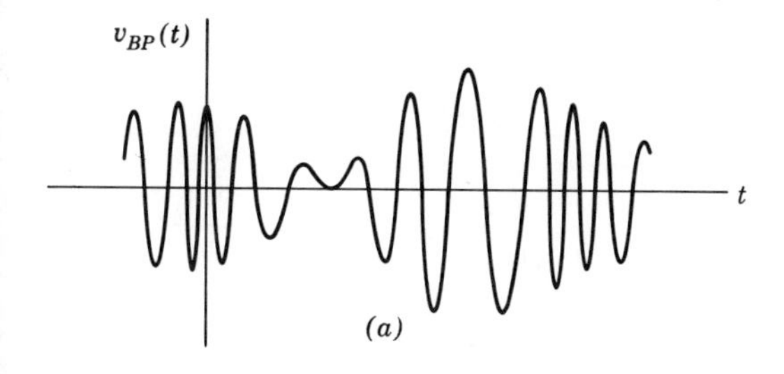

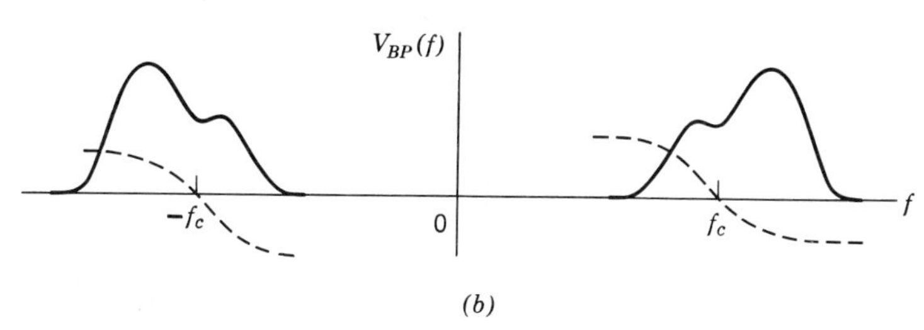

Figure 7.25. A bandpass signal. (*a*) Waveform. (*b*) Spectrum.

which, as sketched in Figure 7.26*a*, is simply the positive-frequency portion of $V_{BP}(f)$ translated down to be centered at $f = 0$. Notice, incidentally, that $v_{LP}(t) = \mathcal{F}^{-1}[V_{LP}(f)]$ may be a *complex* time function since $V_{LP}(f)$ does not necessarily have hermitian symmetry about the origin. We shall assume, however, that the bandpass signal $v_{BP}(t)$ is *real*; then $V_{BP}(-f) = V^*_{BP}(f)$ and it follows that

$$V_{BP}(f) = V_{LP}(f - f_c) + V^*_{LP}(-f - f_c) \tag{9}$$

whose second term is best understood by constructing $V^*_{LP}(-f)$, as in parts (*b*) and (*c*) of Figure 7.26.

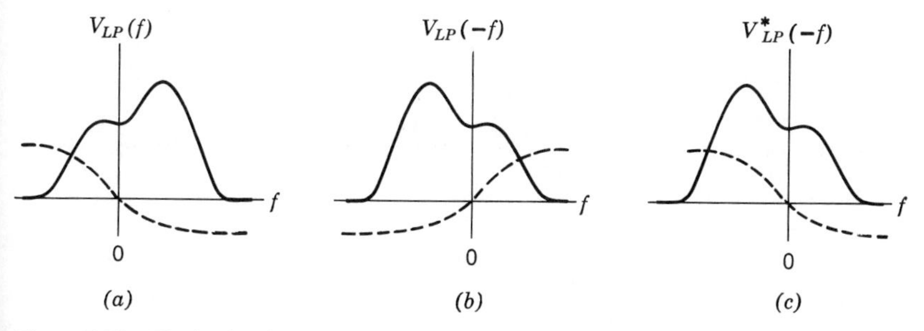

Figure 7.26. Equivalent lowpass spectrum.

Equations (8) and (9) are the *bandpass-to-lowpass* and *lowpass-to-bandpass transformations*, respectively, expressed in the frequency domain. Identical transformations apply to transfer functions as well as to signal spectra. Thus, the relationship

$$H_{BP}(f) = H_{LP}(f-f_c) + H_{LP}^*(-f-f_c) \tag{10}$$

is often used to design bandpass filters from known lowpass configurations. What (10) implies with respect to actual circuit elements is covered in most network theory texts, e.g., Close (1966, pp. 566–568).

Turning to the time domain, the inverse Fourier transform of (9) yields

$$\begin{aligned} v_{BP}(t) &= v_{LP}(t)e^{j\omega_c t} + v_{LP}^*(t)e^{-j\omega_c t} \\ &= \text{Re}\,[2v_{LP}(t)e^{j\omega_c t}] \end{aligned} \tag{11}$$

Comparing the latter with (7) then shows that

$$v_{LP}(t) = \tfrac{1}{2}A_v(t)e^{j\phi_v(t)} \tag{12}$$

and the method behind all this madness begins to emerge. For the lowpass-equivalent signal $v_{LP}(t)$ has the amplitude- and phase-modulation functions of $v_{BP}(t)$ without the notational nuisance of dragging along the carrier term. Similar conclusions apply to the lowpass system's impulse response $h_{LP}(t)$ as contrasted with $h_{BP}(t)$.

Now we are ready to consider bandpass systems analysis. With reference to Figure 7.27*a*, let $X_{LP}(f)$, $H_{LP}(f)$, and $Y_{LP}(f)$ be lowpass equivalents related to the corresponding bandpass functions according to (8); then, using (9) to expand $Y_{BP}(f) = H_{BP}(f)X_{BP}(f)$ we have

$$\begin{aligned} Y_{LP}(f-f_c) + Y_{LP}^*(-f-f_c) &= [H_{LP}(f-f_c) + H_{LP}^*(-f-f_c)] \\ &\quad \times [X_{LP}(f-f_c) + X_{LP}^*(-f-f_c)] \end{aligned}$$

Multiplying out the right-hand side, there are two products which vanish for all f, i.e.,

$$H_{LP}(f-f_c)X_{LP}^*(-f-f_c) = H_{LP}^*(-f-f_c)X_{LP}(f-f_c) = 0$$

because, from (8), $H_{LP}(f-f_c) = 0$ for $f \leq 0$ while $X_{LP}^*(-f-f_c) = 0$ for

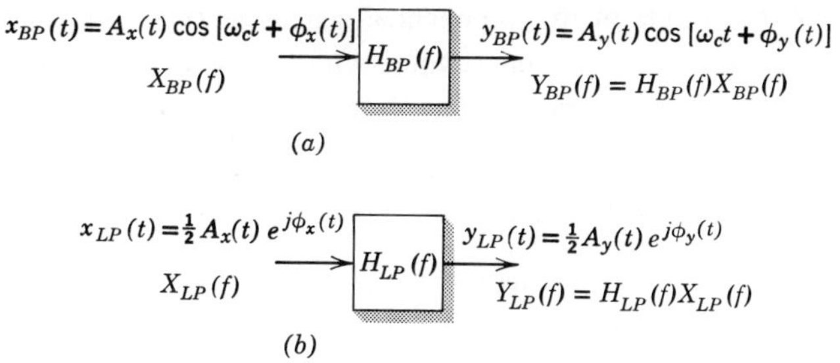

$x_{BP}(t) = A_x(t)\cos[\omega_c t + \phi_x(t)]$ → $H_{BP}(f)$ → $y_{BP}(t) = A_y(t)\cos[\omega_c t + \phi_y(t)]$

$X_{BP}(f)$ $\qquad$ $Y_{BP}(f) = H_{BP}(f)X_{BP}(f)$

(a)

$x_{LP}(t) = \frac{1}{2}A_x(t)\,e^{j\phi_x(t)}$ → $H_{LP}(f)$ → $y_{LP}(t) = \frac{1}{2}A_y(t)\,e^{j\phi_y(t)}$

$X_{LP}(f)$ $\qquad$ $Y_{LP}(f) = H_{LP}(f)X_{LP}(f)$

(b)

Figure 7.27. (*a*) Bandpass system. (*b*) Equivalent lowpass system.

$f \geq 0$, etc. Therefore,

$$Y_{LP}(f-f_c) + Y^*_{LP}(-f-f_c) = H_{LP}(f-f_c)X_{LP}(-f-f_c) + H^*_{LP}(-f-f_c)X^*_{LP}(-f-f_c)$$

and hence

$$Y_{LP}(f) = H_{LP}(f)X_{LP}(f) \tag{13}$$

Equation (13) means that we can use the equivalent lowpass system diagrammed in Figure 7.27*b* rather than the bandpass system. The advantage of doing so is that the calculations are palpably easier to perform because they do not involve the sinusoidal carrier directly. For instance, having determined $Y_{LP}(f)$ from (13), the amplitude and phase modulation of the actual *bandpass* signal are given by

$$A_y(t) = 2|\mathcal{F}^{-1}[Y_{LP}(f)]| \qquad \phi_y(t) = \arg\{\mathcal{F}^{-1}[Y_{LP}(f)]\} \tag{14}$$

since, drawing upon (7) and (12), $y_{BP}(t) = A_y(t)\cos[\omega_c t + \phi_y(t)]$ and $\mathcal{F}^{-1}[Y_{LP}(f)] = y_{LP}(t) = \frac{1}{2}A_y(t)\exp[j\phi_y(t)]$.

Problems

7.1 Assuming $v(t)$ is real, with even and odd parts $v_e(t)$ and $v_o(t)$, respectively, as in Eq. (21), Sect. 5.1, express $|V(f)|$ and $\arg[V(f)]$ in terms of $V_e(f) = \mathcal{F}[v_e(t)]$ and $V_o(f) = \mathcal{F}[v_o(t)]$.

7.2 Find the Fourier transforms of the exponential signals $w(t) = Ae^{at}u(-t)$ and $z(t) = Ae^{-a|t|}$, where $a > 0$. Compare these with Example 7.2 and discuss. *Answer:* $W(f) = A/(a - j2\pi f)$, $Z(f) = 2aA/[a^2 + (2\pi f)^2]$.

7.3 Calculate $V(f)$ for the *gaussian pulse* $v(t) = A \exp[-\pi(t/\tau)^2]$ and demonstrate reciprocal spreading by plotting $v(t)$ and $V(f)$ for two values of τ.

7.4 Suppose all the frequency components outside $|f| < W$ are removed from the signal of Example 7.2. Find the resulting energy E_W and plot E_W/E as a function of W.

7.5 Use Rayleigh's theorem and appropriate transform pairs (see Appendix B.1) to evaluate: (a) $\int_{-\infty}^{\infty} \text{sinc}^2\, a\lambda\, d\lambda$; (b) $\int_{-\infty}^{\infty} \text{sinc}^4\, a\lambda\, d\lambda$; (c) $\int_{-\infty}^{\infty} (a^2 + \lambda^2)^{-1} d\lambda$.

7.6 Find $\mathfrak{F}[v(ct - t_d)]$ in terms of $V(f)$. *Answer:* $|c|^{-1}V(f/c) \exp(-j2\pi f t_d/c)$.

7.7 Using the results of Example 7.1 and appropriate theorems, find the Fourier transforms of the signals in Figure P7.1. Sketch the amplitude spectra.

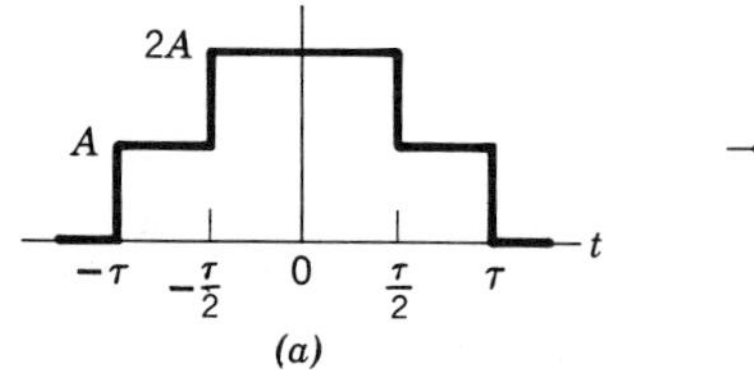

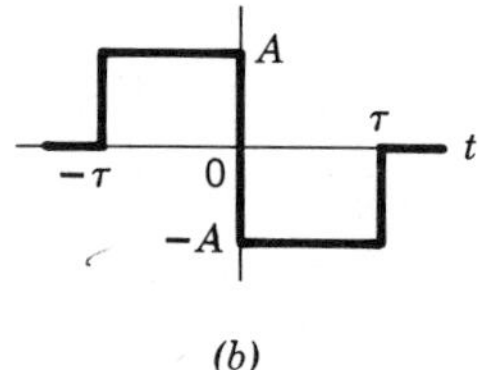

Figure P7.1.

7.8 With the aid of duality, find $z(t)$ corresponding to $Z(f) = (A/W) \exp(-|f|/W)$. *Hint:* see Problem 7.2.

7.9 Derive Eq. (6), Sect. 7.2, by direct Fourier transformation.

7.10 Investigate what happens in Example 7.6 for the special case of $\tau = 1/2f_c$. In particular, sketch the waveform and amplitude spectrum under this condition.

7.11 Generate a new theorem by applying duality to Eq. (9), Sect. 7.2. Check your result directly via the transform integral.

7.12 The nth *moment* of any function, say $z(\lambda)$, is defined as

$$M_z^n = \int_{-\infty}^{\infty} \lambda^n z(\lambda)\, d\lambda$$

(a) If $v(t) \leftrightarrow V(f)$ and $v_0^{(n)} \triangleq d^n v(t)/dt^n|_{t=0}$, etc., show that

$$M_V^n = (j2\pi)^{-n} v_0^{(n)} \qquad \text{and} \qquad M_v^n = (-j2\pi)^{-n} V_0^{(n)}$$

Hint: note from the differentiation theorem that $v^{(n)}(t) = \mathfrak{F}^{-1}[(j2\pi f)^n V(f)]$.

(b) Now expand $v(t)$ in a Taylor series about $t = 0$, expressed in terms of the moments of $V(f)$.

7.13 If $v_1(t) \leftrightarrow V_1(f)$, etc., find $V(f)$ in terms of $V_1(f)$, etc., in each of the following cases: (a) $v(t) = [v_1 * v_2(t)][v_3 * v_4(t)]$; (b) $v(t) = [v_1(t) + v_2(t)] * [v_3(t) + v_4(t)]$; (c) $v(t) = [v_1(t)] * [v_1^2(t)]$.

7.14 Use the convolution theorem to find and sketch the spectrum of the *trapezoidal pulse*, Figure P7.2. Compare with the spectrum of a rectangular pulse having $A = 10$ and $\tau = 6$ millisec and comment.

Answer: $0.06 \operatorname{sinc}(0.002f) \operatorname{sinc}(0.006f)$.

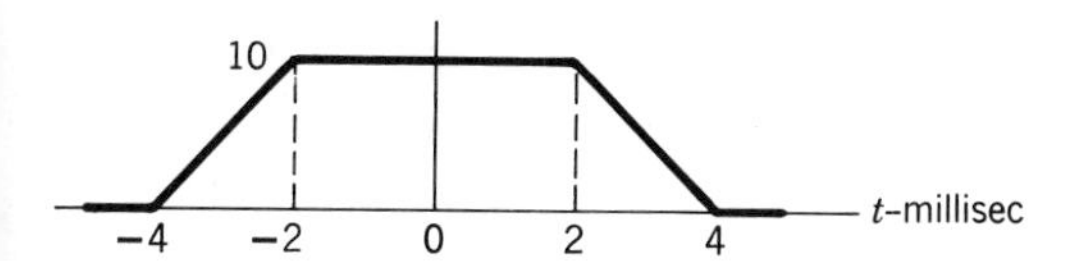

Figure P7.2.

7.15 Without integrating, find the second moment (defined in Problem 7.12) of $z(t) = v * w(t)$ in terms of the moments of $v(t)$ and $w(t)$.

7.16 Further justify Eq. (1), Sect. 7.3, by starting with the transform pair $Ae^{-a|t|} \leftrightarrow 2aA/[a^2 + (2\pi f)^2]$ and letting $a \to 0$. In particular, sketch the frequency function for two values of a.

7.17 Show that Eq. (2), Sect. 7.3, is consistent with the frequency-translation and convolution theorems by taking $w(t) = e^{j\omega_c t}$ in Eq. (12), Sect. 7.2.

7.18 Without integrating, find the Fourier transforms of the waveforms shown in Figure P7.3. Also sketch the amplitude spectra.

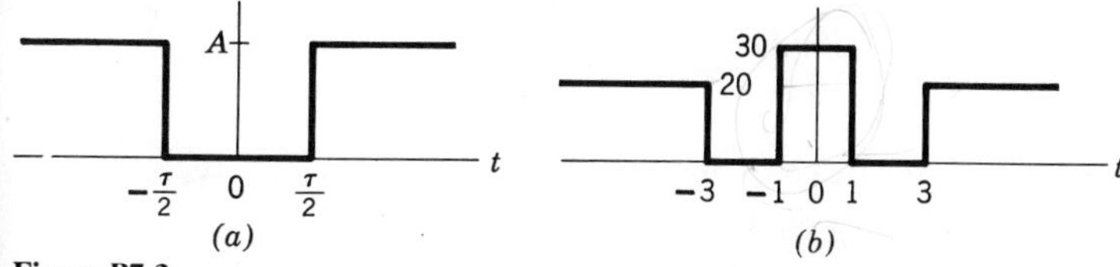

Figure P7.3.

7.19 If a single rectangular pulse is the input to a frequency-modulation system, the modulated signal will be of the form

$$v(t) = \begin{cases} A \cos 2\pi (f_c + \Delta f)t & |t| \leq \tau/2 \\ A \cos 2\pi f_c t & \text{otherwise} \end{cases}$$

Find $V(f)$ and sketch the amplitude spectrum for $f > 0$ assuming $f_c >> \Delta f >> 1/\tau$. *Hint:* decompose $v(t)$ into an eternal cosine wave plus two sinusoidal pulse bursts.

7.20 In view of the differentiation theorem one can write

$$\mathcal{F}[v(t)] = (j2\pi f)^{-n}\mathcal{F}[v^{(n)}(t)]$$

Thus, $V(f)$ can be found more readily in some cases by Fourier transformation of a derivative of $v(t)$, the key being whether or not $v^{(n)}(t)$ consists of simple functions for some reasonable value of n. However, care must be exercised when discontinuities are involved, since they become impulses upon differentiation; simple sketches of $v(t)$, $\dot{v}(t)$, etc., are helpful in this regard. Using this method with $n = 2$, find $V(f)$ for the *parabolic pulse* $v(t) = A[1-(2t/\tau)^2]\Pi(t/\tau)$. *Answer:* $(2A/\pi^2 f^2 \tau)(\text{sinc}\, f\tau - \cos \pi f\tau)$.

7.21 Use the method of the previous problem to find the transform of the *signum function* $\text{sgn}\, t \triangleq 2u(t) - 1$. From your result show that

$$\mathcal{F}[u(t)] = \tfrac{1}{2}\delta(f) + \frac{1}{j2\pi f}$$

and explain why the differentiation trick applied directly to $u(t)$ does not yield the correct answer.

7.22 The *analytic signal* associated with a real signal $v(t)$ is

$$\varphi_v(t) = \tfrac{1}{2}[v(t) + jv_H(t)]$$

where $v_H(t) \triangleq [v(t)] * [1/\pi t]$ is the *Hilbert transform* of $v(t)$. Show that

$$\mathcal{F}[\varphi_v(t)] = V(f)u(f) = \begin{cases} V(f) & f > 0 \\ 0 & f < 0 \end{cases}$$

and develop a similar expression for $\mathcal{F}[\varphi_v^*(t)]$. *Hint:* use the fact that $v(t) = v * \delta(t)$ and apply duality to $\mathcal{F}[u(t)]$ in Problem 7.21.

7.23 Derive Eq. (4), Sect. 6.5, from Eq. (3), Sect. 7.4, and Eq. (4), Sect. 7.3.

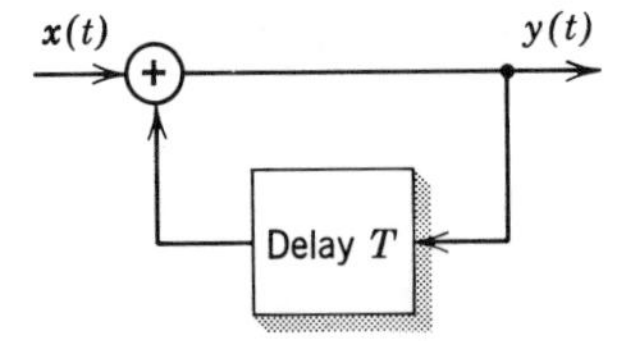

Figure P7.4.

7.24 If $x(t)$ is a pulse applied at $t = 0$ to the system in Figure P7.4, the output is a semi-infinite pulse train $y(t) = \sum_{n=0}^{\infty} x(t-nT)$. Determine the system's frequency response function by finding $Y(f)$ in terms of $X(f)$; convert your answer to closed form.

7.25 A filter is said to be *matched* to a specific input $x(t)$ if it has $h(t) = Kx(-t+t_d)$, where K and t_d are constants. (a) Show that $H(f) = Ke^{-j\omega t_d}X^*(f)$ and, hence, $Y(f) = Ke^{-j\omega t_d}|X(f)|^2$. (b) Using Eq. (7), Sect. 7.4, find $y(t_d)$ in terms of the input energy E_x and show that $y(t)$ has a *maximum* at $t = t_d$. *Hint:* consider the mathematical properties of $|X(f)|^2$.

7.26 The signal $x(t) = (1+\cos 2\pi f_c t)\Pi(f_c t/1000)$ is applied to an RC lowpass filter having $B = 2$ kHz. Taking the approach of Example 7.9, sketch the resulting output waveform when f_c equals: (a) 10 kHz; (b) 100 kHz; (c) 10 MHz.

7.27 Suppose the integrator in Figure 7.15*a* is replaced by an RC lowpass filter. Find the resulting $H(f)$ and determine the best value for B and the restrictions on $X(f)$ so the system will still act essentially as a finite-time integrator.

7.28 An ideal highpass filter may be modeled as shown in Figure P7.5; obtain $h(t)$ therefrom. *Answer:* $C[\delta(t-t_d) - 2f_1 \operatorname{sinc} 2f_1(t-t_d)]$.

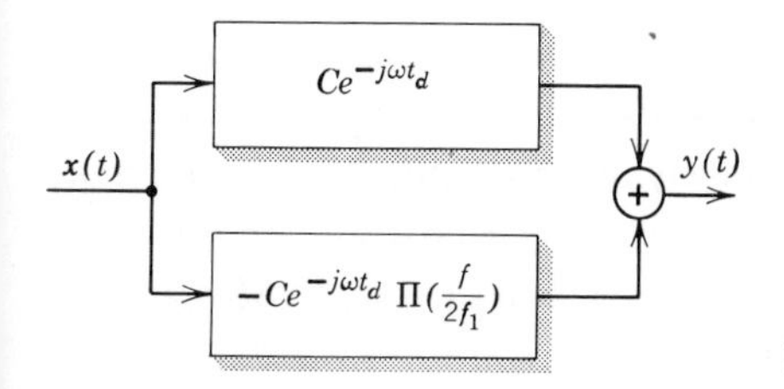

Figure P7.5.

7.29 The circuit in Figure P7.6 is a Butterworth lowpass filter. Find and plot $|H(f)|$ and arg $[H(f)]$.

7.30 Carry out the time-domain analysis of Figure 7.19 in terms of $x(t)$, $h_s(t)$,

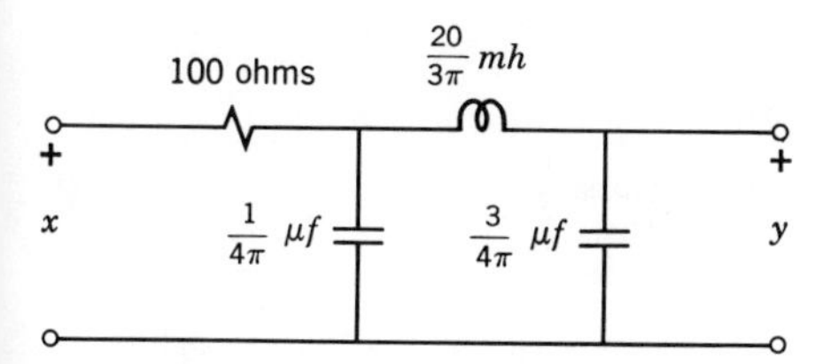

Figure P7.6.

and $h_{eq}(t)$. Can you arrive at an explicit formula for $h_{eq}(t)$ without going to the frequency domain?

7.31 Suppose a transmission system has constant amplitude ratio but rippled phase shift, $\arg [H(f)] = -\omega t_p + \alpha \sin \omega\Delta$. Assuming $\alpha^2 << 1$, show that $y(t)$ contains two echoes.

7.32 A signal, bandlimited in W, is to be transmitted over a system whose $H_s(f)$ has small variations in amplitude and phase. Design a tapped-delay-line equalizer, similar to Figure 7.22, by expanding $[H_s(f)]^{-1}$ in an exponential Fourier series in the frequency domain with period $2W$, valid over $|f| \leq W$, and use the first $2N+1$ terms as an approximation. *Hint:* take $t_d \geq N/2W$.

7.33 Suppose an amplifier has no dynamic elements but is nonlinear, the input-output relation being $y(t) = A_1 x(t) + A_2 x^2(t)$. Demonstrate that equalization of the nonlinear distortion is not possible in general.

7.34 Four voice signals are to be transmitted via a system having usable frequency response over the following bands: 50–60, 65–75, and 80–100 kHz. Design an appropriate transmitter and receiver in the form of Figure 7.23*a*, specifying all numerical parameters. Take account of the fact that, although typical voice signals contain higher frequencies, they can be restricted to a 4-kHz bandwidth without loss of intelligibility.

7.35 Referring to the receiver portion of Figure 7.23*a*, suppose that $y_1(t)$ is multiplied by an *unsynchronized* wave so that $z_1(t) = 2y_1(t) \cos(\omega_1 t + 2\pi\Delta f t + \Delta\theta)$, where Δf and $\Delta\theta$ are slowly and randomly drifting with time. Analyze the resulting filtered output taking $x_1(t) = \cos 2\pi f_\mu t$. In particular, consider what happens when: (a) $\Delta f < W_1 - f_\mu$ and $\Delta\theta = 0$; (b) $\Delta f > W_1 - f_\mu$ and $\Delta\theta = 0$; (c) $\Delta f = 0$ and $|\Delta\theta| = 90°$.

7.36 *Speech scramblers* are sometimes employed to defeat electronic eavesdropping. Analyze the simple scrambler diagrammed in Figure P7.7 by sketching the spectra at points (a) through (d). Assume the filters are ideal, both cutting off at $|f| = f_c$, and that $f_c >> W$. How would you unscramble the scrambled output?

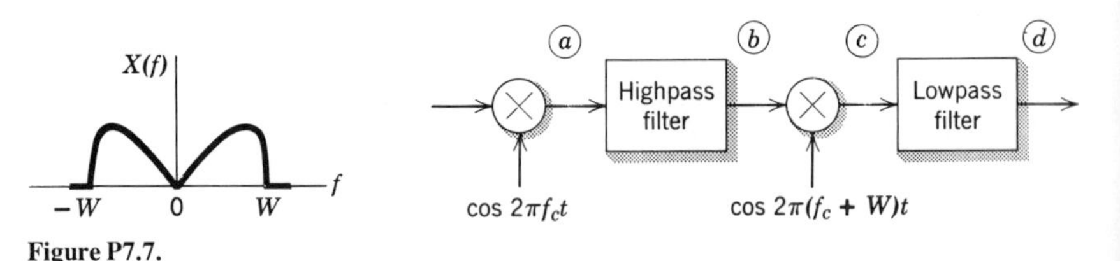

Figure P7.7.

7.37 A special multiplexing technique, good for only two channels, uses as the transmitted signal $x(t) = x_1(t) \cos \omega_c t + x_2(t) \sin \omega_c t$. Show that the system in Figure P7.8 will achieve the demultiplexing.

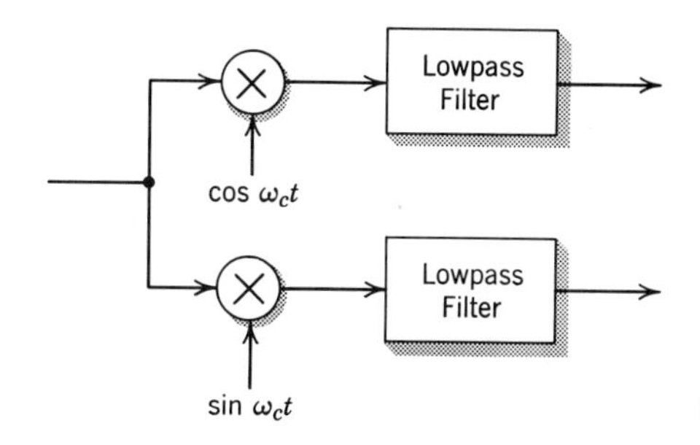

Figure P7.8.

7.38 Prove Eq. (6), Sect. 7.5, taking the approach used to derive Rayleigh's energy theorem.

7.39 The frequency response of a high-Q circuit resonating at f_c is given approximately by

$$H_{BP}(f) = [1 + j2Q(f - f_c)/f_c]^{-1} \qquad f > 0$$

(a) Show that $H_{LP}(f)$ has the form of an RC filter, and use Eq. (11), Sect. 7.5, to find $h_{BP}(t)$. (b) Determine $y_{BP}(t)$ when $x_{BP}(t) = A \cos \omega_c t \, u(t)$.

7.40 Consider the amplitude-modulated signal $x_{BP}(t) = x(t) \cos 2\pi f_c t$, where $x(t)$ is bandlimited in $W \ll f_c$. This signal is transmitted over a system having $H_{BP}(f) = Ce^{j\theta(f)}$ where $\theta(f) = \arg [H(f)]$ is reasonably smooth but not strictly linear. Show that, to a good approximation,

$$y_{BP}(t) = Cx(t - t_{ed}) \cos 2\pi f_c (t - t_{cd})$$

where $t_{cd} = -\theta(f_c)/2\pi f_c$ and $t_{ed} = -(1/2\pi)\, d\theta(f_c)/df$, which are known as the *carrier* and *envelope delays*, respectively. *Hint:* use a Taylor series to obtain a linear approximation for $\theta(f + f_c)$.

8

Analysis of cascade systems: Laplace transforms

In Chapters 6 and 7 we found the Fourier transform and series to be useful tools for investigating signals and systems in the frequency domain. Generally our interest was in the spectrum of the system's zero-state response and we seldom had need to solve for the actual output waveform. Under these conditions, which are characteristic of communication systems engineering, we have essentially a *steady-state* situation for which the frequency-response function $H(f)$ and Fourier-analysis methods are best suited.

In subsequent chapters where we shall be studying control systems, or feedback systems in general, it is imperative that our analytic techniques be capable of handling *transients* – that is, given an excitation that starts at some particular instant and the state of the system at that instant, it is necessary to find the resulting complete-response waveform. Furthermore, many of the signals with which we shall commonly work are not of the energy type and thus do not possess Fourier transforms. Therefore, the mathematical tool that will be most useful for our future work with feedback systems is the *Laplace transform*. This chapter introduces Laplace-transform analysis methods in the context of cascade systems, with the treatment of feedback systems deferred until Chapter 9.

Starting with the definition of the transform, various theorems and the transforms of common functions will be derived. Then the theorems and relationships that are particularly suited to the solution for the zero-input and zero-state responses are derived and illustrated. In contrast to frequency-domain analysis, we are particularly interested in finding the inverse transform which is covered in Section 8.3. The chapter concludes by making use of transform methods to obtain some basic and very general results for multi-input-output systems described by the matrix state-variable equations.

8.1 DEFINITIONS, THEOREMS, AND TRANSFORMS

The Laplace transform of $v(t)$, denoted by $\mathcal{L}[v(t)] = V(s)$, is a function of the complex variable $s = \sigma + j2\pi f = \sigma + j\omega$. It is defined to be the integral

$$V(s) \triangleq \int_0^\infty v(t)e^{-st}dt \tag{1}$$

where $v(t)$ must vanish for all $t < 0$. By writing (1) as

$$V(s) = \int_0^\infty [v(t)e^{-\sigma t}]e^{-j\omega t}dt$$

and comparing it with Eq. (3), Sect. 7.1, noting that $\omega = 2\pi f$, we can see a strong similarity to the Fourier transform. Heuristically, the Laplace transform may be considered as the result of *1*. restricting consideration to functions that are zero for $t < 0$ (thereby accounting for the lower integration limit of zero†), and *2*. introducing the convergence factor $e^{-\sigma t}$ into the integrand. This convergence factor, in effect, makes the product $v(t)e^{-\sigma t}$ into an energy signal even if $v(t)$ itself is not an energy signal, thereby expanding the class of transformable functions.

A function $v(t)$ will have a Laplace transform‡ provided that it is of *exponential order* which, in mathematical terms, requires that there must exist a real number σ_1 for which

$$\lim_{t\to\infty} |v(t)e^{-\sigma_1 t}| = 0$$

For example, $v(t) = Me^{\alpha t}$ is of exponential order for *any* combination of finite M and α. Hence the restrictions for the existence of a Laplace transform are very loose, the principal one being that the function must vanish for all $t < 0$.

†The transform defined by (1) is more accurately referred to as the *one-sided* or *unilateral* Laplace transform to distinguish it from the *two-sided* or *bilateral* Laplace transform in which the range of integration is $-\infty < t < \infty$. The bilateral Laplace transform may be viewed as the parent of both the Fourier transform and the one-sided Laplace transform being more of conceptual than computational value. See LePage (1961, Chap. 10) for further details.

‡See Schwarz and Friedland (1965, Sect. 6.3) for a more rigorous treatment of the subject.

We have shown the lower limit of integration in (1) as being exactly at $t = 0$. In the literature, however, it is often taken as $t = 0-$, so as to include impulses and other singularities that straddle the time origin. Occasionally, $t = 0+$ is used, so as to exclude impulses, etc., from the transform integral. Any of the three lower limits can be adopted provided that one is consistent, particularly with regard to impulses (and higher-order singularities, e.g., doublets). To provide the necessary consistency, we shall define the class of *causal functions* to include all functions that are identically zero for $t \leq 0$; in other words, they exist only for positive values of time. Henceforth, we shall deal exclusively with causal functions, unless otherwise noted. Thus, in contrast to Eq. (2), Sect. 4.2, we define the unit impulse as having the sampling or shifting property

$$\int_0^\infty v(t)\delta(t)\,dt \triangleq v(0+) \tag{2}$$

where

$$v(0+) \triangleq \lim_{\substack{t\to 0\\ t>0}} v(t) \tag{3}$$

i.e., the limit is approached from the right. To achieve this property we need only restrict the class of functions that become impulses in the limit to causal functions, two of which are shown in Figure 8.1; no further modifications in the theoretical development of the impulse are required.

In keeping with the above, the *causal step function* will be defined as

$$u(t) \triangleq \begin{cases} 0 & t \leq 0 \\ 1 & t > 0 \end{cases} \tag{4}$$

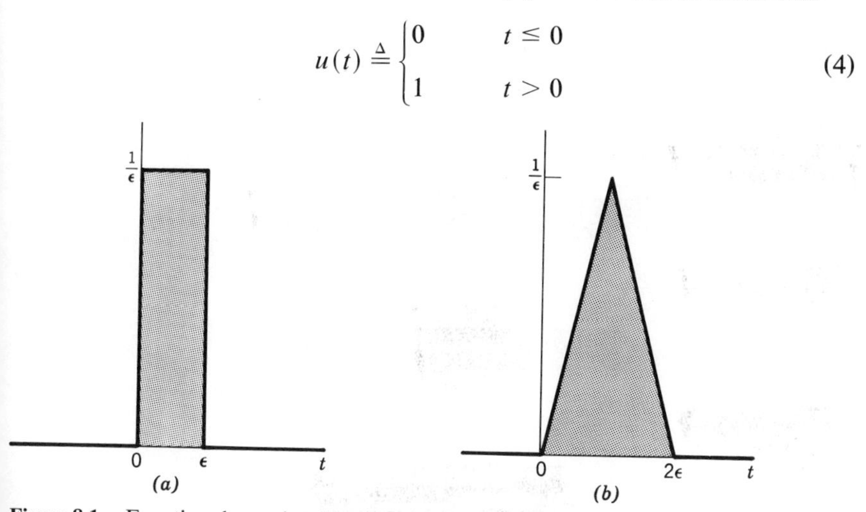

Figure 8.1. Functions becoming the causal impulse as $\epsilon \to 0$.

which is consistent with Eq. (13), Sect. 4.2 – recall that we reserved the right to assign that value to $u(0)$ which best suited the particular application. By defining $u(0)=0$ at this point we are also consistent with the causal impulse in that $du/dt = \delta(t)$. For all functions $v(t)$ that do not contain an impulse or higher-order singularity at $t=0$, the causal version of the function refers to the product $v(t)u(t)$ where $u(t)$ is the causal step. Two common examples of such functions – the exponential and cosine functions – are shown in Figure 8.2. Henceforth, it will be implicit that *all functions under consideration are multiplied by* $u(t)$.

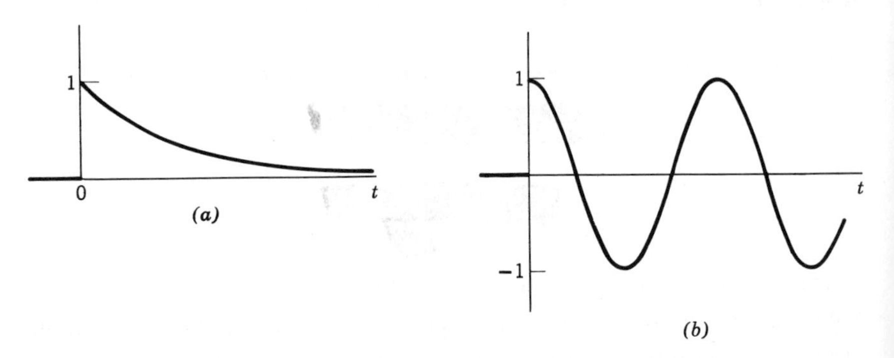

Figure 8.2. Two causal waveforms. (*a*) Decaying exponential. (*b*) Cosine.

Laplace transform theorems

Many Laplace-transform theorems are directly analogous to Fourier-transform theorems with $j\omega$ replaced by s and due account taken of the restriction to causal functions. Three of these are:

Superposition

$$\alpha v(t) + \beta w(t) \leftrightarrow \alpha V(s) + \beta W(s) \tag{5}$$

Time scaling

$$v(ct) \leftrightarrow \frac{1}{c} V\left(\frac{s}{c}\right) \qquad c > 0 \tag{6}$$

Integration

$$\int_0^t v(\lambda)\, d\lambda \leftrightarrow \frac{1}{s} V(s) \tag{7}$$

Proof of these is left for the reader. Other theorems, requiring more attention, are discussed below.

Differentiation

In order to relate the transform of the derivatives and integrals of $v(t)$ to $V(s)$, we insert dv/dt into the transform integral and then integrate by parts, obtaining

$$\mathcal{L}\left[\frac{dv}{dt}\right] = \int_0^\infty \left(\frac{dv}{dt}\right) e^{-st} dt = -\int_0^\infty v(t)(-s)e^{-st} dt + v(t)e^{-st}\Big|_0^\infty$$

$$= s\underbrace{\int_0^\infty v(t)e^{-st}dt}_{V(s)} - v(0) + v(t)e^{-st}\Big|_{t=\infty}$$

By virtue of the fact that $v(t)e^{-st}$ must vanish as $t \to \infty$ in order for the integral expression of $V(s)$ to converge in the first place, the third term above is zero, leaving

$$\frac{dv}{dt} \leftrightarrow sV(s) - v(0) \tag{8}$$

The use of (8) in a recursive manner results in the family of transform formulas

$$\begin{aligned} \frac{d^2v}{dt^2} &\leftrightarrow s^2V(s) - sv(0) - \dot{v}(0) \\ &\vdots \\ \frac{d^nv}{dt^n} &\leftrightarrow s^nV(s) - s^{n-1}v(0) - s^{n-2}\dot{v}(0) - \cdots - v^{(n-1)}(0) \end{aligned} \tag{9}$$

Time delay

The transform of $v(t-t_d)$ will be related to $V(s)$ by a complex phase shift:

$$v(t-t_d) \leftrightarrow e^{-st_d}V(s) \qquad t_d \geq 0 \tag{10}$$

To prove (10), the transform integral is

$$\mathcal{L}[v(t-t_d)] = \int_0^\infty v(t-t_d)e^{-st}dt = \int_0^\infty v(t-t_d)e^{-s(t-t_d)}e^{-st_d}dt$$

which, upon changing the variable of integration to $\lambda = t - t_d$, becomes

$$\mathfrak{L}[v(t-t_d)] = e^{-st_d} \int_{-td}^{\infty} v(\lambda) e^{-s\lambda} d\lambda$$

where the change in the lower limit results from the change of variable. However, because $t_d \geq 0$ and $v(\lambda) = 0$ for $\lambda < 0$, the lower limit of integration can be changed back to zero without affecting the integral, thereby giving (10).

The time-delay theorem is analogous to the Fourier-transform time-shift theorem, which is valid for both positive and negative values of t_d. However, the Laplace-transform version given by (10) has been restricted to $t_d \geq 0$ in order to avoid the possibility of shifting a nonzero portion of the function $v(t)$ to negative values of time. Such a shift would violate the requirement that a function vanish for all $t < 0$ in order to have a Laplace transform.

Next we turn to the evaluation of the transforms of specific functions.

Singularity functions

Substituting the causal impulse into the transform integral (1) and making use of the sampling property (2), it follows that

$$\mathfrak{L}[\delta(t)] = \int_0^{\infty} \delta(t) e^{-st} dt = e^{-st}\Big|_{t=0+} = 1$$

or, in the form of a transform pair,

$$\delta(t) \leftrightarrow 1 \tag{11}$$

The transforms of higher-order singularities can be obtained by applying the differentiation rule (8) to (11). For example, the causal doublet, denoted by $\delta'(t)$, is the derivative of $\delta(t)$ in the sense described in Chapter 4, and, using $\delta(0) = 0$,

$$\delta'(t) \leftrightarrow s \tag{12}$$

Consider, on the other hand, the causal step function which is the integral of the causal impulse; it follows from (7) that

$$u(t) \leftrightarrow \frac{1}{s} \tag{13}$$

This result can also be derived by direct substitution of $u(t)$ into the transform integral. Similarly, the causal ramp function $tu(t)$ is the integral of $u(t)$ and its transform is given by

$$t \leftrightarrow \frac{1}{s^2} \tag{14}$$

By repeated application of the integral rule, one obtains

$$t^n \leftrightarrow \frac{n!}{s^{n+1}} \qquad n = 2, 3, \ldots \tag{15}$$

Causal exponential functions

One of the most useful functions is the causal exponential defined by the product of $e^{at}u(t)$, and written merely as e^{at} (see Figure 8.2a). Substituting e^{at} into (1),

$$\mathfrak{L}[e^{at}] = \int_0^\infty e^{-(s-a)t} dt = \frac{1}{s-a} - \frac{e^{-(s-a)t}}{s-a}\bigg|_{t=\infty}$$

The second term will vanish as $t \to \infty$ if $\mathrm{Re}[s] > a$, which must be the case if the integral is to converge. Hence

$$e^{at} \leftrightarrow \frac{1}{s-a} \tag{16}$$

Because of the dominant role played by the exponential time function in the response of fixed linear systems – recall the preservation of form property – this transform pair will be used extensively in the inversion of transforms to be taken up in Section 8.3. The closely related transforms of the causal sine and cosine functions can be easily evaluated by using their complex exponential forms along with (16) with $a = \pm j\omega$. The results are

$$\sin \omega t \leftrightarrow \frac{\omega}{s^2 + \omega^2} \tag{17}$$

and

$$\cos \omega t \leftrightarrow \frac{s}{s^2 + \omega^2} \tag{18}$$

8.2 SYSTEMS ANALYSIS

Having introduced the Laplace transform and determined a number of its properties, we shall now focus our attention on its use in solving the types of problems that arise in the study of fixed, linear continuous systems of the single-input-output (SIO) type. In particular, it was shown in Chapters 2, 3, and 4 that the mathematical models of such systems can be put in the form

$$a_n y^{(n)} + a_{n-1} y^{(n-1)} + \cdots + a_0 y = b_m x^{(m)} + \cdots + b_0 x \tag{1}$$

However, because the inverse transform will not be discussed until the following section, for the moment we shall establish only the transforms of the various system responses, except for several examples in which the time function can be obtained more or less by inspection.

Since our major interest is in the system's transient response, we shall limit our attention to $t \geq 0$, thereby making the use of the Laplace transform defined in Eq. (1), Sect. 8.1, appropriate. For reasons that will become apparent as we proceed (see Example 8.1) we must provide a consistent framework for the behavior of the system at and on either side of the time origin $t = 0$. Hence, we shall adopt the convention – and that is all it is, a convention – that all *input functions* are zero for $t \leq 0$ and start at $t = 0+$, and all *initial conditions* on state or output variables exist at $t = 0$. In fact, our definition of causal functions, particularly the impulse, was motivated by looking ahead to their use as system input functions subject to the above convention.

Zero-state response

To solve for the transform of the zero-state response, that is, $Y_{zs}(s) \triangleq \mathcal{L}[y_{zs}(t)]$, we transform both sides of (1), thereby obtaining an algebraic equation that is readily solved. It follows that if the state of the system is zero at $t = 0$, the initial value of $y(t)$ and its first $n-1$ derivatives must also vanish at $t = 0$. Note that because of our convention, the input $x(t)$ cannot commence until $t = 0+$ so it cannot possibly affect a causal system at $t = 0$. Transforming the left-hand side and using the derivative theorems, Eqs. (8) and (9), Sect. 8.1, with $y(0) = \dot{y}(0) = \cdots = y^{(n-1)}(0) = 0$, it follows that

$$\mathcal{L}[a_n y^{(n)} + a_{n-1} y^{(n-1)} + \cdots + a_0 y] = (a_n s^n + a_{n-1} s^{n-1} + \cdots + a_0) Y_{zs}(s) \tag{2}$$

Transforming the right-hand side with $x(0) = \dot{x}(0) = \cdots = x^{(m-1)}(0) = 0$,

$$\mathcal{L}[b_m x^{(m)} + \cdots + b_0 x] = (b_m s^m + \cdots + b_0) X(s) \tag{3}$$

Because of (1) and the superposition theorem, Eq. (5), Sect. 8.1, the right-hand sides of (2) and (3) may be equated, yielding

$$(a_n s^n + a_{n-1} s^{n-1} + \cdots + a_0) Y_{zs}(s) = (b_m s^m + \cdots + b_0) X(s)$$

To find the transform of the response to a specific causal input $x(t)$, the preceding algebraic equation can be divided by the polynomial

$$P(s) \triangleq a_n s^n + a_{n-1} s^{n-1} + \cdots + a_0 \tag{4}$$

which happens to be the system's characteristic polynomial, giving

$$Y_{zs}(s) = \frac{b_m s^m + \cdots + b_0}{a_n s^n + a_{n-1} s^{n-1} + \cdots + a_0} X(s) \tag{5}$$

Alternately, the system itself can be characterized by dividing both sides of (5) by $X(s)$, forming $Y_{zs}(s)/X(s)$ which is none other than the system *transfer function* $H(s)$. Thus

$$H(s) = \frac{Y_{zs}(s)}{X(s)} = \frac{b_m s^m + \cdots + b_0}{a_n s^n + a_{n-1} s^{n-1} + \cdots + a_0} \tag{6}$$

and the transform of the zero-state response is just the product of the transfer function and the input signal's transform:

$$Y_{zs}(s) = H(s) X(s) \tag{7}$$

Henceforth, we shall take (6) as a replacement for our prior definition of $H(s)$ as the ratio of input to output when $x(t) = e^{st}$ for all $-\infty < t < \infty$. As might be expected, $H(s)$ is the Laplace transform of the impulse response $h(t)$, a fact that is easily shown by letting $X(s) = \mathcal{L}[\delta(t)] = 1$ in (7).

Zero-input response

When evaluating the transform of the zero-input response, $Y_{zi}(s) \triangleq \mathcal{L}[y_{zi}(t)]$, (1) reduces to the homogeneous equation

$$a_n y^{(n)} + a_{n-1} y^{(n-1)} + \cdots + a_0 y = 0 \tag{8}$$

with the initial conditions $y_0 = y(0)$, $y_0^{(1)} = \dot{y}(0), \ldots, y_0^{(n-1)} = y^{(n-1)}(0)$, at least one of which must be nonzero. Because $y_{zi}(t)$ is the output variable, it and its derivatives may have nonzero values at $t = 0$ according to the convention we have adopted. Thus, the use of the derivative formulas to transform each term in (8) and a regrouping of terms yields the *algebraic* equation

$$(a_n s^n + a_{n-1}s^{n-1} + \cdots + a_0)Y_{zi}(s)$$
$$= a_n[s^{n-1}y_0 + s^{n-2}y_0^{(1)} + \cdots + sy_0^{(n-2)} + y_0^{(n-1)}]$$
$$+ a_{n-1}[s^{n-2}y_0 + s^{n-3}y_0^{(1)} + \cdots + y_0^{(n-2)}]$$
$$\vdots$$
$$+ a_2[sy_0 + y_0^{(1)}]$$
$$+ a_1 y_0 \qquad (9)$$

By defining the n polynomials

$$\begin{aligned} A_0(s) &\triangleq a_1 + a_2 s + \cdots + a_{n-1}s^{n-2} + a_n s^{n-1} \\ A_1(s) &\triangleq a_2 + a_3 s + \cdots + a_n s^{n-2} \\ &\vdots \\ A_{n-2}(s) &\triangleq a_{n-1} + a_n s \\ A_{n-1}(s) &\triangleq a_n \end{aligned} \qquad (10)$$

and regrouping terms once again, (9) may be solved for the transform of the zero-input response:

$$Y_{zi}(s) = \frac{A_0(s)y_0 + A_1(s)y_0^{(1)} + \cdots + A_{n-1}(s)y_0^{(n-1)}}{a_n s^n + a_{n-1}s^{n-1} + \cdots + a_0} = \frac{\sum_{i=0}^{n-1} A_i(s)y_0^{(i)}}{P(s)} \qquad (11)$$

where the numerator is a polynomial in s of degree $n-1$ or less.

Complete response

Using (5) and (11) with the superposition properties of linear systems and the Laplace transform, the transform of the complete response is

$$Y(s) = Y_{zi}(s) + Y_{zs}(s) = \frac{\sum_{i=0}^{n-1} A_i(s)y_0^{(i)} + \left(\sum_{k=0}^{m} b_k s^k\right)X(s)}{P(s)} \qquad (12)$$

where $P(s)$ is the characteristic polynomial.

Example 8.1

As a simple demonstration of the application of transforms to the task of solving for the response of a system, we shall consider the first-order system with input $x(t)$ and output $y(t)$ described by

$$\dot{y} + \alpha y = x \tag{13}$$

Comparing (13) with the general SIO equation (1), we see that the system order is $n = 1$ and

$$a_1 = 1 \qquad b_1 = 0$$

$$a_0 = \alpha \qquad b_0 = 1$$

To find the step response $y_u(t)$ we let $X(s) = \mathfrak{L}[u(t)] = 1/s$ and set $y(0) = y_0 = 0$, in which case (5) reduces to

$$Y_u(s) = \frac{1}{s(s+\alpha)}$$

Although we have not yet taken up the transform-inversion process, the reader may readily verify that $Y_u(s)$ can be rewritten as

$$Y_u(s) = \frac{1}{\alpha}\left(\frac{1}{s} - \frac{1}{s+\alpha}\right)$$

Recalling that $u(t) \leftrightarrow 1/s$ and $e^{-\alpha t} \leftrightarrow 1/(s+\alpha)$, it follows that

$$y_u(t) = \frac{1}{\alpha}(1 - e^{-\alpha t}) \qquad t > 0$$

and is identically zero for $t \leq 0$.

As for the system's zero-input response following the initial condition $y_0 = C$, the only nonzero polynomial in (10) is $A_0(s)$, which becomes $A_0(s) = 1$. Therefore, from (11),

$$Y_{zi}(s) = C\frac{1}{s+\alpha}$$

whence

$$y_{zi}(t) = Ce^{-\alpha t} \qquad t > 0$$

Finally, we shall find the complete response to the input $x(t) = \delta(t) + u(t)$ when the initial state is C. Restating the problem, we are to solve the differential equation

$$\dot{y} + \alpha y = \delta(t) + u(t) \qquad y(0) = C$$

Transforming the equation term by term,

$$sY(s) - C + \alpha Y(s) = 1 + \frac{1}{s}$$

from which

$$Y(s) = \frac{C+1+1/s}{s+\alpha} = \frac{s(C+1)+1}{s(s+\alpha)} = \frac{1/\alpha}{s} + \frac{C+1-1/\alpha}{s+\alpha}$$

Thus, identifying the two terms of $Y(s)$ as the transforms of the causal functions $u(t)$ and $e^{-\alpha t}$ multiplied by the appropriate constants, we can find $y(t)$ for $t > 0$. Combining this information with the knowledge that $y(0) = C$, the complete expression for the output is

$$y(t) = \begin{cases} 0 & t < 0 \\ C & t = 0 \\ \frac{1}{\alpha}u(t) + \left(C+1-\frac{1}{\alpha}\right)e^{-\alpha t} & t > 0 \end{cases} \tag{14}$$

which is sketched in Figure 8.3*a*. We observe that the discontinuities in $y(t)$ due to the initial state and the impulsive input function are clearly distinguishable because of the convention adopted in Section 8.1 of establishing the initial state *before* the input arrives. However, if the establishment of the initial state and the impulse in the input function had been allowed to occur simultaneously, it would not be clear whether (14) or a response that satisfied $y(0+) = C$, as shown in Figure 8.3*b*, was

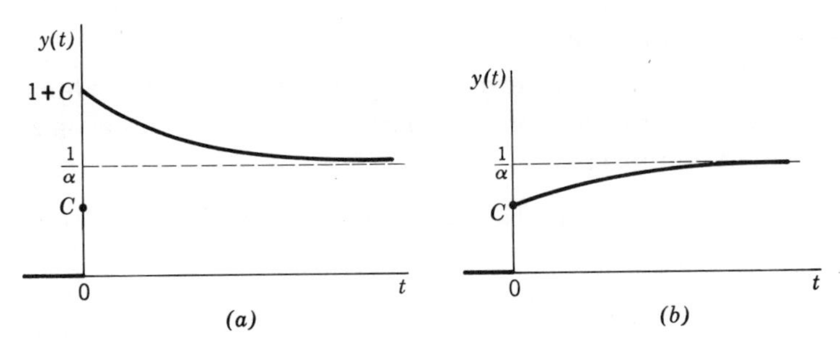

Figure 8.3.

intended. To avoid ambiguities of this type, we shall use the notion of the initial state and insist that it exist at $t = 0$, with all inputs commencing at $t = 0+$.

Time-domain convolution

Because the zero-state response of a system is given in the time domain by $y_{zs}(t) = h * x(t)$ and the corresponding transform relationship has been shown to be $Y_{zs}(s) = H(s)X(s)$, it would seem likely that the transform of the convolution of any two causal functions which are themselves transformable would be the product of their respective transforms. This is indeed the case, as we shall now prove.

Substituting the integral expression for the convolution of $v(t)$ with $w(t)$ into the transform integral gives

$$\mathfrak{L}[v * w(t)] = \int_0^\infty \overbrace{\left[\int_0^t v(\lambda) w(t-\lambda) d\lambda \right]}^{v * w} e^{-st} dt \tag{15}$$

where the limits of the integration over λ are from 0 to t because of the causal nature of both functions. Our next step will be to change the order and limits of integration, the validity of which can be understood by referring to Figure 8.4. The area in the λ-t plane over which the double integration is performed is the shaded semi-infinite region.

The double integration of (15) is summed first along the vertical strips shown in Figure 8.4a and then these strips are summed over $0 \le t < \infty$. Recalling the definition of integration over two variables, the integrand can just as well be summed first along the horizontal strip in part (b), and then over $0 \le \lambda < \infty$. Making this change, (15) becomes

$$\mathfrak{L}[v * w(t)] = \int_0^\infty \int_\lambda^\infty v(\lambda) w(t-\lambda) e^{-st} d\lambda\, dt$$

$$= \int_0^\infty v(\lambda) e^{-s\lambda} \left[\int_\lambda^\infty w(t-\lambda) e^{-s(t-\lambda)} dt \right] d\lambda \tag{16}$$

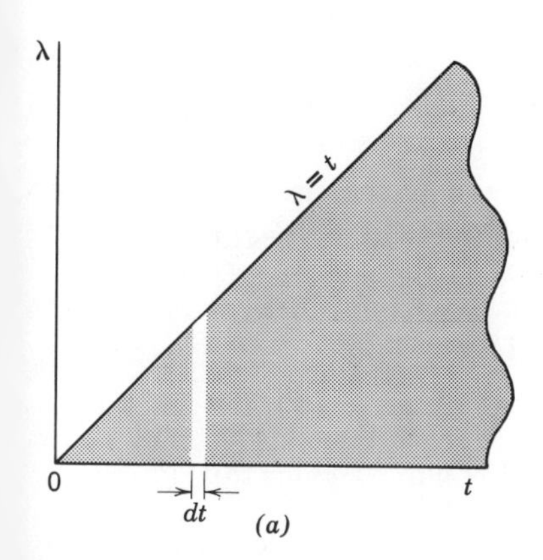

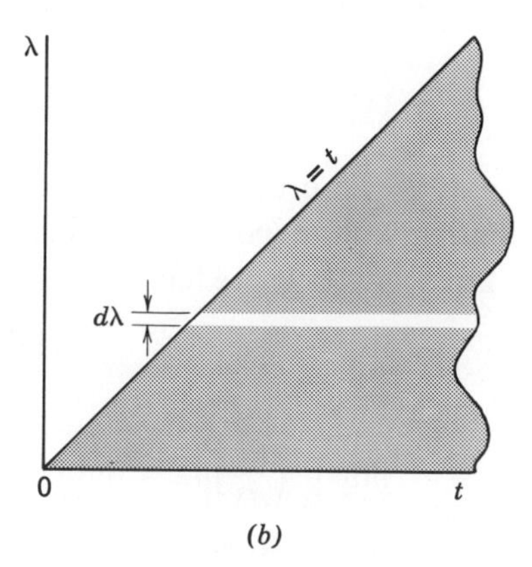

Figure 8.4.

where e^{-st} has been written in the equivalent form $e^{-s(t-\lambda)}e^{-s\lambda}$. Now, making the change of variable $\mu = t - \lambda$,

$$\mathcal{L}[v * w(t)] = \underbrace{\int_0^\infty v(\lambda)e^{-s\lambda}\,d\lambda}_{V(s)} \underbrace{\int_0^\infty w(\mu)e^{-s\mu}\,d\mu}_{W(s)}$$

which, in our shorthand notation, becomes

$$v * w(t) \leftrightarrow V(s)W(s) \tag{17}$$

Final-value theorem

In the analysis of dynamic systems, one often desires to know the value that a variable approaches as $t \to \infty$, assuming, of course, that it does approach a limit. Using the partial-fraction method of inverting Laplace transforms (to be discussed in the following section), it is not difficult to show that $v(t)$ approaches a limit as $t \to \infty$ only if all of the poles of $V(s)$ lie in the left-half plane (LHP), with the possible exception of a simple pole at the origin. A more compact way of phrasing this condition is to say that $sV(s)$ must be *analytic*† on the imaginary axis and in the right-half plane (RHP). The final-value theorem states that when this condition on $sV(s)$ is satisfied, then

$$\lim_{t\to\infty} v(t) = \lim_{s\to 0} sV(s) \tag{18}$$

To derive (18), we start by rewriting the integral for $\mathcal{L}[dv/dt]$ over a finite interval, followed by a limiting process:

$$\lim_{t\to\infty} \int_0^t \left(\frac{dv}{d\lambda}\right)e^{-s\lambda}\,d\lambda = sV(s) - v(0)$$

Next, we take the limit of both sides of the equation as $s \to 0$, with the limiting with respect to s *preceding* the limiting with respect to t. Since λ is finite as $s \to 0$, we have

$$\lim_{s\to 0} e^{-s\lambda} = 1$$

whence

$$\lim_{t\to\infty} \int_0^t \left(\frac{dv}{d\lambda}\right)d\lambda = \lim_{t\to\infty} [v(t) - v(0)] = \lim_{s\to 0} [sV(s) - v(0)]$$

†The reader who is not familiar with such basic concepts of complex-variable theory should now refer to Appendix A.

Because the terms involving $v(0)$ are unaffected by either limiting process they may be canceled; having done so, the final-value theorem results.

Before continuing, let us point out the danger of using this theorem for transforms that do not satisfy the condition for its applicability. To demonstrate, consider

$$V(s) = \frac{2s^2 + \omega^2}{s(s^2 + \omega^2)}$$

Applying the final-value theorem yields

$$\lim_{s \to 0} sV(s) = 1$$

which would indicate that $v(t)$ should approach unity as $t \to \infty$. However, the reader may verify that $V(s)$ is the transform of $(1 + \cos \omega t)$ which has no limit as $t \to \infty$. The fact that its transform has a pair of poles on the imaginary axis $(s = \pm j\omega)$ is the clue that the final-value theorem must *not* be applied, in spite of the temptation to do so. What has been found is the constant portion of the function that exists, along with the undamped oscillation $\cos \omega t$, after any transients have decayed.

Initial-value theorem

When the function $v(t)$ contains a discontinuity at $t = 0$, it is sometimes desirable to ascertain $v(0+)$ directly from the transform $V(s)$, without having to obtain the complete function $v(t)$ by inverting the transform. This can be done by applying the initial-value theorem which states that

$$v(0+) = \lim_{s \to \infty} sV(s) \tag{19}$$

The proof of the theorem starts with the expression for $\mathcal{L}[dv/dt]$ and involves interchange and limiting arguments that are somewhat similar to those used in proving the final-value theorem. The details are left as an exercise for the reader.

8.3 TRANSFORM INVERSION BY PARTIAL FRACTIONS

In contrast to the inversion of Fourier transforms, which is seldom carried out in communication-systems analysis, the inversion of Laplace transforms plays an important role in the study of control systems, primarily because of the emphasis on transient response. Although the inversion integral† provides the underlying connection between the complex-frequency and time domains, an approach based upon the partial-fraction theorem (see Appendix A for its proof) is widely used for computational purposes and will be given primary emphasis here. We shall focus our attention on transforms that are rational functions, that is, the ratio of two polynomials in s. Usually, $V(s)$ will represent the transform of a signal associated with a system, i.e., its input, a state variable, or its output, although in some instances it may be a transfer function – in which case $v(t)$ will be an impulse response. On occasion we shall use the shorthand notation $\mathcal{L}^{-1}[V(s)]$ to denote the inverse transform of $V(s)$. We shall start with the simplest case: transforms having distinct poles.

Distinct poles

Since $V(s)$ is assumed to be a rational function it can be written as

$$V(s) = \frac{b_m s^m + \cdots + b_0}{a_n s^n + a_{n-1}s^{n-1} + \cdots + a_0} \tag{1}$$

which, in factored form, becomes

$$V(s) = \left(\frac{b_m}{a_n}\right) \frac{\prod_{i=1}^{m} (s - z_i)}{\prod_{k=1}^{n} (s - p_k)} \tag{2}$$

The terms $(s - z_i)$ are the factors of the numerator polynomial and the z_i are referred to as the *zeros* of $V(s)$ because $V(z_i) = 0$ for $i = 1, 2, \ldots, m$. The terms $(s - p_k)$ are the factors of the denominator polynomial and are known as the *poles* of the transform. If no two of the p_k are equal to one another, the poles are said to be *distinct*. If two or more of the p_k are equal, that pole is said to be repeated or to have a multiplicity greater than one.

Subject to the restrictions that $m < n$ and the poles are distinct, the

†The function $v(t)$ and its Laplace transform $V(s)$ defined by Eq. (1), Sect. 8.1, satisfy the inversion integral

$$v(t) = \frac{1}{2\pi j} \int_{\sigma_c - j\infty}^{\sigma_c + j\infty} V(s) e^{st} ds$$

where σ_c must be greater than the real parts of all of the poles of $V(s)$. As shown in Schwarz and Friedland (1965, Sect. 6.6), the contour-integral approach to finding $v(t)$ is more fundamental and applicable to a wider class of transforms than the partial-fraction method employed here.

partial-fraction expansion theorem states that $V(s)$ can be written as the summation

$$V(s)=\frac{C_1}{s-p_1}+\frac{C_2}{s-p_2}+\cdots+\frac{C_n}{s-p_n} \tag{3}$$

where the p_k are the poles of $V(s)$ and the C_k are a unique set of constants (perhaps complex). To compute C_1, for example, we multiply both sides of (3) by $(s-p_1)$, yielding

$$(s-p_1)V(s)=C_1+\frac{s-p_1}{s-p_2}C_2+\cdots+\frac{s-p_1}{s-p_n}C_n \tag{4}$$

Because the poles are distinct, evaluating (4) at $s=p_1$ will make each term on the right-hand side vanish, except the constant term C_1. Thus, $C_1=(s-p_1)V(s)|_{s=p_1}$ which generalizes to

$$C_k=(s-p_k)V(s)|_{s=p_k} \qquad k=1,2,\ldots,n \tag{5}$$

Notice that (3) is comprised of the summation of individual terms of the form $1/(s-p_k)$, each of which has the inverse transform

$$\mathfrak{L}^{-1}\left[\frac{1}{s-p_k}\right]=e^{p_k t}$$

It then follows from the transform superposition property that

$$v(t)=\sum_{k=1}^{n} C_k e^{p_k t} \qquad t>0 \tag{6}$$

where the C_k are given by (5) and the p_k are the distinct poles of $V(s)$.

Example 8.2

To demonstrate the inversion of transforms with real, distinct poles, we shall find $v_1(t)$ and $v_2(t)$ corresponding to

$$V_1(s)=\frac{1}{(s+1)(s+3)} \qquad \text{and} \qquad V_2(s)=\frac{s+\alpha}{(s+1)(s+3)}$$

Both transforms have distinct poles $(s_1=-1,\ s_2=-3)$ and the order of the numerator polynomial is less than that of the denominator so we know immediately that both time functions have the form $v(t)=C_1e^{-t}+C_2e^{-3t}$

In the case of $V_1(s)$,

$$C_1 = (s+1)V_1(s)|_{s=-1} = \left.\frac{1}{s+3}\right|_{s=-1} = \frac{1}{2}$$

$$C_2 = (s+3)V_1(s)|_{s=-3} = \left.\frac{1}{s+1}\right|_{s=-3} = -\frac{1}{2}$$

resulting in

$$v_1(t) = \tfrac{1}{2}e^{-t} - \tfrac{1}{2}e^{-3t}$$

On the other hand, the expansion coefficients for $V_2(s)$ are

$$C_1 = (s+1)V_2(s)|_{s=-1} = \left.\frac{s+\alpha}{s+3}\right|_{s=-1} = \frac{\alpha-1}{2}$$

$$C_2 = (s+3)V_2(s)|_{s=-3} = \left.\frac{s+\alpha}{s+1}\right|_{s=-3} = \frac{3-\alpha}{2}$$

giving

$$v_2(t) = \frac{\alpha-1}{2}e^{-t} + \frac{3-\alpha}{2}e^{-3t}$$

Thus, by virtue of its transform having the term $s+\alpha$ as its numerator, rather than unity, the weighting of the two exponential functions in $v_2(t)$ is dependent upon the parameter α, i.e., upon the zero at $s=-\alpha$. For example, if $\alpha = 1$ the zero coincides with the pole at $s=-1$ and $v_2(t)$ reduces to e^{-3t}. Likewise, if $\alpha = 3$ the zero coincides with the pole at $s=-3$ and $v_2(t)$ reduces to e^{-t}.

Complex poles

Invariably, the coefficients of transforms in which we shall be interested will be real, thereby requiring that all complex poles (and zeros) shall occur in complex-conjugate pairs. Thus, if $s=p_1$ is a complex pole of $V(s)$, there must also be a pole, say $s=p_2$, such that $p_2 = p_1^*$. Assuming that its poles are distinct, $V(s)$ may be rewritten as

$$V(s) = \frac{C_1}{s-p_1} + \frac{C_2}{s-p_1^*} + \sum_{k=3}^{N} \frac{C_k}{s-p_k} \tag{7}$$

where it can be shown that $C_2 = C_1^*$ if the coefficients of $V(s)$ are to be real. In this case it is possible to express the *combined* contribution of p_1 and p_2 to $v(t)$ in terms of a cosine wave with an exponentially varying envelope, a form that is usually more convenient to work with in transient analysis than the complex exponentials resulting from a direct application of (6).

Inverting (7) term by term,

$$v(t) = C_1 e^{p_1 t} + C_1^* e^{p_1^* t} + \sum_{k=3}^{n} C_k e^{p_k t}$$

$$= 2|C_1| e^{\sigma_1 t} \cos(\omega_1 t + \arg[C_1]) + \sum_{k=3}^{n} C_k e^{p_k t} \tag{8}$$

where $p_1 = \sigma_1 + j\omega_1$. To simplify the evaluation of C_1, from which $|C_1|$ and $\arg[C_1]$ can be found, we define the rational function

$$V_0(s) \triangleq \left(\frac{b_m}{a_n}\right) \frac{\prod_{i=1}^{m} (s - z_i)}{\prod_{k=3}^{n} (s - p_k)} \tag{9}$$

such that $V(s) = V_0(s)/[(s-p_1)(s-p_1^*)]$. Then, drawing upon the fact that $p_1 - p_1^* = j2\omega_1$,

$$C_1 = \left.\frac{V_0(s)}{(s - p_1^*)}\right|_{s=p_1} = \frac{V_0(p_1)}{j2\omega_1} = \frac{V_0(p_1)}{2\omega_1} e^{-j90°} \tag{10}$$

whence

$$|C_1| = \frac{|V_0(p_1)|}{2\omega_1} \qquad \arg[C_1] = \arg[V_0(p_1)] - 90°$$

which can be substituted into (8). Any additional complex poles in $V(s)$ can be treated in exactly the same fashion.

V(s) not strictly proper ($m \geq n$)

When the number of zeros is less than the number of poles ($m < n$) the transform $V(s)$ is said to be a *strictly proper* rational function. Because the partial-fraction theorem is valid only for such functions, steps must be taken to modify the form of $V(s)$ when $m \geq n$ so as to consist of the sum of a strictly proper function, say $V_1(s)$, and other terms whose inverse transforms are readily identified.

The technique used is to divide the denominator polynomial into the numerator polynomial, such that the quotient consists of the sum of a polynomial of degree $m-n$ and a strictly proper rational function for which the numerator degree will be at least one less than that of the denominator. The inverse transform of the $m-n$ degree polynomial will be a set of singularity functions, i.e., impulse, doublet, etc., and the strictly proper rational function may be inverted as described above.

Example 8.3

To illustrate this technique, consider

$$V(s)=\frac{s^2+6s+8}{s^2+4s+3}$$

which has $m=n=2$ and hence is not strictly proper. Dividing the numerator by the denominator yields the equivalent form

$$V(s)=1+\frac{2s+5}{s^2+4s+3}=1+\frac{2s+5}{(s+1)(s+3)}$$

each term of which can be inverted separately and then added to obtain $v(t)$. Because $\mathcal{L}^{-1}[1]=\delta(t)$, the constant term in $V(s)$ corresponds to a unit impulse at $t=0+$ and the proper-fraction portion of $V(s)$ contributes the exponential terms e^{-t} and e^{-3t}, giving

$$v(t)=\delta(t)+\tfrac{3}{2}e^{-t}+\tfrac{1}{2}e^{-3t}$$

Repeated poles

Because the notation becomes very cumbersome when an arbitrary number of poles are allowed to have arbitrary multiplicities, and because multiplicities greater than 2 are seldom encountered in practice, only the case of a double pole (multiplicity of 2) will be treated. If $p_1=p_2$ but $p_1\neq p_3, p_4, \ldots, p_n$, we say that p_1 is a double pole. If the remaining poles $p_3, p_4, \ldots, p_n$ are distinct, it can be proved† that the transform can be expressed in partial-fraction form as

$$V(s)=\frac{C_{11}}{(s-p_1)^2}+\frac{C_{12}}{s-p_1}+\sum_{k=3}^{n}\frac{C_k}{s-p_k} \tag{11}$$

For $k=3, 4, \ldots, n$, the coefficients C_k are found by applying (5), as before. However, using (5) with $k=1$ yields $C_{11}=\infty$, a clue to the fact that a different approach is needed for the repeated pole.

It is apparent that C_{11} can be found merely by multiplying both sides of

†The proof of the partial-fraction theorem in Appendix A is readily extended to repeated roots at the expense of more complicated notation.

(11) by the factor $(s-p_1)^2$, giving

$$(s-p_1)^2V(s) = C_{11} + (s-p_1)C_{12} + (s-p_1)^2 \sum_{k=3}^{n} \frac{C_k}{s-p_k} \tag{12}$$

and then letting $s = p_1$. Furthermore, if (12) is differentiated with respect to s, we have

$$\frac{d}{ds}[(s-p_1)^2V(s)] = C_{12} + (s-p_1) \sum_{k=3}^{n} C_k\left[\frac{2}{s-p_k} - \frac{(s-p_1)}{(s-p_k)^2}\right]$$

the right-hand side of which reduces to C_{12} when evaluated for $s = p_1$. Thus, the n unknown coefficients in (11) can be evaluated according to

$$C_{11} = (s-p_1)^2V(s)|_{s=p_1} \tag{13a}$$

$$C_{12} = \left\{\frac{d}{ds}[(s-p_1)^2V(s)]\right\}_{s=p_1} \tag{13b}$$

and

$$C_k = (s-p_k)V(s)|_{s=p_k} \qquad k = 3, 4, \ldots n \tag{13c}$$

To find $v(t)$, the first term in (11) can be inverted by using the convolution theorem, Eq. (17), Sect. 8.2, to see that $1/(s-p_1)^2$ must be the transform of e^{p_1t} convolved with itself. Carrying out the convolution integral yields the transform pair

$$te^{p_1t} \leftrightarrow \frac{1}{(s-p_1)^2} \tag{14}$$

Thus, the inverse transform of (11) is given by

$$v(t) = C_{11}te^{p_1t} + C_{12}e^{p_1t} + \sum_{k=3}^{n} C_ke^{p_kt} \tag{15}$$

Interpretation of poles and zeros: modes

By combining the preceding discussion of a transform in terms of its poles and zeros with the transform of the complete response we can obtain some insight into the behavior of fixed linear systems. First, we express the input signal's transform $X(s)$ as a rational function, defining its numerator and denominator polynomials by

$$\frac{N_X(s)}{D_X(s)} \triangleq X(s) \tag{16}$$

The roots of $N_X(s) = 0$ are the *zeros* of $X(s)$ and the roots of $D_X(s) = 0$ are its *poles*.

Then, substituting for $X(s)$ in Eq. (12), Sect. 8.2, the transform of the complete response is

$$Y(s) = \underbrace{\frac{\sum_{i=0}^{n-1} A_i(s) y_0^{(i)}}{P(s)}}_{Y_{zi}(s)} + \underbrace{\frac{\left(\sum_{k=0}^{m} b_k s^k\right) N_X(s)}{P(s) D_X(s)}}_{Y_{zs}(s)} \tag{17}$$

where the transforms of the zero-input and zero-state responses $Y_{zi}(s)$ and $Y_{zs}(s)$, respectively, are readily identified. For reference purposes, we also rewrite the transfer function of the system as

$$H(s) = \frac{\sum_{k=0}^{m} b_k s^k}{P(s)} \tag{18}$$

Clearly, the n roots of the characteristic equation $P(s) = 0$ are the poles of $Y_{zi}(s)$ which, in turn, completely specify the exponential time functions, or *modes*, comprising the zero-input response. The initial values y_0, $\dot{y}_0, \ldots y_0^{(n-1)}$, help determine the numerator of $Y_{zi}(s)$ and, hence, its zeros. As such, they contribute to the partial-fraction expansion coefficients or, equivalently, to the weighting of the modes in the zero-input response. As we shall see in the following example, the initial conditions can be selected so as to excite specific modes and suppress others by causing a zero of $Y_{zi}(s)$ to coincide with a pole that is to be suppressed.

Because the denominator of $Y_{zs}(s)$ is the product of $P(s)$ and $D_X(s)$, the poles of $Y_{zs}(s)$ are the combined poles of $H(s)$ and of $X(s)$. This property implies that the zero-state, or forced, response is composed of the modes of the free response and the time functions comprising the input $x(t)$. Exceptions to this rule occur when *1*. a pole is common to both $H(s)$ and $X(s)$, in which case the multiplicity of the pole is increased, and *2*. a pole of $H(s)$ is canceled by a zero of $X(s)$, or vice versa. Finally, the zeros of $Y_{zs}(s)$ are the combined zeros of $H(s)$ and $X(s)$ and contribute to the weighting of the system modes and the components of $x(t)$ in the forced response.

Example 8.4

Evaluating the transform of the zero-input response of the system which obeys the equation

$$\ddot{y} + 4\dot{y} + 3y = 2\dot{x} + x$$

and has the initial conditions y_0 and $\dot{y}_0$ gives

$$Y_{zi}(s) = \frac{y_0 s + (\dot{y}_0 + 4y_0)}{(s+1)(s+3)} = \frac{y_0(s+4+\alpha)}{(s+1)(s+3)} \tag{19}$$

where $\alpha = y_0/\dot{y}_0$. It is apparent that $Y_{zi}(s)$ has poles at $s=-1$ and -3 and a zero at $s=-(4+\alpha)$. Thus, the two modes comprising the free response are e^{-t} and e^{-3t}. Before evaluating the partial-fraction expansion coefficients of $Y_{zi}(s)$ to find the weighting of the individual modes, we can observe that when $\alpha=-3$ the zero of the numerator falls exactly on the pole at $s=-1$, causing $Y_{zi}(s)$ to be reduced to $y_0/(s+3)$; in other words, that particular ratio of initial states excites only the e^{-3t} mode. Similarly, setting $\alpha=-1$ suppresses the e^{-3t} mode and excites the e^{-t} mode.

The zero-input response is found by carrying out the partial-fraction expansion of (19), yielding, for $t \geq 0$,

$$\begin{aligned} y_{zi}(t) &= \left[\frac{y_0(s+4+\alpha)}{(s+3)}\right]_{s=-1} e^{-t} + \left[\frac{y_0(s+4+\alpha)}{(s+1)}\right]_{s=-3} e^{-3t} \\ &= \left[\frac{y_0(3+\alpha)}{2}\right] e^{-t} - \left[\frac{y_0(1+\alpha)}{2}\right] e^{-3t} \end{aligned}$$

where $\alpha = y_0/\dot{y}_0$. As deduced from pole-zero considerations, the e^{-t} mode is not excited when $\alpha=-3$ and the e^{-3t} mode is absent when $\alpha=-1$.

If the system is now excited by the causal exponential $x(t)=e^{-2t}$ with y_0 and $\dot{y}_0$ specified, the transform of the complete response is

$$[s^2Y(s) - sy_0 - \dot{y}_0] + 4[sY(s) - y_0] + 3Y(s) = 2sX(s) + X(s)$$

where $x(0)=0$ because of the causal property of the input. Rewriting $Y(s)$ so as to separate $Y(s)$ into its zero-input and zero-state components,

$$Y(s) = \underbrace{\frac{(s+4)y_0 + \dot{y}_0}{(s+1)(s+3)}}_{Y_{zi}(s)} + \underbrace{\overbrace{\frac{2(s+\frac{1}{2})}{(s+1)(s+3)}}^{H(s)} \overbrace{\frac{1}{s+2}}^{X(s)}}_{Y_{zs}(s)}$$

Having discussed the zero-input response above, we shall focus our attention on $Y_{zs}(s)$. For the numerical values given, the zero of $H(s)$ is at $s=-\frac{1}{2}$ and does not coincide with the single pole of $X(s)$ at $s=-2$. Furthermore, the pole of $X(s)$ does not coincide with either of the poles of $H(s)$ which are at $s=-1$ and -3. Thus, the forced response y_{zs} will consist of the three exponential terms e^{-t}, e^{-2t}, and e^{-3t}, the coefficients of which are found from a partial-fraction expansion of $Y_{zs}(s)$. Performing the calculations, we find that for $t > 0$

$$y_{zs}(t) = -\tfrac{1}{2}e^{-t} + 3e^{-2t} - \tfrac{5}{2}e^{-3t}$$

which can be added to the previously calculated zero-input response to obtain the complete response $y(t)$.

8.4 STATE-VARIABLE EQUATIONS

Although the state-variable equations and the structure of the zero-input response of multi-input-output (MIO) systems were derived in Chapter 3, no attempt was made to solve these equations in any systematic fashion. However, it is possible to use Laplace transforms, coupled with the results of our study of single-input-output (SIO) systems, to provide insight into both the methods of solution and some of the general properties of such systems. Bear in mind that a definitive study of the relevant theory requires an understanding of linear algebra, e.g., eigenvectors, functions of a matrix, etc., none of which will be attempted here. Furthermore, in practice, calculations are usually carried out on a digital computer because the use of such methods is warranted only when studying systems of at least moderate complexity.

Transfer-function matrix

The Laplace transform of a vector function of time can be defined to be the vector whose elements are the transforms of the elements of the time-function vector.† For example, the n-vector $\mathbf{q}(t)$ and its transform, the vector $\mathbf{Q}(s) \triangleq \mathcal{L}[\mathbf{q}(t)]$, form the transform pair $\mathbf{q}(t) \leftrightarrow \mathbf{Q}(s)$, which is a shorthand notation for the relationship

$$\begin{bmatrix} q_1(t) \\ q_2(t) \\ \vdots \\ q_n(t) \end{bmatrix} \leftrightarrow \begin{bmatrix} Q_1(s) \\ Q_2(s) \\ \vdots \\ Q_n(s) \end{bmatrix} \tag{1}$$

†Boldface capital letters will be used to denote the Laplace transforms of both vectors and matrices. The argument associated with the transform will distinguish it from a constant matrix.

These vector transforms will be applied to the state-variable equations of a fixed MIO system, namely the *state equation*

$$\dot{\mathbf{q}}(t) = \mathbf{Aq}(t) + \mathbf{Bx}(t) \tag{2}$$

and the *output equation*

$$\mathbf{y}(t) = \mathbf{Cq}(t) \tag{3}$$

where $\mathbf{q}$ is the $n \times 1$ state vector, $\mathbf{x}$ is the $p \times 1$ input vector, $\mathbf{y}$ is the $r \times 1$ output vector, and $\mathbf{A}$, $\mathbf{B}$, and $\mathbf{C}$ are constant matrices of appropriate dimension.

Using Eq. (8), Sect. 8.1, with (1) it follows that $\mathcal{L}[\dot{\mathbf{q}}(t)] = s\mathbf{Q}(s) - \mathbf{q}(0)$, where $\mathbf{q}(0)$ is the initial state of the system and is also an $n \times 1$ vector. Therefore, transforming each term in (2) yields the following *algebraic* equation for the transform of the state-vector:

$$s\mathbf{Q}(s) - \mathbf{q}(0) = \mathbf{AQ}(s) + \mathbf{BX}(s) \tag{4}$$

where $\mathbf{X}(s) = \mathcal{L}[\mathbf{x}(t)]$. The transform of the zero-state response vector $\mathbf{q}_{zs}(t)$ can be found by setting $\mathbf{q}(0) = \mathbf{0}$ in (4) and collecting the terms involving $\mathbf{Q}_{zs}(s) \triangleq \mathcal{L}[\mathbf{q}_{zs}(t)]$, giving

$$s\mathbf{Q}_{zs}(s) - \mathbf{AQ}_{zs}(s) = \mathbf{BX}(s)$$

Introducing the $n \times n$ identity matrix $\mathbf{I}$ in order to combine the terms involving $\mathbf{Q}_{zs}$, we have

$$(s\mathbf{I} - \mathbf{A})\mathbf{Q}_{zs}(s) = \mathbf{BX}(s) \tag{5}$$

which, when premultiplied by the inverse† of $(s\mathbf{I} - \mathbf{A})$, results in

$$\mathbf{Q}_{zs}(s) = (s\mathbf{I} - \mathbf{A})^{-1}\mathbf{BX}(s) \tag{6}$$

The transform of the zero-state response of the output vector $\mathbf{y}_{zs}(t)$ is obtained by transforming the output equation (3) and substituting (6) to yield

$$\mathbf{Y}_{zs}(s) = \mathbf{C}\overbrace{(s\mathbf{I} - \mathbf{A})^{-1}\mathbf{BX}(s)}^{\mathbf{Q}_{zs}(s)} \tag{7}$$

Recalling that the scalar transfer function $H(s)$ was defined so as to satisfy $Y_{zs}(s) = H(s)X(s)$ for SIO systems, it is reasonable to define the

†The temptation to divide matrices rather than use the inverse operator must be resisted.

transfer-function matrix as

$$\mathbf{H}(s) \triangleq \mathbf{C}(s\mathbf{I}-\mathbf{A})^{-1}\mathbf{B} \tag{8}$$

which has r rows and p columns and depends only upon the system matrices **A**, **B**, and **C**. Having made this definition, (7) becomes

$$\mathbf{Y}_{zs}(s) = \mathbf{H}(s)\mathbf{X}(s) \tag{9}$$

Characteristic values and poles

The inverse of the matrix $(s\mathbf{I}-\mathbf{A})$ used in forming $\mathbf{H}(s)$ in (8) can be written in terms of its adjoint matrix – not to be confused with the "adjoint system" of optimal-control theory – and its determinant as

$$(s\mathbf{I}-\mathbf{A})^{-1} = \frac{\text{Adj}\,[s\mathbf{I}-\mathbf{A}]}{|s\mathbf{I}-\mathbf{A}|} \tag{10}$$

The denominator $|s\mathbf{I}-\mathbf{A}|$ will be recognized as the system's characteristic polynomial, first introduced in Section 3.5 and its roots are known as the characteristic values (eigenvalues) of the matrix **A**. Because $(s\mathbf{I}-\mathbf{A})$ has the rather special form

$$(s\mathbf{I}-\mathbf{A}) = \begin{bmatrix} s-a_{11} & -a_{12} & \cdots & -a_{1n} \\ -a_{21} & s-a_{22} & \cdots & -a_{2n} \\ \vdots & \vdots & & \vdots \\ -a_{n1} & -a_{n2} & \cdots & s-a_{nn} \end{bmatrix} \tag{11}$$

it follows that Adj $[s\mathbf{I}-\mathbf{A}]$ will be an $n \times n$ matrix, each element of which is a polynomial in s of degree $n-1$ or less. Hence, the n^2 elements of $(s\mathbf{I}-\mathbf{A})^{-1}$ will be rational functions of s, each having the nth-degree polynomial $|s\mathbf{I}-\mathbf{A}|$ as its denominator.

Because the matrices **B** and **C** do not depend upon the variable s, the $r \times p$ elements of the transfer-function matrix $\mathbf{H}(s)$ will also be functions of s, each having the polynomial $|s\mathbf{I}-\mathbf{A}|$ as its denominator – at least before any factors that are common to both numerator and denominator are canceled. To be more explicit, substitution of (10) into (8) gives

$$\mathbf{H}(s) = \frac{\mathbf{C}\,\text{Adj}\,[s\mathbf{I}-\mathbf{A}]\mathbf{B}}{|s\mathbf{I}-\mathbf{A}|} \tag{12}$$

where the element $H_{ij}(s)$ is the transfer function $Y_i(s)/X_j(s)$.

Hence, all input-output transfer functions associated with the system will have as their poles the characteristic values of $\mathbf{A}$ and these characteristic values will be the p_k in the exponential terms $e^{p_k t}$ comprising the system's response modes. If the numerator of a particular $H_{ij}(s)$ should have a root that coincides with one of the characteristic values of $\mathbf{A}$, then the corresponding mode will not appear in the response of $y_i(t)$ to the impulsive input $x_j(t) = \delta(t)$. This fundamental relationship is just one manifestation of the fact that once the dynamics of a fixed linear system are put into state-variable form, the very powerful analytical methods of linear algebra and computational capabilities of the digital computer can be brought to bear on the problems of analysis and synthesis.

State-transition matrix

Having extended the transfer-function notion to MIO systems, we shall consider the zero-input response of the state vector, denoted by $\mathbf{q}_{zi}(t)$, which results from the initial state $\mathbf{q}(0)$. Returning to (4) and setting $\mathbf{X}(s) = \mathbf{0}$ we obtain, after a slight rearrangement of terms,

$$(s\mathbf{I}-\mathbf{A})\mathbf{Q}_{zi}(s) = \mathbf{q}(0) \tag{13}$$

where $\mathbf{Q}_{zi}(s) \triangleq \mathcal{L}[\mathbf{q}_{zi}(t)]$. As before, $\mathbf{Q}_{zi}(s)$ may be solved for by using the inverse of $(s\mathbf{I}-\mathbf{A})$, yielding

$$\mathbf{Q}_{zi}(s) = (s\mathbf{I}-\mathbf{A})^{-1}\mathbf{q}(0) \tag{14}$$

Taking the inverse Laplace transform of both sides of (14), the zero-input response vector must be expressable in the form

$$\mathbf{q}_{zi}(t) = \mathbf{F}(t)\mathbf{q}(0) \qquad t \geq 0 \tag{15}$$

where

$$\mathbf{F}(t) \triangleq \mathcal{L}^{-1}[(s\mathbf{I}-\mathbf{A})^{-1}] = \mathcal{L}^{-1}\left[\frac{\text{Adj}\,[s\mathbf{I}-\mathbf{A}]}{|s\mathbf{I}-\mathbf{A}|}\right] \tag{16}$$

In words, (15) says that in the absence of any inputs the state at $t = 0$ is transformed into the state at any time $t \geq 0$ according to the matrix $\mathbf{F}(t)$ which, from (16), depends strictly on $\mathbf{A}$ and t.

Actually, (15) is part of a more general relationship in which the state at *any* time t (positive or negative) can be related to the state at any other time t_0 by

$$\mathbf{q}_{zi}(t) = \mathbf{\Phi}(t-t_0)\mathbf{q}(t_0) \qquad -\infty < t < \infty \tag{17}$$

provided that all inputs are zero during the interval between t_0 and t. The matrix function $\mathbf{\Phi}(t)$, which provides the connection between the states at any two times, is known as the *state-transition matrix* and is equal to the exponential function $e^{\mathbf{A}t}$ which is *defined* by the infinite series

$$e^{\mathbf{A}t} \triangleq \mathbf{I} + \mathbf{A}t + \mathbf{A}^2 \frac{t^2}{2!} + \mathbf{A}^3 \frac{t^3}{3!} + \cdots \tag{18}$$

The fact that

$$\mathbf{\Phi}(t) = e^{\mathbf{A}t} \tag{19}$$

can be proved by showing that the infinite series given in (18) satisfies the matrix differential equation $\dot{\mathbf{\Phi}}(t) = \mathbf{A}\mathbf{\Phi}(t)$ with the initial condition $\mathbf{\Phi}(0) = \mathbf{I}$, which implies that $\mathbf{\Phi}(t)\mathbf{q}(0)$ must satisfy (2) with $\mathbf{x}(t) = \mathbf{0}$. [See Zadeh and Desoer (1963, Chap. 5) for the proof and for the other numerous properties of the exponential function.] The relevant point here is that the matrix $\mathbf{F}(t)$ in (15) must be the state-transition matrix $\mathbf{\Phi}(t)$ over the interval $t \geq 0$.

Example 8.5

For illustrative purposes, we shall find the transfer function and state-transition matrix of the second-order system discussed in Example 3.3, although it is a bit like using a cannon to shoot a humming bird. Rewriting the state and output equations in matrix form, we have

$$\frac{d}{dt}\begin{bmatrix} q_1 \\ q_2 \end{bmatrix} = \underbrace{\begin{bmatrix} 1 & 0 \\ 1 & -3 \end{bmatrix}}_{\mathbf{A}} \begin{bmatrix} q_1 \\ q_2 \end{bmatrix} + \underbrace{\begin{bmatrix} 1 \\ 0 \end{bmatrix}}_{\mathbf{B}} x(t)$$

$$y = \underbrace{\begin{bmatrix} -\frac{1}{4} & 1 \end{bmatrix}}_{\mathbf{C}} \begin{bmatrix} q_1 \\ q_2 \end{bmatrix}$$

from which the matrices **A**, **B**, and **C** are readily identified.

In order to find the transfer-function matrix $\mathbf{H}(s)$ – in this case a scalar – we write

$$s\mathbf{I} - \mathbf{A} = \begin{bmatrix} s-1 & 0 \\ -1 & s+3 \end{bmatrix}$$

from which

$$|s\mathbf{I} - \mathbf{A}| = (s-1)(s+3)$$

and

$$\text{Adj}\,[s\mathbf{I}-\mathbf{A}] = \begin{bmatrix} s+3 & 0 \\ 1 & s-1 \end{bmatrix}$$

Substituting into (8) and using (10), the scalar transfer function is

$$\mathbf{H}(s) = \frac{[-\tfrac{1}{4} \quad 1]\begin{bmatrix} s+3 & 0 \\ 1 & s-1 \end{bmatrix}\begin{bmatrix} 1 \\ 0 \end{bmatrix}}{(s-1)(s+3)}$$

$$= \frac{-\tfrac{1}{4}(s-1)}{(s-1)(s+3)} = -\frac{1}{4}\frac{1}{(s+3)}$$

which is in agreement with the calculations in Example 3.3. As before, we note that $\mathbf{H}(s)$ has a zero which coincides with its pole at $s = 1$, thereby eliminating the mode e^t from the impulse response of the output, although it is present in the state variables q_1 and q_2.

As for the state-transition matrix, we can use (16) to get an analytical result; or, if a computer is handy and numerical values of $\mathbf{\Phi}(t)$ computed at discrete points in time are satisfactory, (18) can be used. Pursuing the former,

$$\mathbf{F}(t) = \mathcal{L}^{-1}\left[\frac{\begin{bmatrix} s+3 & 0 \\ 1 & s-1 \end{bmatrix}}{(s-1)(s+3)}\right] = \mathcal{L}^{-1}\begin{bmatrix} \dfrac{1}{s-1} & 0 \\ \dfrac{1}{(s-1)(s+3)} & \dfrac{1}{s+3} \end{bmatrix}$$

Inverting the transform matrix term by term and noting that although $\mathbf{F}(t)$ is defined only for $t \geq 0$, $\mathbf{\Phi}(t) = \mathbf{F}(t)$ over this interval but is defined for $-\infty < t < \infty$, it follows that

$$\mathbf{\Phi}(t) = \begin{bmatrix} e^t & 0 \\ \tfrac{1}{4}(e^t - e^{-3t}) & e^{-3t} \end{bmatrix}$$

Having obtained the state-transition matrix, the zero-input response of $\mathbf{q}(t)$ can be written in terms of the initial state from (15) as

$$q_1(t) = q_1(0)e^t$$

$$q_2(t) = \tfrac{1}{4}(e^t - e^{-3t})q_1(0) + e^{-3t}q_2(0)$$

Now, employing the output equation (3),

$$y_{zi}(t) = -\tfrac{1}{4}q_1(t) + q_2(t) = [-\tfrac{1}{4}q_1(0) + q_2(0)]e^{-3t}$$

Complete response

A time-domain expression can be found for $\mathbf{q}_{zs}(t)$ by observing that the matrix $(s\mathbf{I}-\mathbf{A})^{-1}$ and the state-transition matrix form a Laplace-transform pair, i.e.,

$$\mathbf{\Phi}(t) \leftrightarrow (s\mathbf{I}-\mathbf{A})^{-1}$$

Thus the frequency-domain multiplication in (6) becomes a time-domain convolution of the matrices $\mathbf{\Phi}(t)$ and $\mathbf{Bx}(t)$, which can be written as

$$\mathbf{q}_{zs}(t) = \int_0^t \mathbf{\Phi}(t-\lambda)\mathbf{Bx}(\lambda)\,d\lambda \tag{20}$$

where the three matrices must appear in the sequence shown. Although the notation used in (20) may appear to be rather overpowering at first glance, the equation merely says that any element of the vector $\mathbf{q}_{zs}(t)$ is the integral of the corresponding row of the matrix product $\mathbf{\Phi}(t-\lambda)\mathbf{Bx}(\lambda)$.

When $\mathbf{q}_{zs}(t)$ is added to the zero-input response of the state vector as given by (15), and the output equation (3) is included, we have the complete response of both the state and output vectors:

$$\mathbf{q}(t) = \mathbf{\Phi}(t)\mathbf{q}(0) + \int_0^t \mathbf{\Phi}(t-\lambda)\mathbf{Bx}(\lambda)\,d\lambda \tag{21}$$

$$\mathbf{y}(t) = \mathbf{Cq}(t) \tag{22}$$

Example 8.6

To find the complete response of the system discussed in Example 8.5 to a unit step with an arbitrary initial state, we let $\mathbf{x}(t)$ be the scalar $u(t)$. Substituting the appropriate matrices from the previous example into (21), the state vector is

$$\begin{bmatrix} q_1 \\ q_2 \end{bmatrix} = \overbrace{\begin{bmatrix} e^t & 0 \\ \frac{1}{4}(e^t - e^{-3t}) & e^{-3t} \end{bmatrix}}^{\mathbf{\Phi}(t)} \begin{bmatrix} q_1(0) \\ q_2(0) \end{bmatrix}$$

$$+ \int_0^t \overbrace{\begin{bmatrix} e^{(t-\lambda)} & 0 \\ \frac{1}{4}(e^{(t-\lambda)} - e^{-3(t-\lambda)}) & e^{-3(t-\lambda)} \end{bmatrix}}^{\mathbf{\Phi}(t-\lambda)} \overbrace{\begin{bmatrix} 1 \\ 0 \end{bmatrix}}^{\mathbf{B}} u(\lambda)\,d\lambda$$

In component form, the first row reduces to

$$q_1(t) = q_1(0)e^t + e^t \int_0^t e^{-\lambda}\,d\lambda = e^t[q_1(0) + 1 - e^{-t}]$$

and $q_2(t)$ may be found in a similar manner, the detailed calculations being omitted. Substituting q_1 and q_2 into the output equation, the complete response is found to be

$$y(t) = \underbrace{[-\tfrac{1}{4}q_1(0) + q_2(0)]e^{-3t}}_{y_{zi}(t)} + \underbrace{\tfrac{1}{12}(e^{-3t} - 1)}_{y_{zs}(t)}$$

where the zero-input and zero-state components are readily identified. We notice that whenever the output variable is computed, the mode e^t never appears, in keeping with the fact that that particular mode is unobservable.

Problems

8.1 Derive the following transform relationships from Section 8.1: (a) Superposition, Eq. (5); (b) Scale change, Eq. (6); (c) Integration, Eq. (7).

8.2 Derive the transforms of the causal functions $\sin \omega t$ and $\cos \omega t$ as given by Eqs. (17) and (18), Sect. 8.1.

8.3 Prove the initial-value theorem.

8.4 (a) Derive the transform relationships

$$e^{at}v(t) \leftrightarrow V(s-a)$$

$$tv(t) \leftrightarrow -\frac{dV(s)}{ds}$$

(b) Derive the transform pairs

$$t \sin \omega t \leftrightarrow \frac{2\omega s}{(s^2+\omega^2)^2}$$

$$t^2 e^{-\alpha t} \leftrightarrow \frac{2}{(s+\alpha)^3}$$

8.5 Use Laplace transforms to derive the following expressions for the zero-state response of a second-order system: (a) Eq. (9b), Sect. 3.3; (b) Eq. (11), Sect. 3.3. In both cases, take $t_0 = 0$.

8.6 Find the transfer function $Y(s)/X(s)$ for the system discussed in Example 3.3.

8.7 Using Eq. (7), Sect. 8.2, and the definition of steady-state response given in Sect. 3.3, find an expression for the steady-state response in terms of the system's transfer function $H(s)$ for the following causal inputs: (a) $x(t) = e^{\alpha t}$; (b) $x(t) = \cos \omega t$; (c) $x(t) = t$. *Answer:* (a) $y_P(t) = H(\alpha)e^{\alpha t}$.

8.8 Find the Laplace transforms of the functions shown in Figure P8.1.

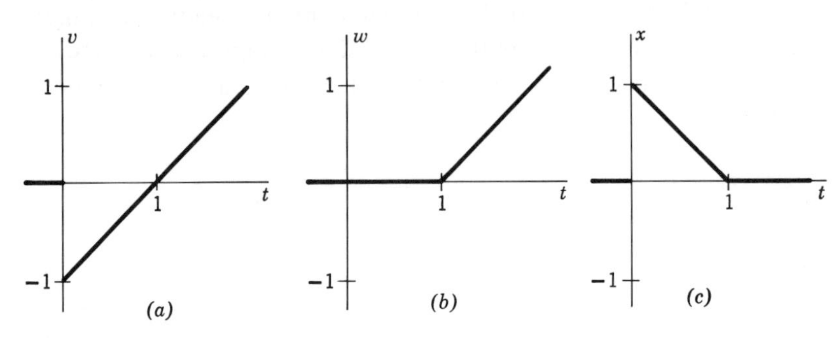

Figure P8.1

8.9 Find and sketch the functions $v(t)$ and $w(t)$ corresponding to the transforms given below. Where applicable, apply the initial- and final-value theorems.

$$V(s) = \frac{2s^2 + 14s + 3}{s^2 + 5s + 4} \qquad W(s) = \frac{3s^2 + 10s + 26}{s^3 + 4s^2 + 13s}$$

Answer: $v(t) = 2\delta(t) - 3e^{-t} + 7e^{-4t}$.

8.10 (a) Consider the "causal periodic function" $v_T(t) = \sum_{m=0}^{\infty} v(t - mT)$, where $v(t)$ is zero outside the interval $0 \le t < T$ and has the transform $V(s)$. Show that

$$v_T(t) \leftrightarrow \frac{V(s)}{1 - e^{-sT}}$$

(b) Find the transform of the causal sawtooth wave shown in Figure P8.2.

(c) Find the transform of $\cos(2\pi t/T)[u(t) - u(t-T)]$.

8.11 It is known that the transform $F(s)$ has a pair of distinct poles at $s = 0$ and -1 and a single zero at $s = 1$ and that $\lim_{t\to\infty} f(t) = 10$. Find $F(s)$ and $f(t)$. *Answer:* $f(t) = 10u(t) - 20e^{-t}$, $t > 0$.

8.12 Find and sketch the impulse response of the system described by

$$y^{(4)} + 6y^{(3)} + 13y^{(2)} + 12y^{(1)} + 4y = x^{(4)} + 6x^{(3)} + 13x^{(2)} + 12x^{(1)} + 3x$$

Note: The characteristic polynomial is $P(s) = (s+1)^2(s+2)^2$.

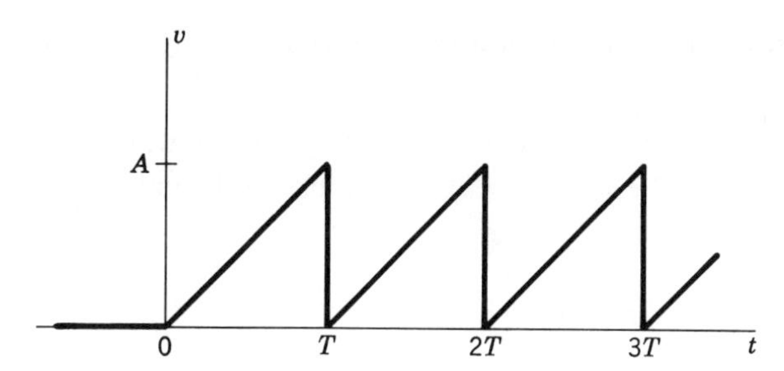

Figure P8.2

8.13 A fixed linear system initially at rest is excited by the input signal $x(t) = (1-e^{-t})u(t)$ yielding the output $y(t) = \frac{1}{4}u(t) - e^{-t} + \frac{1}{2}te^{-2t} + \frac{3}{4}e^{-2t}$. Find the system's impulse response and transfer function. *Answer:* $H(s) = 1/(s+2)^2$.

8.14 Given the input $x(t) = 10u(t)$ and the impulse response $h(t) = (1-t) \times [u(t) - u(t-1)]$, use Laplace transforms to find $H(s)$, $Y(s)$, and $y(t)$.

8.15 A system's transfer function is

$$H(s) = \frac{(s^2+5s+4)(s+3)}{s^4+2.5s^3+s^2}$$

(a) Find the step and ramp components of $h(t)$. (b) Find $h(0+)$ and $\dot{h}(0+)$. (c) Sketch $h(t)$ with a degree of accuracy that conveys the important features of its behavior.

8.16 Find the sinusoidal component of the function $y(t)$ whose transform is

$$Y(s) = \frac{s^4+2s^3+s^2+3s+6}{(s^2+4)(s^3+6s^2+4s+6)}$$

Hint: Use a method that avoids having to find the complete function $y(t)$.

8.17 (a) Compute or obtain by simulation the step response of the second-order system with transfer function

$$H(s) = \left(\frac{1}{\alpha}\right)\frac{s+\alpha}{s^2+s+1}$$

for $\alpha = 0.2$, 1, and 10. (b) Relate the system's step response to the step and impulse responses of a system whose transfer function is $1/(s^2+s+1)$.

8.18 (a) Show that the inverse transform of $Y_{zi}(s)$ as given by Eq. (11), Sect. 8.2, with $n = 2$ and $a_n = 1$ reduces to the expression for $y_{zi}(t)$ given by Eq. (13), Sect. 3.1, with $t_0 = 0$. *Hint:* Because $s^2 + a_1 s + a_0 = (s-p_1)(s-p_2)$ it follows that $a_1 = -p_1 + p_2$ and $a_0 = (p_1 p_2)$.

(b) Derive the corresponding result for a repeated root, namely Eq. (16), Sect. 3.1.

8.19 For each of the sets of transfer functions and inputs given below, evaluate the output $y(t)$ and check the value of $y(0+)$ by applying the initial-value theorem:

(a) $H(s) = \dfrac{s+3}{s(s+2)}$ $\qquad x(t) = 3u(t)$

(b) $H(s) = \dfrac{s+1}{s+2}$ $\qquad x(t) = e^{-2t}$

(c) $H(s) = \dfrac{6}{(s+2)(s+3)}$ $\qquad x(t) = \sin 2t$

Answer: (a) $y(t) = \frac{9}{2}t - \frac{3}{4}u(t) + \frac{3}{4}e^{-2t}$, $y(0+) = 0$.

8.20 One method of characterizing first-order systems is to say that a transfer function of the form $F_1(s) = A/s$ represents a *nonself-regulatory* process, while one of the form $F_2(s) = A/(\tau s + 1)$ represents a *self-regulatory* process. Explain why this distinction is appropriate when viewed in terms of their respective step responses and give a physical example of each type.

8.21 Repeat Example 3.1 using Laplace transforms and taking $t_0 = 0$.

8.22 Repeat Example 3.2 using Laplace transforms.

8.23 Use Laplace transforms to do Problem 3.4.

8.24 Use Laplace transforms to do Problem 3.5.

8.25 Use Laplace transforms to do Problem 4.2.

8.26 Use Laplace transforms to do Problem 4.17.

8.27 Do part (b) of Problem 4.19 using Laplace transforms. Explain why part (a) cannot be done this way.

8.28 Figure P8.3 depicts a proof mass connected to a base or frame by a parallel spring-viscous damper combination. The variable $y(t)$ represents the relative displacement and can be measured in a variety of ways and will be considered as the output of the device. The variable $x(t)$ represents the location of the base or frame relative to an inertial reference frame.

(a) Show that the transfer function relating the transforms $X(s) \triangleq \mathcal{L}[x(t)]$ and $Y(s) \triangleq \mathcal{L}[y(t)]$ is

$$H(s) = \frac{Y(s)}{X(s)} = \frac{Ms^2}{Ms^2 + Bs + K}$$

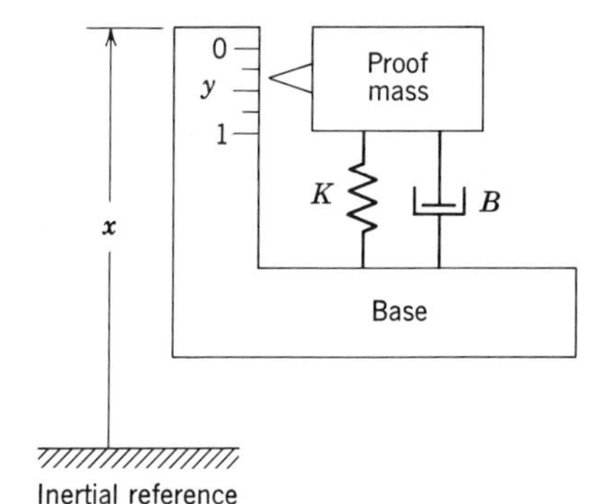

Figure P8.3

(b) If a small mass and a stiff spring are used with moderate damping ($\zeta \approx 1$) the unit is an accelerometer. For what range of frequencies will $y(t)$ be proportional to $\ddot{x}(t)$? What are its transfer function as an accelerometer and constant of proportionality relating y and $\ddot{x}$? (c) If a large mass and a very soft spring are used with light damping ($\zeta \approx 0$) the unit is a seismograph, having an output proportional to the displacement of the base during an earthquake. Give the range of frequencies for which the device will behave in this manner, with the appropriate transfer function and constant of proportionality.

8.29 Find the state-transition matrix for the system treated in Example 3.1; use the state vector $\mathbf{q}^T = [y \quad \dot{y} \quad \ddot{y}]$.

8.30 (a) Write state equations for the circuit in Figure P8.4, taking the inductor current and the capacitor voltage as state variables, as indicated in the figure. Find the state-transition matrix $\mathbf{\Phi}(t)$.

(b) Find the response of both state variables if $i(t) = u(t)$, $i_L(0) = 1$, and $v_c(0) = \frac{3}{2}$.

(c) Find the output voltage $e(t)$ if the circuit has no initial stored energy and $i(t) = (1 + \sin t)u(t)$.

8.31 (a) Show that defining state variables such that $q_2 = \dot{q}_1$, $q_3 = \dot{q}_2, \ldots, q_n = \dot{q}_{n-1}$ implies that each row of $\mathbf{\Phi}(t)$ is the derivative of the row directly above

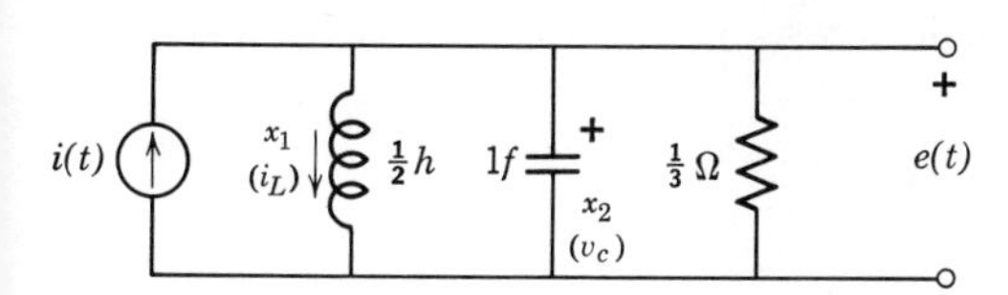

Figure P8.4

it. What is the corresponding relationship between the rows of Adj $[s\mathbf{I}-\mathbf{A}]$?
(b) Show that the ith column of the state-transition matrix is the state vector which results when $q_i(0)=1$ with the remaining initial states equal to zero.

8.32 The uniform stick of length L and mass M shown in Figure P8.5 is held by a frictionless pivot at its lower end. A torque $\tau(t)$ can be applied about the pivot by an electric motor.

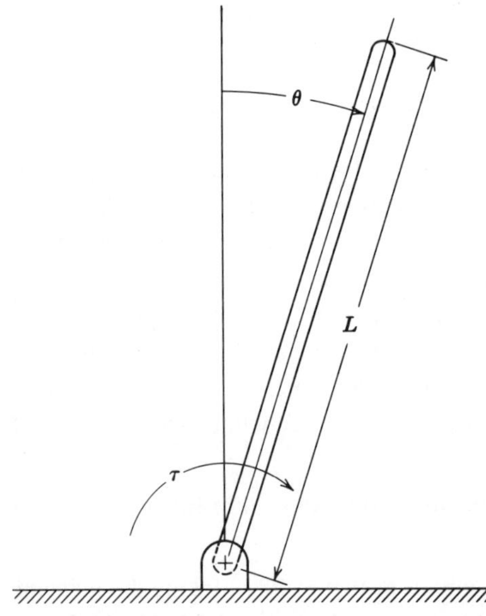

Figure P8.5

(a) Show that the system equation is $(ML^2/3)\ddot{\theta}-(MgL/2)\theta=\tau(t)$.
(b) Taking θ and $\dot{\theta}$ as the two state variables, write the system equation in state-variable form and find the characteristic values of $\mathbf{A}$.
(c) Find the state-transition matrix $\mathbf{\Phi}(t)$ and an expression for $\theta(t)$ resulting from the initial state $[\theta_0, \dot{\theta}_0]$ with $\tau=0$. What relationship between θ_0 and $\dot{\theta}_0$ will result in $\theta(t)\to 0$ as $t\to\infty$, with $\tau=0$?
(d) Defining the output variable $y=\theta+k\dot{\theta}$, find the transfer function $Y(s)/\tau(s)$ by applying Eq. (8), Sect. 8.4.

8.33 The behavior of an airplane flying at a constant speed of 250 ft/sec can be described by the differential equations

$$\ddot{\theta}+0.32\dot{\theta}+0.14\dot{\alpha}+0.42\alpha=-1.40\eta$$

$$\dot{\alpha}=\dot{\theta}-0.42\alpha$$

As indicated in Figure P8.6, the angles θ and α are the pitch angle and angle-of-attack, respectively, and η is the elevator deflection, all measured in radians.

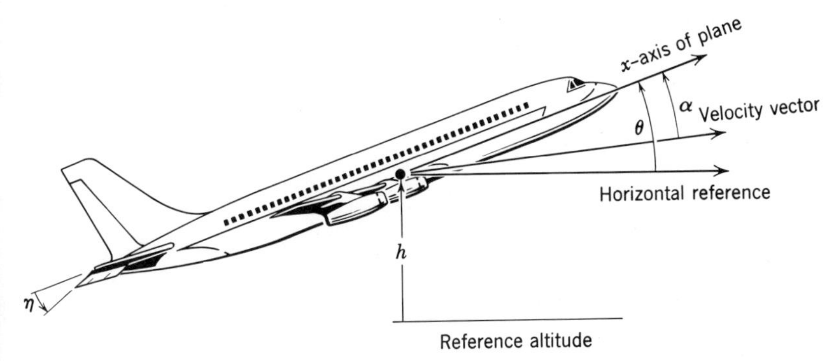

Figure P8.6

(a) Taking θ, $\dot{\theta}$, and α as the states, write the state-variable equations in matrix form. What are the characteristic values of **A**? Find the transfer function $\theta(s)/\eta(s)$ and explain the significance of the characteristic values in terms of the response of θ to a step change in η.

(b) Expand the model to include the incremental altitude h as a state variable, in addition to θ, $\dot{\theta}$, and α. Find the characteristic values of the system matrix and the transfer function $H(s)/\eta(s)$.

(c) Repeat part (b) using $\dot{\theta}$, α, h, and $\dot{h}$ as the state variables. Why is this an equally valid representation?

9

Feedback systems

It was pointed out in the first chapter that control systems usually have a closed-loop configuration, the outputs being fed back to modify the inputs. This use of feedback is feasible because, unlike most communication systems, the input and output points of a control system are generally in close physical proximity. More significant, however, is the fact that the use of feedback can accomplish certain desirable results that would not be readily possible with an open-loop system. For this reason, feedback loops are frequently designed into various electronic circuits as well as automatic control systems.

Some of the potential benefits (and liabilities) of feedback will be discussed in this chapter, but our main purpose here is to develop the special techniques required to handle conveniently the analysis of closed-loop systems. It is perhaps not too surprising that the study of feedback is somewhat more involved than the study of cascade systems. This occurs because the input and output signals of all subsystems connected within any closed loop are interrelated, thereby preventing independent analysis of the individual blocks such as was done in Chapter 7.

Generally, the material presented here is applicable to any system whose mathematical model takes the form of a fixed linear feedback configuration. These occur not only in automatic control and electronics, where the feedback is introduced by design, but also in such diverse fields as physiology, economics, ecology, etc., where feedback is inherent in the natural laws governing the process. However, consistent with the major thrust of this book, our terminology, conventions, and illustrations will be taken in the context of control-systems engineering.

9.1 TRANSFER FUNCTIONS OF FEEDBACK SYSTEMS

9.1 TRANSFER FUNCTIONS OF FEEDBACK SYSTEMS

Usually, a process or plant to be controlled has one or more inputs that can be manipulated in some prescribed manner so as to affect its response. Likewise, the process is usually affected by other inputs that cannot be manipulated, often fluctuating in a random fashion. Because they have undesired effects on the behavior of the process, they are referred to as disturbance inputs. The objective of the control-system designer is to monitor the performance of the process via suitable transducers and combine these signals with externally-generated reference signals that are then used as inputs to the controller in order to generate the manipulated inputs to the process. The result is a feedback or closed-loop system.

Although real systems are usually nonlinear and of the multi-input-output (MIO) nature, they can often be analyzed in terms of linear single-input-output (SIO) mathematical models by restricting one's attention to an incremental model for which superposition holds. Furthermore, systems of high order with a variety of feedback paths frequently may be treated as interconnected lower-order systems, each having a single feedback path, by neglecting the connections between weakly-coupled modes. Thus, the analytical techniques to be presented in this and the following two chapters are devoted to "single-loop" systems of the form shown in Figure 9.1, but, by means of judicious modeling, they may be applied to many types of truly MIO systems. In fact, the system shown in the figure is not strictly SIO in that the process, and hence the complete system, is allowed to have two inputs; however, we shall use superposition to treat each input separately.

To put Figure 9.1 in more quantitative form, each block can be represented by its transfer function – the Laplace-transform version,

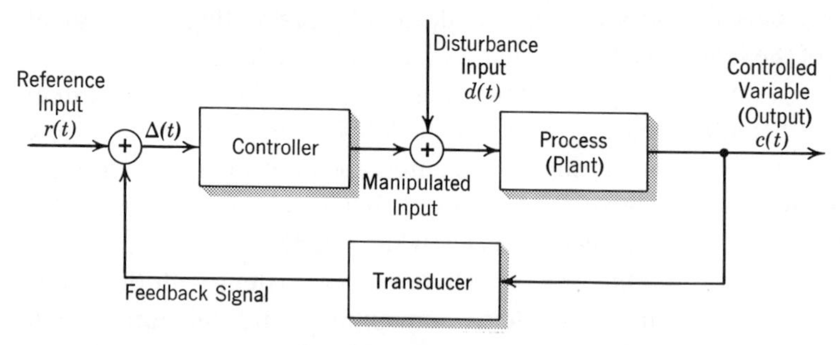

Figure 9.1. General structure and nomenclature for a single-loop control system.

since we shall be very much interested in transient responses – resulting in Figure 9.2. We have adopted the terminology used in the figure in deference to long-standing conventions in the analysis of single-loop control systems. Specifically, $H(s)$ will henceforth be associated exclusively with the *feedback* path from the controlled variable to the reference signal,

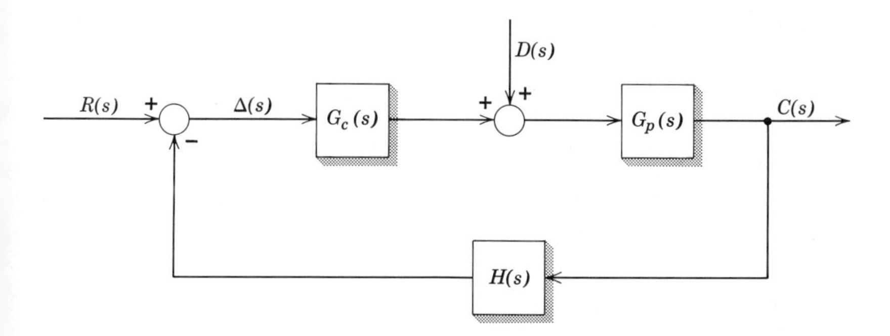

Figure 9.2. Transform model of Figure 9.1.

while $G(s)$, either with or without subscripts, will denote a transfer function comprising the *forward* path from the reference signal to the controlled variable.

Particular mention should be made of the summing junction in Figure 9.2 where the minus sign has been associated with the feedback signal. Most of the time – but not always – the feedback signal is *subtracted* from the reference signal when the two are combined, producing the *difference* signal

$$\Delta(t) = \mathcal{L}^{-1}[R(s) - H(s)C(s)] \tag{1a}$$

Since we can certainly have a minus sign associated with $H(s)$ and a plus sign with the summing junction, the choice is obviously just a matter of convention. In those cases in which the objective of the system is to make the output $c(t)$ follow the reference signal $r(t)$, it is appropriate to define the *error* as

$$\varepsilon(t) \triangleq r(t) - c(t) \tag{1b}$$

which, when $H(s) = 1$, is the difference signal $\Delta(t)$. However, it should be kept in mind that unless $H(s) = 1, \Delta(t) \neq \varepsilon(t)$.

Reference transfer function

If the disturbance input in Figure 9.2 is set to zero and the forward-path transfer functions $G_c(s)$ and $G_p(s)$ are combined according to

$$G(s) \triangleq G_c(s)G_p(s) \tag{2}$$

the single-input system of Figure 9.3*a* results.

The transfer function from the reference input $r(t)$ to the output $c(t)$ is referred to as the *reference transfer function* and will be denoted by $T_R(s)$. To solve for $T_R(s)$ we write the pair of transform equations

$$\Delta(s) = R(s) - H(s)C(s) \qquad C(s) = G(s)\Delta(s)$$

and then eliminate $\Delta(s)$ to obtain

$$[1+G(s)H(s)]C(s) = G(s)R(s)$$

which leads to the desired result:

$$T_R(s) = \left.\frac{C(s)}{R(s)}\right|_{D(s)=0} = \frac{G(s)}{1+G(s)H(s)} \tag{3}$$

A special case of (3) which is of particular interest arises when $H(s) = 1$ and is often referred to as a *unity-feedback system*, Figure 9.3*b*. In this

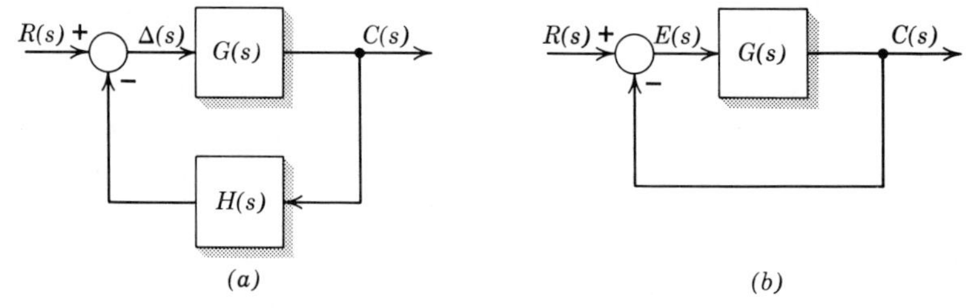

Figure 9.3. System diagram without disturbance input, $G(s) = G_c(s)G_p(s)$. (*a*) General form, $H(s) \neq 1$. (*b*) Unity feedback, $H(s) = 1$.

situation the signal $\Delta(t)$ is the error and (3) reduces to

$$T_R(s) = \frac{G(s)}{1+G(s)} \tag{4}$$

Situations in which a unity-feedback model is appropriate generally arise when the transducer has a flat frequency response insofar as its input $c(t)$ is concerned. Although a system's block diagram may not appear to be of the unity-feedback form when initially derived, it is often possible and convenient to manipulate the diagram into that form, in which case the error $\varepsilon(t)$ appears explicitly.

Disturbance transfer function

Similarly, we obtain the *disturbance transfer function* $T_D(s)$ by setting the reference input to zero in Figure 9.2 yielding Figure 9.4, and then

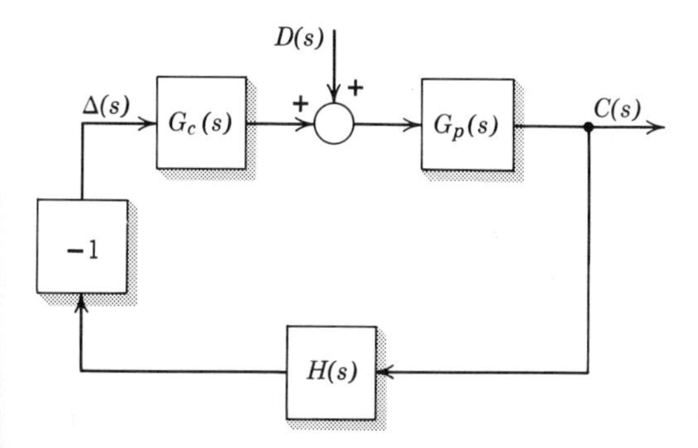

Figure 9.4. System diagram without reference input.

solving for $C(s)/D(s)$. From the revised block diagram,

$$\Delta(s) = -H(s)C(s) \qquad\qquad C(s) = G_p(s)[D(s) + G_c(s)\Delta(s)]$$

from which $\Delta(s)$ can be eliminated to give

$$T_D(s) = \left.\frac{C(s)}{D(s)}\right|_{R(s)=0} = \frac{G_p(s)}{1+G_c(s)G_p(s)H(s)} \tag{5}$$

It is of interest to note from (2) that the denominator of $T_D(s)$ is precisely the same as that of $T_R(s)$. Although neither of these transfer functions is written as a rational function so as to clearly display their respective poles and zeros, they do have the same poles as one another when written in the form of (4) and (5). However, when the numerator and denominator polynomials of a transfer function are written in *factored* form, one or more of the poles may coincide with zeros and cancel one another. In case of such a cancellation, the particular transfer function in question will have fewer than the original number of poles. The proper interpretation of this situation is that the mode corresponding to the missing pole is not observable insofar as that particular output variable is concerned; it is incorrect to conclude that the system order is the number of transfer-function poles remaining *after* the cancellation.

In the absence of any such cancellations, all transfer functions associated with a given closed-loop system, single loop or otherwise, will have the same poles – it is only in their zeros that they will differ. Since pole-zero cancellations do not arise routinely in practice, we shall assume henceforth that common factors are not present. Should common factors exist, the more fundamental form of the characteristic equation given by Eq. (12), Sect. 3.5, must be used, rather than setting the denominator of either $T_R(s)$ or $T_D(s)$ to zero.

Multiple-loop systems and block-diagram algebra

Often a feedback system will have more than one feedback path in which case some manipulations must be performed on the original block diagram before the transfer functions $T_R(s)$ or $T_D(s)$ can be found from (3) or (5). For example, suppose the original block diagram of a system has the structure shown in Figure 9.5, and it is desired to find $T_R(s)$ in terms of the individual block transfer functions G_1, G_2, H_1, and H_2.

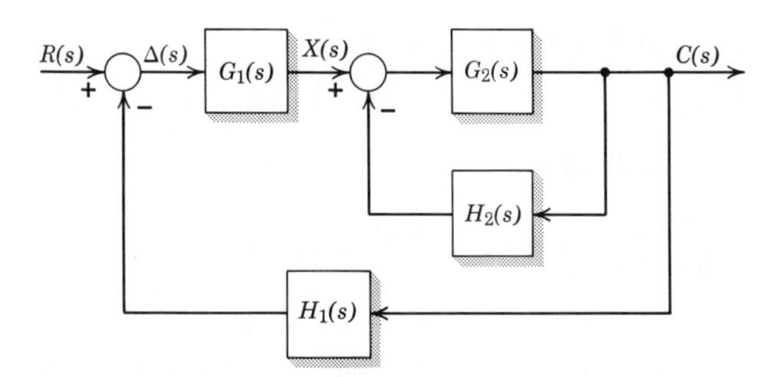

Figure 9.5. A two-loop system.

Basically, two alternatives present themselves, both involving the elimination of the inner loop in order to obtain a single-loop configuration which fits the form of Figure 9.3*a*. If (3) is used to find $C(s)/X(s)$, the transfer function of the inner loop, then Figure 9.6*a* can be drawn. Now, applying (3) once again, the desired transfer function results:

$$T_R = \frac{C}{R} = \frac{G_1G_2}{1 + G_2H_2 + G_1G_2H_1} \tag{6}$$

The second approach is to expand the inner loop to give the single-loop equivalent in Figure 9.6*b*, to which (3) can be applied, yielding (6). The intermediate step is shown in Figure 9.7.

The process by which we have found the transfer functions of the above closed-loop configurations is commonly referred to as *block-diagram algebra.* In reality, it is nothing more than the manipulation of a set of algebraic transform equations in order to eliminate the transforms of all signals except those of the input – either reference or disturbance – and the output. The only role played by the block diagram in this process is that it is a convenient means of representing the various algebraic equations, rather than writing them out explicitly. While this tool is useful, it is a mistake to place too much physical significance on the block diagram itself. For example, the feedback block in Figure 9.7 whose transfer function is H_2/G_1 does not refer to any physical portion of the original system. Rather, it represents the result of manipulating the pair of equations $\Delta = R - H_1C$ and $C = G_2(G_1\Delta - H_2C)$, corresponding to Figure 9.5, into the single equation

$$C = G_1G_2\left[R - H_1C - \left(\frac{H_2}{G_1}\right)C\right]$$

which is represented pictorially by Figure 9.7.

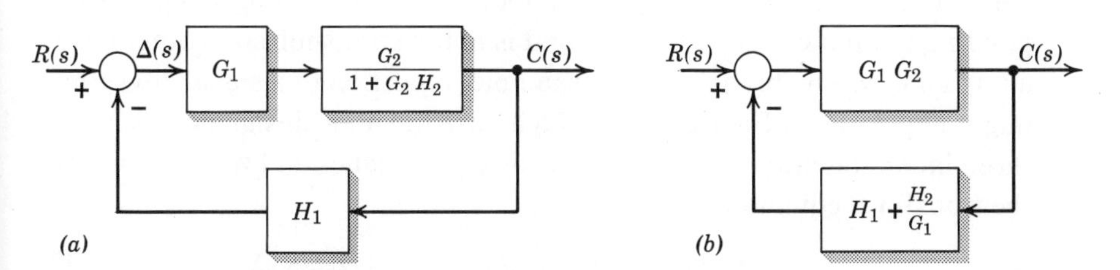

Figure 9.6. Single-loop equivalents of the two-loop system.

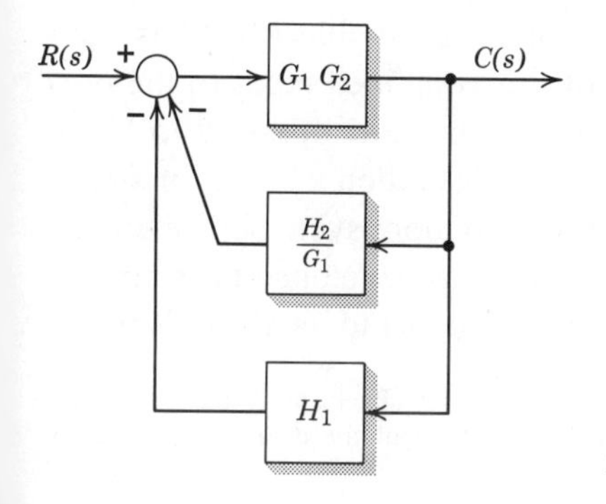

Figure 9.7.

Thus the reader is urged to think in terms of the *equations* that the block diagrams represent, rather than to attach any special significance to the block diagrams themselves. The most commonly encountered block-diagram operations are summarized in Figure 9.8.

One consequence of the fact that block-diagram reduction is actually just the manipulation of transform equations is that there is no need to concern ourselves with questions of interactions between physical components, i.e., loading effects, once the original block diagram has been properly established. However, it is imperative that the original derivation of the system's differential equations and the corresponding block diagram take loading effects into account. Otherwise, any subsequent analysis will be carried out with incorrect transfer functions.

Signal flow graphs

An alternate method of representing the transform equations of a fixed linear system is the *signal flow graph*. Basically, the method consists of drawing a set of interconnected *nodes*, which represent the signals of the system. The lines connecting the nodes are *branches* with which both *directions* and *transmittances* are associated. The convention that has been adopted is that the signal at a node is the sum of the signals corresponding to all incoming branches and is in turn transmitted to other nodes along all outgoing branches. For example, a portion of a signal flow graph might appear as in Figure 9.9 in which case the node designated as X_3 has three incoming branches and two departing branches and would represent the transform equation

$$X_3(s) = T_{13}(s)X_1(s) + T_{23}(s)X_2(s) + T_{33}(s)X_3(s) \tag{7}$$

For the single-loop feedback system of Figure 9.2, the signal flow graph is shown in Figure 9.10, where the intermediate variable $W(s)$ has been defined as representing the transform of the combined input to the process. One characteristic that flow graphs do have, as opposed to block diagrams, is that they lend themselves to presentation as a set of simultaneous equations of the type in (7) and thus to more systematic methods for their reduction. One such method, which was developed for simplifying moderately complicated signal flow graphs by hand, is Mason's rule.†

†See, for instance, Dorf (1967, pp. 46–51). Mason's rule can also be applied to the reduction of block diagrams, thereby avoiding repeated redrawing of the diagram in complex problems.

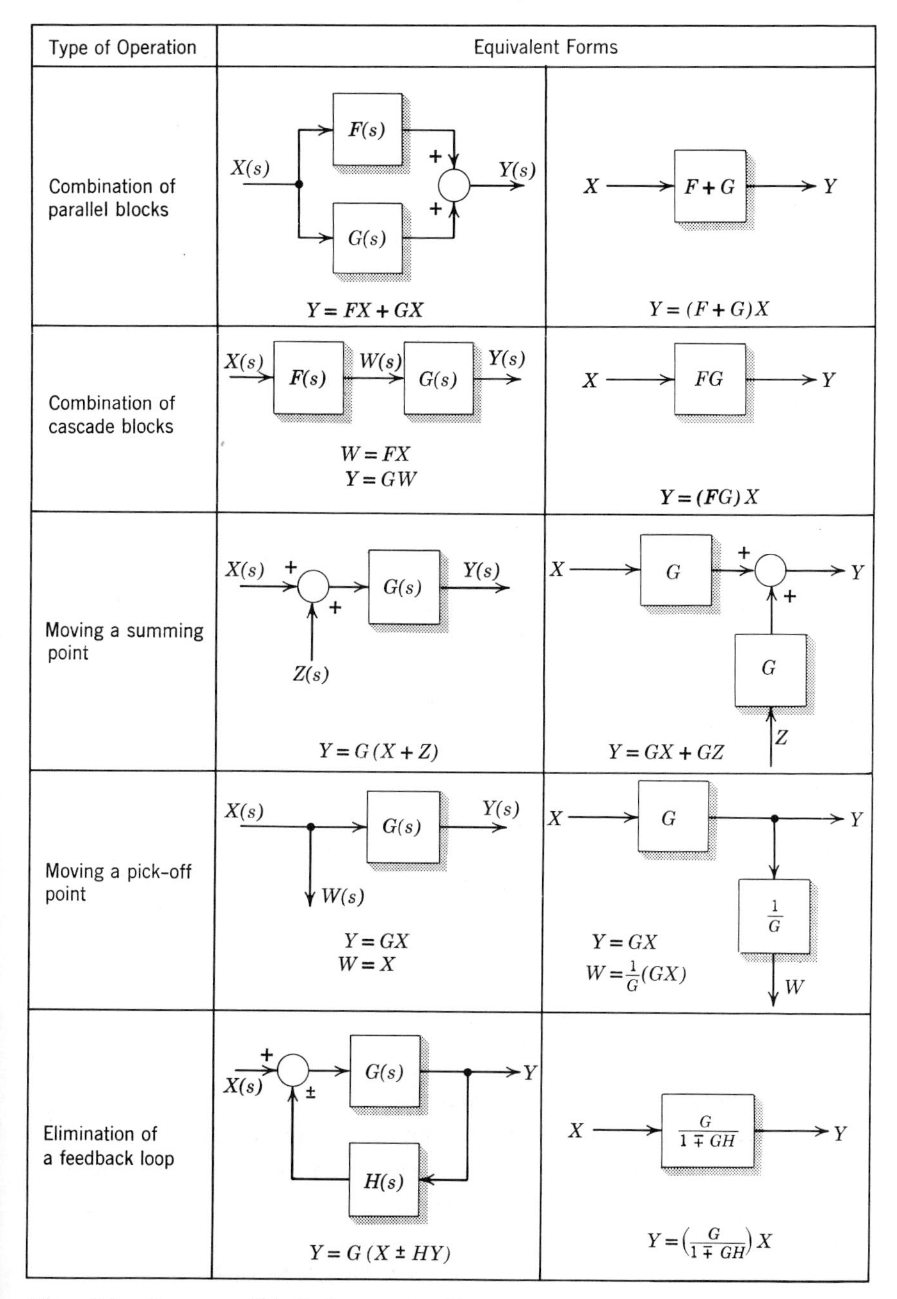

Figure 9.8. Summary of block-diagram operations.

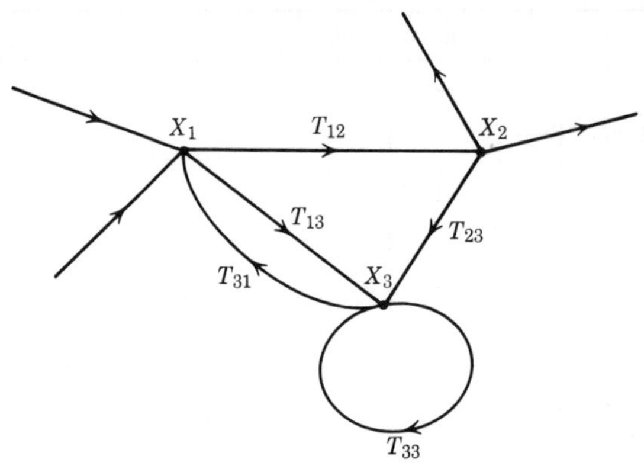

Figure 9.9. A portion of a signal flow graph.

However, the use of digital computers tends to make other approaches, based on the state-variable form of the system equations, more attractive. In essence, results that can be obtained with signal flow graphs can also be found using block diagrams or the transform equations directly and vice versa. Therefore, we shall not pursue the flow-graph formulation further.

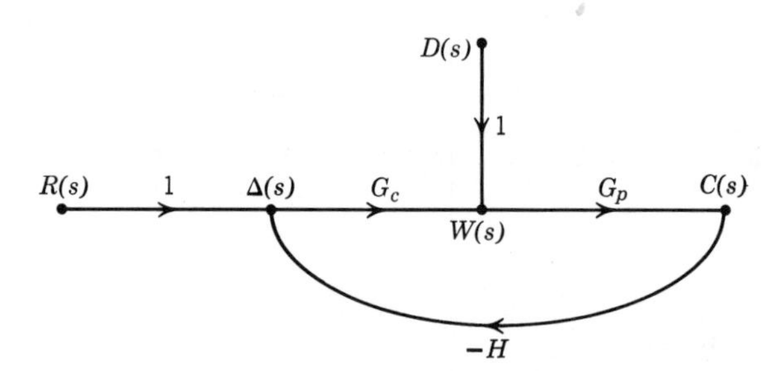

Figure 9.10. Flow-graph equivalent of the single-loop system in Figure 9.2.

Advantages and disadvantages of feedback

With the help of the transfer-function equations, we are now in a position to consider briefly some of the advantages and disadvantages of feedback systems. For this purpose, let us take the simple configuration shown in Figure 9.11 where the controller is just a high-gain wideband amplifier, i.e., $G_c(s) = A$ with $|A| >> 1$, and there is unity feedback.

Substituting $G(s) = AG_p(s)$ and $H(s) = 1$ into (3) yields, for the

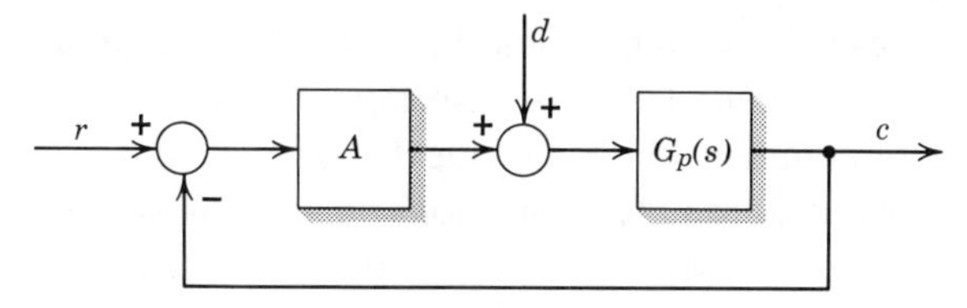

Figure 9.11.

reference transfer function,

$$T_R(s) = \frac{AG_p(s)}{1+AG_p(s)} \approx 1$$

assuming that $|AG_p(s)| \gg 1$. Under these conditions, all the dynamics of the process (and the controller) have been corrected for by the feedback and hence $c(t)$ *directly follows* the reference signal $r(t)$. In addition, since neither A nor $G_p(s)$ appear in $T_R(s)$, the system's operation will be *insensitive* to any variations of these parameters as long as $|AG_p(s)| \gg 1$. Turning to the disturbance transfer function, it is given by

$$T_D(s) = \frac{G_p(s)}{1+AG_p(s)} \approx \frac{1}{A}$$

Thus, $|T_D(s)| \ll |T_R(s)|$, which can be interpreted as meaning that the response to any disturbance $d(t)$ is greatly reduced compared with the response to any reference $r(t)$ of comparable strength.

While we have derived these qualitative results for a particular case, similar conclusions hold for most any well-designed feedback system. To summarize then, the potential advantages of feedback are threefold:

1. The dynamics or bandwidth limitations of the forward path can be substantially modified, presumably for improved performance.

2. The sensitivity to parameter variations in the forward path can be substantially reduced.

3. The effects of disturbance inputs, including noise, can be much reduced.

But these advantages are not achieved without paying a price. Aside from the cost of additional hardware, the feedback system's response will be very sensitive to parameter variations of the elements in the feedback path. Furthermore, since the poles of the closed-loop system can be relocated in the s-plane, there is the danger of having one end up in the right-half plane, in which case the system would be *unstable*. Hence, an important aspect of feedback-system design is ensuring that the resulting system will be stable, not only under nominal or ideal conditions but also in the presence of anticipated parameter variations and perhaps even failure of individual components. Consequently, the designer must take care to insure that by adding a feedback path he is not creating more problems than he is solving. Some of the techniques presented in Chapters 10 and 11 are addressed to this very point.

Example 9.1 A feedback amplifier

Feedback is often used to advantage in communications engineering – or electronic instrumentation in general – in designing amplifiers in accordance with some set of specifications. Illustrating this point, suppose that a given amplifier has as its transfer function $T(s) = A/[1 + (s/2\pi B)]$ so its frequency-response function is

$$T(f) = T(s)|_{s=j2\pi f} = \frac{A}{1 + j(f/B)}$$

Therefore, A is the low-frequency gain and B the half-power bandwidth. Suppose further that we have more than enough gain but the bandwidth is inadequate for the input signals in question. To overcome this defect, let us try the configuration of Figure 9.12, where a feedback loop with a sign inversion and a gain of $1/\alpha$ has been added around the original amplifier. We assume that $1 < \alpha << A$.

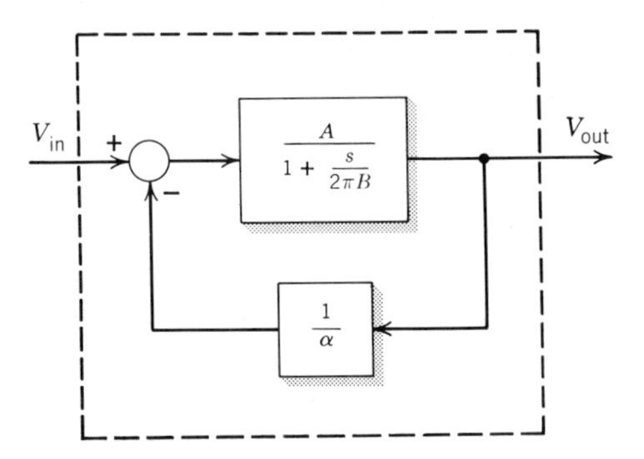

Figure 9.12.

Applying (3), the feedback amplifier is described by

$$T_F(s) = \frac{T(s)}{1+\dfrac{1}{\alpha}T(s)} = \frac{A}{1+\dfrac{s}{2\pi B}+\dfrac{A}{\alpha}}$$

$$\approx \frac{\alpha}{1+\dfrac{\alpha}{A}\dfrac{s}{2\pi B}} \qquad \frac{A}{\alpha} >> 1$$

Therefore

$$T_F(f) \approx \frac{A_F}{1+j(f/B_F)}$$

where

$$A_F = \alpha \qquad B_F = \frac{A}{\alpha}B$$

so the introduction of feedback has increased the bandwidth by a factor of A/α and reduced the gain by α/A. Thus, the design specifications may be met if α can be set at the desired low-frequency gain while satisfying $A/\alpha >> 1$. As a further advantage of the feedback amplifier, it should be mentioned that α is usually implemented with a resistive network, whereas A depends directly upon the characteristics of the active device; this means that $A_F = \alpha$ (independent of A) is much less likely to vary with temperature, aging, etc., than A.

Many more exotic designs are possible with feedback amplifiers. Angelo (1968, Chap. 19) discusses some of them, while Bode (1945) is devoted entirely to the subject.

9.2 RESPONSE OF FEEDBACK SYSTEMS

In the previous section we showed how the transfer functions $T_R(s)$ and $T_D(s)$ could be found in terms of the transfer functions of the individual blocks within the loop. Thus, since we already have a substantial array of tools for the analysis of cascade systems, one might conclude that there is nothing more to be done with the analysis of feedback systems. To the contrary, there are a number of additional techniques that will both simplify the analysis and provide more insight in the design of feedback systems than can be gained by a direct application of the cascade-system methods.

In terms of manpower and insight, the efficient analysis of a feedback system requires that the important features of its response as a closed-loop system be readily ascertainable from the characteristics of its component subsystems, e.g., $G_c(s)$, $G_p(s)$, and $H(s)$ in Figure 9.2, rather than from its closed-loop transfer functions $T_R(s)$ and $T_D(s)$. Because of the interaction among all subsystems in the closed loop, a direct analysis of $T_R(s)$ or the corresponding closed-loop impulse response is impractical in all but the simplest of systems. In fact, we shall demonstrate shortly that the impulse-response and convolution methods, which were so useful in analyzing cascade systems, are of very limited value, compared with transform methods, for the study of feedback systems. Hence, it is no accident that the vast majority of our work in this and the two subsequent chapters will be done in the frequency domain – either complex ($s = \sigma + j\omega$) or real ($j\omega$).

Poles and zeros of $T_R(s)$

Rewriting the closed-loop transfer function $T_R(s)$ of the single-input system shown in Figure 9.3*a*, namely

$$T_R(s) = \frac{C(s)}{R(s)} = \frac{G(s)}{1+G(s)H(s)} \tag{1}$$

we observe that the form of $T_R(s)$ given in (1), while perfectly valid and useful for some purposes, does not give any indication as to the locations of its poles or zeros. By finding the poles and zeros of $T_R(s)$ we shall be able to characterize the behavior of the closed-loop system's zero-input response and its response to a reference input. For example, in terms of the foregoing discussion on the advantages and disadvantages of feedback, we shall be able to verify that the closed-loop system is asymptotically stable, a fact that affords some peace-of-mind even if the response is not all that is desired. However, to determine these closed-loop poles and zeros we must have $T_R(s)$ expressed as the ratio of two factored polynomials in s, i.e., as a factored rational function. Since the transfer functions $G(s)$ and $H(s)$ comprising $T_R(s)$ are generally rational functions themselves, (1) must be modified in order to have its poles and zeros explicitly available.

Assuming both $G(s)$ and $H(s)$ to be rational functions of s, they can each be expressed as ratios of their numerator and denominator polynomials of the form

$$\frac{N_G(s)}{D_G(s)} \triangleq G(s) \tag{2}$$

and

$$\frac{N_H(s)}{D_H(s)} \triangleq H(s) \tag{3}$$

For instance, if the forward-path transfer function $G(s)$ has k poles, then $D_G(s)$ can be written as

$$D_G(s) = a_k s^k + a_{k-1} s^{k-1} + \cdots + a_0$$

or, in factored form,

$$D_G(s) = a_k[(s-p_1)(s-p_2)\cdots(s-p_k)]$$

where $p_1, p_2, \ldots, p_k$ are the poles of $G(s)$. Substituting (2) and (3) into (1),

$$T_R(s) = \frac{N_G/D_G}{1+(N_G N_H)/(D_G D_H)}$$

which reduces to

$$T_R(s) = \frac{N_G D_H}{D_G D_H + N_G N_H} \tag{4}$$

Having obtained the reference transfer function in rational form, we are ready to make two rather important observations.

First, the finite *zeros* of $T_R(s)$ are those values of s (excluding $s = \infty$) for which either $N_G(s)$ or $D_H(s)$ vanishes. But (2) indicates that the values of s for which $N_G(s) = 0$ are precisely the zeros of the forward-path transfer function $G(s)$ and (3) shows that the values of s for which $D_H(s) = 0$ are the poles of the feedback-path transfer function $H(s)$. Second, the *poles* of $T_R(s)$ are the roots of the equation

$$D_G(s)D_H(s) + N_G(s)N_H(s) = 0 \tag{5}$$

which, because its left-hand side is the denominator polynomial of the transfer function, must be the characteristic equation for the closed-loop system shown in Figure 9.3*a*. As such, the roots of (5) determine the modes of the closed-loop system's zero-input response and, consequently, the stability of the closed-loop system. However, because the denominator is the *sum* of two polynomials rather than their product, the poles of $T_R(s)$ will not be related to the poles and zeros of $G(s)$ and $H(s)$ in a straightforward manner.

The fact that the closed-loop poles of a feedback system do not have a direct relationship to the open-loop poles and zeros tends to complicate the analysis. However, it does mean that the designer has the capability of making the closed-loop system behave quite differently from the open-loop system. In order to make effective use of this ability it is necessary to have methods for analyzing the effects of parameter changes without having to solve (4) for $T_R(s)$ or (5) for the closed-loop poles each time a parameter is changed. Hence, much of our subsequent work will be directed toward developing analytical tools that will allow us to take advantage of this unusual property of feedback systems.

Example 9.2

The block diagram in Figure 9.13 represents an integrator whose transfer function is $1/s = \mathcal{L}[u(t)]$, around which a negative-feedback path with an adjustable gain of K has been placed. Our objective is to demonstrate the manner in which the gain K affects the closed-loop transfer function $T_R(s)$.

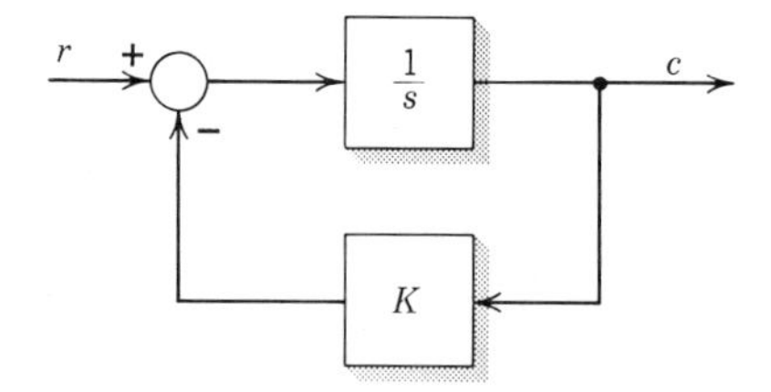

Figure 9.13.

Comparing the block diagram with Figure 9.3a, it is apparent that $G(s) = 1/s$ and $H(s) = K$. Substituting these open-loop transfer functions into (4), the reference transfer function can be written as the rational function

$$T_R(s) = \frac{1}{s+K} \tag{6}$$

which has a single pole at $s = -K$ and has no finite zeros.†

Because the integrator's transfer function $(1/s)$ has a pole at $s = 0$ and $H(s)$ has no poles at all, we say that there is one open-loop pole at the origin of the s-plane – imagine, if you will, that the feedback path was physically opened and $C(s)/R(s)$ evaluated. However, with the loop closed and $K > 0$ the system transfer function has its pole at $s = -K$, which is on the negative real axis. Thus, the addition of the feedback path

†Strictly speaking, $T_R(s)$ has a single zero at $s = \infty$; however, we generally refer only to those poles and zeros in the finite s-plane.

has caused the transfer-function pole to shift from the origin to the left in the s-plane as K is increased from zero to some positive value. In the process of raising K the system has been changed from a pure integrator to a first-order low-pass filter whose half-power bandwidth is K rad/sec. For negative values of K the pole will be on the positive real axis, resulting in an unstable system. By treating the gain K as a parameter that is varied from 0 to $+\infty$, we can draw the locus of the closed-loop pole in the complex plane that is shown in Figure 9.14.

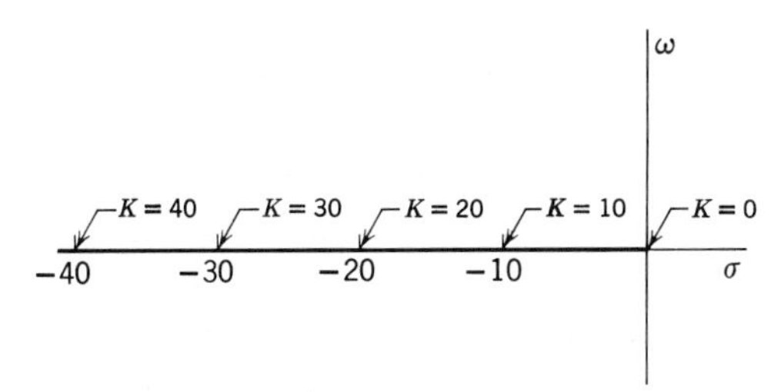

Figure 9.14.

Methods for rapidly constructing more complicated root-locus plots will be presented in Chapter 10.

Poles and zeros of $T_D(s)$

Starting with Eq. (5), Sect. 9.1, and defining the rational functions

$$\frac{N_{G_c}(s)}{D_{G_c}(s)} \triangleq G_c(s) \qquad \frac{N_{G_p}(s)}{D_{G_p}(s)} \triangleq G_p(s)$$

it follows that the rational-function form of the disturbance transfer function is

$$T_D(s) = \frac{N_{G_p} D_{G_c} D_H}{D_{G_c} D_{G_p} D_H + N_{G_c} N_{G_p} N_H} \tag{7}$$

To better compare the two closed-loop transfer functions, we recognize that because of Eq. (2), Sect. 9.1, it is necessary for $N_{G_c} N_{G_p} = N_G$ and $D_{G_c} D_{G_p} = D_G$. Rewriting (7), we get

$$T_D(s) = \frac{N_{G_p} D_{G_c} D_H}{D_G D_H + N_G N_H} \tag{8}$$

which can be compared directly with $T_R(s)$ on the basis of their closed-loop poles and zeros.

Because the numerator of (8) is the product $N_{G_p}(s)D_{G_c}(s)D_H(s)$ it follows that the zeros of $T_D(s)$ are composed of: *1*. the zeros of $G_p(s)$, the process transfer function; *2*. the poles of $D_{G_c}(s)$, the controller transfer function, and *3*. the poles of $H(s)$, the transducer transfer function. By examining the block diagram of the system, we can see that this result is consistent with our findings for $T_R(s)$ in that for a disturbance input the feedback path consists of both $H(s)$ and $G_c(s)$, whereas the forward path consists of only $G_p(s)$.

On the other hand, the polynomials comprising the denominators of $T_R(s)$ and $T_D(s)$ are identical, implying that both transfer functions have precisely the same poles. In fact, if any pair of inputs and outputs was considered in addition to those two sets, one would find that the corresponding closed-loop transfer function would have as its denominator the characteristic polynomial $P(s) = D_G D_H + N_G N_H$. This situation is a consequence of the fact that the characteristic polynomial is a property of the closed-loop configuration and not of the particular variables that happen to be designated as the input and output.

Because their poles are the same, the only differences between $T_R(s)$ and $T_D(s)$ are in their zeros, i.e., numerator polynomials, except for a multiplicative constant. In fact it would be rather unfortunate if this difference did not exist because we certainly do not want the output to respond to disturbance inputs in the same way that it responds to reference inputs – generally we would prefer $T_R(s)$ to be unity, so the output would follow the reference input exactly, and $T_D(s)$ to be zero, so the output would be unaffected by disturbances. As it turns out, the only difference between the two transfer functions is in the way the controller transfer function $G_c(s)$ enters the two numerator polynomials, suggesting that in the selection of a controller transfer function the relationship of its zeros to its poles will play an important role in the relationship of $T_R(s)$ to $T_D(s)$.

Example 9.3

If the simple feedback system of the previous example is rearranged as shown in Figure 9.15*a*, with a disturbance input included, the respective transfer functions are found to be $T_R(s) = K/(s + K)$ and $T_D(s) = 1/(s + K)$, the only difference being the presence of the factor K in the numerator of $T_R(s)$. Although both transfer functions have the same pole, $s = -K$, and both exhibit the frequency response characteristic of a first-order lowpass filter, see Figure 9.15*b*, the low-frequency gains differ by the factor K. Thus, by increasing K the ability of the output to follow low frequency reference inputs is not affected, but its susceptability to low frequency disturbances is reduced, both effects usually being desirable. The bandwidth associated with both transfer functions increases linearly with K.

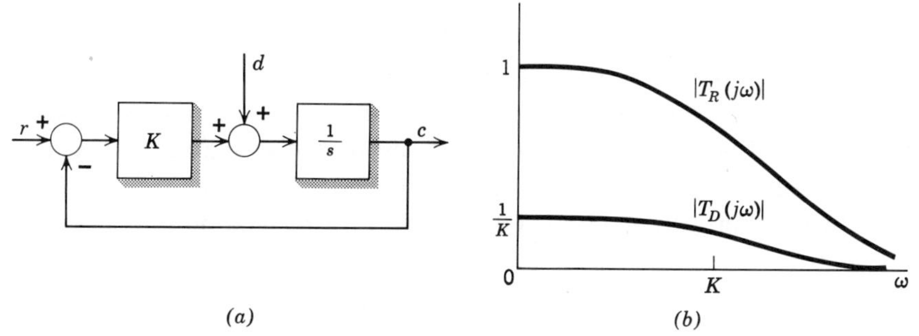

Figure 9.15.

Time-domain analysis

One of our principal themes in earlier chapters was that when certain analytical operations are carried out in the real or complex-frequency domains there are corresponding operations in the time domain and vice versa, e.g., the dual relationship between multiplication and convolution. One might ask why our discussion of the input-output characteristics of a closed-loop system has been carried out exclusively in the complex-frequency domain, i.e., in terms of transforms and transfer functions. Actually, there is a very good reason for working in the frequency domain: it is virtually impossible to reduce even a single feedback loop to cascade form by working exclusively in the time domain.

To convince the reader of this statement, we consider the time-domain equations representing the unity-feedback system in Figure 9.3*b*:

$$\varepsilon(t) = r(t) - c(t) \tag{9}$$

and

$$c(t) = g(t) * \varepsilon(t) \tag{10}$$

where $g(t)$ is the impulse response of the forward path. Substituting (9) into (10) and using the associative property of convolution gives

$$c(t) + [g(t) * c(t)] = g(t) * r(t) \tag{11}$$

Because a function convolved with the unit impulse is the function itself, (11) may be written as

$$[\delta(t)+g(t)] * c(t) = g(t) * r(t)$$

which is merely a shorthand notation for the rather nasty integral equation

$$\int_0^\infty [\delta(t-\lambda)+g(t-\lambda)]c(\lambda)d\lambda = \int_0^\infty g(t-\mu)r(\mu)d\mu \qquad (12)$$

Because the input $r(\mu)$ is inside one integral and the output $c(\lambda)$ is inside another integral, the direct solution of (12) in the time domain is certainly inpractical, if not impossible. In fact, if confronted with an equation such as (12), one should transform it into the complex frequency domain, thereby obtaining the algebraic Laplace-transform equation

$$[1+G(s)]C(s) = G(s)R(s)$$

which leads directly to Eq. (4), Sect. 9.1.

Hence, most of the work to follow that deals with feedback systems will be done in the frequency domain, either real or complex, rather than the time domain, with care taken to relate our *answers* to the time-domain behavior of the system. For example, if a particular reference input $r(t)$ is specified for the unity-feedback system discussed above, we can write

$$c(t) = \mathfrak{L}^{-1}\left[\frac{G(s)R(s)}{1+G(s)}\right]$$

which can then be evaluated as described in Section 8.3 to find $c(t)$.

9.3 APPLICATION: SATELLITE ATTITUDE CONTROL

At this point, we shall undertake a detailed investigation of an illustrative control problem which is continued in Chapters 10, 11, and 13 as we broaden our analytic capabilities. There are at least three reasons for doing this: *1*. to relate our somewhat abstract mathematical concepts to an actual system, *2*. to introduce the basic control laws which have been found to be suitable for the vast majority of control problems, and *3*. to provide the reader with some worked-out examples supplementing our theoretical presentation. In this section we shall first highlight the important physical aspects of the system, taking a number of liberties with the practical aspects so as to avoid undue complication which would tend to obscure our main objectives; and then, we shall examine some basic control strategies.

The physical system and its mathematical model

A telescope is rigidly fixed to a satellite† which is in orbit around the earth, Figure 9.16*a*. It is desired to be able to point the telescope axis (Z) directly at a dim target star whose line of sight is within several degrees of the line of sight to a relatively bright reference star.

The satellite will be subjected to a variety of disturbance torques from such sources as micrometeorites, aerodynamic drag, solar pressure, gravity-gradient torques, and the rotation of various satellite components, e.g., pumps, motors, and antennas. It will be assumed that the vehicle has initially been aligned so that the axis of the telescope is pointing at the target star and the vehicle is not spinning. Because of the above-mentioned disturbance torques, it will be necessary to provide a feedback control system in order to maintain the satellite very close to the desired attitude while photographs requiring long time exposures (1 or 2 minutes) are being taken. In addition to counteracting the effects of the disturbance torques acting on the vehicle, the control system will be called upon to sight the telescope on other target stars, subject to commands received from ground stations.

To accomplish this objective we may construct – on paper – a control system consisting of sensors to measure the angular deviation of the telescope axis from the line to the reference star and electrically operated gas jets, which will provide the control torques. The task of the control-system designer is to decide how the signals from the sensors are to be used in order to generate the electrical signals that operate the gas jets. However, before proceeding further, we must develop a mathematical model of the system, which will account for the important aspects of the response yet be simple enough so that it can be treated analytically.

Figure 9.16*b* shows the set of axes and angles used in the solution of the problem. The X, Y, Z frame is fixed with respect to the vehicle and coincides with the axes of the satellite about which control torques may be exerted by the gas jets. So long as the angular velocity and the error angles of the vehicle are small, angular motions about the three axes are *essentially uncoupled* and may be *controlled independently*; thus, we may design

†See Perkins and Cruz (1969) for a more faithful and comprehensive treatment of a satellite attitude control system.

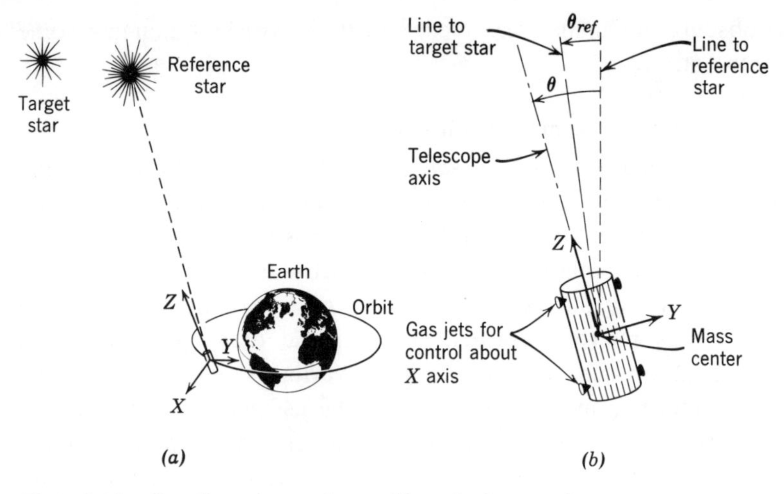

Figure 9.16. Coordinate system for satellite attitude control.

each of the three attitude-control loops separately, which is an important simplification. Because it is small (less than 10 degrees) the angle between the line to the reference star and the telescope axis may be resolved into two components, one a rotation about the X axis and the other a rotation about the Y axis. We shall restrict our attention to the rotation about the X axis, described by the angle θ in Figure 9.16*b*.

The block diagram of the system, as initially proposed, is given in Figure 9.17 and is assumed to consist of the following elements:

1. An *input transducer* that provides an electrical reference signal whose voltage is $K_\theta \theta_{ref}(t)$. This signal will be generated from a signal transmitted from earth and converted to electrical form in the satellite.

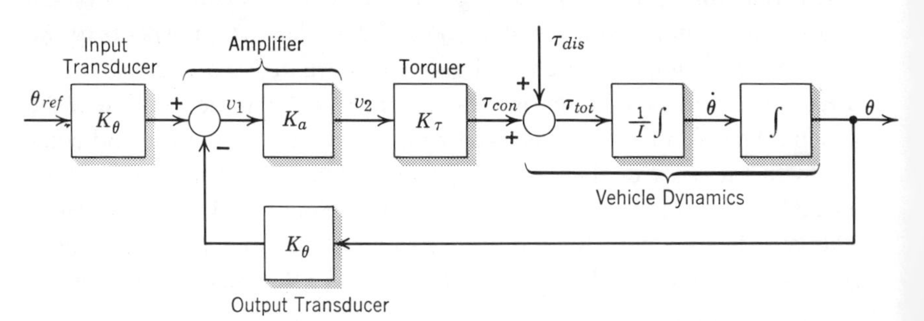

Figure 9.17. Block diagram model of satellite attitude system with proportional control.

2. An *output transducer*, e.g., star tracker, that measures the deviation of the telescope axis from the line to the reference star. Its output is a voltage equal to $K_\theta \theta(t)$.

3. A *differencing amplifier* that provides a voltage $v_2(t)$ proportional to $v_1(t)$, the difference of the voltages from the input and output transducers. The amplifier must provide sufficient power to actuate the gas jets of the torquers. However, since the source of power is not directly involved in the analysis of the control system, it is customary to omit it from the block diagram (as part of a complete system study such factors as power drain and saturation would be important).

4. A set of linear *gas jets*† that is assumed to exert a torque τ_{con} about the X axis equal to $K_\tau v_2$ but exerts no net force. We have neglected the dynamics of the gas jets by assuming that the opening time is short compared with the important time constants of the system. The moment arms of the jets from the vehicle's center of mass are taken into account in determining the gain K_τ. As in item 3, the source of the energy provided by the jets, i.e., the gas stored in tanks, is omitted from Figure 9.17.

5. The *vehicle dynamics*, namely the rotation of the satellite about its X axis, that are governed by the differential equation

$$I\ddot{\theta} = \tau_{con} + \tau_{dis} = \tau_{tot} \tag{1}$$

where I is the moment of inertia about the X axis and the total torque $\tau_{tot}(t)$ about the X axis is the sum of the control torque $\tau_{con}(t)$ and the disturbance torques $\tau_{dis}(t)$. Since θ and $\dot{\theta}$ are the logical choices for state variables, the vehicle dynamics may be modeled by two integrators in series with a gain of $1/I$. Although the moment of inertia will vary over the life of the satellite because of the consumption of fuel, it presumably changes very slowly compared with the response of the vehicle. Thus we are justified in treating it as a constant but should ascertain how sensitive the performance of the final system design is to anticipated changes in I.

Proportional control

We shall begin the design process by analyzing the response of the basic feedback system in Figure 9.17, which involves a control torque that is

†Usually a combination of on-off gas jets and reaction wheels would be employed.

proportional to the pointing error, namely

$$\tau_{con}(t) = K_a K_\tau K_\theta \overbrace{[\theta_{ref}(t) - \theta(t)]}^{\varepsilon(t)} \tag{2}$$

As a first step, we set the disturbance torque equal to zero and evaluate the closed-loop reference transfer function. Using Eq. (3), Sect. 9.1, with $G_c(s) = K_a K_\tau$, $G_p(s) = 1/Is^2$, and $H(s) = K_\theta$, and then multiplying the result by the input-transducer gain K_θ, we find

$$T_R(s) = \frac{\theta(s)}{\theta_d(s)} = K_\theta \left[\frac{K_a K_\tau / Is^2}{1 + K_a K_\tau K_\theta / Is^2} \right]$$

$$= \frac{K_a K_\tau K_\theta}{Is^2 + K_a K_\tau K_\theta}$$

By dividing through by the moment of inertia I and defining the parameter

$$\omega_n \triangleq \left(\frac{K_a K_\tau K_\theta}{I} \right)^{1/2} \tag{3}$$

we see that $T_R(s)$ takes on the familiar form of the transfer function of an undamped, simple harmonic oscillator with unity low-frequency gain:

$$T_R(s) = \frac{\omega_n^{\ 2}}{s^2 + \omega_n^{\ 2}} \tag{4}$$

Comparing $T_R(s)$ with $\mathfrak{L}\,[\sin \omega t]$ as given by Eq. (17), Sect. 8.1, the impulse response $h_1(t)$ is found to be

$$h_1(t) = \omega_n \sin \omega_n t$$

The response to a step-function change in θ_{ref} of magnitude $\bar{\theta}$ radians, with θ and $\dot{\theta}$ initially zero, is

$$\theta(t) = \bar{\theta} \int_0^t h_1(\lambda)\, d\lambda = \bar{\theta}(1 - \cos \omega_n t)$$

which is sketched in Figure 9.18*a*. Alternately, we might have multiplied $T_R(s)$ by $\bar{\theta}/s = \mathfrak{L}\,[\bar{\theta} u(t)]$ and inverted the transform. The corresponding pointing error, denoted by $\varepsilon(t)$ is

$$\varepsilon(t) = \bar{\theta} - \theta(t) = \bar{\theta} \cos \omega_n t$$

and, because of the proportional control law assumed, the control torque is

$$\tau_{con}(t) = K_a K_\tau K_\theta \varepsilon(t)$$

which, using the parameter ω_n defined by (3), can be written as

$$\tau_{con}(t) = I\omega_n{}^2\theta \cos \omega_n t$$

and is plotted in Figure 9.18*b*, along with $\varepsilon(t)$.

It should be obvious that the above control system does not provide a satisfactory response because the oscillations following a change in θ_{ref} do not decay with time. In terms of the locations of the poles of

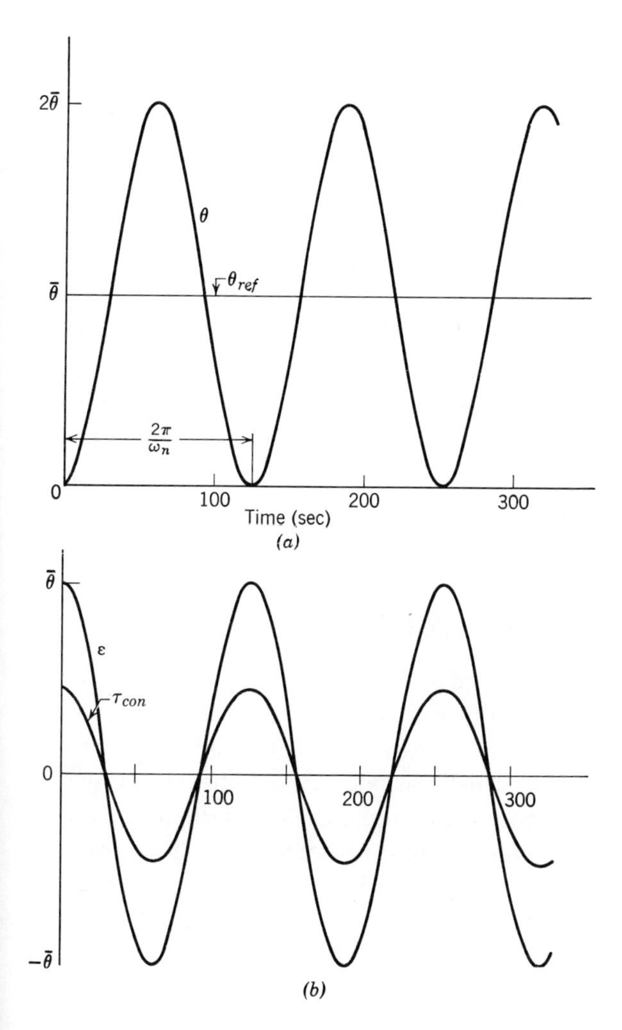

Figure 9.18. Response with proportional control.

$T_R(s)$ or, equivalently, the roots of the characteristic equation in the s-plane, (4) indicates that they lie on the imaginary axis, thereby accounting for the presence of undamped oscillations in the step response. Our next task is to find a means of assuring that the oscillations are damped out which, when viewed in the frequency domain, is equivalent to shifting these poles from the imaginary axis into the LHP.

Proportional-plus-derivative control

One way of approaching a solution to this problem which has considerable *conceptual* merit is to recognize from Figure 9.18*b* that the control torque $\tau_{con}(t)$ and the pointing error $\varepsilon(t)$ are oscillating exactly *in phase* – as a consequence of the proportional control law any oscillations must be exactly in phase. Thus, the control torque does not reverse sign until the error has reached zero, although the derivative of the error was negative during the preceding $0.5\ \pi/\omega_n$ seconds.

If the controller had been able to anticipate the fact that the error was diminishing and consequently reversed its sign before $\varepsilon(t)$ reached zero, the overshoot might have been reduced. Certainly the negative peak of the error should have been less than $\bar{\theta}$ radians in magnitude, which, when applied to successive cycles, would imply that the oscillations should eventually damp out. One way of introducing anticipation into the controller would be to make the control torque at some time t_0 be proportional to the value of the pointing error that would exist α seconds in the future, i.e., to make

$$\tau_{con}(t_0) = K_a K_\tau K_\theta \varepsilon(t_0 + \alpha) \tag{5}$$

for all t_0 and some as yet unspecified α.

While the control law described by (5) might provide the desired response in principle, it does not represent a practical solution due to the requirement that it predict the error α seconds in the future. Because such a device is *noncausal* it is physically unrealizable. We therefore seek a realizable device, at least in an idealized sense, which will approximate the input-output relationship defined by (5). As a guide in selecting the realizable controller, we expand $\varepsilon(t_0+\alpha)$ in a Taylor series about the point $t = t_0$, getting

$$\varepsilon(t_0+\alpha) = \varepsilon(t_0) + \dot{\varepsilon}(t_0)\alpha + \frac{\ddot{\varepsilon}(t_0)}{2!}\alpha^2 + \cdots$$

If we retain the first two terms of the expansion as an approximation to $\varepsilon(t_0+\alpha)$ and denote the result as $\hat{\varepsilon}(t_0)$, then

$$\hat{\varepsilon}(t_0) \triangleq \varepsilon(t_0)+\dot{\varepsilon}(t_0)\alpha \tag{6}$$

which differs from the exact expression only in terms of order α^2 and higher. Thus, for any value of t, $\hat{\varepsilon}(t)$ can be synthesized by taking the proper linear combination of $\varepsilon(t)$ and $\dot{\varepsilon}(t)$, namely by a "proportional-plus-derivative" (P + D) operation. Taking the Laplace transform of (6) and solving for the transfer function relating $E(s) \triangleq \mathfrak{L}[\varepsilon(t)]$ to $\hat{E}(s) \triangleq \mathfrak{L}[\hat{\varepsilon}(t)]$, it follows that

$$\frac{\hat{E}(s)}{E(s)} = 1+\alpha s$$

Defining the P + D control law according to

$$\tau_{con}(t) = K_a K_\tau K_\theta \underbrace{[\varepsilon(t)+\alpha\dot{\varepsilon}(t)]}_{\hat{\varepsilon}(t)} \tag{7}$$

the block diagram of the system becomes that shown in Figure 9.19.

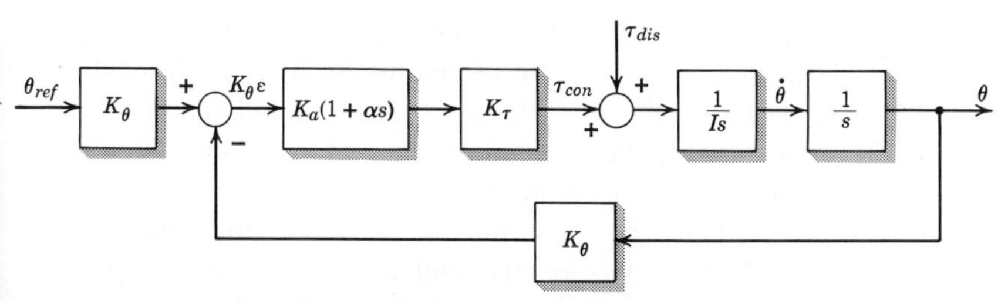

Figure 9.19. System diagram for proportional-plus-derivative (P + D) control.

By reducing the system block diagram and using (3), the closed-loop transfer function with a P + D controller is found to be

$$T_R(s) = \frac{\omega_n^2(1+\alpha s)}{s^2+\alpha\omega_n^2 s+\omega_n^2} \tag{8}$$

A comparison of (8) with (4) reveals two major changes in the reference transfer function of the closed-loop system due to the inclusion of the derivative term. The most important change is that the denominator of (8) contains a term proportional to s, which means that α may be selected so that the poles of $T_R(s)$ will lie in the LHP rather than on the imaginary axis. The time-domain implication of this shift of the closed-loop poles is that the oscillations in Figure 9.18 will be damped out. In fact, the closed-loop poles can be placed *anywhere* in the complex plane (provided they are complex conjugates when not real) by adjusting α and ω_n.

The other difference in the two cases is that $T_R(s)$ now has a closed-loop zero at $s=-1/\alpha$ whereas the previous version had no zeros in the finite s-plane. As a consequence, for a step-function change in θ_{ref} the torque $\tau_{con}(t)$ must contain an impulse at $t=0+$ because of the differentiating action of the P+D controller. Alternatively, an evaluation of $\tau_{con}(s)$ using (7) and (8) with $\theta_{ref}(s)=\bar{\theta}/s$ will give the impulsive term in the control torque. Although this torque cannot be an ideal impulse in practice, it is possible for it to contain a pulse of finite amplitude that is short enough compared with the time constants of the vehicle to be approximated by an impulse. The response of the system with the P+D controller is shown in Figure 9.20 for a step-function input of magnitude $\bar{\theta}$, along with the resulting control torque and pointing error; the essential feature is the damped response, in contrast to the undamped oscillations of Figure 9.18.

The fact that the introduction of the derivative term in the control law resulted in damped transients can also be explained in terms of energy dissipation. One way of making the oscillation decay is to dissipate the energy that caused it in the first place; another method is to provide a means of storing the angular momentum in a rotating wheel (reaction wheel) within the satellite, without dissipating the energy. However, if energy is to be dissipated by the control system there must be a torque dependent upon the derivative of $\theta(t)$ and when τ_{con} is strictly proportional to $\varepsilon(t)$ no such dissipative torque exists, thus allowing any transients to persist indefinitely in the form of undamped oscillations. However, the use of the P+D control law does introduce energy dissipation due to the presence of the term $\alpha\dot{\varepsilon}(t)$ in (7) and thus accounts for the decaying transients shown in Figure 9.20.

A third means of explaining the observed changes in the system's behavior is to note the change in the poles of the closed-loop transfer function which, of course, are the roots of the characteristic polynomial. The characteristic polynomial corresponding to a strictly proportional control law is the denominator of (4), namely,

$$P_1(s)=s^2+\omega_n{}^2=(s-j\omega_n)(s+j\omega_n)$$

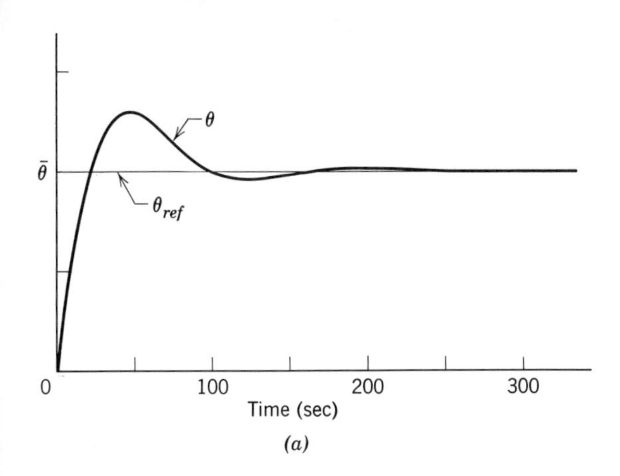

(a)

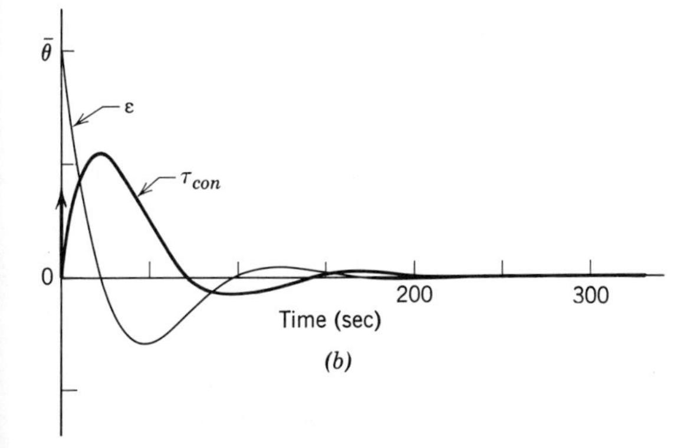

(b)

Figure 9.20. Response with P + D control.

On the other hand, the characteristic polynomial corresponding to a P + D controller is, from (8),

$$\begin{aligned} P_2(s) &= s^2 + \alpha\omega_n{}^2 s + \omega_n{}^2 \\ &= \left[s + \left(\frac{\alpha}{2} - j\sqrt{\frac{1}{\omega_n{}^2} - \frac{\alpha^2}{4}}\right)\omega_n{}^2\right]\left[s + \left(\frac{\alpha}{2} + j\sqrt{\frac{1}{\omega_n{}^2} - \frac{\alpha^2}{4}}\right)\omega_n{}^2\right] \end{aligned}$$

the roots of which are both in the LHP provided only that $\alpha > 0$. Graphically, the addition of the derivative term has caused the system's closed-loop poles to shift from the imaginary axis to the points indicated in Figure 9.21, which happen to lie on the arc of a circle of radius ω_n centered at the origin. This graphical interpretation will be discussed further in the next chapter.

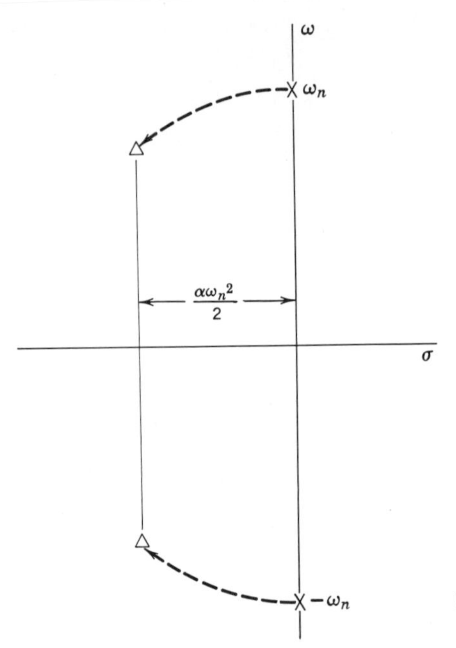

Figure 9.21.

Effects of disturbance torques

Thus far, our attention has been directed toward the control system's ability to make the satellite follow the reference pointing angle, θ_{ref}. An equally important task for the system is that of reducing the effects of disturbance torques upon the vehicle's orientation. For example, the collision of a micrometeorite with the vehicle will exert an impulsive force that will cause an instantaneous change in the angular momentum about one or more of the vehicle's three axes. If a component about the X axis exists, we may consider this effect to be either an impulsive disturbance torque or a step change in the angular rate $\dot{\theta}$, which may be thought of as providing an initial value of $\dot{\theta}$. When a satellite is maintained in an inertially fixed orbit the gravity-gradient torque will act as a sinusoidal disturbance input at twice the orbital frequency. Since a circular near-earth orbit has a period of approximately 90 minutes, the gravity-gradient torque will have a period of 45 minutes, which is generally much greater than the longest time constant of the closed-loop system. Therefore, the effects of this particular disturbance can be considered by examining the steady-state response to a constant disturbance torque.

The torques mentioned above result from sources external to the satellite. However, internal motions of the vehicle may also affect its orientation since the total angular momentum of the satellite must remain constant in the absence of external torques. Therefore, if an electric motor whose spin axis is not orthogonal to the X axis undergoes a change in angular velocity, the corresponding change in the angular momentum of its rotor will be imparted to the remainder of the vehicle and must be counteracted by the control system. If θ_{ref} is set to zero, the pointing error is given by $\varepsilon(t) = -\theta(t)$ and the objective is to maintain $\theta(t) = 0$, regardless of what $\tau_{dis}(t)$ might be.

We shall now proceed with the analysis to see how close we can come to this goal with the system structure of Figure 9.22. Reducing the block diagram, the disturbance transfer function is

$$T_D(s) = \frac{1/I}{s^2 + \omega_n^2 \alpha s + \omega_n^2} \tag{9}$$

By finding the inverse transform of (9), we obtain the response to an *impulsive* disturbance torque, as due, for instance, to micrometeorite impact.

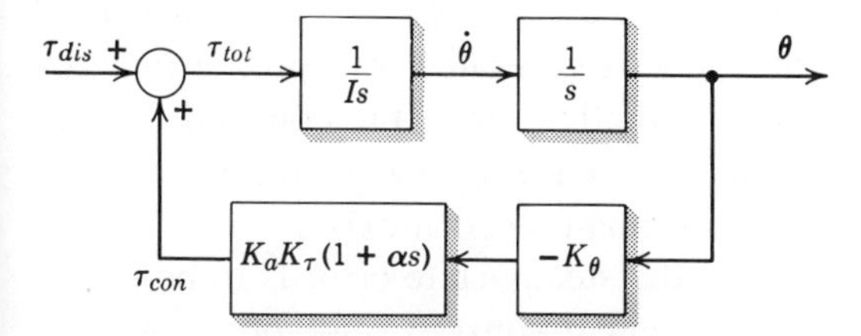

Figure 9.22. Model for disturbance torque, P + D control.

To assess the vehicle's steady-state disturbance behavior, the transform of its response to a step disturbance torque is found and the final-value theorem applied. Letting $\tau_{dis}(t) = Cu(t)$ so $D(s) \triangleq \mathfrak{L}[\tau_{dis}] = C/S$, it follows from (9) that

$$\theta(s) = \frac{C/I}{s(s^2 + \omega_n^2 \alpha s + \omega_n^2)} \tag{10}$$

Provided that $\alpha > 0$ the final-value theorem may be applied to $\theta(s)$ in order to find the steady-state component of the pointing angle; the result is

$$\theta_{ss} = \frac{C}{I\omega_n^2} = \frac{C}{K_a K_\tau K_\theta} \tag{11}$$

Thus, the magnitude of the steady-state offset angle is proportional to the magnitude of the disturbance torque and inversely proportional to the low-frequency gain of the feedback path in Figure 9.22. Furthermore, because the gravity-gradient torque varies at twice the orbital frequency, which is essentially at $\omega = 0$ insofar as $|T_D(j\omega)|$ is concerned, the resulting pointing angle will be an oscillation at the same frequency with an amplitude that is inversely proportional to the product $K_a K_\tau K_\theta$.

Thus, if the satellite specifications require that it be capable of having no pointing error while subjected to a gravity-gradient torque, or any other constant or very low-frequency torque for that matter, it is not possible to meet this objective with a P + D control law because the product $K_a K_\tau K_\theta$ must certainly remain finite. The obvious conclusion is that the *structure* of the control system must be modified.

Integral control

If the telescope axis of the satellite is to point in a fixed direction over an interval of time the total torque, $\tau_{tot} = \tau_{con} + \tau_{dis}$ must be zero during the entire interval, in addition to having the proper initial conditions. In other words, a control torque must be generated to counteract or "buck out" any disturbance torque occurring during the interval. We have just seen, for example, that if the control torque is generated by a proportional or P + D control law acting upon the pointing error, the constant control torque required to counteract a constant disturbance torque can be generated only if a constant, nonzero pointing error exists.

The obvious conclusion is that if the steady-state error is to be maintained at zero over a finite time interval, a nonzero value of τ_{con} will be required even when $\theta = \theta_{ref}$. While we cannot hope to achieve this condition for all times with a nonconstant disturbance torque, it can be done in the steady-state by making the control torque depend on the *integral* of the pointing error, in addition to the error and its derivative. We shall attempt to verify the validity of this assertion by returning to the P + D control law (7) and appending a term proportional to the integral of $\varepsilon(t)$ such that

$$\tau_{con}(t) = K_a K_\tau K_\theta \left[\varepsilon(t) + \alpha \frac{d\varepsilon}{dt} + \beta \int_0^t \varepsilon(\lambda)\, d\lambda \right] \tag{12}$$

which will be designated as a P + I + D controller. Because of the presence of the integral term, the possibility now exists for having $(\tau_{con})_{ss} \neq 0$

even when $\varepsilon_{ss} = 0$ because the integral term in (12) involves the area under the curve of $\varepsilon(t)$, which in general will not vanish when $\varepsilon_{ss} = 0$.

The transfer function of the P + I + D controller, namely $\tau_{con}(s)/E(s)$ can be obtained by transforming (12) to give

$$\frac{\tau_{con}(s)}{E(s)} = K_a K_\tau K_\theta \left(1 + \alpha s + \frac{\beta}{s}\right) = K_a K_\tau K_\theta \left(\frac{\alpha s^2 + s + \beta}{s}\right)$$

The block diagram of the complete system is shown in Figure 9.23 which, when simplified, yields the disturbance transfer function

$$T_D(s) = \frac{s/I}{s^3 + \omega_n{}^2 \alpha s^2 + \omega_n{}^2 s + \beta \omega_n{}^2} \tag{13}$$

Reevaluating the steady-state pointing angle due to a constant disturbance torque of magnitude C,

$$\theta_{ss} = \lim_{s \to 0} s\left(\frac{C}{s}\right) \frac{s/I}{(s^3 + \omega_n{}^2 \alpha s^2 + \omega_n{}^2 s + \beta \omega_n{}^2)} = 0$$

provided that the transients decay. Likewise, the reader can readily show that the error caused by the gravity-gradient torque, while not precisely zero, will be significantly smaller than that achieved with the P + D controller.

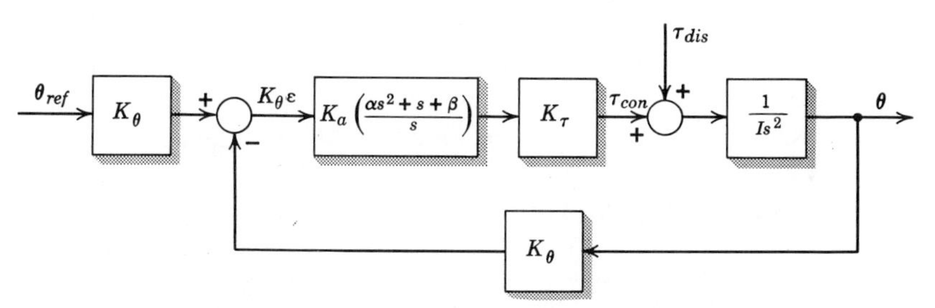

Figure 9.23. System diagram for proportional-plus-integral-plus-derivative (P + I + D) control.

However, before we can attain this improved steady-state response we must be assured that the transients will indeed decay, which is to say that the poles of $T_D(s)$, or the roots of the closed-loop characteristic equation, must all lie in the LHP. Inspection of the characteristic equation

$$P_3(s) = s^3 + \omega_n{}^2 \alpha s^2 + \omega_n{}^2 s + \beta \omega_n{}^2 = 0 \tag{14}$$

reveals that the task of ascertaining a combination of α, β, and ω_n for which all three of its roots lie in the LHP† is not likely to be a simple task, at least with the analytical tools at our disposal. Until now we have encountered only quadratic polynomials or cubic polynomials that were readily factored, e.g., the denominator of (10). Even if we could guarantee that the three roots of (14) were in the LHP, it would be desirable to be able to determine, with a minimum of work, how the three parameters should be chosen in order to place the roots in certain portions of the s-plane or to obtain certain characteristics for the closed-loop frequency response functions $T_R(j\omega)$ and $T_D(j\omega)$, e.g., bandwidth or rejection bands.

Since there are well-known methods for the numerical solution for the roots of high-degree polynomial equations, we could start guessing values of α, β, and ω_n and compute the corresponding roots. Although such an approach might, with a little luck, solve our immediate problems, its value in reaching the ultimate goal of being able to *design* feedback systems, rather than just analyze them, would be minimal.

Rather than working with the *closed-loop* transfer functions directly, our thrust in the next two chapters will be to develop methods for determining the essential features of the closed-loop response by using only the open-loop characteristics of the system. In Chapter 10 the root-locus technique, which is a powerful method for determining how the variation of a gain can affect the locations in the s-plane of the poles of a closed-loop transfer function, is discussed. Then, in Chapter 11, methods that use the open-loop frequency-response are developed. Before going on, however, we shall show how the state-variable formulation can be used to give a concise model for the feedback control of MIO systems.

9.4 STATE-VARIABLE FORMULATION

The basic notion of feedback control can readily be extended to multi-input-output (MIO) systems. The analytic means by which this extension is possible is the vector-matrix formulation of the system equations presented in earlier chapters. Thus, we start by repeating the state and output equations for an nth-order system with p inputs and r outputs:

$$\dot{\mathbf{q}} = \mathbf{A}\mathbf{q} + \mathbf{B}\mathbf{x} \tag{1}$$

$$\mathbf{y} = \mathbf{C}\mathbf{q} \tag{2}$$

where the time dependence of the variables $\mathbf{q}$, $\mathbf{x}$, and $\mathbf{y}$ is implicit.

†As in the second-order case with the P + D control law, the three roots can be placed anywhere in the s-plane by adjusting the three parameters α, β, and ω_n.

Now recall that the general model of single-loop systems presented in Figure 9.1 included two classes of inputs to the process: *1*. those inputs that can be manipulated in a predetermined manner and, hence, are available for controlling the process; and *2*. those inputs that are not the result of any premeditated action on the part of the controller, i.e., disturbance inputs. We can impart this distinction to the state-variable formulation merely by considering the p-element input vector $\mathbf{x}(t)$ to be partitioned into the vectors $\mathbf{v}(t)$ and $\mathbf{w}(t)$ where $\mathbf{v}(t)$ is a p_v-element vector of *control* inputs and $\mathbf{w}(t)$ is a p_w-element vector of *disturbance* inputs, with $p_v + p_w = p$. If the matrix $\mathbf{B}$ is partitioned in a corresponding manner into the $n \times p_v$ matrix $\mathbf{B}_v$ and the $n \times p_w$ matrix $\mathbf{B}_w$, the state equation (1) may be rewritten as

$$\dot{\mathbf{q}} = \mathbf{A}\mathbf{q} + \overbrace{[\mathbf{B}_v | \mathbf{B}_w]}^{\mathbf{B}} \overbrace{\left[\frac{\mathbf{v}}{\mathbf{w}}\right]}^{\mathbf{x}} = \mathbf{A}\mathbf{q} + \mathbf{B}_v\mathbf{v} + \mathbf{B}_w\mathbf{w} \tag{3}$$

As the final preliminary step, we assume that each of the output variables $y_i(t)$, $i = 1, 2, \ldots r$ has a desired value $(y_i)_{des}$ which may be a constant or an arbitrary function of time, written more compactly as the r-element vector $\mathbf{y}_{des}(t)$.

Now we are ready to introduce feedback into the MIO system by deciding in precisely what manner the control input $\mathbf{v}$ is to depend upon the measured output $\mathbf{y}$ and the desired output $\mathbf{y}_{des}$. Certainly the simplest choice is to take

$$\mathbf{v} = \mathbf{K}(\mathbf{y}_{des} - \mathbf{y}) \tag{4}$$

where $\mathbf{K}$ is a $p_v \times r$ matrix of constant feedback gains, commonly referred to as the *gain matrix*. Although (4) may appear at first glance to be nothing more than the proportional control law studied in the previous section, it is really far more general because we are working with the state-variable formulation of the system, rather than specific scalar variables. For instance, if both a variable and its derivative are among the measurable outputs, then the proportional-plus-derivative control law is automatically included in (4). Continuing in the same vein, if the integral of the variable in question is also an output, the P + I + D controller is yet another special case. In fact, (4) is so general that design techniques for choosing the elements of the K matrix are a topic of current research.

Having settled upon the above control law, we can perform a little matrix manipulation to rework the equations of the closed-loop system into the form of (1) to which the analytical methods of Section 8.4 can be directly applied. Using the output equation (2) to express $\mathbf{v}$ in terms of $\mathbf{y}_{des}$ and the state $\mathbf{q}$ we have

$$\mathbf{v} = \mathbf{K}\mathbf{y}_{des} - \mathbf{K}\mathbf{C}\mathbf{q}$$

which can then be substituted into (3), yielding, after a bit of rearranging,

$$\dot{\mathbf{q}} = \underbrace{(\mathbf{A} - \mathbf{B}_v\mathbf{K}\mathbf{C})}_{\tilde{\mathbf{A}}}\mathbf{q} + \mathbf{B}_v\mathbf{K}\mathbf{y}_{des} + \mathbf{B}_w\mathbf{w} \tag{5}$$

Comparing (5) with (3), it is apparent that the introduction of feedback has resulted in two changes in the state equations.

The first difference between the two state equations is that the input term $\mathbf{B}_v\mathbf{v}(t)$ has been changed to $\mathbf{B}_v\mathbf{K}\mathbf{y}_{des}(t)$, thereby expressing the manipulated input directly in terms of the desired output, rather than the more general $\mathbf{v}(t)$.

Second, and most important, the original $\mathbf{A}$ has been modified by the subtraction of the $n \times n$ matrix $\mathbf{B}_v\mathbf{K}\mathbf{C}$, which involves the feedback gains that we are free to choose. Denoting the *modified* $\mathbf{A}$ by $\tilde{\mathbf{A}} \triangleq \mathbf{A} - \mathbf{B}_v\mathbf{K}\mathbf{C}$ we can conclude that the n modes of the zero-input response of the feedback system will be of the form $e^{\tilde{s}_k t}$ where the s_k are the characteristic values of $\tilde{\mathbf{A}}$, i.e., the solutions of $|s\mathbf{I} - \tilde{\mathbf{A}}| = 0$. Since the characteristic values of the open-loop system, s_k, are the roots of $|s\mathbf{I} - \mathbf{A}| = 0$ the introduction of feedback has shifted the system's characteristic values or characteristic-equation roots from the set s_k to the set $\tilde{s}_k$, $k = 1, 2, \ldots, n$.

We have been implicitly assuming that all of the characteristic values are subject to modification by the introduction of the feedback signals. In reality, this may not be the case as the structure of the system embodied in $\mathbf{A}$ and $\mathbf{B}$ may be such that one or more of the modes of the open-loop system is completely unaffected by any of the control inputs. Such a mode is said to be *uncontrollable* and the corresponding characteristic value will not be shifted by the addition of feedback, no matter what control law is used. If a mode is well damped, the fact that it is uncontrollable is probably of no consequence; however, a lightly damped mode could be troublesome and an unstable mode disastrous if it were uncontrollable. The reader is referred to Schwarz and Friedland (1965, Sect. 11.7) for details and to Problem 9.20 for an illustration.

Employing the graphical techniques first used in Figure 3.16, the systems with and without feedback are represented in Figure 9.24 so as to clearly indicate the feedback paths that have been added in going from (1) to (5). The structural similarity between part (*b*) of the figure and the single-loop feedback system of Figure 9.1 is readily apparent, to say the least.

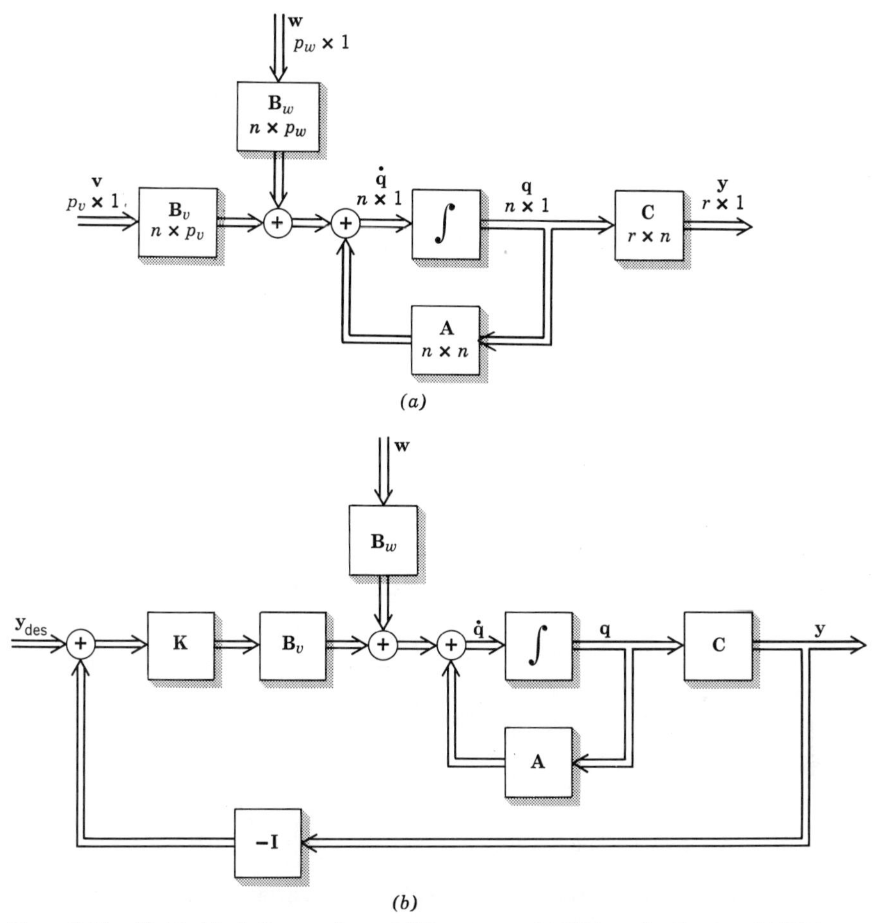

Figure 9.24. Matrix block diagram for an MIO system. (a) Without feedback. (b) With feedback.

Example 9.4

To illustrate the use of the state-variable formulation, consider the attitude-control problem introduced in the previous section. Defining the pointing angle and its derivative to be the two components of the state vector $\mathbf{q}(t)$ and the control and disturbance torques to be the two components of the input vector $\mathbf{x}(t)$, Eq. (1), Sect. 9.3, may be rewritten in the form of (3) as

$$\frac{d}{dt}\begin{bmatrix}\theta\\ \dot{\theta}\end{bmatrix}=\underbrace{\begin{bmatrix}0 & 1\\ 0 & 0\end{bmatrix}}_{\mathbf{A}}\underbrace{\begin{bmatrix}\theta\\ \dot{\theta}\end{bmatrix}}_{\mathbf{q}}+\underbrace{\left[\begin{array}{c|c}0 & 0\\ 1/I & 1/I\end{array}\right]}_{\mathbf{B}}\underbrace{\begin{bmatrix}\tau_{con}\\ \hline \tau_{dis}\end{bmatrix}}_{\mathbf{x}} \tag{6}$$

where $\mathbf{v}=\tau_{con}$, $\mathbf{w}=\tau_{dis}$, both reducing to scalars in this case. Assuming that each of the states can be measured ideally (without sensor dynamics or errors) with gains a_1 and a_2, respectively, the equation for the output vector defined by (2) becomes

$$\underbrace{\begin{bmatrix}y_1\\ y_2\end{bmatrix}}_{\mathbf{y}}=\underbrace{\begin{bmatrix}a_1 & 0\\ 0 & a_2\end{bmatrix}}_{\mathbf{C}}\underbrace{\begin{bmatrix}\theta\\ \dot{\theta}\end{bmatrix}}_{\mathbf{q}} \tag{7}$$

To introduce a control law that makes use of both outputs, we define the desired output vector to be

$$\mathbf{y}_{des}=\begin{bmatrix}a_1\theta_{des}\\ a_2\dot{\theta}_{des}\end{bmatrix}=\mathbf{C}\underbrace{\begin{bmatrix}\theta_{des}\\ \dot{\theta}_{des}\end{bmatrix}}_{\mathbf{q}_{des}}$$

in which case (4) becomes

$$\mathbf{v}=\tau_{con}=\underbrace{[k_1 \quad k_2]}_{\mathbf{K}}\underbrace{\begin{bmatrix}a_1 & 0\\ 0 & a_2\end{bmatrix}}_{\mathbf{C}}\begin{bmatrix}\theta_{des}-\theta\\ \dot{\theta}_{des}-\dot{\theta}\end{bmatrix} \tag{8}$$

Substituting for the control torque in (6), the state equation of the closed-loop system becomes

$$\frac{d}{dt}\begin{bmatrix}\theta\\ \dot{\theta}\end{bmatrix}=\begin{bmatrix}0 & 1\\ 0 & 0\end{bmatrix}\begin{bmatrix}\theta\\ \dot{\theta}\end{bmatrix}+\begin{bmatrix}0\\ 1/I\end{bmatrix}[k_1 \quad k_2]\begin{bmatrix}a_1 & 0\\ 0 & a_2\end{bmatrix}\begin{bmatrix}\theta_{des}-\theta\\ \dot{\theta}_{des}-\dot{\theta}\end{bmatrix}+\begin{bmatrix}0\\ 1/I\end{bmatrix}\tau_{dis}$$

or, upon combining terms involving the state variables so as to explicitly show the modified system matrix $\tilde{\mathbf{A}}$,

$$\frac{d}{dt}\begin{bmatrix}\theta\\ \dot{\theta}\end{bmatrix}=\underbrace{\begin{bmatrix}0 & 1\\ \dfrac{-a_1k_1}{I} & \dfrac{-a_2k_2}{I}\end{bmatrix}}_{\tilde{\mathbf{A}}}\begin{bmatrix}\theta\\ \dot{\theta}\end{bmatrix}+\begin{bmatrix}0 & 0\\ \dfrac{a_1k_1}{I} & \dfrac{a_2k_2}{I}\end{bmatrix}\begin{bmatrix}\theta_{des}\\ \dot{\theta}_{des}\end{bmatrix}+\begin{bmatrix}0\\ \dfrac{1}{I}\end{bmatrix}\tau_{dis}$$

The characteristic values of $\tilde{\mathbf{A}}$ must satisfy the equation

$$|s\mathbf{I}-\tilde{\mathbf{A}}| = \begin{vmatrix} s & -1 \\ \dfrac{a_1 k_1}{I} & s+\dfrac{a_2 k_2}{I} \end{vmatrix} = 0$$

namely

$$s^2 + \frac{a_2 k_2}{I} s + \frac{a_1 k_1}{I} = 0 \tag{9}$$

Thus, if $k_1 > 0$ and $k_2 = 0$, the closed-loop characteristic values will be

$$s = \pm j\sqrt{\frac{a_1 k_1}{I}}$$

which corresponds to proportional feedback and results in undamped oscillatory modes in the satellite's response to either a change in the desired pointing angle or a disturbance torque. However, if the sensor gains (a_1, a_2) and the controller gains (k_1, k_2) are all positive, the roots of (9) will be in the LHP and the closed-loop system will be asymptotically stable. We recognize this case as the P+D controller, for which the characteristic polynomial was found to be

$$P(s) = s^2 + \alpha\omega_n{}^2 s + \omega_n{}^2$$

where $\omega_n{}^2 = K_a K_\tau K_\theta / I$. Thus, if $a_1 k_1 = K_a K_\tau K_\theta$ and $a_2 k_2 = \alpha K_a K_\tau K_\theta$, the characteristic values of $\tilde{\mathbf{A}}$ will coincide with the roots of $P(s)$.

Hence, the control law defined by (8) includes *both* the proportional and P+D systems. However, if a P+I+D control law is to be considered, an additional dynamic element is required in the model to obtain the integral action of the controller. Thus, the two-element state vector must be expanded to include the integral of $\theta(t)$ as a third state.

Having expanded the model, the P+I+D system can be analyzed in the same fashion, with the modified system matrix now being of dimension 3×3 and having three characteristic values. Similarly, the dynamics of either the controller or the sensors can be included in the system model by appending additional state variables to $\mathbf{q}$, along with corresponding rows and columns to $\mathbf{A}$ and $\tilde{\mathbf{A}}$, etc. Although one would soon lose interest in continuing calculations by hand as more dynamic elements are added to the system model, the digital computer could be used for the analysis with no reprogramming required in going from case to case because the structure of the problem, as represented by the matrix equations, is unchanged.

Problems

9.1 Describe a system involving feedback that arises in connection with each of the items listed below. Attempt to identify the plant or process, the inputs (both reference and disturbance), and the feedback path(s).

(a) Around the house
(b) Economics
(c) Sociology
(d) Ecology
(e) Automobiles
(f) Human physiology

9.2 Reduce the block diagrams in Figure P9.1 to find the reference and disturbance transfer functions $T_R(s)$ and $T_D(s)$, respectively. The results should be

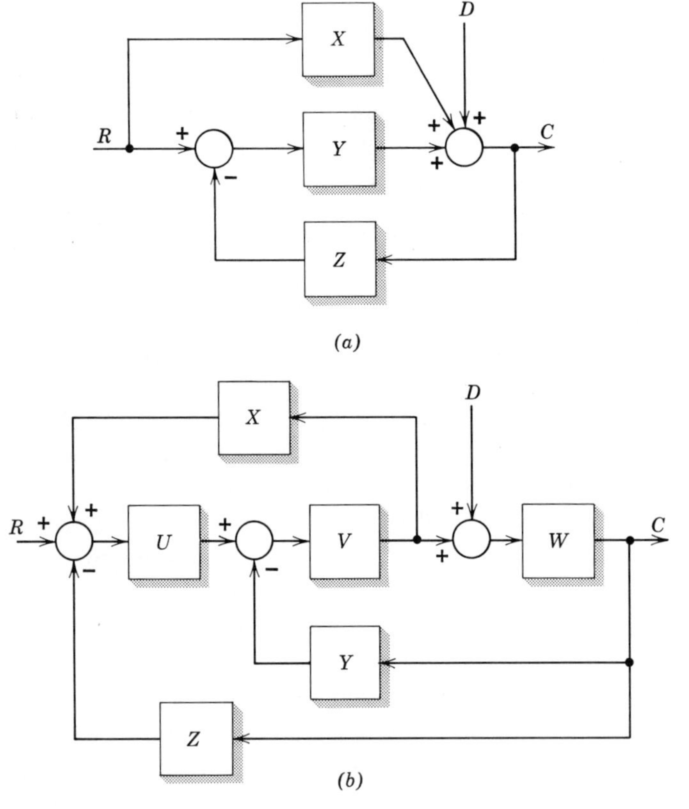

Figure P9.1

in the form $T(s) = F_1(s)/F_2(s)$ where F_1 and F_2 involve only *sums* and *products* (no quotients please) of the individual block transfer functions $U(s), V(s), \ldots, Z(s)$. Comment on the relationship between the denominators of the reference and disturbance transfer functions of a system.

Answer: (a) $T_R(s) = \dfrac{X+Y}{1+YZ}, \quad T_D(s) = \dfrac{1}{1+YZ}$

9.3 (a) Find the four transfer functions associated with the system shown in Figure P9.2, with $F(s) = G(s) = 0$.

(b) Select the transfer functions $F(s)$ and $G(s)$ such that $y_1(t)$ will depend only on $x_1(t)$ and $y_2(t)$ will depend only on $x_2(t)$.

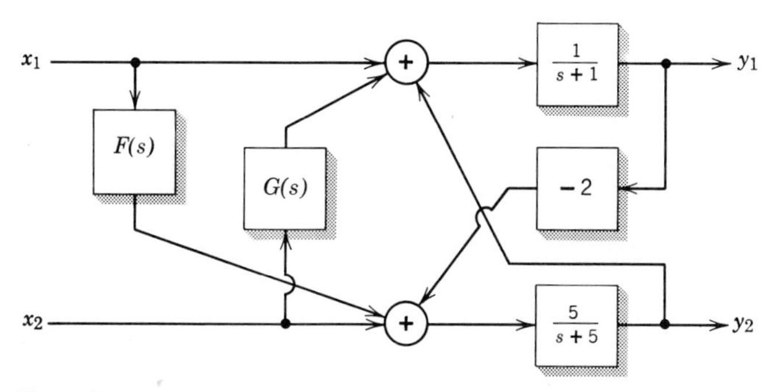

Figure P9.2

9.4 A feedback system which has the structure of Figure 9.2 is known to have the reference and disturbance transfer functions

$$T_R(s) = \frac{10K(s+5)}{s(s+4)(s+5)+10K}$$

and

$$T_D(s) = \frac{10(s+4)(s+5)}{s(s+4)(s+5)+10K}$$

Identify the individual transfer functions $G_c(s)$, $G_p(s)$, and $H(s)$.

9.5 For each of the simulation diagrams in Figure P9.3 verify that the diagram satisfies the differential equation $\ddot{y} + a_1\dot{y} + a_0 y = b_1\dot{x} + b_0 x$. For the diagram in part (b) of the figure evaluate β_0 and β_1 in terms of the a's and b's. For the diagram in part (c) of the figure evaluate the λ's and α's, assuming that λ_1 and λ_2 are real. How are the λ's related to the characteristic roots of the system?

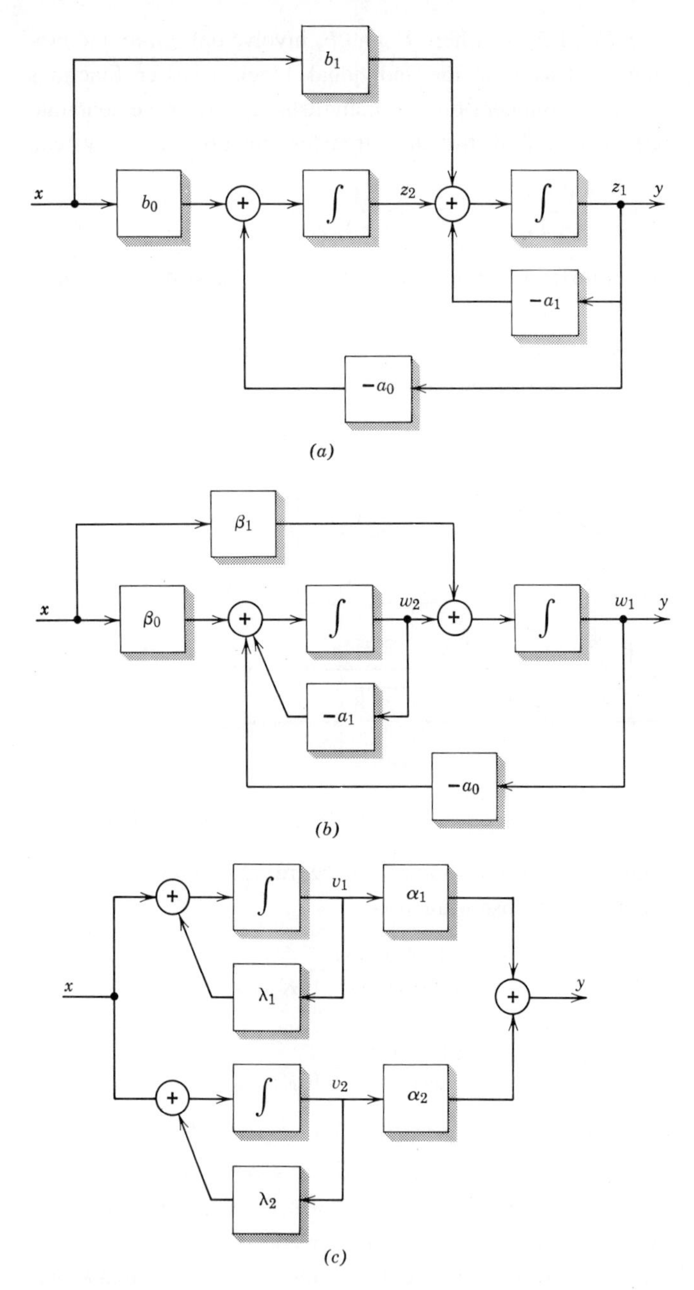

b_1
x
b_0
+
$\int$
z_2
+
$\int$
z_1
y
$-a_1$
$-a_0$
(a)
β_1
x
β_0
+
$\int$
w_2
+
$\int$
w_1
y
$-a_1$
$-a_0$
(b)
+
$\int$
v_1
α_1
x
λ_1
+
y
+
$\int$
v_2
α_2
λ_2
(c)

Figure P9.3

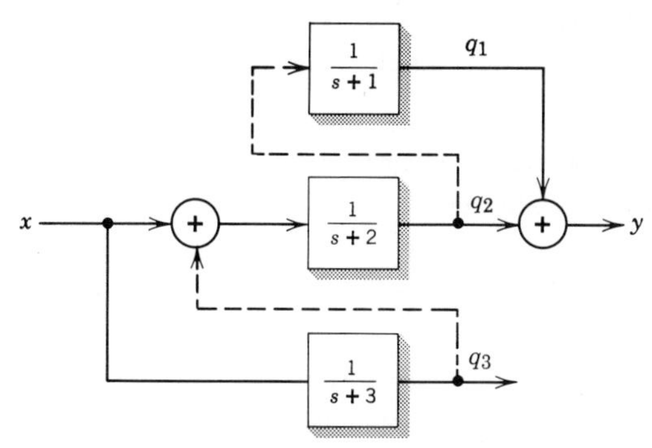

Figure P9.4

9.6 (a) For the system shown in Figure P9.4, considering the dashed connections to be *open*, write the state and output equations, taking q_1, q_2, and q_3 as the state variables. Find the transfer function $Y(s)/X(s)$ and comment on its form. (b) Repeat part (a), considering the dashed connections to be *closed*.

9.7 A feedback system is formed by using n identical blocks with transfer function $F(s)$ and the feedback gain $1/K$ where K is adjustable between 0.10 and 10.0 (see Figure P9.5).

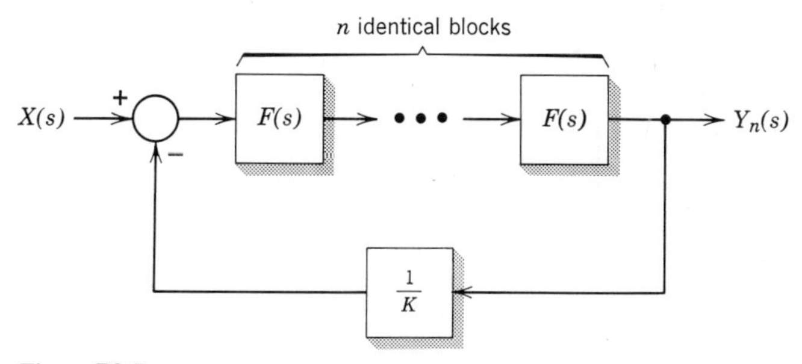

Figure P9.5

(a) Assuming that $F(s)$ can be represented by the constant gain A, where $A \geq 100$, show that

$$T_n \triangleq \frac{Y_n}{X} \approx K\left(1 - \frac{K}{A^n}\right)$$

Comment on the relative roles played by the forward- and feedback-path elements in determining the static gain T_n.
(b) A more accurate representation of $F(s)$ might be as a first-order low pass filter with a DC gain A and a time constant τ such that $F(s) = A/(\tau s + 1)$.

Find the closed-loop transfer function $T_n(s) \triangleq Y_n(s)/X(s)$ for $n = 1$ and 2, and show that in both cases the system is stable. Compare the closed-loop DC gain with that found in part (a).

9.8 The unity feedback system in Figure 9.3b, for which the error is given by $\varepsilon = r - c$, is known to be asymptotically stable.

(a) Show that if $G(s)$ does not have a pole at $s = 0$, then the step input $Au(t)$ will result in the constant steady-state error $\varepsilon_{ss} = A/[1 + G(0)]$ while a ramp input will cause an unbounded error.

(b) Show that if $G(s)$ has a single pole at $s = 0$ such that $G(s) = \hat{G}(s)/s$ where $\hat{G}(0)$ is finite and nonzero, then a step input will result in zero steady-state error, while the ramp input Bt will cause the steady-state error $\varepsilon_{ss} = B/[\hat{G}(0)]$.

(c) Derive comparable results for the situation where $G(s)$ has a double pole at $s = 0$.

9.9 The operational amplifier in the configuration used in analog computers can be modeled as in Figure P9.6. The amplifier draws negligible current ($I_f = -I_i$) and may be considered as having the transfer function $T_a(s) = -A/(\tau s + 1)$ where A is on the order of 10^5 volts/volt and the time constant τ will be taken as 0.10 sec.

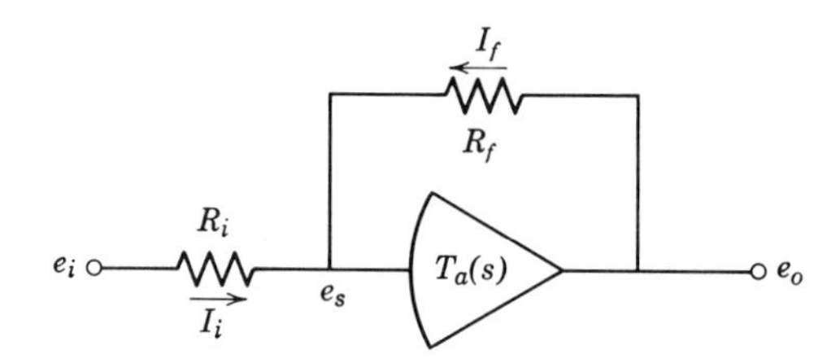

Figure P9.6

(a) Construct a static or DC model by taking $T_a(s) = -A$ and show that the voltage gain of the unit is, to a good approximation, $e_o/e_i = -R_f/R_i$. Find the error in this relationship resulting from the finite value of A.

(b) Using the dynamic model of the amplifier, find the transfer function $T_R(s) \triangleq E_o(s)/E_i(s)$ and make a Bode plot of both $|T_R(j\omega)|$ and $|T_a(j\omega)|$ on the same axes. Comment on the relationship between the DC gains and bandwidths.

(c) Take $R_f = R_i = 1\ k\Omega$ and evaluate the response to a unit step. Calculate the steady-state value and the rise time t_r, defined as the time required for the step response to reach 90% of its final value.

(d) Evaluate separately the effects on the steady-state step response of 10% increases in the values of A, R_f, and R_i. Comment.

9.10 Instead of using the P + D control law of Eq. (7), Sect. 9.3, in the forward path for control of the satellite considered in Section 9.3, one can introduce damping by feeding back the angular rate such that

$$v_1 = K_\theta(\theta_{ref} - \theta) - K_{\dot{\theta}}\dot{\theta}$$

$$\tau_{con} = K_\tau K_a v_1$$

(a) Draw a simulation diagram of the system and find the closed-loop reference and disturbance transfer functions.

(b) Compare the zeros of $\theta(s)/\theta_{\text{ref}}(s)$ with those of $T_R(s)$ in Eq. (8), Sect. 9.3, and find the value of $K_{\dot{\theta}}$ such that their closed-loop poles coincide.

9.11 (a) Find the steady-state errors $\varepsilon = r - c$ of the systems shown in Figure P9.7 to the reference inputs $r_1 = A_1u(t)$, $r_2 = A_2tu(t)$, and $r_3 = A_3t^2u(t)$.

(b) Evaluate and compare the steady-state errors to a step disturbance input $d = Bu(t)$.

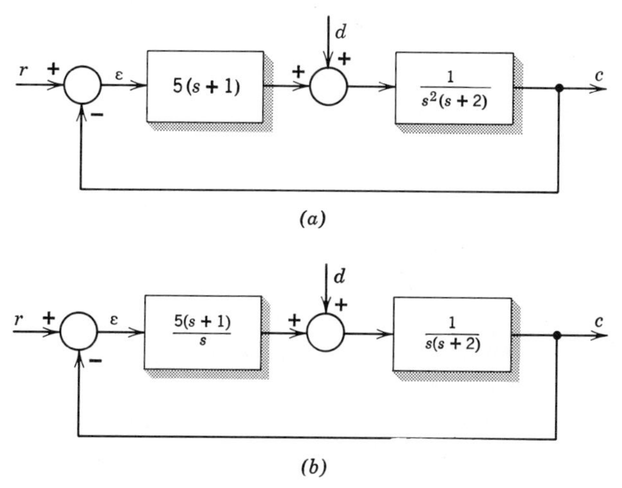

Figure P9.7

9.12 Consider the system in Figure 9.2 with $H(s) = 0$ and $G_p(s)$ satisfying $G_p(0) \neq 0$, that is, a nonzero DC gain.

(a) Show that when $G_c(s)$ represents a proportional controller, the steady-state error resulting from a constant disturbance input will be a nonzero constant and the error due to a ramp input will grow without bound.

(b) When $G_c(s)$ represents a P + I controller, a constant disturbance results in zero steady-state error and a ramp disturbance will result in a constant steady-state error inversely proportional to the gain of the integral term in the control law. Prove the validity of this assertion.

9.13 An unstable process has the transfer function $F_p(s) = 1/[(s-1)(s+5)]$. If the output is passed through a filter with transfer function $F_f(s) = (s-1)/(s+3)$ ideally the unstable mode e^t will not appear in the filter output signal.
(a) Show that if the cancellation is not exact, for example, the filter zero is at $s = 1 + \epsilon$ instead of $s = 1$, the unstable mode will appear in the filter output. Find the impulse response of the process and filter combination with inexact cancellation.
(b) Explain why it should be acceptable to cancel a mode that corresponds to a LHP pole, even with inexact cancellation.
(c) Show that feeding back the output of the process and subtracting it from the input will stabilize the system provided sufficient gain is introduced in the forward path. Find the minimum value for stability.

9.14 A small DC motor driving an output shaft having negligible inertia and friction through a 10:1 gear ratio can be made to follow an electrical signal by attaching a potentiometer (pot) to the output shaft and subtracting the resulting signal from the input signal representing the desired shaft orientation. As indicated in Figure P9.8a, the motor parameters are $K_s = 60$ rad/sec/volt and $\tau_m = 0.40$ sec; the angle of the motor shaft is ψ rad. The gain of the pot in the forward path is adjustable in the interval $0 \leq K_g \leq 1$ volts/volt.
(a) Find the closed-loop transfer function $T_R(s) \triangleq \theta(s)/E_{in}(s)$, treating K_g as a parameter.
(b) Plot the locus of the closed-loop roots in the s-plane for $0 \leq K_g \leq 1$. (See Figure 9.14 for an example). Comment on the character of the step response attainable by adjusting the pot whose gain is K_g.
(c) Redraw the block diagram with the angles ψ and θ expressed in degrees, making the necessary adjustments in the other system parameters. What effect does this change have on $T_R(s)$?
(d) Show that the addition of a tachometer driven by the output shaft having the transfer function $K_t s$, as indicated in Figure P9.8*b*, will allow a higher damping ratio than before for the same value of ω_n. Determine the values of K_t and K_g that will yield closed-loop roots having $\zeta = 0.5$ and $\omega_n = 5$ rad/sec.

9.15 Reconsider the DC motor presented in Problem 9.14, using the state-variable formulation presented in Section 9.4. Take ψ and $\dot{\psi}$ as the two state

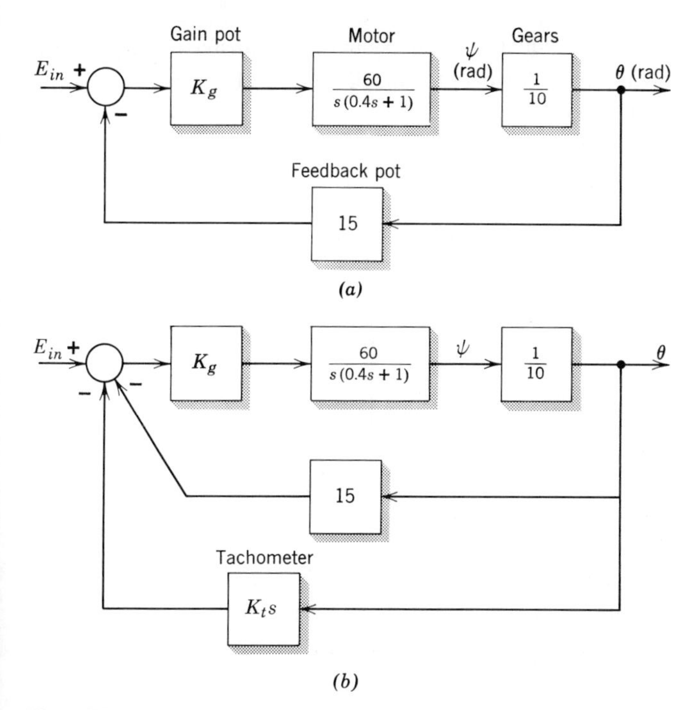

Figure P9.8

variables, θ and $\dot{\theta}$ as the output variables, and consider $E_{in}/15$ as representing θ_{des}.

(a) Using angular feedback ($K_t = 0$), write the state equation, first in the form of Eqs. (1) and (2), Sect. 9.4, and then in the form of Eq. (5), Sect. 9.4. Evaluate the characteristic equation and identify each matrix and vector involved. *Note*: A disturbance torque is not included in the model, so $\mathbf{w} = 0$.

(b) Repeat part (a), adding the tachometer feedback. Using the gains $K_g = 1$ volt/volt and $K_t = 2$ volts/rad/sec, find the state-transition matrix of the closed-loop system. *Hint*: You should obtain characteristic roots of -10 and -22.5.

9.16 For each of the simulation diagrams in Figure P9.3, *1*. write the state and output equations in the form of Eq. (1), Sect. 9.4, *2*. find the state-transition matrix, denoting the two characteristic roots as s_1 and s_2, and *3*. evaluate $H(s)$ using Eq. (12), Sect. 8.4. What relationships between parameters will insure that all three diagrams will yield the same transfer function?

9.17 Write the state-variable equations and evaluate the transfer-function matrix $\mathbf{H}(s)$ shown in Figure P9.9, where the blocks denoted by $a, b, \ldots, f$ are all gains, that is, no dynamics.

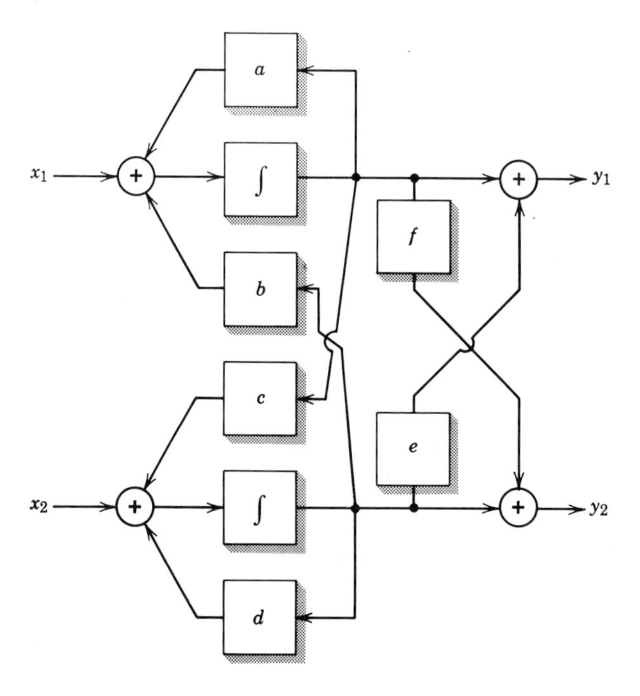

Figure P9.9

9.18 Extend Example 9.4 to include an integral term in the control law. In defining the desired output vector $\mathbf{y}_{des}$, set $(\int \theta dt)_{des}$ and $\dot{\theta}_{des}$ to zero, retaining only θ_{des} as the reference input.

(a) Show explicitly the state, output, and control equations in matrix form.

(b) Find $[s\mathbf{I} - \tilde{\mathbf{A}}]^{-1}$ and compare the characteristic equation with Eq. (14), Sect. 9.3.

(c) Evaluate the reference and disturbance transfer functions and show that there is zero steady-state pointing error to a constant disturbance torque.

9.19 The inherently unstable inverted stick described in Problem 8.32 can be stabilized by sensing θ with a potentiometer and $\dot{\theta}$ with a tachometer and making the motor torque τ obey the control law $\tau = -(\alpha\theta + \beta\dot{\theta})$.

(a) Show that if $\beta = 0$, the best that can be achieved is marginal stability and find the minimum value of α required.

(b) Show that with both $\alpha > 0$ and $\beta > 0$, an asymptotically stable system can be had. What relationships must exist between α, β, and the stick parameters to yield a critically damped system with $\omega_n = 2\sqrt{g/L}$?

(c) Express the feedback system's equations in state-variable form and show how the characteristic values of **A** have been modified by the feedback.

9.20 Consider the system described by the equation

$$\dot{\mathbf{x}} = \begin{bmatrix} -4 & -1 & -2 \\ -2 & -3 & -2 \\ 2 & 2 & 1 \end{bmatrix} \mathbf{x} + \begin{bmatrix} 2 \\ 1 \\ -1 \end{bmatrix} u$$

(a) Show that the characteristic values of the system are $s_i = -1, -2$, and -3.

(b) Show that the mode corresponding to $s = -1$ is uncontrollable. *Hint:* Using Eq. (12), Sect. 8.4, find the three transfer functions $X_i(s)/U(s)$, $i = 1, 2, 3$ and observe that each one has a zero which cancels the unstable mode. Then explain why this implies that the system is uncontrollable.

10

s-Plane analysis: Root locus

By now the reader hopefully has been convinced of the advantages of knowing the roots of a feedback system's characteristic equation in terms of its open-loop parameters (poles, zeros, gains, etc.). Accordingly, we shall proceed to establish analytic methods for finding these roots using either the complex-frequency domain (the s-plane) or the real-frequency domain, as covered in this and the next chapter, respectively.

The principal vehicle used in the s-plane is the root locus, which yields the exact locations of all of the system's roots. A less involved method, giving only a yes-or-no answer to the question of a system's stability, is based upon Routh's criterion. Both are considered here. To demonstrate the utility of these tools and to give the reader a little practice before turning him loose, we shall conclude by revisiting the satellite-attitude control problem introduced in the previous chapter.

10.1 THE ROOT-LOCUS METHOD

In a nutshell, the root-locus is a plot in the s-plane of all possible locations that the roots of a closed-loop system's characteristic equation can have as a specific parameter is varied, usually from zero to infinity. Because of the central role played by the characteristic equation we shall start by giving a variety of different forms that it can take, all of which will be useful in understanding the root locus and in deriving rules for rapidly plotting it.

The characteristic equation

Recall from Section 9.2 that the basic single-loop feedback system represented by the block diagram in Figure 10.1 has the characteristic

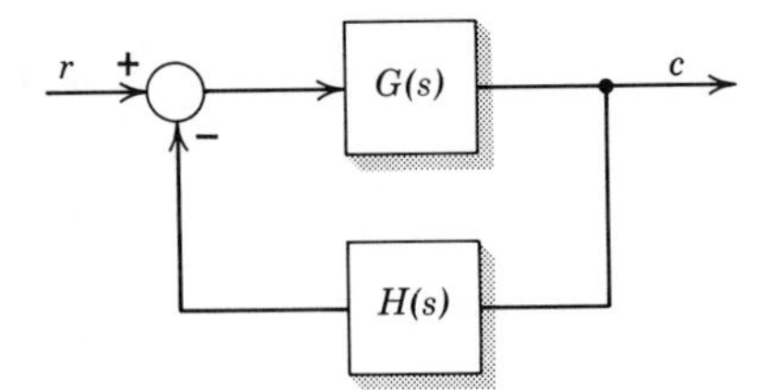

Figure 10.1.

equation

$$D_G(s)D_H(s) + N_G(s)N_H(s) = 0 \tag{1}$$

where $N_G(s)/D_G(s) \triangleq G(s) \triangleq G_c(s)G_p(s)$ and $N_H(s)/D_H(s) \triangleq H(s)$. An alternate form of (1) can be obtained by dividing it by the polynomial $D_G(s)D_H(s)$, resulting in

$$1 + \frac{N_G(s)N_H(s)}{D_G(s)D_H(s)} = 0 \tag{2}$$

which can be written more compactly as

$$1 + G(s)H(s) = 0 \tag{3}$$

Although (3) could perhaps have been obtained by inspection, the reader should note that its left-hand side is not a polynomial; however, the values of s for which it is satisfied will be the same as those for which (1) is satisfied provided that the numerator and denominator polynomials in (2) have no common factors.

In anticipation of the equations we shall need in constructing the root locus, we rewrite the rational function $G(s)H(s)$ as a ratio of *factored polynomials* such that

$$G(s)H(s) = K \underbrace{\frac{\prod_{i=1}^{m} (s - z_i)}{\prod_{k=1}^{n} (s - p_k)}}_{F(s)} \tag{4}$$

in which case the characteristic equation can be written as

$$1 + KF(s) = 0 \tag{5}$$

where

$$F(s) \triangleq \frac{\prod_{i=1}^{m}(s-z_i)}{\prod_{k=1}^{n}(s-p_k)} \tag{6}$$

The parameter K is the constant necessary for the coefficients of the highest powers of s in both numerator and denominator of $F(s)$ to be unity and it is necessarily real. It is referred to as the *root-locus gain* and will play an important role in our development and subsequent use of the root locus. From (6), it is apparent that $F(s)$ has the combined poles and zeros of the individual transfer functions of all of the dynamic elements contained within the closed-loop. Furthermore, because of the rational-function form of $F(s)$ we can write

$$|F(s)| = \frac{\prod_{i=1}^{m}|s-z_i|}{\prod_{k=1}^{n}|s-p_k|} \tag{7}$$

and

$$\arg[F(s)] = \sum_{i=1}^{m}\arg[s-z_i] - \sum_{k=1}^{n}\arg[s-p_k] \tag{8}$$

where the terms $(s-z_i)$ may be represented by vectors from the zeros z_i to the point s at which $F(s)$ is being evaluated and likewise for the poles. A simple example corresponding to $n=m=2$ is shown in Figure 10.2. For instance, both the magnitude and argument of (s_0-z_1) can be measured on the figure, giving $|s_0-z_1|=2$ and $\arg[s_0-z_1]=90°$. Carrying out the process for the other zero and the two poles and substituting into (7) and (8)

$$|F(s_0)| = \frac{2.00\times 3.20}{2.25\times 3.65} = 0.778$$

$$\arg[F(s_0)] = (90+38)-(26+56) = 46°$$

The other forms of the characteristic equation that we shall need can be derived from (5) and (6) and are

$$\prod_{k=1}^{n}(s-p_k) + K\prod_{i=1}^{m}(s-z_i) = 0 \tag{9}$$

$$F(s) = -\frac{1}{K} \tag{10}$$

$$\frac{1}{K}\prod_{k=1}^{n}(s-p_k) + \prod_{i=1}^{m}(s-z_i) = 0 \tag{11}$$

the last being contingent upon $K \neq 0$.

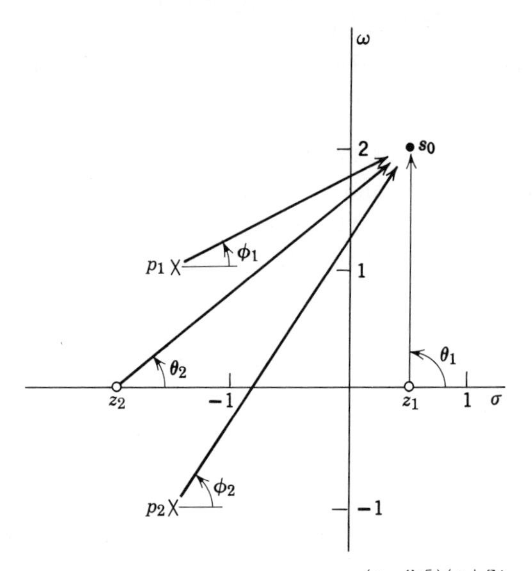

Figure 10.2. Evaluating $F(s) = \dfrac{(s-0.5)(s+2)}{(s+1.5-j)(s+1.5+j)}$ at $s = 0.5 + j2$.

Angle and magnitude criteria

We assume that we know the open-loop transfer function $G(s)H(s)$ – this will be needed in *factored* form – and that the root-locus gain K can be thought of as potentially variable from 0 to ∞ (sometimes from $-\infty$ to 0 or from $-\infty$ to $+\infty$). Because (10) is a valid form of the characteristic equation in that its roots are the roots of (1), the problem of finding the roots of the characteristic equation (closed-loop poles) is equivalent to finding the solutions to the algebraic equation

$$F(s) = -\frac{1}{K}$$

where K is real. Expressing both $F(s)$ and K in polar form gives yet another version of the characteristic equation, namely

$$\underbrace{|F(s)|e^{j\arg[F(s)]}}_{F(s)} = \underbrace{\frac{1}{|K|}e^{jq180^\circ}}_{-1/K} \tag{12}$$

where q is an *odd integer* if $K > 0$ and an *even integer* if $K < 0$.

In the light of (12), any value of s that is to be a closed-loop pole must simultaneously satisfy the *angle criterion*

$$\arg [F(s)] = q\,180° \tag{13}$$

and the *magnitude criterion*

$$|F(s)| = \frac{1}{|K|} \tag{14}$$

The existence of these two requirements suggests that the roots of the system's closed-loop characteristic equation corresponding to specific parameter values may be obtained by the following sequence of operations:

1. Find *all* values of s that satisfy the angle criterion, (13), the plot in the s-plane of these values being the *root locus*.

2. Find the specific values of s on the root locus that satisfy the magnitude criterion, (14), or, as an alternative, find the value of K that will cause a specific point on the locus to be a root.

Although this two-stage process seems perhaps like a roundabout way of solving the problem, it turns out to have several important advantages over other methods. For one thing, by breaking it into the two parts the task of finding the roots of the characteristic equation is greatly simplified, as we shall see shortly. For another, the method has the attractive feature that by applying the angle criterion first we are finding *all* of the possible root locations for any $K \neq 0$ (the case $K = 0$ is easily included, although the magnitude criterion is undefined there). Because the root-locus gain K usually depends on a physical parameter that is readily adjusted, e.g., an amplifier gain, we can make effective use of the root locus as a design tool to suggest those values of gain that yield preferred closed-loop roots. Finally, if we cannot find a value of gain that yields satisfactory roots, e.g., stable and well damped, an examination of the root-locus plot will usually provide insight as to how other system parameters should be varied or how the structure of the system should be altered, such as by adding filters or feeding back different signals.

Example 10.1

To demonstrate the above notions with a very elementary example, we shall draw the root locus for the system shown in Figure 10.3*a* for which $F(s) = 1/s$ and locate the closed-loop root(s) for the specific value

$K = 10$. Substituting for $F(s)$ in (13) and (14), the angle and magnitude criteria become $\arg[F(s)] = -\arg[s]$ and $|F(s)| = 1/|s|$, respectively.

Considering the angle criterion with $K > 0$, q is an odd integer. Consequently those points that lie on the root locus are those values of s for which $\arg[F(s)]$ is an odd multiple of 180°, e.g., ±180°, ±540°. However, this cannot happen unless $\arg[s]$ is also an odd multiple of 180°, which is to say that s is restricted to points on the negative-real axis. Hence, the root locus for the system with $K > 0$ must be as shown in Figure 10.3*b*.

Having obtained the root locus for $K > 0$, we can identify the specific closed-loop root corresponding to $K = 10$ by applying the magnitude criterion, obtaining $|s| = 10$. Since $s = -10$, denoted by the triangle in Figure 10.3*c*, is the only point in the entire s-plane that satisfies both the angle and the magnitude criteria, it must be the single root of the closed-loop system. A glance back to Example 9.2 will substantiate the validity of the above conclusions.

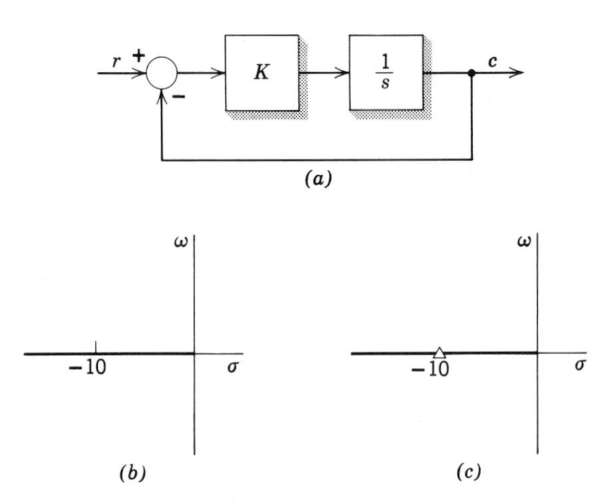

Figure 10.3. First-order system. (*a*) Block diagram. (*b*) 180° locus. (*c*) Root location, $K = 10$.

Construction of the 180° root locus

The procedure followed in the previous example of solving analytically for $|F(s)|$ and $\arg[F(s)]$ is not at all practical for other than trivial problems and will not be pursued further. Rather, a number of rules exist that allow one to *sketch* certain salient features of the root locus with only a modest effort, even for rather complex systems. Also, a device called a

"Spirule"† has been developed which, with a bit of practice, can be used to obtain a more accurate *plot* of the root locus, should it be required. Finally, digital-computer programs‡ have been written for achieving high accuracy and relieving the engineer of the plotting burden.

We shall present the rules for constructing a number of the important features of the root locus, deferring their derivations until the following section. Because the angle criterion requires that arg $[F(s)]$ be an *odd* multiple of 180° when $K > 0$, we refer to the locus corresponding to $K \geq 0$ as the *180° locus*. When $K < 0$ the angle criterion must yield an *even* multiple of 180°, e.g., 0°, ±360°; hence the designation *0° locus* is used for $K \leq 0$. In the interest of simplicity, we shall work with the 180° locus for now and show the relatively minor modifications necessary for the 0° locus later.

It will be assumed that the n poles and m zeros of $F(s)$ are distinct, that no poles coincide with zeros, and that $m \leq n$ (the case of repeated poles and/or zeros will be taken up in Section 10.3). The basic rules for the construction of a 180° root locus ($K \geq 0$) are given below.

1. The locus has exactly n branches, where a branch is the path formed by any one root as K is varied continuously from zero to infinity.

2. The locus is symmetric with respect to the real axis of the s-plane.

3. Any point on the real axis is on the root locus if the total number of real poles and zeros to the right of that point is odd.

4. As K increases from 0, the n branches of the root locus depart from the poles of $F(s)$, one branch per pole.

5. As $K \to \infty$, m of the branches of the locus approach the finite zeros of $F(s)$, one branch per zero.

6. If $m < n$, there are $n-m$ branches of the root locus that approach infinity as $K \to \infty$; furthermore, they approach infinity asymptotic to the $n-m$ straight lines that intersect the real axis at the point

$$\sigma_0 = \frac{\sum_{k=1}^{n} p_k - \sum_{i=1}^{m} z_i}{n-m} \tag{15}$$

†Available from your bookstore or directly from The Spirule Co., 9728 El Venado, Whittier, California. See D'Azzo and Houpis (1966, App. D.) for a description of its use.

‡See Ash and Ash (1968).

and form the angles ψ_ν with the real axis where

$$\psi_\nu = \frac{180° + \nu 360°}{n - m} \qquad \nu = 0, 1, \ldots, (n - m - 1) \tag{16}$$

7. The angle with which the locus departs from a complex open-loop pole p_J is given by

$$\phi_J = \sum_{i=1}^{m} \arg\,[p_J - z_i] - \sum_{\substack{k=1 \\ k \neq J}}^{n} \arg\,[p_J - p_k] + q180° \tag{17}$$

where q is an odd integer.

8. The angle at which the locus arrives at a complex open-loop zero z_J is given by

$$\theta_J = -\sum_{\substack{i=1 \\ i \neq J}}^{m} \arg\,[z_J - z_i] + \sum_{k=1}^{n} \arg\,[z_J - p_k] + q180° \tag{18}$$

where q is an odd integer.

Having applied the above rules to a given problem, enough of the root locus is usually known so that the remainder can be sketched to within a reasonable approximation. As a further refinement, one can always select a test point $\hat{s}$ that looks as if it should be close to the locus and evaluate the argument of $F(\hat{s})$. If $\arg\,[F(\hat{s})] = \pm 180°$, then $\hat{s}$ is on the locus. If not, a nearby point is selected and the argument reevaluated until the result is sufficiently close to $\pm 180°$ to consider the test point as being on the locus. It is in this phase of the graphical process that the Spirule is of greatest assistance, allowing for the rapid measurement and addition or subtraction of the angles of vectors to the test point $\hat{s}$ from the poles and zeros of $F(s)$.

Example 10.2

As an example that is only slightly less trivial than Example 10.1, we shall find the root locus corresponding to

$$F(s) = \frac{1}{s(s+2)}$$

where $0 \leq K \leq \infty$. As $F(s)$ has two poles (at $s = 0$ and -2) and no zeros, $n = 2$ and $m = 0$. Applying rules *5* and *6*, the locus has two branches, one emanating from $s = 0$ and the other from $s = -2$ for $K \approx 0$ and both going to infinity as $K \to \infty$. Applying rule *6*, the two branches approach infinity at angles of $\pm 90°$ with respect to the real axis, asymptotic to a vertical line that intersects the real axis at

$$\sigma_0 = \frac{p_1 + p_2}{n - m} = \frac{0 - 2}{2 - 0} = -1$$

Only that portion of the real axis between 0 and -2 is on the locus because *1*. there are no real poles or zeros of $F(s)$ to the right of $s = 0$; *2*. there is a single real pole to the right of $s = -2$; and *3*. there are two real poles to the right of all points to the left of $s = -2$ (rule *3*).

Drawing what we know at this point about the locus leads to the heavy lines shown in Figure 10.4*a*, which accounts for high and low values of K; however, the behavior of the locus for intermediate values of K during the transition from the real axis to the vertical asymptotes is as yet unknown. If we select any test point $\hat{s}$ on the vertical asymptote (Figure 10.4*a*) we see that $\phi_1 + \phi_2 = 180°$, which means that $\arg [F(\hat{s})] = -180°$ and that the test point must lie on the locus. Since the angle criterion is satisfied for any point on the vertical asymptote, the complete root locus must appear as in part (*b*).

Having found the root locus, we can apply the magnitude criterion if we wish to know the root locations for a specific nonnegative value of K. Alternatively, suppose we desire that the point $s_1 = -1 + j\sqrt{3}$ shown in part (*c*) be a root of the characteristic equation. We can use the magnitude

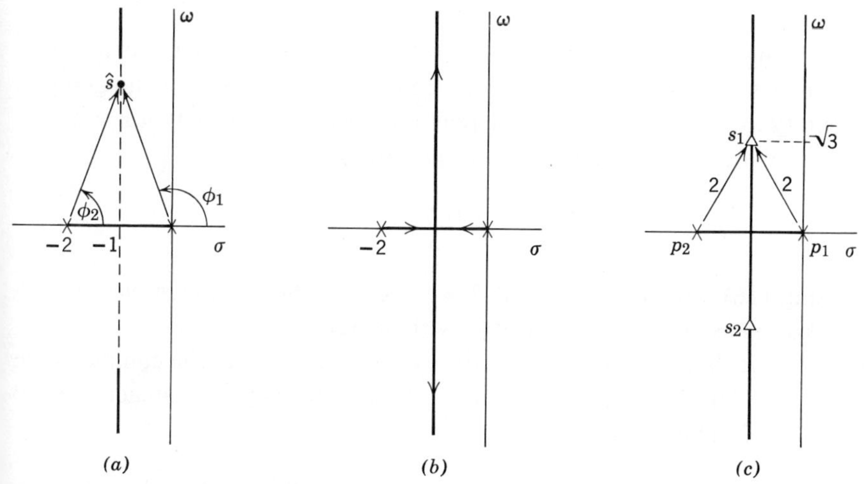

Figure 10.4. (*a*) Applying angle criterion at s. (*b*) Complete locus. (*c*) Applying magnitude criterion at s_1.

criterion to solve for the required value of K. Constructing the vectors from p_1 and p_2 to s_1, both of which are of length two, it follows that

$$|F(s_1)| = \frac{1}{|s_1 - p_1| \times |s_1 - p_2|} = \frac{1}{2 \times 2} = \frac{1}{4}$$

which, from (14), indicates that K must be equal to 4 if s_1 is to be a closed-loop root. Because of the symmetry condition about the real axis, $s_2 = 1 - j\sqrt{3}$ will be the other root corresponding to $K = 4$.

For this simple example we can verify the result by directly computing the roots of $1 + KF(s) = 0$ with $K = 4$. Substituting for $F(s)$ gives $1 + [4/s(s+2)] = 0$ which, in polynomial form, becomes $s^2 + 2s + 4 = 0$ and has the required roots.

Example 10.3

The feedback system shown in Figure 10.5a represents a control system for which the process is unstable due to the open-loop pole at $s = 1$. By plotting the root locus we can find the range of the gain α such that all three closed-loop poles lie in the LHP. From the block diagram

$$G(s)H(s) = \frac{5\alpha}{(s-1)(s^2+2s+5)} = \underbrace{5\alpha}_{K}\underbrace{\left[\frac{1}{(s-1)(s+1-j2)(s+1+j2)}\right]}_{F(s)}$$

Applying the basic root-locus rules with $n = 3$ and $m = 0$, there are three branches emanating from the three poles of $F(s)$ for $K \approx 0$ and going to infinity as $K \to \infty$ ($\alpha \to \infty$). The real-axis segment to the left of $s = 1$ is on the root locus and the large-gain asymptotes intersect at

$$\sigma_0 = \frac{1 + (-1+j2) + (-1-j2)}{3} = -\frac{1}{3}$$

Using (16) with $\nu = 0$, 1, and 2 we see that the large-gain asymptotes make angles of 60°, 180°, and 300° with the real axis.

To calculate the angle of departure as the locus leaves the complex pole at $p_2 = -1 + j2$ we use (17) with $q = 1$ (actually, any odd integer will do) and $J = 2$. Thus,

$$\phi_2 = -\{\underbrace{\arg[p_2 - p_1]}_{135^\circ} + \underbrace{\arg[p_2 - p_3]}_{90^\circ}\} + 180^\circ = -45^\circ$$

Because of the symmetry property of the root locus, the angle of departure from p_3 is $\phi_3 = -\phi_2 = +45°$.

Figure 10.5*b* shows the complete 180° root locus of the system, where the segments of the two branches leaving the complex poles have been determined for intermediate values of α by checking the angle criterion at several test points. For instance, $s_1 = j\sqrt{3}$ turns out to be the point at which the upper branch crosses into the RHP and application of the magnitude criterion gives

$$K_1 = \frac{1}{|F(s)|_{s=s_1}} = 2 \times 1.05 \times 3.80 = 8.00$$

where the lengths of the three vectors can be obtained graphically from the root locus. Relating the "cross-over" value of the root-locus gain K to the adjustable loop gain α, $\alpha_1 = 8.00/5 = 1.80$ since $K = 5\alpha$.

For very low values of the gain there will be a root on the positive real axis between the origin and $s = 1$. To find the value of K, denoted by K_2, at which the real branch of the locus crosses into the LHP we apply

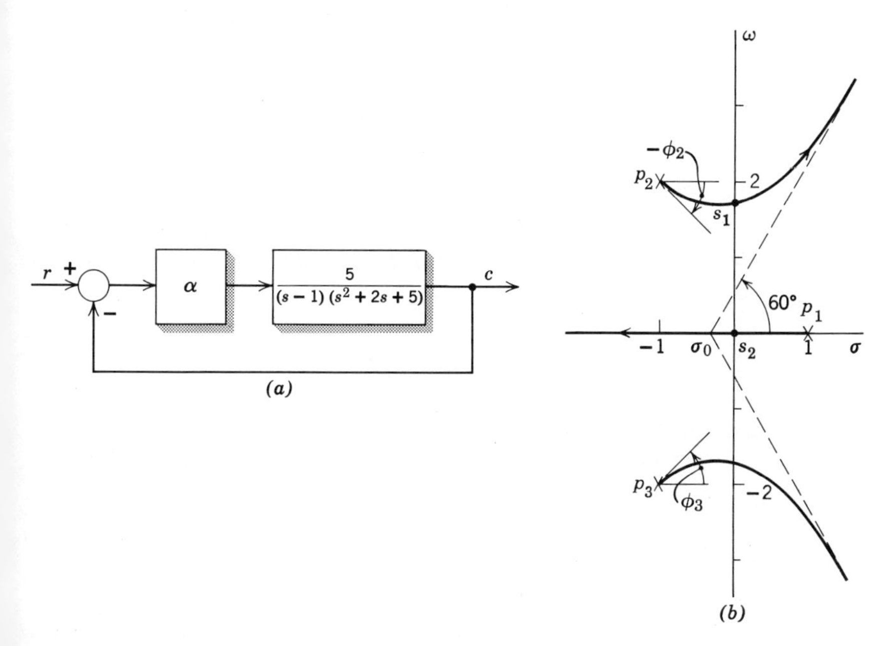

Figure 10.5. A third-order system without zeroes. (*a*) Block diagram. (*b*) Root locus.

the magnitude criterion at the point $s_2 = 0$, obtaining

$$K_2 = \frac{1}{|F(0)|} = 1 \times \sqrt{5} \times \sqrt{5} = 5$$

whence $\alpha_2 = 1.00$.

Combining the two cross-over values of α and using the root-locus plot as the basis for inferring the range of α that will result in all three roots being in the LHP, we see that the closed-loop system will be stable for $\alpha_2 < \alpha < \alpha_1$ or, using numerical values, for $1.00 < \alpha < 1.80$.

Example 10.4

As a final example and an indication of the manner in which feedback signals can be used to improve the stability characteristics of a system, we shall find the root locus for the system of the previous example when $H(s)$ is changed from unity to $2s+1$ and determine the limits on α for stability. The alteration of $H(s)$ is equivalent to adding a feedback signal of $-2\dot{c}(t)$ to the unity-feedback system of Figure 10.5a, resulting in Figure 10.6a. Writing the transfer function $G(s)H(s)$ directly from the block diagram and then rearranging so as to identify K and $F(s)$, we have

$$G(s)H(s) = \underbrace{10\alpha}_{K}\underbrace{\left[\frac{s+0.5}{(s-1)(s^2+2s+5)}\right]}_{F(s)}$$

Although the poles of $F(s)$ have not been affected, there is now a zero at $z_1 = -0.5$ so $m = 1$, thereby reducing the number of large-gain asymptotes from three to two. Applying rule 6, they make angles of $\pm 90°$ with the real axis and pass through the point

$$\sigma_0 = \frac{[1+(-1+j2)+(-1-j2)]-[-0.5]}{3-1} = -\frac{1}{4}$$

The real-axis portion of the locus lies between the pole at $s = 1$ and the zero at $s = -0.5$. Finally, the angles of departure from the complex poles can be computed. For example, a branch leaves p_2 with an angle of

$$\phi_2 = \underbrace{\arg[p_2 - z_1]}_{114°} - \underbrace{\arg[p_2 - p_1]}_{135°} - \underbrace{\arg[p_2 - p_3]}_{90°} + 180° = 69°$$

Having the above information at our disposal, we can sketch the salient features of the root locus, resulting in Figure 10.6b, or at least a close

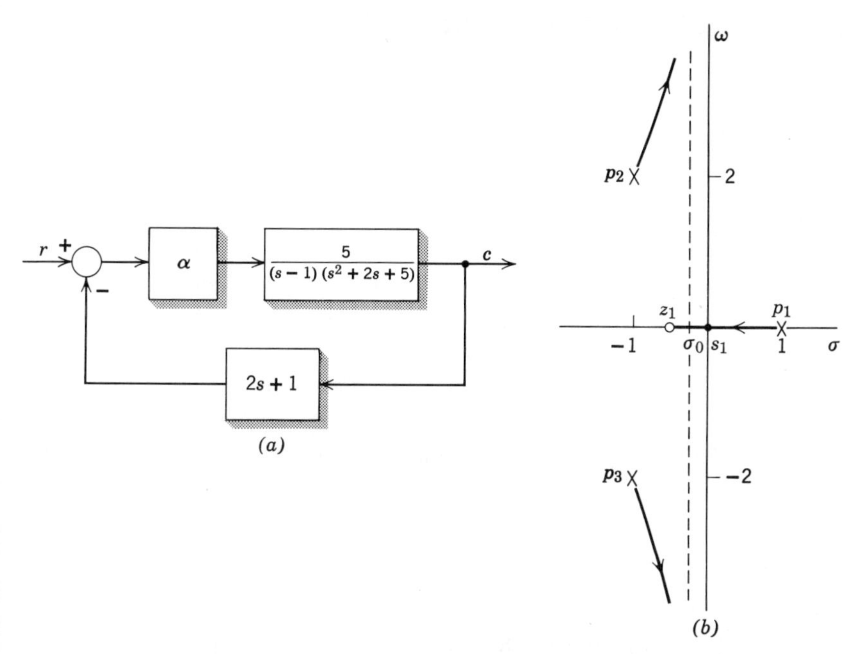

Figure 10.6. A third-order system with one zero. (*a*) Block diagram. (*b*) Root locus.

approximation thereof. Comparing the locus with Figure 10.5*b*, we see that the complex branches emanating from p_2 and p_3 no longer enter the RHP as α is increased. Thus, the addition of the zero – in physical terms, the addition of the rate-feedback signal – has had the effect of keeping the complex branches of Figure 10.5*b* in the LHP. Hence, the modified system will be stable for all values of $\alpha > \alpha_1$ which, upon applying the magnitude criterion with $s = 0$ and using the relationship $K = 10\alpha$, turns out to be $\alpha_1 = 1.00$.

10.2 DERIVATION OF BASIC ROOT-LOCUS RULES

In this section the basic rules introduced in the previous section for constructing plots of the 180° root locus will be derived.

Number of branches (Rule *1*)

From the version of the characteristic equation given in Eq. (9), Sect. 10.1, as the *sum* of polynomials of degree n and m, it is apparent that its degree is n when the condition $m \leq n$ is satisfied. Hence there will be exactly n roots of the characteristic equation for any given value of K, each of which yields one branch of the root locus as K is varied from 0 to ∞. For a specific value of K, the magnitude criterion will be satisfied for one and only one point on each of the n branches.

Symmetry (Rule 2)

Because the characteristic polynomial has strictly real coefficients, any complex roots must occur in complex-conjugate pairs. Therefore, when a complex root of the characteristic equation exists for a specific value of K, there must be another root symmetric to it with respect to the real axis, for the *same* value of K. Extending the argument from a specific value of K to an interval, it follows that if a locus contains a branch in the upper half of the complex plane, its image in the lower half of the complex plane must also be a branch of the locus.

Real-axis portions (Rule 3)

This rule is derived by applying the angle criterion at a real test point $\hat{s}$ which may lie anywhere along the real axis of the s-plane except at a pole or zero of $F(s)$. Each pole or zero of $F(s)$ must fall into one of the following three categories, corresponding to the three parts of Figure 10.7, which immediately determines its contribution to arg $[F(s)]$.

Case (a). Complex pairs of poles or zeros make no *net* contribution to arg $[F(s)]$ because the sum of the two arguments always equals 360°.

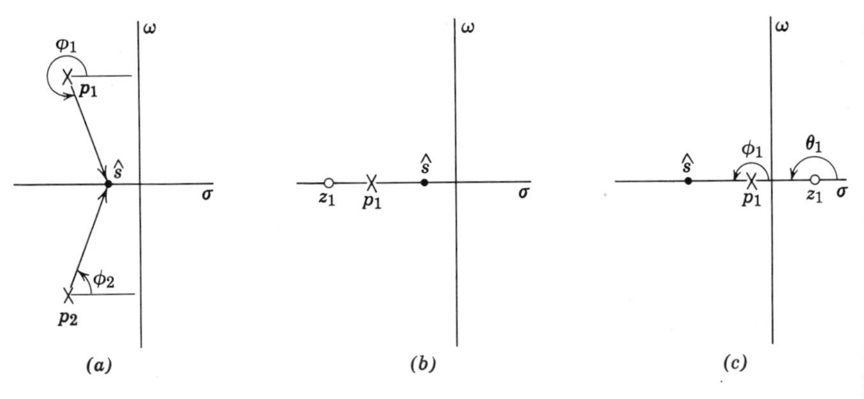

Figure 10.7. (a) $\phi_1 + \phi_2 = 360°$. (b) $\phi_1 = \theta_1 = 0°$. (c) $\phi_1 = \theta_1 = 180°$.

Case (*b*). Real poles or zeros to the *left* of $\hat{s}$ make no contribution to arg $[F(\hat{s})]$ because the argument of $(\hat{s}-p_k)$ or $(\hat{s}-z_i)$ is zero.

Case (*c*). Real poles or zeros to the *right* of $\hat{s}$ will contribute $-180°$ or $+180°$, respectively, to arg $[F(\hat{s})]$.

Since all possibilities arising from finite poles and zeros have been discussed, we conclude that only those poles and zeros lying on the real axis to the *right* of $\hat{s}$ need be considered. Furthermore, any combination of real poles and zeros to the *right* of $\hat{s}$, such that the total number is *odd*, will result in a net value of $+180°$ for arg $[F(\hat{s})]$ and, hence, a point on the 180° locus.

Behavior for $K \approx 0$ (Rule 4)

When $K = 0$ the version of the characteristic equation given by Eq. (9), Sect. 10.1, reduces to $(s-p_1)(s-p_2)\cdots(s-p_n)=0$ from which the roots are readily identifiable, namely $s = p_k$, $k = 1, 2, \ldots n$. But these values of s are precisely the poles of $F(s)$, i.e., the open-loop poles of the system. Hence, as K increases from 0, the branches of the root locus will emanate from the open-loop poles.

When $K = 0$ the loop is no longer closed and the zero-input response of the system will be that of the individual elements or subsystems, acting without the presence of feedback. Thus, if K is decreased to 0, the roots of the closed-loop system will approach the poles of the dynamic elements comprising the system, i.e., the poles of $G(s)$ and $H(s)$.

Behavior as $K \to \infty$ (Rule 5)

Starting with yet another form of the characteristic equation, namely Eq. (11), Sect. 10.1, it follows that when $K \to \infty$ with s remaining finite, the terms involving p_k vanish. Thus the equation reduces to

$$(s-z_1)(s-z_2)\cdots(s-z_m)=0$$

the roots of which obviously are $s = z_i$, $i = 1, 2, \ldots, m$. Thus, the m zeros of $F(s)$ in the finite plane will be the terminating points of m of the n branches of the locus as $K \to \infty$. The remaining *n-m* branches will approach infinity in the manner described by rule 6, the derivation of which follows.

Branches approaching infinity (Rule 6)

When $|s|$ is much larger than the magnitude of any of its poles or zeros, the function $F(s)$ may be approximated by the new function $\hat{F}(s)$ – not to be confused with $F(\hat{s})$ – defined by

$$\hat{F}(s) \triangleq \frac{1}{s^r} \tag{1}$$

where $r = n - m$. In other words, to a first approximation, the root locus for very large values of s should look like the root locus of a system which has r open-loop poles at the origin and no zeros. By applying the root-locus construction rules which have already been derived, we can see that the locus for such a system will consist of r branches that are straight lines emanating from the origin at equally spaced angles of $360°/r$. If r is odd, then one of these branches will be the negative real axis and if r is even the real axis cannot lie on the locus associated with $\hat{F}(s)$. Figure 10.8 shows the root-locus plots corresponding to $\hat{F}(s)$ for $r = 3$ and 4.

We shall now prove that the branches of the exact locus which approach infinity do so asymptotic to the r branches of the locus of $\hat{F}(s)$ after they are shifted to the right or left by the proper amount. Shifting the approximate locus defined by (1) to the right by the amount σ_0 is equivalent to constructing the root locus of

$$\hat{F}(s - \sigma_0) = \frac{1}{(s - \sigma_0)^r} \tag{2}$$

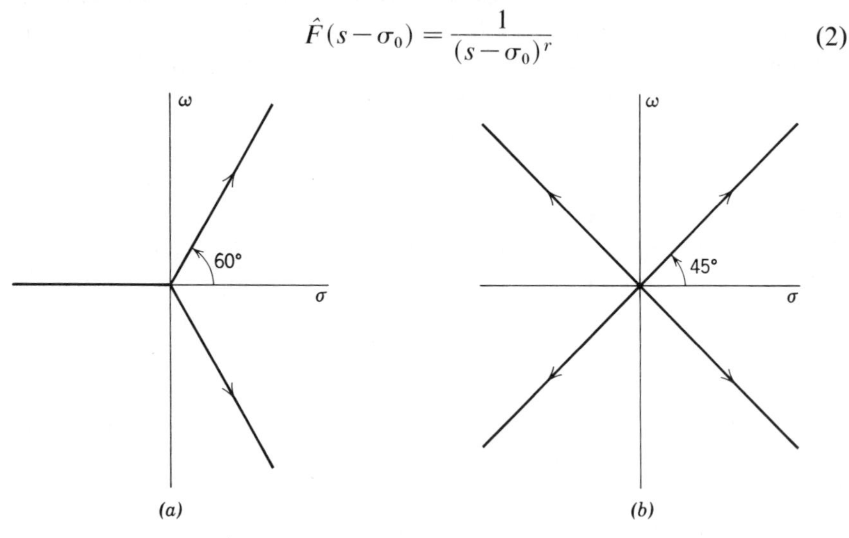

Figure 10.8. Locus of $\hat{F}(s) = 1/s^r$. (a) $r = 3$. (b) $r = 4$.

which has the root-locus plots shown in Figure 10.9, again, for $r=3$ and 4.

The characteristic equation for this second approximation may be written as $1+K\hat{F}(s-\sigma_0)=0$ or, using (2) to substitute for $\hat{F}(s-\sigma_0)$,

$$1+K\frac{1}{(s-\sigma_0)^r}=0$$

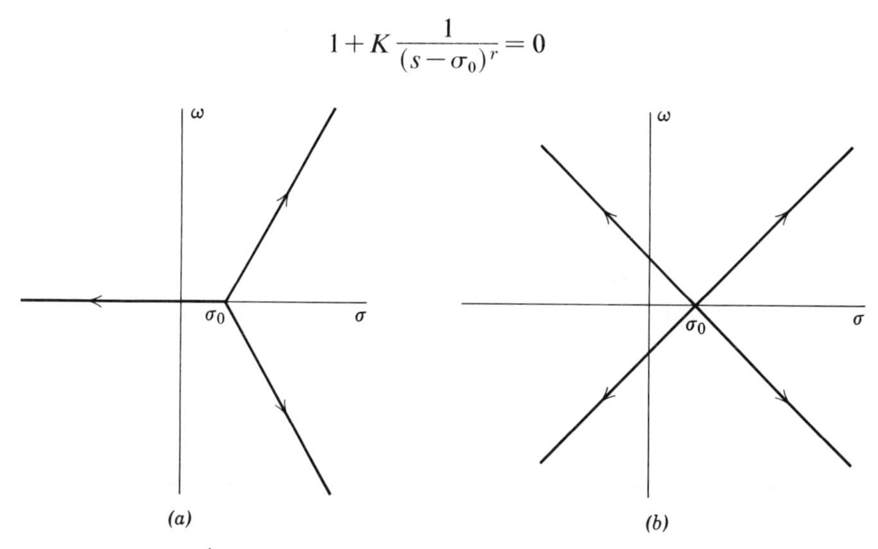

Fig 10.9. Locus of $\hat{F}(s-\sigma_0)$. (a) $r=3$. (b) $r=4$.

Replacing $(s-s_0)^r$ by its binomial-series expansion, the approximate characteristic equation becomes

$$1+K\left[\frac{1}{s^r-r\sigma_0 s^{r-1}+\cdots+(-1)^r(\sigma_0)^r}\right]=0 \tag{3}$$

Next we manipulate the characteristic equation of the actual system into the form of (3) and then choose σ_0 so as to make the two equations agree in as many of the high-order powers of s as possible. The reason for seeking agreement in the high-order powers rather than low-order powers is that we are approximating the behavior of the locus as $|s|\to\infty$ and these are obviously the terms that will dominate. With the numerator and denominator of $F(s)$ written in polynomial form, the characteristic equation of the original system becomes

$$1+K\overbrace{\left[\frac{s^m+b_{m-1}s^{m-1}+\cdots+b_0}{s^n+a_{n-1}s^{n-1}+\cdots+a_0}\right]}^{F(s)}=0 \tag{4}$$

where, from the theory of polynomials†, it is known that

$$b_{m-1} = -\sum_{i=1}^{m} z_i \qquad \text{and} \qquad a_{n-1} = -\sum_{k=1}^{n} p_k$$

If the division of the two polynomials in (4) is carried out through the first and second high-order terms and $n-m$ is replaced by r, we see that the characteristic equation of the actual system is

$$1+K\left[\frac{1}{s^r-(b_{m-1}-a_{n-1})s^{r-1}+\cdots}\right]=0 \tag{5}$$

Comparing the characteristic-equation expansion corresponding to the shifted approximation (3) with that of the exact system (5), it is apparent that the two can be made to agree in the two highest powers of s by choosing

$$\sigma_0 = \frac{b_{m-1}-a_{n-1}}{r} \tag{6}$$

When the above expressions for b_{m-1} and a_{n-1} in terms of the poles and zeros of $F(s)$ are substituted into (6), along with $r=n-m$, Eq. (15), Sect. 10.1, results.

The quantity σ_0 is often referred to as the "center of mass" of the open-loop poles and zeros, this interpretation coming about by considering each pole to be a unit positive mass and each zero to be a unit negative mass. An analogy using positive and negative unit charges can also be devised which might be more pleasing to electrical engineers than the mass analogy.

Angles of departure and arrival (Rules *7* and *8*)

The manner in which the root-locus branches depart from a simple real pole of $F(s)$ is covered by rule *3*. When a branch leaves a simple complex pole its angle of departure is defined to be the angle which the tangent to the locus at the pole makes with the real axis, e.g., ϕ_J in Figure 10.10*a*. Although the expression for ϕ_J given by Eq. (17), Sect. 10.1, can be derived rigorously by expanding $F(s)$ in a Laurent series about the pole p_J, a simpler intuitive argument will be presented instead.

We define the point s_0 to lie on the 180° root locus very close to the pole

†For a summary of polynomial properties and additional references, see D'Azzo and Houpis (1966, App. B.).

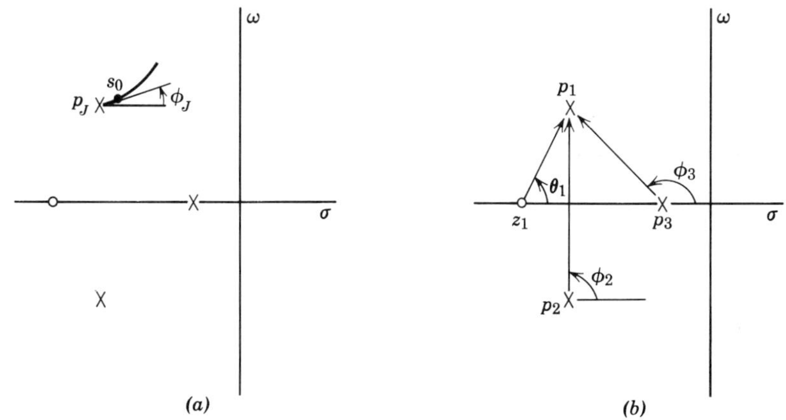

Figure 10.10.

p_J, as indicated in part (a) of the figure. Because s_0 is on the locus we know that $F(s_0)$ must satisfy the angle criterion

$$\arg\,[F(s_0)] = q180^\circ \tag{7}$$

where q is an odd integer. However, we cannot evaluate $\arg\,[F(s_0)]$ because we do not know the precise location of s_0 — all we know at the moment is that it is on the locus. Since $|s_0 - p_J|$ is small, we can use p_J in place of s_0 for evaluating the angular contributions from all of the poles and zeros of $F(s)$ *except* the pole at p_J. The construction used to find the angle of departure from a complex pole is shown in Figure 10.10b for a typical pole-zero pattern. However, as indicated in part (a), the angular contribution of the vector from p_J to the test point s_0 must be $\phi_J = \arg\,[s_0 - p_J]$. When the approximate expressions for the angles are substituted into the angle criterion it follows that

$$\arg\,[F(s_0)] \approx \sum_{i=1}^{m} \arg\,[p_J - z_i] - \sum_{\substack{k=1 \\ k\neq J}}^{n} \arg\,[p_J - p_k] - \phi_J$$

Then, using (7) to substitute for $\arg\,[F(s_0)]$ and solving for ϕ_J, we obtain Eq. (17), Sect. 10.1.

The branches of the root locus that terminate on the complex zeros of $F(s)$ for $K = \infty$ will have tangents that define their angles of arrival. To derive Eq. (18), Sect. 10.1, one makes the same approximations used above with the angle of departure, with the exception that θ_J, the contribution of the Jth zero, is now the unknown.

10.3 ADDITIONAL ROOT-LOCUS PROPERTIES

Having presented and proved the foregoing basic rules for the construction of a root locus we shall conclude the formal presentation of the topic with a discussion of several additional features that are of particular value in control-systems work.

Intersection of branches

If two real-axis branches of a locus are approaching one another for increasing values of K, with no intervening poles or zeros of $F(s)$, they will meet at some point. At this meeting both will undergo abrupt 90° changes of direction, one moving into the upper half of the s-plane and the other into the lower, as in Figure 10.4*b*. Likewise, two complex branches may come together at some point on the real axis as K increases and then move to the right and left along the real axis for further increases in K. While the intersection of root-locus branches is not limited to pairs of branches or to intersections on the real-axis, these are the most common cases in practice. Unfortunately, determining the locations of intersection points generally requires somewhat more effort than that required by the other construction rules.

To gain insight into the conditions under which multiple branches can intersect at a single point and to determine the general features of such an intersection, we assume that s_0 is a point on the 180° locus and expand $F(s)$ in a Taylor series about s_0. Since $F(s)$ is a rational function of s it will be analytic at all points in the s-plane except at its poles, $s = p_k$, $k = 1, 2, \ldots, n$. Thus, within the circle of convergence (Appendix A)

$$F(s) = F(s_0) + F'(s_0)(s - s_0) + \tfrac{1}{2}F''(s_0)(s - s_0)^2 + \cdots \tag{1}$$

where $F'(s_0)$ denotes dF/ds evaluated at $s = s_0$. Because s_0 lies on the root locus, arg $[F(s_0)]$ must satisfy the angle criterion, which is to say that $F(s_0)$ must be real and negative. If the value of K which corresponds to the point s_0 on the root locus is denoted by K_0, it follows from the magnitude criterion that

$$F(s_0) = -\frac{1}{K_0} \tag{2}$$

For the purpose of identifying other root-locus points in the neighborhood of s_0 we consider a small circle of radius ρ centered at s_0 and seek to determine those points on the circle which satisfy the angle criterion.

First, let us assume that $F'(s_0)$ does not vanish and write the difference between the testpoint and s_0 in polar form as $s - s_0 = \rho e^{j\psi}$. Then, taking ρ small enough so that the constant term in (1) dominates the term which is linear in ρ which, in turn, dominates all higher-order terms, we are justified in writing

$$F(s) \approx -\frac{1}{K_0} + \rho F'(s_0) e^{j\psi} \tag{3}$$

In addition to depicting the s-plane relationships in Figure 10.11*a*, the complex numbers in (3) can be represented in a separate complex plane, part (*b*), referred to as the F-plane, in which we show $F(s)$ as a vector whose magnitude and angle are functions of s, namely $|F(s)|$ and arg $[F(s)]$. As required by the angle criterion, the vector representing $F(s_0)$ lies on the negative-real axis of the F-plane and remains fixed during the following discussion.

If the test point s is to be on the 180° locus, $F(s)$ must certainly be real and negative per the angle criterion. But we can see from either (3) or Figure 10.11*b* that as a consequence $\rho F'(s_0) e^{j\psi}$, which is the difference

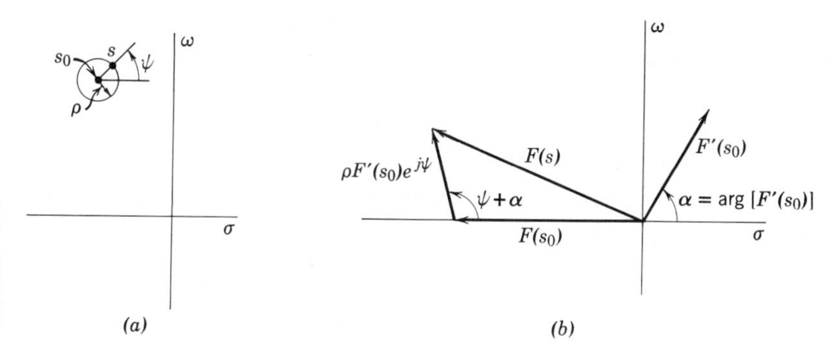

Figure 10.11. Mapping of $F(s)$ near s_0, a point on the locus. (*a*) s-plane. (*b*) F-plane.

between $F(s)$ and $F(s_0)$, must be *real* if the angle criterion is to be satisfied. Since ψ goes from 0° to 360° as the test point is moved around the circle in Figure 10.11*a* and both ρ and $F'(s_0)$ are constants, there will be only two values of ψ in this interval for which the angle criterion will be satisfied. One of these, say ψ_1, will result in $|F(s)| > |F(s_0)|$ when $\psi_1 + \arg\ [F'(s_0)] = 180°$, which corresponds to the branch of the root locus entering the circle at a value of $K_1 < K_0$.

The other value of ψ for which the angle criterion is satisfied will be $\psi_2 = \psi_1 + 180°$, resulting in $|F(s)| < |F(s_0)|$ and corresponding to the branch leaving the circle at a value of $K_2 > K_0$. Thus, if $F'(s)$ does not vanish at a point that is known to lie on the root locus, there is only one branch passing through that point, as in Figure 10.12*a*. In other words, $F'(s_0) = 0$ is a necessary condition for two branches to intersect at the point s_0.

To explore this point further, suppose that $F'(s_0) = 0$ but $F''(s_0) \neq 0$. Returning to (1) and placing the appropriate restriction on ρ, the Taylor-series expansion for $F(s)$ yields

$$F(s) \approx -\frac{1}{K_0} + \frac{1}{2}\rho^2 F''(s_0) e^{j2\psi} \tag{4}$$

The important difference between (3) and (4) is that the latter involves the term $e^{j2\psi}$ rather than $e^{j\psi}$. As a consequence there will be four values of ψ in the interval $0° \leq \psi < 360°$ for which the angle criterion is satisfied as the test point is moved around the circle in Figure 10.11*a*. Therefore, two branches of the locus must enter the circle and two must leave it. Furthermore, as ψ is increased from 0° to 360° the values of ψ for which the angle criterion is satisfied will be separated by 90°, with $|F(s)|$ alternating between being greater than and less than $|F(s_0)|$. Thus, the character of the root locus in the vicinity of s_0 must be as shown in Figure 10.12*b*.

Extending the above arguments to an arbitrary number of intersections, if s_0 is on the 180° locus and the first $\nu - 1$ derivatives of $F(s)$ vanish at $s = s_0$, then ν branches intersect at s_0 with entering and departing branches alternating and separated by angles of $(180/\nu)°$. The case $\nu = 3$ is shown in Figure 10.12*c*.

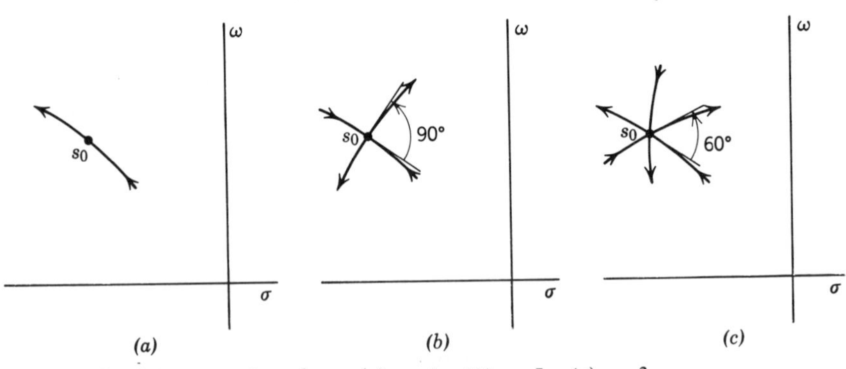

Figure 10.12. Intersecting branches. (*a*) $\nu = 1$. (*b*) $\nu = 2$. (*c*) $\nu = 3$.

The situation most often encountered in practice is the intersection of two branches, for which we have shown that

$$\left.\frac{dF}{ds}\right|_{s=s_0} = 0 \tag{5}$$

is a *necessary* condition, where s_0 is the intersection point on the root locus. However, the mere fact that $F'(s)$ vanishes for some values of s does not imply that two or more branches intersect unless the point at which $F'(s) = 0$ is on the root locus.

Finally, provided that s_0 does not coincide with a zero of $F(s)$, the condition for the intersections of loci can also be expressed as

$$\frac{d}{ds}\left(\frac{1}{F(s)}\right)_{s=s_0} = 0 \tag{6}$$

This form follows from the fact that

$$\frac{d}{ds}\left(\frac{1}{F(s)}\right) = -\frac{1}{F^2(s)}\frac{dF}{ds}$$

which vanishes for those values of s satisfying (5) provided $F(s) \neq 0$. The version given by (6) may be easier to apply than (5) in some cases.

Example 10.5

As an illustration of the manner in which branches intersect, we apply the rules given in Section 10.2 to construct the 180° locus for the system shown in Figure 10.13a, for which

$$F(s) = \frac{s+1}{s(s+0.5)}$$

Based upon the knowledge that the real-axis portions of the locus are the segments $-0.5 \leq s \leq 0$ and $s \leq -1$, we may infer that as K increases the two branches meet at some point $-0.5 < s_1 < 0$ and depart from the real axis for $K = K_1$ at angles of $\pm 90°$. Furthermore, the two branches must come together for some larger value of K, say $K = K_2$, at the point $s_2 < -1$.

We can solve for s_1 and s_2 by finding those solutions of (5) or (6) that lie on the locus. Differentiating $F(s)$,

$$\frac{dF}{ds} = -\frac{s^2+2s+0.5}{s^2(s+0.5)^2}$$

which vanishes for

$$s_1 = -\left(1 - \frac{1}{\sqrt{2}}\right) = -0.293 \qquad \text{and} \qquad s_2 = -\left(1 + \frac{1}{\sqrt{2}}\right) = -1.707$$

Because both s_1 and s_2 lie on the 180° locus, they must be the points at which the two branches leave from and arrive at the real axis. If we had chosen to differentiate $1/F(s)$ instead, we would have obtained

$$\frac{d}{ds}\left(\frac{1}{F(s)}\right) = \frac{s^2 + 2s + 0.5}{(s+1)^2}$$

which also vanishes for the values of s_1 and s_2 found above. The complete root locus is given in Figure 10.13*b*. The combination of a pair of poles and a single zero is one that arises often, and it can be shown analytically that the branches in the upper and lower halves of the s-plane always comprise a circle centered at the zero of $F(s)$ – observe that the mid-point between s_1 and s_2 is indeed $s = -1$, the location of the zero (Prob. 10.18).

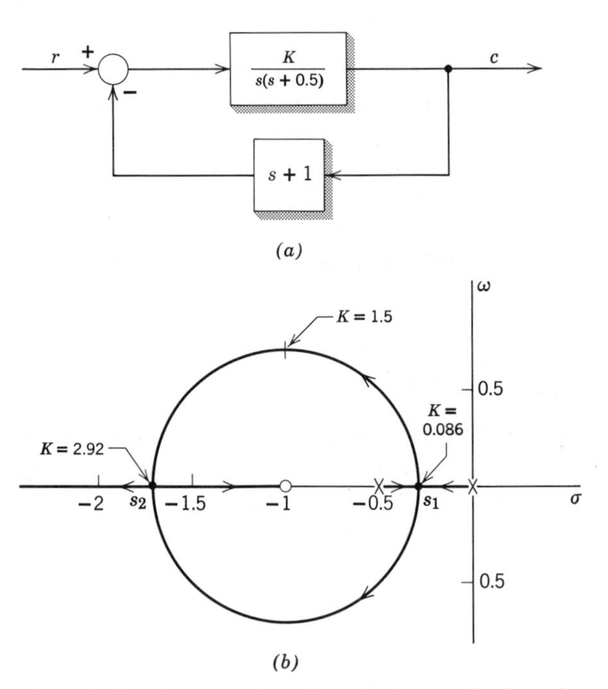

Figure 10.13. A system whose locus has intersecting branches. (*a*) Block diagram. (*b*) 180° locus.

By applying the magnitude criterion the values of K corresponding to the points s_1 and s_2 can be computed. For example, at the point s_1 at which the locus breaks away from the real axis,

$$F(s_1) = \frac{1/\sqrt{2}}{(-1+1/\sqrt{2})(-0.5+1/\sqrt{2})} = -11.64$$

so $K_1 = 1/|F(s_1)| = 0.086$.

Repeated poles and zeros of *F*(s)

When the basic root-locus construction rules were presented in Section 10.2, the assumption was made that all poles or zeros of $F(s)$ were distinct, i.e., not repeated. This restriction can be relaxed by making only minor adjustments in the rules.

Briefly, a pole of $F(s)$ which is repeated r times will be the starting point of r branches and should be counted r times when determining the real-axis portion of the locus, the large-gain asymptotes, and the angles of departure and arrival. The behavior of these branches in the vicinity of the repeated pole for $K \approx 0$ is governed by the preceding discussion on the intersection of branches – these happen to intersect for $K = 0$. Comparable relationships hold for a repeated zero of $F(s)$.

Example 10.6

The adjustments in the application of the root-locus construction rules necessitated by a repeated pole can be demonstrated by obtaining the locus for

$$F(s) = \frac{s+2}{s(s+1)^2}$$

By inspection, $F(s)$ has a repeated pole at $s = -1$ of multiplicity $r = 2$, along with a single pole at the origin and a zero at $s = -2$. Counting the repeated pole twice, it follows that $m = 1$ and $n = 3$; hence the locus has three branches. For $K \approx 0$ one branch emanates from $s = 0$ and two branches must start at the repeated pole. By applying the real-axis rule, we see that all points on the real axis between $s = 0$ and $s = -2$ have an odd number of poles and/or zeros to their right.

Because $n - m = 2$ and

$$\sigma_0 = \frac{[0+2(-1)]-[-2]}{3-1} = 0$$

there are two branches that approach infinity asymptotic to the imaginary axis. If the point at which the branches break away from the real axis is evaluated by solving $dF/ds = 0$, one finds that the breakaway point must satisfy $s^3+4s^2+4s+1=0$ which has as its three solutions $s_1=-0.38$, $s_2=-1$, and $s_3=-2.62$. Because s_3 does not lie on the 180° locus it need not be considered further. The fact that $F'(s)$ vanishes for $s=-1$ is a result of the fact that there is a repeated pole of $F(s)$ at that point in keeping with the discussion on intersection of branches. Thus, by the process of elimination, $s_1=-0.38$ must be the breakaway point we seek. The complete root locus is shown in Figure 10.14.

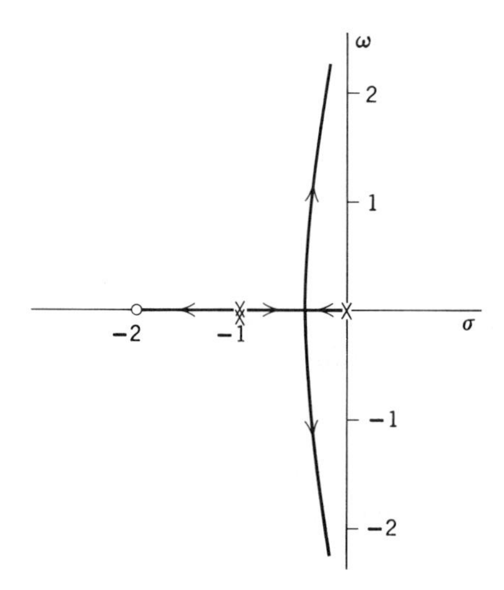

Figure 10.14. Root locus with a double pole.

Sum of the roots

When the number of open-loop poles exceeds the number of open-loop zeros by 2 or more, i.e., $m \le n-2$, the sum of the roots of the characteristic equation happens to be independent of K. This result can be derived by considering the form of the characteristic equation given in Eq. (9), Sect. 10.1, which becomes

$$s^n + a_{n-1}s^{n-1} + \cdots + a_0 + K[s^m + b_{m-1}s^{m-1} + \cdots + b_0] = 0$$

where the b's and a's do not depend on K. Assuming $m \le n-2$, we collect like powers of s to get

$$s^n + a_{n-1}s^{n-1} + \cdots + (a_m + K)s^m + \cdots + (a_0 + Kb_0) = 0 \tag{7}$$

Appealing to the well-known theorem that the sum of the roots of a polynomial of degree n is the negative of the coefficient of the term s^{n-1}, it follows that the sum of the roots is $-a_{n-1}$, which is certainly independent of K. Furthermore, using $K=0$ shows that this sum is the sum of the poles of $F(s)$.

Referring to Figures 10.5*b* and 10.6*b* we see that in both cases the number of poles of $F(s)$ exceeds the number of zeros by at least 2. In Figure 10.5*b*, the real parts of the roots on the two complex branches must approach $+\infty$ as $K \to \infty$ at such a rate that for any given value of K the sum of the roots is equal to -1, since $p_1+p_2+p_3=-1$. Thus, for the specific value of K at which the two complex roots lie on the imaginary axis, the real root must be at $s=-1$. On the other hand, in Figure 10.6*b*, the real parts of the complex roots must remain finite as $K \to \infty$ because the real root moves to the left only as far as $-\frac{1}{2}$ as $K \to \infty$. Since the sum of the open-loop poles is -1, it follows that the real part of the asymptotes must be $-\frac{1}{4}$, which agrees with the value of σ_0 computed in Example 10.4.

The root locus in Figure 10.14 is yet another situation in which the sum of the roots is constant. On the other hand, that of Figure 10.13*b* is an example in which $m=n-1$ and the sum of the roots is not independent of K, moving to the left as K increases.

The zero-degree locus

So far, our attention has been devoted to systems that are characterized by both a minus sign at the feedback summing junction and a positive value of the root-locus gain K. However, if the feedback summing junction has a plus sign, the characteristic equation, as given by the block-diagram rules in Figure 9.8, is $1-G(s)H(s)=0$. Substituting $KF(s)$ for $G(s)H(s)$, the characteristic equation can be written as

$$F(s)=+\frac{1}{K}=\frac{1}{|K|}e^{jq\pi} \tag{8}$$

where q is now an *even* integer. Likewise, if the feedback sign is negative but the gain K is negative, (8) is applicable, with q even. It follows that the only change to be made in the angle criterion is that in Eq. (13), Sect. 10.1, q must be even (zero is considered to be even) and the magnitude criterion is unaffected. Hence, the designation "0° locus" is used to distinguish both of these cases from the more common 180° locus.

Because their derivations did not involve the angle criterion, rules *1*, *2*, and *4* of Section 10.1 are unchanged. Rule *5* requires only the insertion of "As $K \to -\infty$" in place of "As $K \to \infty$" and rule *3* must be modified by replacing "odd" with "even"—a moment's thought will indicate that all points on the real axis will lie on either the 0° locus or the 180° locus.

In rule *6* "as $K \to -\infty$" should be inserted for "as $K \to \infty$" and Eq. (16), Sect. 10.1, should be changed to

$$\psi_\nu = \frac{\nu 360^\circ}{n-m} \qquad \nu = 0, 1, \ldots, (n-m-1)$$

The reader may verify that there is no change needed in Eq. (15), Sect. 10.1, which is the equation for the intersection of the asymptotes. Finally, rules *7* and *8* need only be modified by restricting q to be even rather than odd.

Example 10.7

To demonstrate the application of the 0°-locus rules, the locus for the system of Example 10.5 with $K \leq 0$ will be constructed. With reference to Figure 10.13*b*, there will be two branches on the 0° locus that will be those segments of the real axis not belonging to the 180° locus: $-1 \leq \sigma \leq -0.5$ and $\sigma \geq 0$. In this case no other points lie on the 0° locus as there are no complex poles or zeros and the real roots never merge. The 0° locus is shown in Figure 10.15*a*; if the 180° portion in Figure 10.13*b* is combined with the 0° locus, Figure 10.15*b* results. The reader should note that the combination of the two portions forms a continuous locus as K goes from $-\infty$ to $+\infty$, with the change from the 0° locus to the 180° locus occurring at $K = 0$, at which point the roots coincide with the open-loop poles.

10.4 ROUTH'S CRITERION

In the initial stages of a design the situation often arises where one only needs to know whether a system is stable, rather than the precise values of all of the characteristic-equation roots. By applying *Routh's criterion* we can answer this limited question far more easily than by solving for the roots of an open-loop system directly or by drawing a root locus in the case of a feedback system. In fact, we can often use Routh's criterion as one of the steps in constructing the root locus of a feedback system. One final feature of Routh's criterion is that it may be applied to a system whose elements are expressed in literal rather than numerical form.

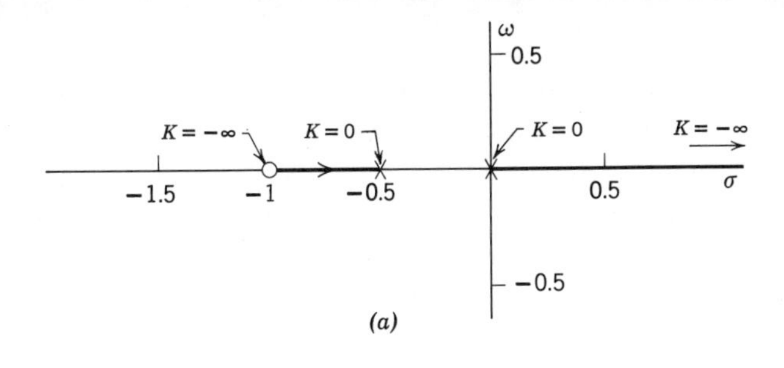

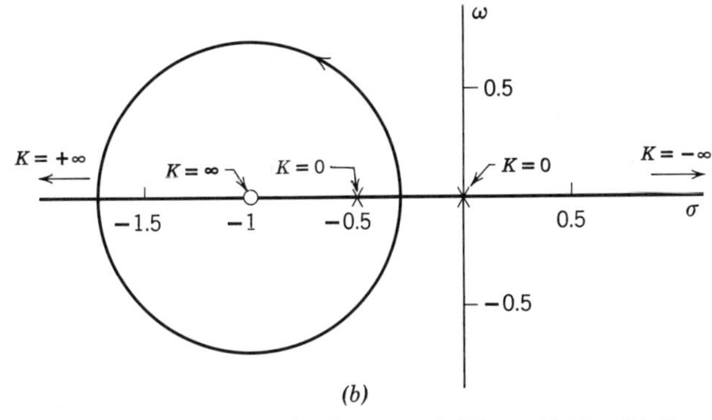

Figure 10.15. (*a*) 0° locus for the system in Figure 10.13. (*b*) Complete locus, $-\infty < K < \infty$.

It is well known that if all the roots of

$$\alpha_n s^n + \alpha_{n-1} s^{n-1} + \cdots + \alpha_0 = 0 \tag{1}$$

are to be in the LHP it is necessary, *but not* sufficient, that all of the coefficients α_i be positive. However, we seek a necessary *and* sufficient condition in the sense that if the condition is not satisfied we shall be assured that at least one of the roots of the polynomial does not lie in the LHP. If (1) is constructed so as to be the characteristic equation, we shall have a test for asymptotic stability.

Such a test was presented by Routh in 1874 and an equivalent test was derived independently by Hurwitz in 1895; both forms are often referred to as the *Routh-Hurwitz criterion.*† Here we shall present the version given by Routh; the proof, which is rather involved, is omitted.‡ Routh's formulation states that certain quantities, $n+1$ in number and known as

†A simplification of the Routh-Hurwitz criterion which could be useful for high-order systems is the Liénard-Chipart test. See Lindorff (1965, App. A) for details.

‡See Guillemin (1956, pp. 395–409).

the *Routh series*, must all be positive if all the roots of (1) are to lie in the LHP. Furthermore, if all elements of the series are nonzero, the number of roots in the RHP is equal to the number of changes in sign encountered in going through the series in order. We shall show the manner in which the elements of the Routh series are computed and then give several results in a form particularly suited to the analysis of feedback systems.

In order to systematize the evaluation of the Routh series, the results of intermediate calculations are arranged along with the elements of the series in the *Routh array* which, in its general form, is

$$
\begin{array}{|c|cccc}
\hline
A_n & B_n & C_n & D_n & \cdots \\
A_{n-1} & B_{n-1} & C_{n-1} & D_{n-1} & \cdots \\
A_{n-2} & B_{n-2} & \cdot & \cdot & \\
\cdot & \cdot & \cdot & \cdot & \\
\cdot & \cdot & C_4 & & \\
\cdot & \cdot & 0 & & \\
A_2 & B_2 & 0 & & \\
A_1 & 0 & 0 & & \\
A_0 & 0 & 0 & &
\end{array}
$$

Routh series

The set $A_n, A_{n-1}, \ldots, A_0$ comprising the first column is the Routh series, and each element of it must be positive if all of the roots of (1) are to lie in the LHP. The B's, C's, etc., are the results of intermediate calculations and, with the exception of B_2, will be of no further use in assessing the stability of the system. There will always be $n+1$ rows in the array and the number of columns will be $(n+2)/2$ if n is even and $(n+1)/2$ if n is odd. As indicated in the partial array shown above, the column of B's will terminate with B_2, the column of C's with C_4, etc. The reason for this will be clear after the algorithm for computing the elements is given.

To start the process, the first two rows of the array are merely the coefficients of the polynomial in (1), arranged so that

$$
\begin{array}{lll}
A_n = \alpha_n & B_n = \alpha_{n-2} & C_n = \alpha_{n-4} \cdots \\
A_{n-1} = \alpha_{n-1} & B_{n-1} = \alpha_{n-3} & C_{n-1} = \alpha_{n-5} \cdots
\end{array}
$$

The filling out of the first two rows continues until α_0 is reached and it will fall in the second row if n is odd and in the first row if n is even, in which case the corresponding element in the second row is set to zero. All entries to the right of the column containing α_0 are considered to be zeros.

At this point, the entries in the third row are calculated using the elements in the first and second rows according to

$$A_{n-2} = \frac{A_{n-1}B_n - A_nB_{n-1}}{A_{n-1}} \qquad B_{n-2} = \frac{A_{n-1}C_n - A_nC_{n-1}}{A_{n-1}} \tag{2}$$

and so forth. The reader should be able to convince himself that the third row will have one fewer nonzero elements than the two rows above it. Once the third row is completed, the fourth and subsequent rows may be computed using the generalizations of (2):

$$A_{i-1} = \frac{A_iB_{i+1} - A_{i+1}B_i}{A_i} \qquad B_{i-1} = \frac{A_iC_{i+1} - A_{i+1}C_i}{A_i} \tag{3}$$

and similarly for C_{i-1}, D_{i-1}, etc., until the process ends. With the evaluation of each pair of rows the number of nonzero elements in a row is reduced by one, thus causing the process to terminate eventually. The next to last row will contain only the element A_1 and the last row only the element A_0. In fact, it is readily shown that if the sequence of operations given above is properly executed, the last nonzero element in each column of the Routh array will be α_0.

From an inspection of (3) it is obvious that the process can not proceed if $A_i = 0$ at any stage in the calculation. This is not generally a problem because we know from the statement of Routh's criterion that if one of the A_i is zero, at least one of the roots of (1) will not lie in the LHP. There are methods for continuing the calculations when this happens; see, for instance, Cannon (1967, Sect. 11.12).

Having shown how to construct the Routh series we now state the results that are of greatest utility in the study of feedback systems.

1. For all the roots of the characteristic equation to lie in the LHP, i.e., for the system to be asymptotically stable, it is both necessary and sufficient that each of the $n+1$ elements of the Routh series, $A_n, A_{n-1}, \ldots, A_0$, be positive.

2. If $A_0 = 0$ and the remaining n elements of the Routh series A_n, $A_{n-1}, \ldots, A_1$, are positive, then the characteristic equation has a single root at $s = 0$ and the remaining $n-1$ roots are in the LHP (marginal stability).

3. If $A_1 = 0$ and the remaining n elements of the Routh series,

$A_n, \ldots, A_2, A_0$, are positive, then the characteristic equation has a pair of imaginary roots at

$$s = \pm j\sqrt{\frac{A_0}{A_2}} \tag{4}$$

and the remaining $n-2$ roots are in the LHP (marginal stability).

While these rules allow us to quickly check the system's stability, they also are useful in constructing root loci and in parameter selection.

In the process of constructing the root locus for a feedback system, condition *3* can be used to locate the two points at which a pair of complex branches cross the imaginary axis if the other roots are in the LHP for the cross-over gain K_{co}. Because the A_i will be functions of the root-locus gain K, we can find K_{co} by solving the set of equations

$$A_1(K_{co}) = 0$$
$$A_i(K_{co}) > 0 \qquad i = 0, 2, \ldots, n$$

Having found K_{co}, the values of s_{co} at which the locus crosses the imaginary axis may be found by using (4) with the values of A_0 and A_2 corresponding to $K = K_{co}$:

$$s = \pm j\sqrt{\frac{A_0(K_{co})}{A_2(K_{co})}} \tag{5}$$

The above requirements also constitute a particularly useful stability test in control-system design if the characteristic equation of the closed-loop system is such that the A_i can be obtained in analytical form in terms of one or two unspecified parameters, e.g., gains, time constants. Thus, when there are two such parameters, say k_1 and k_2, the designer can construct curves in a two-dimensional *parameter space* corresponding to the solutions of the $n+1$ equations

$$A_i(k_1, k_2) = 0 \qquad i = 0, 1, \ldots, n$$

If a region in the k_1, k_2 plane exists such that each of the $A_i(k_1, k_2)$ is positive for all values of k_1 and k_2 in that region, we know that the closed-loop system can be made asymptotically stable by selecting the k_1 and k_2 corresponding to *any* point in that region (see Problem 10.26).

Example 10.8

To determine whether all the roots of

$$4s^5 + 6s^4 + 9s^3 + 2s^2 + 5s + 4 = 0$$

lie in the LHP, we begin by writing the first two rows of the Routh array directly from the polynomial. Using (3) to compute the remaining entries yields

$$
\begin{array}{|lll}
\hline
4 & 9 & 5 \\
6 & 2 & 4 \\
\frac{23}{3} = \frac{6\times 9 - 2\times 4}{6} & \frac{7}{3} = \frac{6\times 5 - 4\times 4}{6} & 0 \\
\frac{4}{23} = \left(\frac{23}{3}\times 2 - \frac{7}{3}\times 6\right)\frac{3}{23} & 4 = \left(\frac{23}{3}\times 4 - 0\times 6\right)\frac{3}{23} & 0 \\
-174 = \left(\frac{4}{23}\times\frac{7}{3} - 4\times\frac{23}{3}\right)\frac{23}{4} & 0 & \\
4 = \frac{-174\times 4 - 0\times 4/23}{-174} & &
\end{array}
$$

where the calculations leading to each entry have been included.

The Routh series, being the first column of the array, is

$$4 \qquad 6 \qquad 23/3 \qquad 4/23 \qquad -174 \qquad 4$$

which contains the two sign reversals $4/23 \rightarrow -174$ and $-174 \rightarrow 4$. Thus we have ascertained that the equation has two roots that do not lie in the LHP, but we have no idea as to their specific locations—we only know that if the equation were a characteristic equation, the system would be unstable.

Example 10.9

In Example 10.3 the root locus was drawn for a feedback system that was stable only when a system parameter, designated as α, was within a finite interval. Using Routh's criterion we can readily compute the stability range for α and the points at which the complex branches of the locus cross into the RHP.

From an inspection of the block diagram in Figure 10.5a we can see that when the closed-loop characteristic equation is written in the form of Eq. (1), Sect. 10.1, it is

$$(s-1)(s^2+2s+5)+5\alpha = s^3+s^2+3s+5(\alpha-1) = 0$$

The Routh array is

$$\begin{array}{|cc} \hline 1 & 3 \\ 1 & 5(\alpha-1) \\ 8-5\alpha & 0 \\ 5(\alpha-1) & 0 \end{array}$$

and the Routh series is

$$1 \quad 1 \quad 8-5\alpha \quad 5(\alpha-1)$$

The last two terms of the series will be positive if and only if $1 < \alpha < \frac{8}{5}$, which agrees with the results of the root-locus analysis in Example 10.3. Furthermore, if $\alpha = \frac{8}{5}$ then $A_1 = 0$ and the conditions for the existence of a pair of imaginary roots are satisfied. Using (5) with $A_0 = 3$ and $A_2 = 1$ gives $s_{co} = \pm j\sqrt{3}$, which are the points in Figure 10.5*b* at which the complex branches enter the RHP.

10.5 APPLICATION

In order to demonstrate the manner in which the analytical techniques presented in this chapter might be applied to the design of a feedback system, let us reconsider the satellite-attitude control system discussed in Section 9.3. Usually, most of the system parameters are fixed before the design of the control system begins, leaving at the discretion of the control-system designer only the form and parameter values of the control law; sometimes he is free to select or modify transducers also. For the sake of argument, we shall assume the following parameters are fixed:

$$\text{Angle sensor gain:} \quad K_\theta = 0.20 \text{ volts/rad}$$

$$\text{Torque constant:} \quad K_\tau = 0.10 \text{ ft lb/volt}$$

$$\text{Moment of inertia:} \quad I = 10.0 \text{ slug ft}^2$$

Hence, if we use the P+I+D control law of Eq. (12), Sect. 9.3, the amplifier gain K_a (volts/volt), the derivative gain α (seconds), and the integral gain β (seconds^{-1}) are to be selected. To complete the problem statement, certain requirements will be placed on the behavior of the closed-loop system such as specifying the locations of closed-loop poles in the s-plane and bounds upon steady-state pointing errors caused by disturbance torques. Since the control laws to be discussed are not physically realizable they should be thought of as idealizations of the actual characteristics that can be obtained in practice, with their approximation by actual networks postponed until Chapter 11.

Proportional control

It was shown in Section 9.3 that the proportional control law

$$\tau_{con} = K_a K_\tau K_\theta [\theta_{ref} - \theta] \tag{1}$$

resulted in a closed-loop system whose response to any input was unsatisfactory because of the presence of an undamped oscillatory mode. Before going on to more satisfactory control laws, we can use a simple root-locus plot to verify that this will indeed be the case. Substituting the fixed parameter values and using (1), the system's block diagram becomes that shown in Figure 10.16a. If we restrict our interest for the moment to a determination of the roots of the closed-loop characteristic equation, the block diagram can be simplified by eliminating all inputs and outputs and by reordering any blocks within the loop – recall that we are concerned only with the properties of the product $G(s)H(s)$. Thus, for the construction of the root locus the block diagram becomes the simplified form

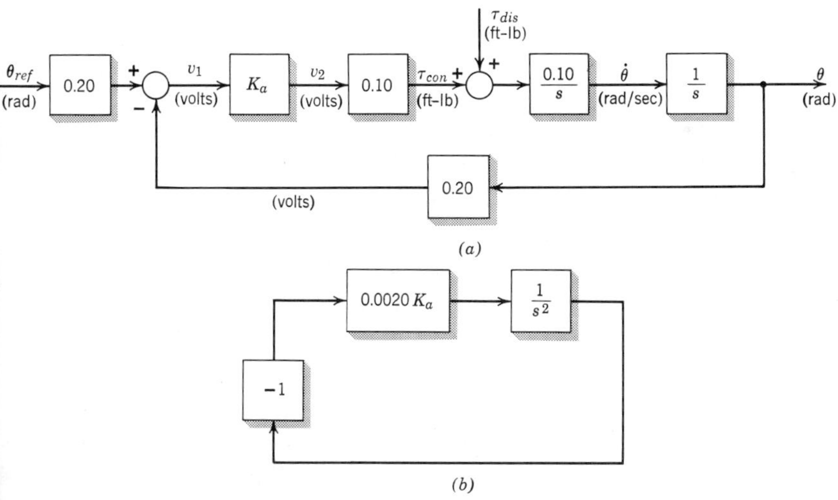

Figure 10.16. Satellite attitude system with proportional control. (a) Complete diagram. (b) Reduced loop.

of Figure 10.16*b*. Identifying $F(s) = 1/s^2$ and the root-locus gain as $K_{rl} = 0.0020K_a$, we can immediately obtain the 180° root locus for the system, shown in Figure 10.17*a* (the use of the 180° rather than the 0° locus is dictated by the presence of the minus sign at the feedback summing junction, assuming that $K_a > 0$). Because the locus remains on the imaginary axis for all $K_a > 0$, it is apparent that any value of amplifier gain will result in an undamped oscillatory mode. Hence, with the proportional control law, K_a can be selected only on the basis of the resulting frequency of oscillation.

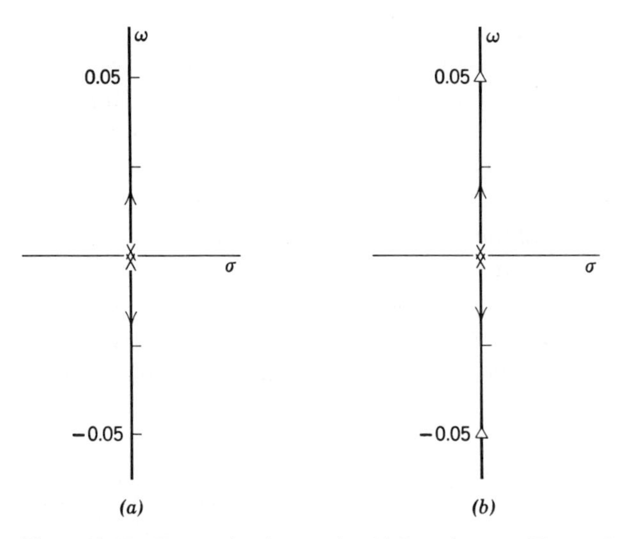

Figure 10.17. Proportional control. (*a*) Root locus. (*b*) Root locations, $K_a = 1.25$.

For example, if K_a is to be chosen so as to yield a natural frequency of $\omega_n = 0.05$ rad/sec, the root-locus magnitude criterion, Eq. (14), Sect. 10.1, must be satisfied at the points $s = \pm j0.05$. Carrying out the calculation

$$\frac{1}{K_{rl}} = \frac{1}{0.0020K_a} = |F(s)|_{s=j0.05} = \frac{1}{(0.05)^2} = 400$$

results in $K_a = 1/(400 \times 0.0020) = 1.25$ volts/volt and the pair of closed-loop roots denoted by the triangles in Figure 10.17*b*. Fixing the amplifier gain at this value and using a bit of block-diagram algebra, Figure 10.16*a* may be redrawn as shown in Figure 10.18.

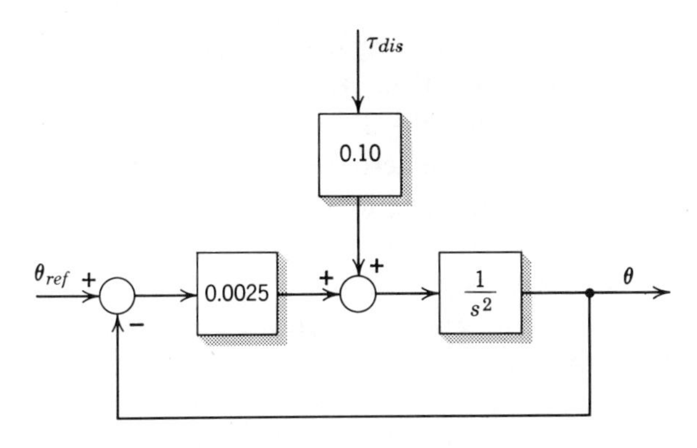

Figure 10.18. Proportional control, $K_a = 1.25$.

Proportional-plus-derivative control

We showed in Section 9.3 that the oscillatory mode can be made to damp out by using the P + D control law represented by the transfer function $1 + \alpha s$ in the forward path (Figure 10.19). A typical specification might be to select α such that the damping ratio of the closed-loop roots is $\zeta = 0.5$, with K_a remaining unchanged from its present value of 1.25.

There are many ways in which root-locus plots can be used to select the value of α and to verify that the specifications are met. For example, we might write

$$G(s)H(s) = 0.0025\left(\frac{1+\alpha s}{s^2}\right) = \underbrace{0.0025\alpha}_{K_{rl}}\underbrace{\left(\frac{s+1/\alpha}{s^2}\right)}_{F(s)} \tag{2}$$

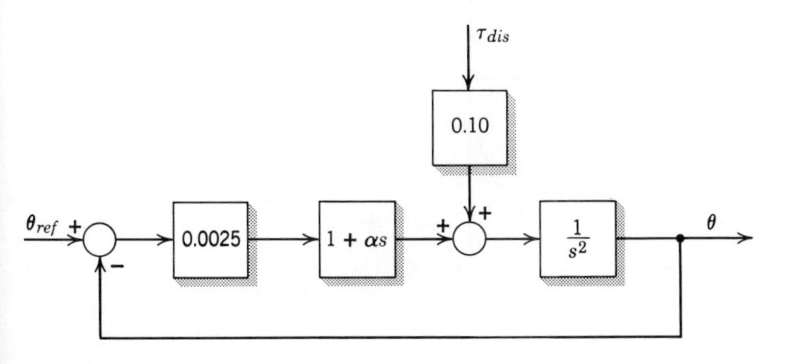

Figure 10.19. Block diagram with P + D control.

and note that $F(s)$ has a double pole at the origin and a zero at $s = -1/\alpha$ and that the root-locus gain is $K_{rl} = 0.0025\alpha$. However, if we use this formulation to construct the locus, it has the disadvantage that both the root-locus gain K_{rl} and the location of the zero depend on α. Thus, one must first guess a trial value for α, draw the locus, and then locate the roots corresponding to the specific value of α. In the likely event that the roots corresponding to K_{rl} do not satisfy the requirement that $\zeta = 0.50$, another value of α must be chosen, another locus drawn and the process repeated, until the designer either gets a satisfactory answer or runs out of patience. Two such plots, for $\alpha = 20$ and 40, are shown in Figure 10.20, with the closed-loop roots corresponding to $K_a = 1.25$ denoted. As the locus for $\alpha = 20$ happens to intersect the line corresponding to $\zeta = 0.5$ for $K_a = 1.25$, we can conclude that $\alpha = 20$ is the value that satisfies the specifications. However, there was no way to know this a priori.

As an alternative to the trial-and-error method used above, it is not difficult to manipulate the block diagram into a form in which the parameter α appears only in the root-locus *gain* and not as a pole or zero of $F(s)$. Having achieved this form, the selection of α to meet the damping-ratio specification is greatly simplified. The required sequence of block-diagram manipulations is shown in Figure 10.21. First, we draw the loop

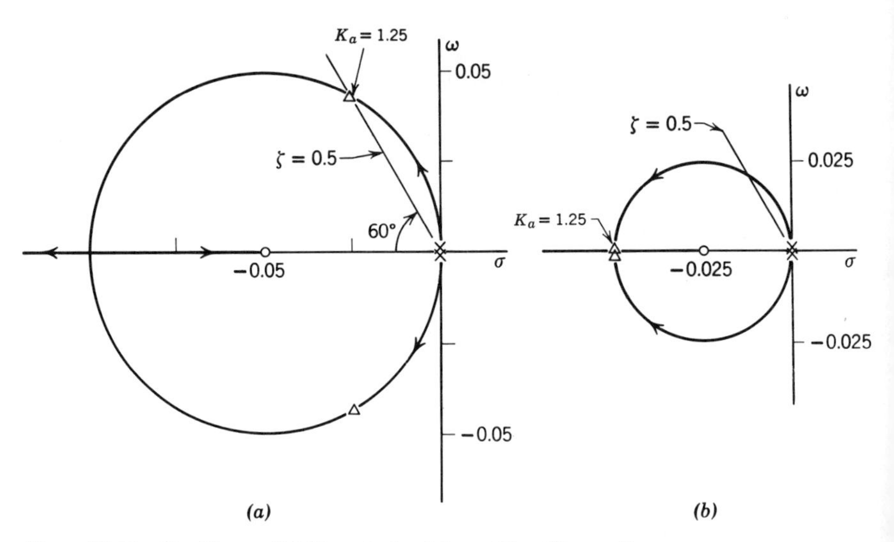

Figure 10.20. Root locus, P + D control. (*a*) $\alpha = 20$. (*b*) $\alpha = 40$.

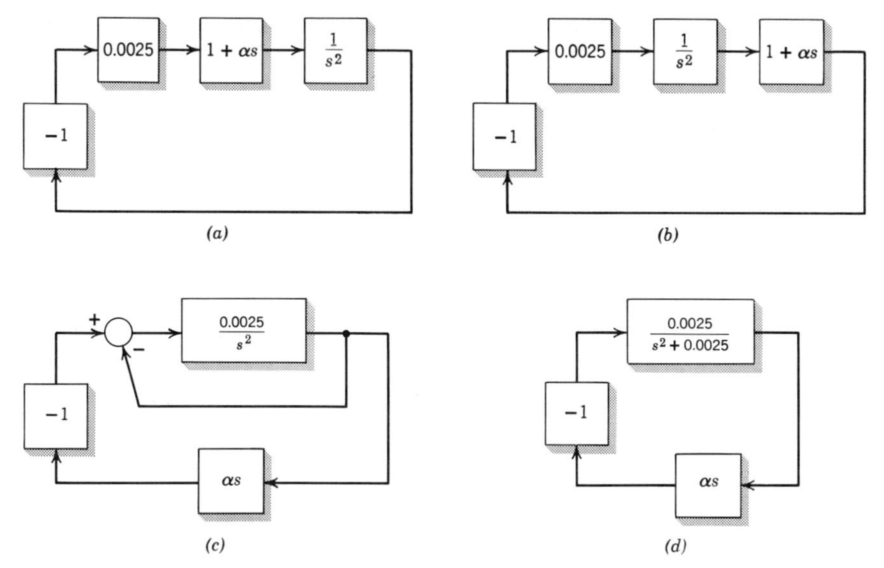

Figure 10.21. Manipulations leading to α as a root-locus gain.

diagram corresponding to Figure 10.19 as shown in part (*a*), ignoring the input and output designations. Since the characteristic equation does not depend on the ordering of the individual blocks, we are entitled to interchange the locations of the blocks whose transfer functions are $1+\alpha s$ and $1/s^2$ – the reason for doing so will become apparent in a moment – resulting in part (*b*). Next, we separate the block with transfer function $1+\alpha s$ into two parallel paths, one with unity transfer function and the other with αs, leading to part (*c*). Finally, the inner loop can be recognized as the proportional feedback system or, equivalently, the P + D system with $\alpha = 0$.

Reformulating the open-loop transfer function based upon Figure 10.21*d*, we write

$$G(s)H(s) = \alpha s\left(\frac{0.0025}{s^2+0.0025}\right) = \underbrace{0.0025\alpha}_{K_{rl}}\underbrace{\left(\frac{s}{s^2+0.0025}\right)}_{F(s)} \tag{3}$$

Thus, in the revised form, $F(s)$ has a zero at the origin and an imaginary pair of poles at $s = \pm j0.050$, all of which are independent of α, which appears only in the root-locus gain. Now, referring to Figure 10.22 which

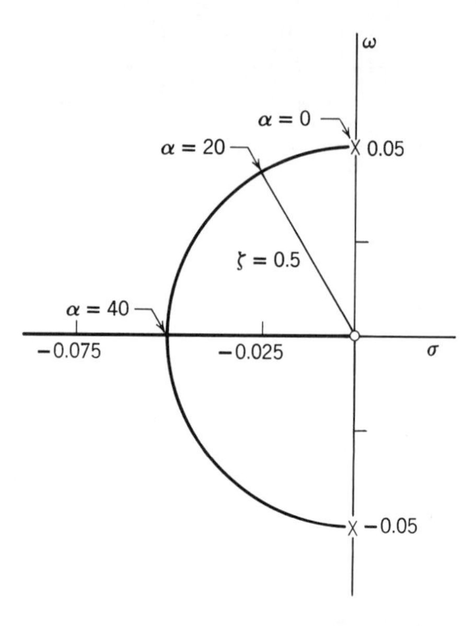

Figure 10.22. Root locus for P + D control showing dependence on α.

is the 180° root locus, we can readily obtain the closed-loop poles for any nonnegative value of α merely by using the magnitude criterion to calibrate a few points along the locus directly in terms of α. It should be clear that $\alpha = 20$ is the value that will result in the closed-loop roots with the proper damping ratio.

It is of interest to note that the roots corresponding to $\alpha = 0$ are at the closed-loop roots of the proportional feedback system. Hence, even if the function $F(s)$ in (3) could not have been factored by inspection, the root locus for the proportional system would have given the open-loop poles for the root locus of Figure 10.22. We shall use this same technique again in selecting an integral controller.

Proportional-integral-derivative control

The control law derived above will result in an asymptotically stable system which, as discussed in Section 9.3, will exhibit a constant pointing error when subjected to a constant disturbance torque. The addition of an integral term to the controller will eliminate this steady-state pointing error, provided that the parameter values are selected so that the resulting system (now third order) is still asymptotically stable.

The inclusion of the integral term, with a gain of β, will result in the block diagram of Figure 10.23 where, as before, $K_a = 1.25$ and $\alpha = 20$.

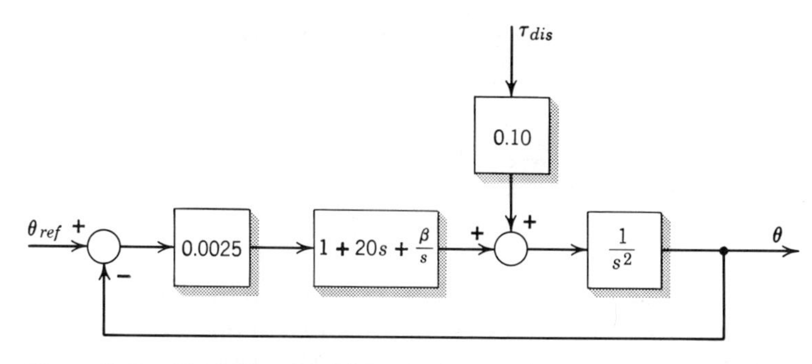

Figure 10.23. Block diagram with P + I + D control.

The steps necessary to isolate the parameter β in the root-locus gain without having it affect the poles or zeros of $F(s)$ are shown in the sequence of block diagrams in Figure 10.24. From the form of the inner loop in part (*b*), we recognize that the poles of the transfer function of the upper block in part (*c*) are the roots of the system having the P + D

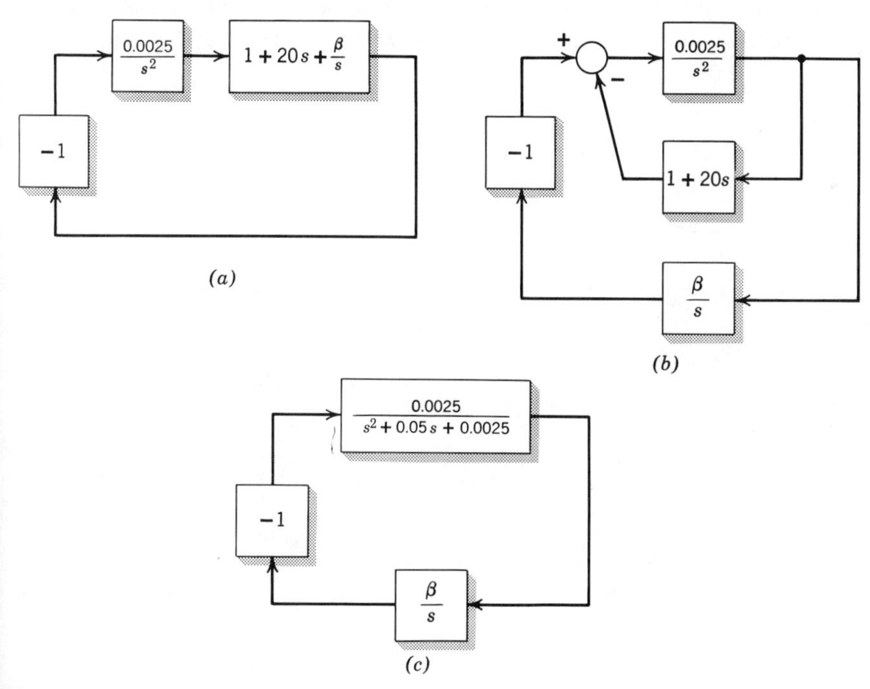

Figure 10.24. Manipulations leading to β as a root-locus gain.

controller with $\alpha = 20$; the pole at $s = 0$ of the transfer function of the lower block results from the integration operation in the controller.

Thus

$$G(s)H(s) = \underbrace{0.025\beta}_{K_{rl}} \underbrace{\left(\frac{1}{s}\right)\left(\frac{1}{s^2 + 0.05s + 0.0025}\right)}_{F(s)} \tag{4}$$

from which the root locus of Figure 10.25*a* is obtained by the application of the root-locus construction rules. Noting that two of the branches cross over into the RHP as β is increased, it is worthwhile to apply Routh's criterion to evaluate the upper limit on the integrator gain. First we write the characteristic polynomial $P(s)$ in the form of Eq. (1), Sect. 10.1, with β as a free parameter, namely,

$$\begin{aligned} P(s) &= s(s^2 + 0.05s + 0.0025) + 0.0025\beta \\ &= s^3 + 0.05s^2 + 0.0025s + 0.0025\beta \end{aligned}$$

Evaluating the Routh series and applying the condition for marginal stability, we find that all of the roots will be in the LHP if and only if $0 < \beta < 0.05$.

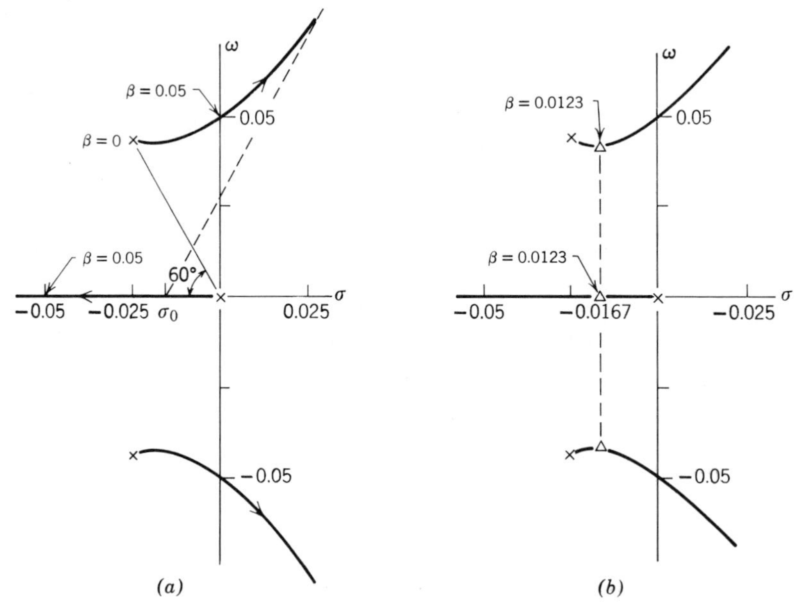

Figure 10.25. P + I + D control. (*a*) Root locus. (*b*) Root locations, $\beta = 0.0123$.

Since the number of poles exceeds the number of zeros by at least two, the sum of the roots will be independent of β. Thus it is reasonable to select β such that all three roots have the same real part, i.e., lie on a vertical line. A simple calculation will show that that line must pass through $s = -0.0167$, as indicated in Figure 10.25*b*. Measuring the lengths of the vectors to $s = -0.0167$ from the three poles, we find $|F(s)|_{s=-0.0167} = 32{,}400$. Applying the magnitude criterion, $1/K_{rl} = 1/0.025\beta = 32{,}400$ whence $\beta = 0.0123$.

The block diagram of the complete system, with proportional, derivative, and integral terms in the controller, is shown in Figure 10.26. The pointing

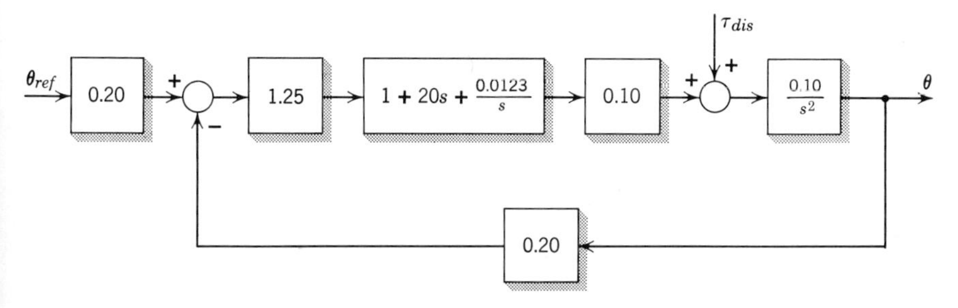

Figure 10.26. Complete block diagram, P + I + D control.

error in response to a ramp shift to a new value of θ_{ref} is shown in Figure 10.27*a*; from the closeness of the curves it is apparent that little has been sacrificed in the satellite's ability to follow changes in θ_{ref}. On the other hand, the all-important role of the integral term in reducing steady-state pointing errors caused by disturbance torques is demonstrated in part (*b*) of the figure.

Sensitivity to parameter variations

Because of the innate perversity of inanimate objects, a physical system never exactly agrees with the fixed linear mathematical model that may have been used in its design. The fact that the model itself is only an approximation to the real thing; tolerances in construction, deterioration due to aging, wear, corrosion, etc. are all factors contributing to such differences. Thus, the system designer should undertake a *sensitivity analysis* of the effects of parameter variations upon the performance of the overall system. Speaking strictly qualitatively, one of the principal

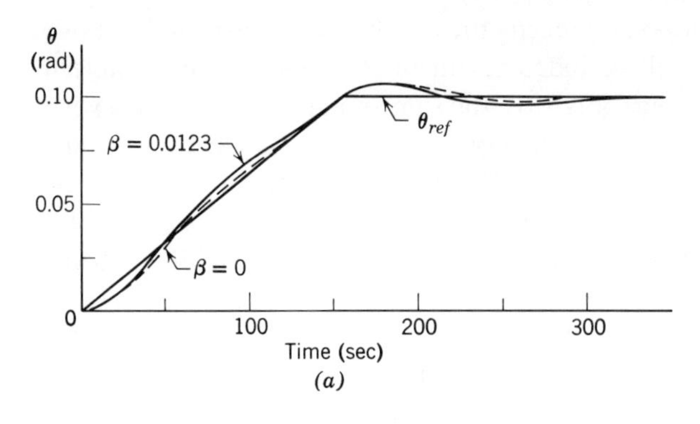

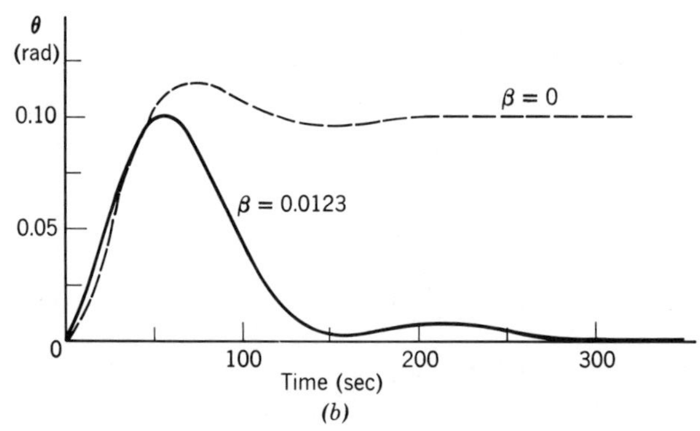

Figure 10.27. Simulated system responses, P + I + D control. (*a*) Change of reference input. (*b*) Step disturbance input.

advantages of using feedback is that the closed-loop performance can be made relatively insensitive to a wide range of parameter variations. In concluding this chapter, we shall demonstrate the use of the root locus and Routh's criterion to investigate the system's sensitivity to parameter variations.

Consider, for instance, the satellite's moment of inertia (I) which will certainly vary over a period of time as fuel is expended for attitude control and is likely to be affected by the deployment or reorientation of antennas. To investigate the effects of such variations the block diagram of Figure 10.26 may be redrawn with the moment of inertia left in literal form and all other parameters expressed by their design values. Doing this and expressing the controller transfer function as a rational function results in Figure 10.28.

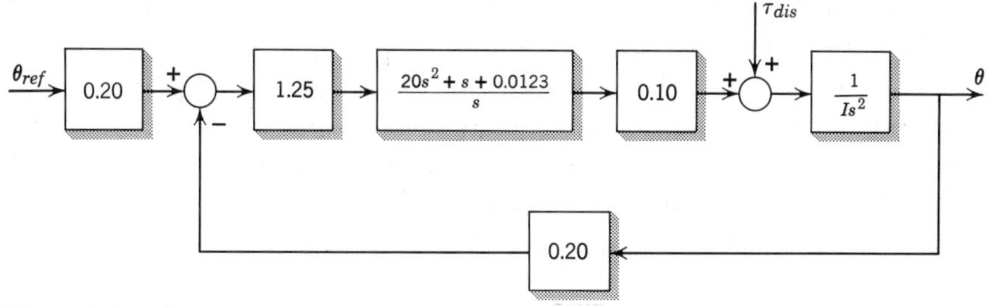

Figure 10.28. System diagram with moment of inertia as a parameter.

Writing the characteristic polynomial,

$$\begin{aligned} P(s) &= (Is^2)s + (1.25)(20s^2 + s + 0.0123)(0.10)(0.20) \\ &= Is^3 + 0.50s^2 + 0.025s + 0.000308 \end{aligned}$$

from which the Routh array can be found:

I	0.025
0.50	0.000308
$0.025 - 0.000616\,I$	0
0.000308	0

Since all terms in the first column will be positive for $0 < I < 40.6$ slug ft^2 we conclude that the only limitation insofar as the system's stability is concerned is that the moment of inertia must not undergo more than a fourfold increase – a rather improbable situation unless the vehicle undergoes a drastic change in its geometry.

Although not important from the standpoint of stability, variations in I of, say, $\pm 20\%$ will affect the transient response of the system by causing a shifting of the closed-loop poles. To assess these effects quantitatively, we can construct a root locus with I treated as the gain parameter – actually the gain will be $1/I$ because the moment of inertia is in the denominator of the loop transfer function.

Referring to the block diagram of Figure 10.28, we write

$$\begin{aligned} GH(s) &= (1.25)(0.10)(0.20)\left(\frac{20s^2 + s + 0.0123}{s}\right)\left(\frac{1}{Is^2}\right) \\ &= \underbrace{\frac{0.50}{I}}_{K_{rl}} \underbrace{\left(\frac{s^2 + 0.05s + 0.000616}{s^3}\right)}_{F(s)} \end{aligned}$$

from which the root-locus gain and $F(s)$ are readily identified†. The 180° root locus – we are certainly interested only in positive values of I – can be drawn as in Figure 10.29*a* and points along the locus can be calibrated in terms of I by applying the magnitude criterion. We note that the roots for $I = 10$ coincide with those of Figure 10.25*b* for $\beta = 0.0123$ because both situations correspond to the complete set of nominal parameters. For a 50% decrease or a 100% increase in I $(5 \leq I \leq 20)$, the

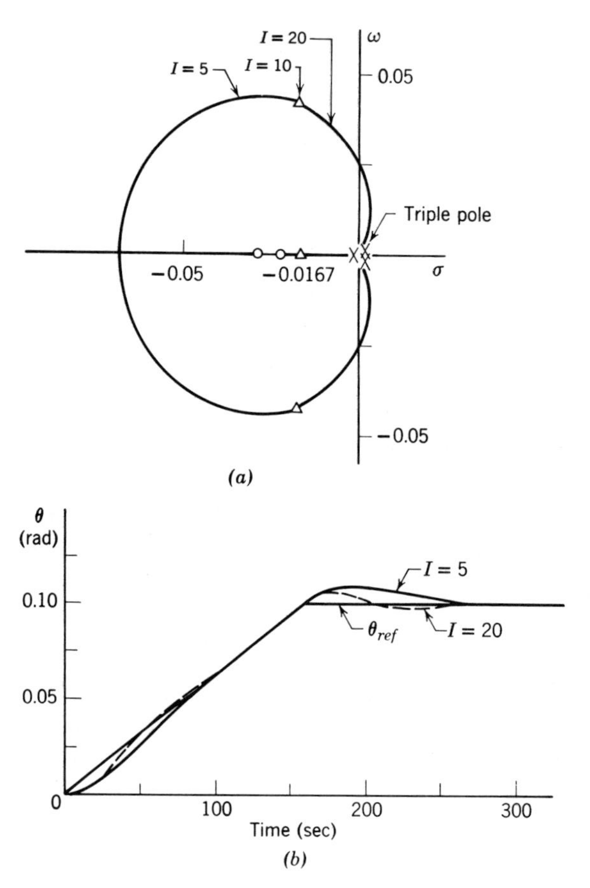

Figure 10.29. Variations in moment of inertia (I). (*a*) Root locus. (*b*) Response to change of reference input.

†If we do not mind having more zeros than poles in $F(s)$, we could also construct an equivalent root-locus problem with the gain proportional to I by working with $1/GH(s)$ rather than with $GH(s)$. See Problem 10.15.

system roots will be confined to the portions of the locus between the points indicated by $I = 5$ and $I = 20$. The corresponding responses to a change in θ_{ref} are shown in Figure 10.29*b* and substantiate the claim that the performance of the closed-loop system is relatively insensitive to variations in I.

To determine the effects of changes in I on the steady-state pointing error, we can repeat the steady-state analysis carried out in Section 9.3, with I as a free parameter and their nominal values assigned to the other parameters. However, recall that the principal result of adding the integral term to the controller was to drive the steady-state pointing error to *zero* when the satellite was subjected to a constant disturbance torque. This being the case, the value of I, or any other parameter for that matter, will not affect the steady-state error caused by a constant disturbance torque. If it is desired to analyze the steady-state effects of other inputs, such as a ramp disturbance, one can return to Eq. (13), Sect. 9.3, and apply the final-value theorem.

For the system we have been studying the components to which the response is most sensitive are likely to be the angular transducers that generate the voltages corresponding to $\theta_{ref}(t)$ and $\theta(t)$. In practice, the gains of these two devices will not be exactly 0.20 volts/degree and they probably will not be equal. Thus, the designer should analyze the effects of variations in the two gains upon the transient and steady-state behavior of the system. The results of such a study will show that if each gain changes by the same factor, such that their ratio is still unity, the effect on the response will not be important. However, if the ratio of the gains differs from unity, there will be a corresponding steady-state pointing error whenever $\theta_{ref} \neq 0$.

Problems

10.1 Assuming that $F(s)$ has the form of Eq. (6), Sect. 10.1, evaluate arg $[F(s)]$ and $|F(s)|$ at the points labeled s_0 for the pole-zero patterns shown in Figure P10.1.

10.2 For the open-loop transfer function $GH(s) = K(s+1)/s^2(s+2)$ plot the root locus for $K \geq 0$. The plot should include the large-gain asymptotes,

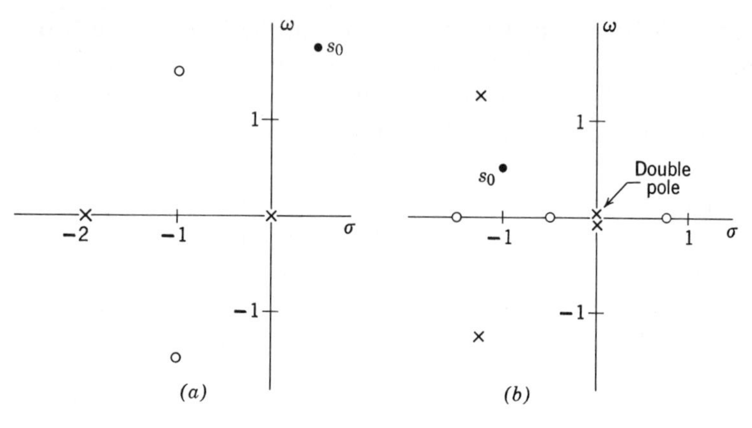

Figure P10.1

angles of departure and arrival computed and shown to within $\pm 5°$, and at least one point on the locus located by trial-and-error application of the angle criterion.

10.3 Repeat Problem 10.2 for $GH(s) = K(s+1)/s^4$.

10.4 Repeat Problem 10.2 for

$$GH(s) = \frac{K(s+1)^2}{(s^2+4)(s+2-j2)(s+2+j2)}$$

Find the approximate value of K for which the branches intersect.

10.5 Repeat Problem 10.2 for

$$GH(s) = \frac{K(s+0.5)}{(s^2-1)(s^2+2s+5)}$$

Find the values of K for which the loci cross the imaginary axis. Is there a value of K for which the system can be made asymptotically stable?

10.6 Repeat Problem 10.2 for $GH(s) = K(s+1)/(s+10)(s^2+4)$.

10.7 Repeat Problem 10.2 for

$$GH(s) = \frac{K(s+5)}{s[(s+2)(s+1)+4]}$$

10.8 Repeat Problem 10.2 for

$$GH(s) = \frac{K(s^2+1)(s^2+9)}{s(s^2+4)(s^2+16)}$$

10.9 Repeat Problem 10.2 for

$$GH(s) = \frac{K(s^2+4)(s^2+9)}{s(s^2+1)(s^2+16)}$$

10.10 (a) Sketch the 180° root locus for the system having

$$G(s) = \frac{K(s+3)}{s(s^2+4s+5)} \qquad H(s) = \frac{1}{s+1}$$

(b) For what values of K do the branches of the root locus leave or meet the real axis? (c) Find the closed-loop roots for $K = 1$ and write the transfer function $T_R(s)$ as a ratio of *factored* polynomials.

10.11 Answer the following questions about the system shown in Figure P10.2, using root-locus sketches to support your answer. (a) Explain why the system will be unstable for all $K > 0$ when $G_1(s)$ represents a P+I controller. (b) Explain why a P+I+D controller can be found such that the system will be stable for some range of $K > 0$.

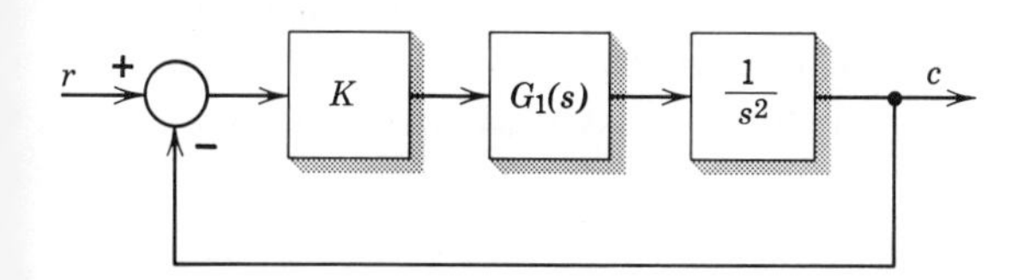

Figure P10.2

10.12 Assume that the system shown in Figure P10.3 must have $K \geq 4000$ in order to meet certain steady-state error requirements.

(a) Show that it will be unstable for any such value of K if $H(s) = 1$.

(b) Find a feedback transfer function $H(s)$ such that $\lim_{s \to 0} H(s) = 1$, that is, the feedback gain is unity for low frequencies, and the system is stable for $K = 4000$.

(c) Determine the closed-loop roots and transfer function.

10.13 The 180° root locus for $F(s) = 1/[s(s+2)]$ has two branches that intersect at $s = -1$. (a) Evaluate the expansion of Eq. (3), Sect. 10.3 at $s = -1/2$ and draw the s- and F-plane plots corresponding to Figure 10.11. Identify the angles ψ_1 and ψ_2 referred to in the text. (b) Evaluate the expansion of Eq. (4), Sect. 10.3 at the intersection point and draw the appropriate s- and F-plane plots.

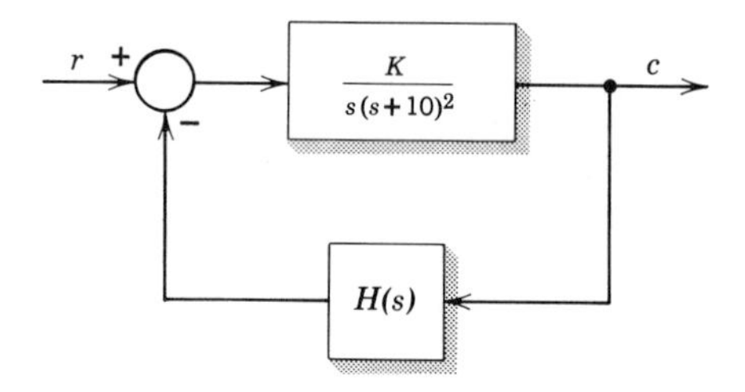

Figure P10.3

10.14 A control system with $G(s) = K/[(s+1)(s+3)]$ and $H(s) = 1+0.5s$ has had the gain selected so as to make the dominant root be $s=-1.5$. (a) Find K and the location of the second root. Why are we justified in referring to $s=-1.5$ as the *dominant* root? (b) Unfortunately, when the control system was put into service, the gain K was of the wrong sign. Describe the kind of behavior that most likely resulted.

10.15 Prove that the root locus corresponding to the loop transfer function $GH(s)$ is identical to that corresponding to $1/GH(s)$, where the root-locus gains are the reciprocals of one another. Note that the poles and zeros are interchanged. Demonstrate the validity of this assertion as applied to the system discussed in Example 10.5.

10.16 (a) List and justify the changes that must be made in the root-locus rules listed in Section 10.1 in order to handle functions $F(s)$ having more zeros than poles, that is, $m > n$? (b) Use your revised rules to construct the 180° locus for $F(s) = s^2(s+10)/(s+1)$.

10.17 The 180° and 0° root loci are merely two members of an infinite set of loci that can be drawn for a given function $F(s)$. For $F(s) = (s-1)/(s+1)$ draw on one set of axes the loci for angle criteria of 0, 45, 90, 135, and 180 degrees.

10.18 Prove that the root locus of a system having a double real pole and a single real zero must consist of a circle centered at the zero and passing through the double pole, in addition to the real-axis portion. *Hint:* Assume a test point s_0 anywhere on the circle except where it crosses the real axis and show that the angle criterion is satisfied by using the vector representation of $F(s_0)$. *Extension:* Allow the poles to be real and distinct or, more generally, complex conjugates.

10.19 Adapt the root locus to find the roots of the following equations:
(a) $2s^3+s^2-4s+9=0$. *Hint:* The desired roots are also the roots of the equation $1+K(s^2+9)/(2s^3-4s)=0$ when $K=1$.

(b) $s^4 + 4s^3 + 2s^2 + 4s + 1 = 0$.

(c) $4s^5 - s^4 + 2s^2 + 5 = 0$.

10.20 The root locus can be drawn for a feedback system involving a time delay of T sec. by noting that $|e^{-sT}| = e^{-\sigma T}$ and arg $[e^{-sT}] = -\omega T$, where e^{-sT} is the transfer function of the time delay and $s = \sigma + j\omega$. (a) Draw the root locus for the system having $G(s) = K/s$ and a delay of T sec. in the feedback path. Show that the system will be stable for $0 < K < 0.5\pi/T$. (b) Repeat for $G(s) = K/s^2$ and show that the system is unstable for all $K > 0$.

10.21 A feedback system whose open-loop transfer function is $\overline{GH}(s) = 4/[s(s+a)]$ has a nominal value of $\bar{a} = 2$ but is subject to occasional shifts of up to $\pm 50\%$. Thus the actual $GH(s)$ can be written as $GH(s) = 4/[s(s+2+\Delta)]$ where $-1 \leq \Delta \leq 1$.

(a) Formulate a root-locus problem such that Δ appears only as the root-locus gain and the poles and zeros are fixed. *Answer:* $\dfrac{\Delta s}{s(s+2)+4} = -1$.

(b) Draw the root locus corresponding to the results of part (a) and indicate the range over which the closed-loop roots can move for $\pm 50\%$ changes in the value of a.

(c) Construct the locus of the original system for one value of $a \neq 2$ and compare the root locations with those predicted in part (b).

10.22 Repeat Problem 10.21 for the open-loop transfer function

$$GH(s) = \frac{75(s+b)}{s(s+1)(s+10)}$$

where $b = 3 + \Delta$, with $-2 \leq \Delta \leq 2$.

10.23 Consider the system model shown in Figure P10.4 where $D(s)$ is the denominator polynomial of $G(s)$. (a) Explain why a root locus can not be drawn for arbitrary variations in the parameter α. (b) Assuming that α has a nominal value of $\bar{\alpha}$ so $\alpha = \bar{\alpha} + \Delta\alpha$ and that the increment $\Delta\alpha$ is "small," show that the movement of the system's roots in the s-plane can be found

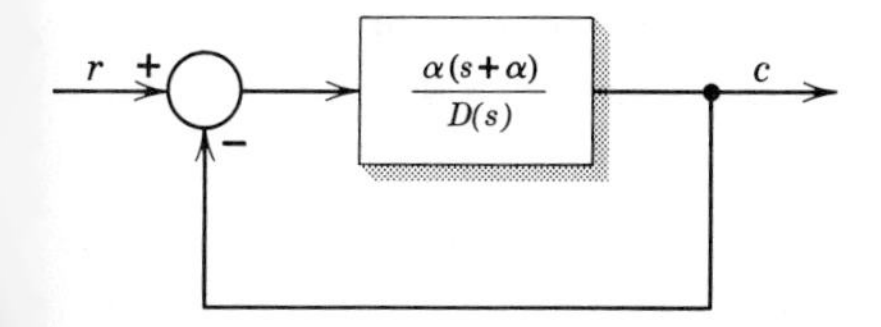

Figure P10.4

by drawing the root locus corresponding to the equation

$$1+\frac{\Delta\alpha(s+2\bar{\alpha})}{D(s)+\bar{\alpha}(s+\bar{\alpha})}=0$$

10.24 (a) Draw the root locus of the system shown in Figure P10.5 with $H(s)=1$ (unity feedback) and determine the maximum value of K for stability. What is the frequency of oscillation when K is at its upper limit?

(b) Let $H(s)=s+2$ (P + D feedback) and draw the root locus. Find the value of K such that the damping ratio of the complex roots is $\zeta=0.707$ and locate all the closed-loop roots.

(c) Repeat part (b) with $H(s)=(s+2)/(s+20)$ (a lead network).

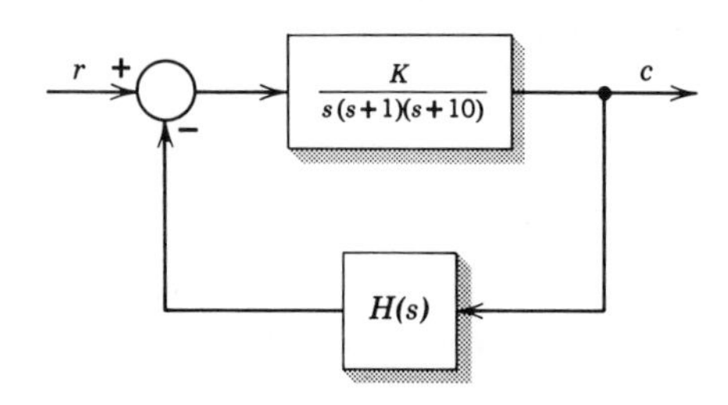

Figure P10.5

10.25 Use Routh's criterion to find the value of K such that the system in Figure P10.3 has imaginary roots for $H(s)=1$. At what value(s) of s does the locus enter the RHP?

10.26 The system shown in Figure P10.6 represents a third-order plant with unity feedback and a P + I control law.

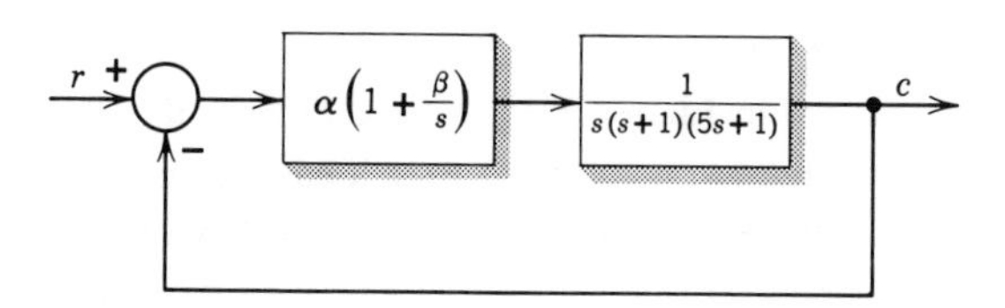

Figure P10.6

(a) Use Routh's criterion to determine all combinations of the controller gains α and β that will insure a stable system. Indicate these points in the α,β plane.

(b) Pick a combination of α and β (both >0) that lies on the stability boundary and draw the root locus with α as the gain. Apply the magnitude criterion to find the value of α giving marginal stability and compare it with the selected value.

10.27 Find values for the gains K and τ such that the system in Figure P10.7 will meet the specifications listed below. The error is defined as $\varepsilon = r - c$.

1. A unit ramp reference input produces a steady-state error not exceeding 0.40 in magnitude.
2. A unit step disturbance input produces a steady-state error less than 0.20 in magnitude.
3. Transients decay at least as rapidly as $e^{-0.5t}$.
4. $K \leq 150$.

Show that your selection (not unique) meets all criteria. Include a root-locus plot and estimate the closed-loop roots.

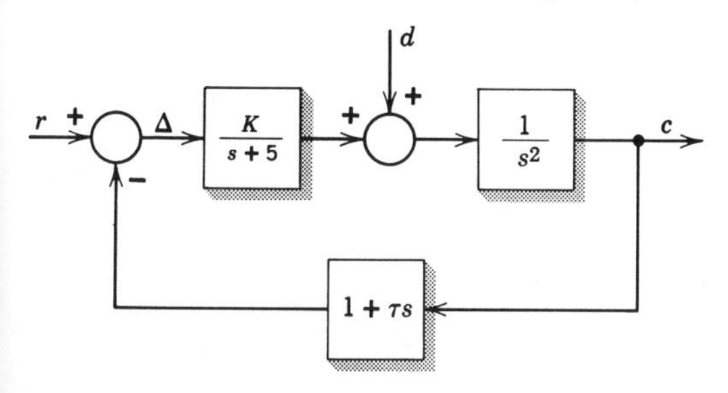

Figure P10.7

10.28 Repeat parts (a) and (b) of Problem 9.19 in which the inverted stick was stabilized, this time using root-locus techniques. In part (a), draw the 180° root locus, with α as the gain parameter. Apply the magnitude criterion to find the value of α that gives $\omega_n = 2\sqrt{g/L}$. In part (b), set α to the value found above and take β as the root-locus gain. Apply the magnitude criterion at the point yielding critically damped roots. *Answer:* (b) $\alpha = 11MgL/6$, $\beta = (4ML^2/3)\sqrt{g/L}$.

10.29 To make the airplane described in Problem 8.33 easier to fly the pitch angle (θ) and its derivative ($\dot{\theta}$) can be sensed and used to generate a portion of the desired elevator signal (η).

(a) Let $\eta = 0.5\dot{\theta} + \eta'$, where η' is now considered to be the input to the airplane with a pitch-rate feedback loop and find the transfer function $H(s)/\eta'(s)$. Use a root-locus plot and simulation diagram to show the effect on the plane's dynamics.

(b) Add a second loop by letting $\eta' = K\theta + \eta''$, where η'' is the input to the airplane with both feedback loops present. Repeat the analysis of part (a) and select a value for K to yield a damping ratio of $\zeta = 0.5$ for the complex roots. Present an argument as to why the input η'' is approximately proportional to the desired rate of climb $(\dot{H})$.

(c) Obtain a computer simulation of the airplane making a change in altitude under the three configurations analyzed.

11

Frequency-response analysis: Nyquist and Bode plots

The frequency-response techniques we found to be so useful for the analysis of cascade systems in Chapters 6 and 7 can be adapted to the study of closed-loop systems. In fact, the use of such methods to analyze feedback systems preceded the introduction of the root locus by more than 15 years. The basis of all frequency-response approaches to closed-loop systems is the *Nyquist criterion*, which leads to a variety of graphical techniques, two of which will be taken up here. One might question the need for studying such topics in light of the root locus and state-variable analysis. However, we have always assumed the existence of a linear mathematical model, either in the form of differential equations or the transfer functions of individual components. Unfortunately, it may not be practical to obtain a mathematical model of the system with the degree of precision required for the construction of a root locus or the solution of the state-variable equations. Likewise, there is little to be gained from a complete root locus when we are interested only in knowing how the natural frequency and damping ratio of a pair of dominant closed-loop poles depend on a gain parameter, rather than the exact locations of all of the closed-loop poles.

Basically, the frequency-response methods focus one's attention on the conditions that must exist in order for a feedback system to have one or more roots of its characteristic equation in the RHP, with the objective of assuring that the system parameters will not be allowed to satisfy those conditions. Because the boundary between asymptotically stable and unstable behavior usually corresponds to a pair of roots on the imaginary axis (a single root at $s = 0$ also defines the boundary but is much less common) and thus to the presence of undamped oscillations in the system's response, we shall begin by examining the requirements for such oscillations to exist. These conditions will be formalized as the Nyquist-stability test which is carried out by constructing a Nyquist plot.

Only after gaining some experience in using these tools will we prove the Nyquist criterion from which they stem. Following that, several topics that are particularly suited to the design of feedback systems, namely gain and phase margins, and Bode plots will be taken up and then applied to the satellite-attitude control problem studied in Chapters 9 and 10.

11.1 OSCILLATIONS IN FEEDBACK SYSTEMS

Consider the single-loop system shown in Figure 11.1*a* which has an undamped oscillatory mode at the frequency $\omega = \omega_{osc}$, with the remainder of its modes being damped. Following an input of finite duration or a nonzero initial state, every signal within the system will become a sinusoidal oscillation at ω_{osc} once the asymptotically stable modes have decayed. To examine this steady-state situation, we can model the system by only the closed-loop diagram in Figure 11.1*b* where $KF(s) = G(s)H(s)$, as used throughout Chapter 10, and the block with gain -1 accounts for the minus sign of the feedback summing junction.

Now suppose we mentally open the loop, thereby considering $x(t)$ to be an input signal and $y(t)$ the corresponding output. Clearly, the closed-loop system can be oscillating only if conditions are such that $x(t)$ and $y(t)$ are sinusoids with $\omega = \omega_{osc}$ and have identical amplitude and phase. Intuitively, then, if we construct the system shown in Figure 11.1*c* and excite it with a sinusoid at the frequency ω_{osc} and an amplitude such that

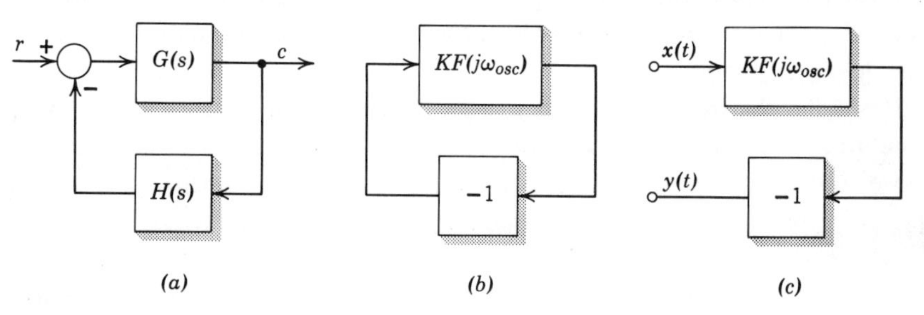

Figure 11.1.

the output has the same amplitude and a phase shift of $-360°$ (or any integer multiple of 360°) the input and output can be connected with $y(t)$ now providing the excitation signal $x(t)$. Such a connection will result in the system of Figure 11.1*b* operating in the oscillatory mode, which is precisely the condition whose existence we wish to check.

Thus, the problem of detecting the possibility of oscillations can be reduced to one of finding those frequencies and positive values of K for which $x(t) = y(t)$. Letting $x(t) = A \sin \omega_{osc}t$, the signal appearing at the output in Figure 11.1*c* will be

$$y(t) = -KA|F(j\omega_{osc})| \sin (\omega_{osc}t + \arg [F(j\omega_{osc})])$$

Absorbing the minus sign as an odd multiple of 180° in the argument of the sine function and equating the input and output gives

$$\overbrace{A \sin \omega_{osc}t}^{x(t)} = \overbrace{KA|F(j\omega_{osc})| \sin (\omega_{osc}t + q180° + \arg [F(j\omega_{osc})])}^{y(t)}$$

where q is any odd integer. Hence, we obtain the following two conditions which must be satisfied simultaneously:

$$\arg [F(j\omega_{osc})] = q180° \qquad q \text{ odd} \tag{1}$$

$$\frac{1}{K} = |F(j\omega_{osc})| \tag{2}$$

Because (1) does not involve the adjustable gain K, it can be solved, generally by trial and error, for all possible frequencies at which the system might oscillate (there may be multiple solutions). Then, for each value of ω_{osc} satisfying (1), a unique value of K may be found which will satisfy (2). If the reader refers back to the angle and magnitude criteria for the 180° root locus, Eqs. (13) and (14), Sect. 10.1, respectively, he will see that (1) is just the condition for the points $s = \pm j\omega_{osc}$ to be on the root locus and (2) is the requirement on K for those points to actually be roots of the closed-loop characteristic equation. Hence, the conditions for the existence of an undamped oscillatory mode must be closely related to the behavior of the system's root locus on the imaginary axis of the s-plane. With this in mind, it should not be surprising that the conditions necessary to insure stability, e.g., the maximum and minimum values of K, can be precisely determined from a knowledge of only the *frequency-response* characteristics of the *open-loop* system. In fact, the frequency-response function $GH(j\omega)$ need not even be obtained in factored form. This means that experimental frequency-response data can be used directly in a stability analysis, without a knowledge of the individual poles and zeros of $G(s)$ and $H(s)$.

As the first step in our investigation, let us interpret (1) and (2) in a second complex plane, the F-plane.

The *F*-plane

Because $F(s)$ is a complex function of the complex variable s, we can consider the functional relationship as a mapping† of any point in the s-plane (the domain) into a unique point in the F-plane (the range). Thus, $F(j\omega)$ is the locus of points in the F-plane into which the positive imaginary axis of the s-plane maps. Figure 11.2 shows the respective planes, and the locus into which the upper half of the s-plane imaginary axis maps is sketched for a typical function $F(s)$, where the arrows indicate the corresponding directions of travel. Because the functions $F(s)$ with which we are dealing have real coefficients, evaluating the function at the complex conjugate of its argument yields the complex conjugate of the function value, i.e., $F(s^*) = F^*(s)$. In geometric terms, the lower half of the imaginary axis in the s-plane will map into the image relative to the real axis of the locus in Figure 11.2*b*.

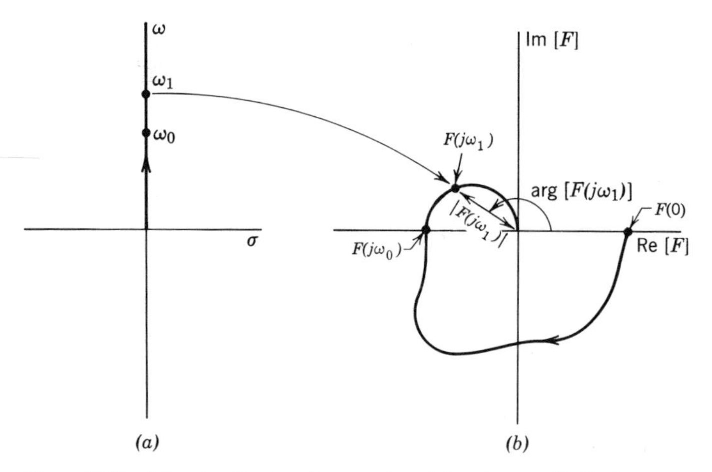

Figure 11.2. Mapping the imaginary axis. (*a*) s-plane. (*b*) F-plane.

†See Appendix A for further discussion.

Having obtained the plot of $F(j\omega)$, it is a simple matter to ascertain whether steady-state oscillations can possibly exist in the feedback system for which $KF(s)$ represents the open-loop transfer function by graphically checking for solutions of (1) and (2). First, if the phase or argument condition (1) is to be satisfied for finite ω, then the plot of $F(j\omega)$ must cross the negative real axis of the F plane because these are the only points in the F-plane possessing arguments that are odd multiples of 180°. In Figure 11.2*b*, for instance, $F(j\omega_0)$ is the only such point, and ω_0 is the only candidate for the frequency of oscillation. Second, in order for such an oscillatory mode to exist, the gain K must equal the reciprocal of the magnitude of $F(j\omega_0)$.

Thus, based on just the information contained in Figure 11.2*b*, we can conclude that the system represented by $KF(s)$ will not exhibit an undamped oscillatory mode if $K \neq 1/|F(j\omega_0)|$. However, we cannot as yet say anything about the *stability* of the feedback system; all we know at the moment is how to ensure that its zero-input response is not marginally stable. In the following section we shall present a test that uses the plot of $F(j\omega)$, plus a little additional information, to provide a rigorous stability test. Before getting to the theory, let us demonstrate the mechanics of making and interpreting such a plot.

Example 11.1

The second-order system diagrammed in Figure 11.3*a* has

$$F(s) = \frac{1}{(s+1)(s+2)}$$

whence

$$F(j\omega) = \frac{1}{(j\omega+1)(j\omega+2)} = \frac{(2-\omega^2)-j3\omega}{\omega^4+5\omega^2+4}$$

Identifying its real and imaginary parts, $F(j\omega)$ can be plotted in the F-plane for $\omega \geq 0$, resulting in Figure 11.3*b*. To assist in the plotting, one can readily see that $F(0) = \frac{1}{2} = 0.5e^{j0°}$ and $F(j\omega) \to 0$ as $\omega \to \infty$. Since $F(s)$ has two poles and no finite zeros, the argument of $F(j\omega)$ will approach $-180°$ as $\omega \to \infty$, with each pole contributing $-90°$ in the limit. Hence, the plot of $F(j\omega)$ must approach the origin of the F-plane tangent to the negative real axis. Furthermore, it is apparent that the argument of $F(j\omega)$ cannot reach $-180°$ for any finite value of ω, implying that the plot of $F(j\omega)$ does not cross the negative real axis.

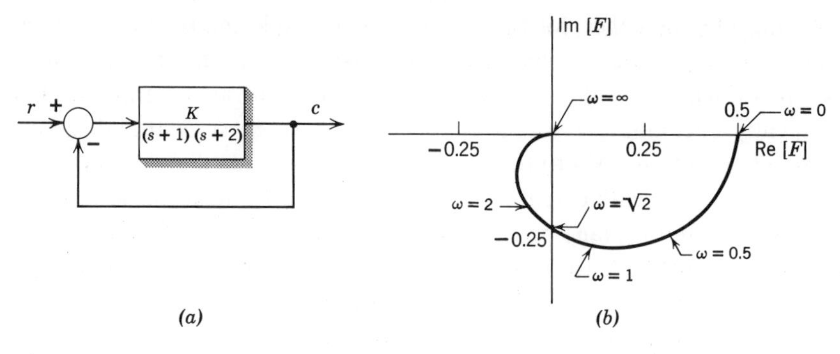

Figure 11.3. A second-order system. (*a*) Block diagram. (*b*) Locus of $F(j\omega)$, $\omega \geq 0$.

As a consequence of the above, there are no finite values of ω for which the phase criterion (1) will be satisfied; therefore it is not possible to find a positive value of K for which the closed-loop system will have an oscillatory mode. Although we can say no more at the moment, a sketch of the root locus will indicate that the system is asymptotically stable for all $K > 0$.

Example 11.2

As an illustration of a system that does have the potential for oscillatory behavior, consider the block diagram of Figure 11.4*a*, which is the system of the previous example with the addition of a pole at $s = -3$ in the feedback transfer function – perhaps a lowpass transducer having a bandwidth of 3 rad/sec.

From the block diagram we see that $F(s) = 1/[(s+1)(s+2)(s+3)]$ whence $F(0) = \frac{1}{6}$ and $F(s) \to 0$ as $s \to \infty$. Rather than writing $F(j\omega)$ and trying to find its real and imaginary parts analytically, it is easier to construct the s-plane diagram corresponding to $F(s)$ and draw the vectors from the poles $s = -1$, -2, and -3 to an arbitrary point on the imaginary axis. Doing so yields Figure 11.4*b*, from which we conclude that

$$\arg\,[F(j\omega)] = -\left(\arctan \omega + \arctan \frac{\omega}{2} + \arctan \frac{\omega}{3}\right) \tag{3}$$

which starts at 0° for $\omega = 0$ and decreases monotonically toward −270° as $\omega \to \infty$. Similarly, the s-plane plot shows that $|F(j\omega)|$ decreases monotonically from $\frac{1}{6}$ to zero as ω goes from zero to infinity.

The only other point on the plot of $F(j\omega)$ of particular *quantitative* interest is the point at which the curve crosses the negative real axis, i.e., at which $\arg[F(j\omega)] = -180°$. Denoting this frequency as ω_0 and using (3) we have

$$\arctan \omega_0 + \arctan \frac{\omega_0}{2} + \arctan \frac{\omega_0}{3} = 180°$$

which can be solved by trial and error to yield $\omega_0 = 3.32$ rad/sec. Having found ω_0, $|F(j\omega_0)|$ can be computed directly and is 1/60. The plot of $F(j\omega)$ for $\omega \geq 0$ is shown in Figure 11.4*c* and differs from that of the

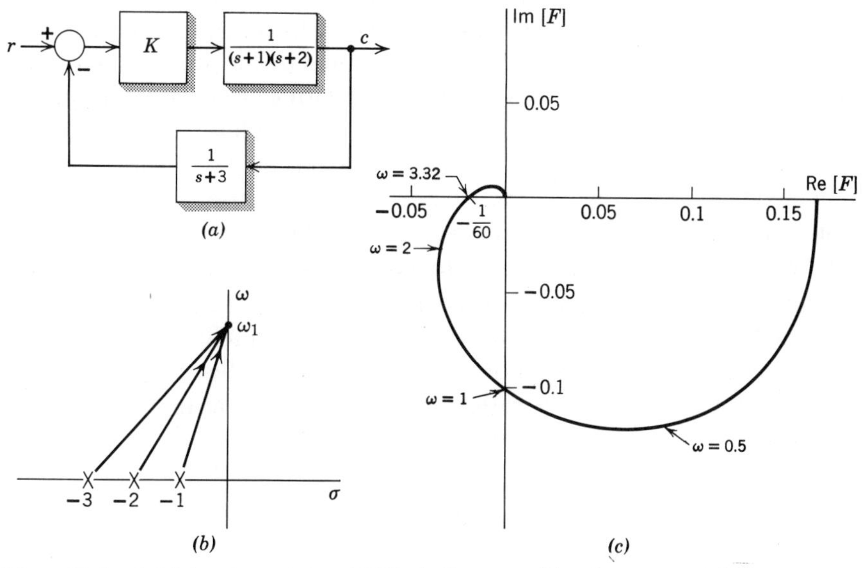

Figure 11.4. A third-order system. (*a*) Block diagram. (*b*) *s*-plane construction. (*c*) Locus of $F(j\omega)$, $\omega \geq 0$.

previous example primarily in that it crosses the negative real axis. Thus, if $K = 1/|F(j\omega_0)| = 60$, the closed-loop system will have an oscillatory mode with a frequency of 3.32 rad/sec, which is to say that it will have a pair of poles at $s = \pm j\omega_0 = \pm j3.32$.

11.2 THE NYQUIST STABILITY TEST

The Nyquist stability test is an *analysis* tool for determining whether a closed-loop system will be asymptotically stable which, with some modest additions, can be used in the *synthesis* of feedback systems to guide the selection of both control laws and parameter values. Before applying the stability test one must construct the Nyquist plot corresponding to the dynamic portion of the open-loop transfer function, $F(s)$.

Nyquist plots

The Nyquist plot is a simple extension of the plot of $F(j\omega)$ discussed in the previous section. Specifically, we define the s-plane contour Γ_s shown in Figure 11.5*a*, which consists of a finite semicircle in the RHP of radius r and the connecting portion of the imaginary axis, such that the entire RHP is enclosed as $r \to \infty$. For any value of r the function $F(s)$ will map Γ_s into a closed contour in the F-plane, denoted by Γ_F, provided only that Γ_s does not pass directly through any singularities (typically imaginary-axis poles) of $F(s)$. A system's Nyquist plot is the limit as $r \to \infty$ of the F-plane contour into which Γ_s is mapped, as indicated in Figure 11.5*b*. It is conventional to traverse the s-plane contour in a clockwise direction, thus defining the direction of travel around the Nyquist plot (note the arrows). If $F(s)$ should happen to possess one or more poles on the imaginary axis of the s-plane, we merely incorporate a semicircular detour of infinitesimal radius around the pole, passing either to its left or right; no modifications are required in the stability test itself. (See Example 11.5.)

Comparing Figure 11.5*b* with Figure 11.2*b*, we note that the plot of

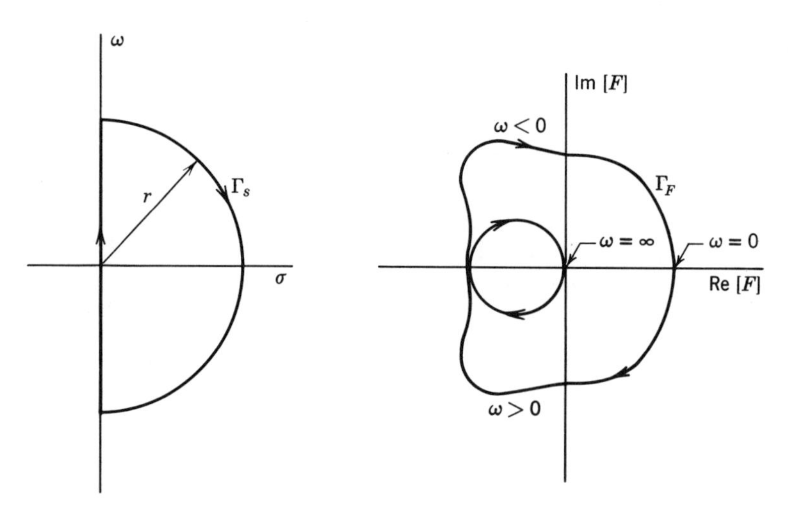

Figure 11.5. (*a*) s-plane contour. (*b*) Typical Nyquist plot.

$F(j\omega)$ for $\omega \geq 0$ contains the essential features of the complete Nyquist plot due to *1.* the symmetry property of $F(s)$ and *2.* the fact that the entire semicircle in Figure 11.5a maps into the origin of the F-plane for this particular example. For this reason, the somewhat heuristic discussion of Section 11.1 regarding the existence of sustained oscillations gave almost all of the information needed for a complete determination of stability.

Before stating the full requirements for stability, we define the number of encirclements of a point s_0 made by a closed contour. Consider the vector from s_0 to a point s_C which is allowed to move around the contour. The number of counterclockwise encirclements is $1/2\pi$ times the net increase in the argument of the vector as s_C makes one full circuit of the contour in the positive sense. For example, a net decrease of 4π radians would imply -2 counterclockwise or $+2$ clockwise encirclements.

The Nyquist stability test

A feedback system whose open-loop transfer function $KF(s) = G(s)H(s)$ has n_p poles within Γ_s as $r \to \infty$ will be stable *if and only if* its Nyquist plot Γ_F encircles the point $-1/K$ in a counterclockwise sense exactly n_p times.

Rather than elaborate further or proceed to its derivation, we shall demonstrate the application of the test.

Example 11.3

To ascertain the range of K for which the second-order feedback system discussed in Example 11.1 will be stable, we draw the complete Nyquist plot and indicate a typical location of the point $-1/K$ for $K > 0$. Noting that $F(s) \to 0$ along the semicircular portion of Γ_s as $r \to \infty$ and that the Nyquist plot must be symmetric about the real axis of the F-plane, Figure 11.3b leads directly to Figure 11.6a. Inspection of the block diagram in Figure 11.3a reveals an absence of poles of $F(s)$ within the s-plane contour Γ_s; hence, $n_p = 0$. Applying the stability test, we require that the point $-1/K$ not be encircled by Γ_F if the feedback system is to be stable. For any finite, positive value of K, $-1/K$ will lie to the left of the entire contour, thereby precluding its encirclement and ensuring a stable system for $K > 0$.

If we desire to consider *negative* values of K we merely allow the point $-1/K$ to lie on the positive real axis of the F-plane, noting the two possibilities shown in Figures 11.6b and c, depending on the value of K relative

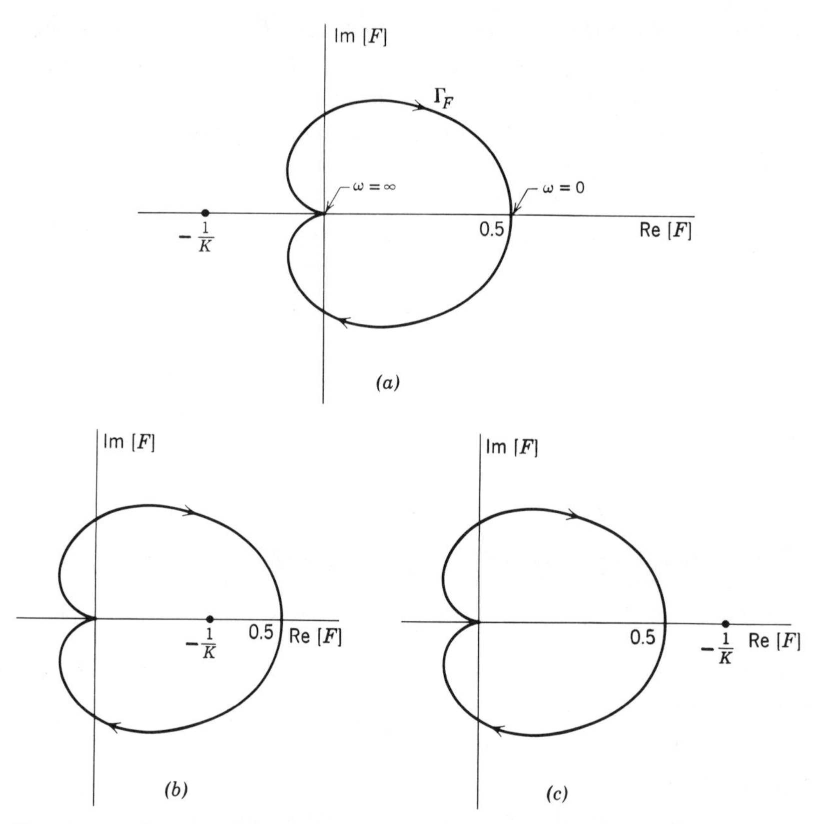

Figure 11.6. Complete Nyquist plot developed from Figure 11.3. (*a*) $K > 0$ (stable). (*b*) $K < -2$ (unstable). (*c*) $-2 < K < 0$ (stable).

to -2. In part (*b*) the Nyquist plot makes one clockwise encirclement of the point $-1/K$ and the system is unstable since $n_p = 0$. In part (*c*), $-1/K$ again lies outside the contour, implying stability. Taking the three parts of Figure 11.6 together, the conclusion is reached that the system will be stable for $-2 < K < \infty$. The reader may verify this result by making a sketch of the root locus or by applying Routh's criterion.

Example 11.4

Having plotted $F(j\omega)$ over positive ω in Example 11.2 for the third-order system shown in Figure 11.4*a*, it is a simple matter to obtain the complete Nyquist plot in Figure 11.7. Because $F(s)$ has all three of its poles in the LHP, $n_p = 0$, and the Nyquist stability test indicates that the closed-loop

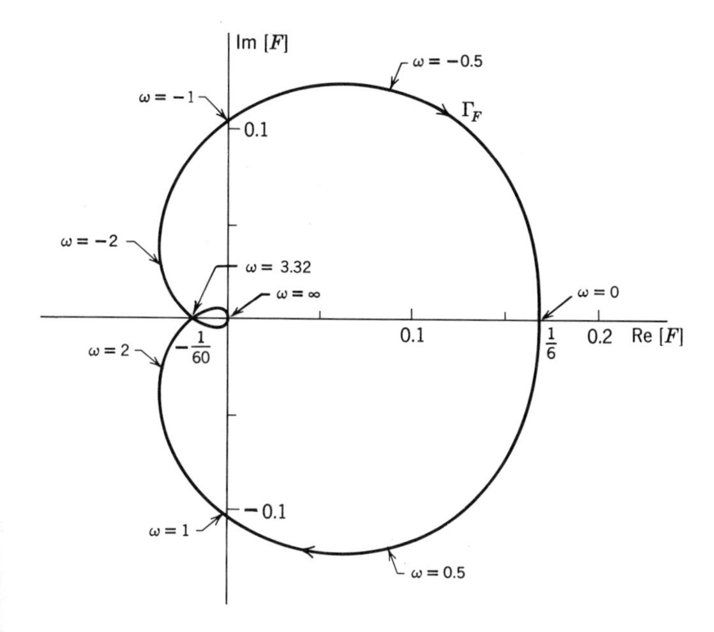

Figure 11.7. Complete Nyquist plot developed from Figure 11.4.

system will be stable if and only if Γ_F does not encircle the point $-1/K$. This will be the case if $-1/K < -\frac{1}{60}$ or $-1/K > \frac{1}{6}$. Thus, the closed-loop system is stable for $-6 < K < 60$. For $K > 60$ there are two clockwise encirclements of the point $-1/K$ and for $K < -6$ there is a single clockwise encirclement. Later, it will be shown that these situations indicate a pair of RHP poles in the former case and a single RHP pole in the latter.

Next, let us consider the results of applying the other two stability tests that we covered in the last chapter, namely, the root-locus and Routh's criterion. Writing the characteristic equation from the block diagram with K as a parameter and constructing the Routh array leads to the Routh series

$$1 \qquad 6 \qquad \frac{60-K}{6} \qquad 6+K$$

each term of which must be positive in order to achieve asymptotic stability. Clearly, the requirement on K is that $-6 < K < 60$. Furthermore, when $K = 60$, the next-to-last term in the series vanishes, with the remainder positive, which is the condition for a stable system having imaginary roots at

$$s = \pm j \sqrt{\frac{6+K}{6}}\Bigg|_{K=60} = \pm j\sqrt{11} = \pm j3.32$$

as follows from Eq. (5), Sect. 10.4. Hence, the Routh array allows us to make an independent evaluation of the point on the Nyquist plot at which $\arg[F(j\omega)] = -180°$ in terms of magnitude and frequency.

Finally, the 180° $(K > 0)$ and 0° $(K < 0)$ root loci can be sketched and are shown in Figure 11.8. From part (*a*) we conclude that the system will

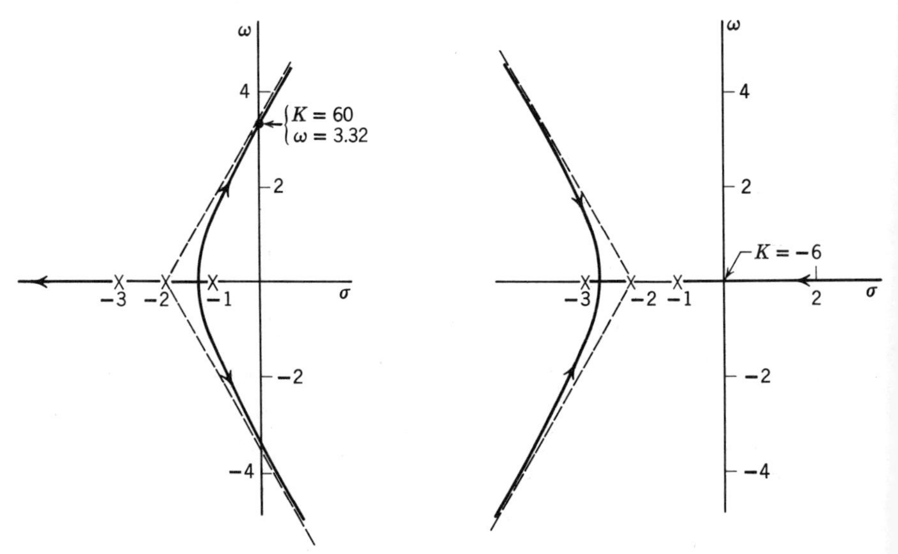

Figure 11.8. Root locus for system in Figure 11.4. (*a*) 180° locus $(K > 0)$. (*b*) 0° Locus $(K < 0)$.

be asymptotically stable for $0 \leq K < 60$, with two branches crossing into the RHP at $s = \pm j3.32$ for $K = 60$. From part (*b*) it is apparent that the system is asymptotically stable for $-6 < K \leq 0$, with a real root crossing into the RHP at $s = 0$ for $K = -6$.

From the application of these three methods to the same problem we can draw several conclusions regarding their relative complexities and the information they yield. For instance, the use of Routh's criterion involves looking only at the three points in Figure 11.8 at which one of the root-locus branches crosses from the LHP into the RHP. Its application takes the least amount of effort but provides the least amount of information about the overall properties of the closed-loop system. The Nyquist criterion expands the point of view to include the entire imaginary axis of the s-plane and, using related topics, some of which are covered in Section 11.4, can furnish a good deal of information about the closed-loop frequency response. Finally, the root locus involves a consideration of the entire s-plane which, in general, will provide the greatest insight of the three into the response of the closed-loop system, but at a greater expenditure of effort or imposing more constraints upon the mathematical model, e.g., requiring $F(s)$ in factored form.

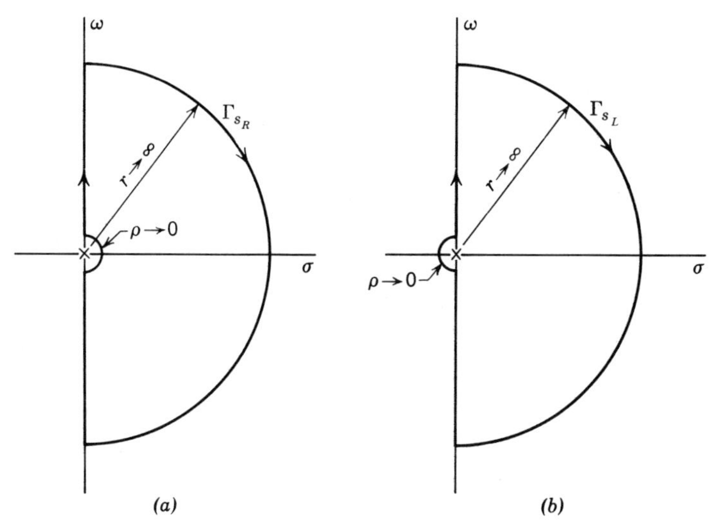

Figure 11.9. Detouring around a pole at the origin.

Imaginary-axis poles

Because the contour Γ_s must avoid all poles of $F(s)$, modifications in the contour are required when $F(s)$ has one or more poles on the imaginary axis. The basic trick, of course, is to take a small detour around the pole, two alternatives being depicted in Figure 11.9 for a pole at $s = 0$. It is this case we shall examine in some detail, and generalize therefrom.

To determine the behavior of the Nyquist plot in the F-plane when Γ_s is close to the pole at the s-plane origin, we substitute $s = \rho e^{j\theta}$ into $F(s)$ and establish the relationship between $\theta \triangleq \arg\,[s]$ and $\arg\,[F]$ as $\rho \to 0$ (because $s = 0$ is a pole we already know that $|F(s)| \to \infty$). Assuming that $F(s)$ has a single pole at the origin, we write $F(s) = \tilde{F}(s)/s$ where

$\tilde{F}(0)$ is real and nonzero. Then, letting $s = \rho e^{j\theta}$, the small semicircle in the s-plane maps approximately into the semicircle

$$F(s) \approx \frac{1}{\rho e^{j\theta}} \tilde{F}(0) = \frac{\tilde{F}(0)}{\rho} e^{-j\theta} \tag{1}$$

where $-90° < \theta < 90°$ if the contour Γ_{s_R} is chosen and $90° < \theta < 270°$ for Γ_{s_L}.

Letting $\rho \to 0$ in order to include the entire right-half plane within the s-plane contour, $|F(s)| \to \infty$ and, because arg $[F]$ contains the additive term $-\theta$, Γ_F undergoes a $180°$ arc in exactly the opposite direction of Γ_s. Hence, if Γ_{s_R} is used (Figure 11.9*a*) Γ_F will undergo a clockwise change of $180°$. When n_p is evaluated in the process of applying the Nyquist stability test, the pole of $F(s)$ at $s = 0$ is not within Γ_{s_R}.

Alternately, if Γ_{s_L} is selected, the pole at $s = 0$ will lie within the contour (Figure 11.9*b*) and n_p is at least equal to 1. Although the Nyquist plots for the two options will be somewhat different from one another, the difference in the respective values of n_p will ensure that the stability test yields the identical conclusions regarding stability.

If $F(s)$ has a repeated pole at the origin, the same technique is used. For example, if a double pole is present, the term $1/s^2$ is factored out of $F(s)$ before letting $\rho \to 0$. It will turn out that $|F(s)|$ contains the factor $1/\rho^2$ and arg $[F(s)]$ has the term -2θ. Thus the radius of Γ_F still goes to ∞ (even faster than before) but arg $[F]$ will undergo a $360°$ change, again in the opposite sense to that of s. If Γ_{s_L} is selected, then n_p is at least equal to 2.

If $F(s)$ has a pair of imaginary poles at $s = \pm j\omega_1$, the s-plane contour must make two detours, as in Figure 11.10. In the vicinity of the upper pole, s is defined as $s = j\omega_1 + \rho e^{j\theta_1}$ and on the lower semicircle we take $s = -j\omega_1 + \rho e^{j\theta_2}$. Although the calculations are likely to be somewhat more complicated than for a pole at $s = 0$, the method is precisely the same.

Example 11.5

The system shown in Figure 11.11*a* has a single pole at the origin. To demonstrate the preceding arguments we shall apply the Nyquist stability test, using both of the contours in Figure 11.9. From the block diagram, $F(s) = 1/[s(s+2)(s+3)]$. Because there are three more poles than zeros, $F(s) \to 0$ as $s \to \infty$ and, as $s \to \infty$ along the positive imaginary

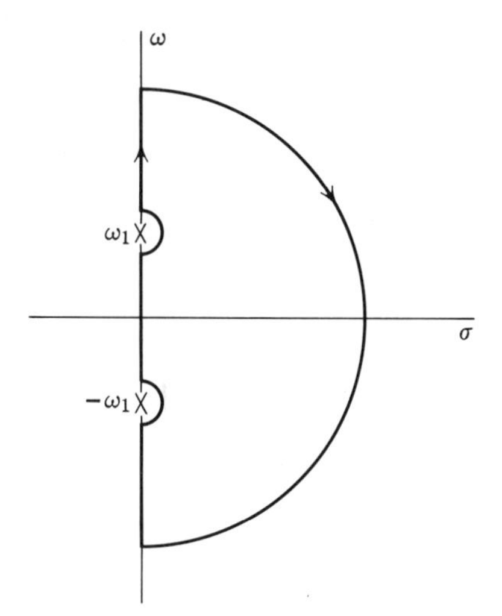

Figure 11.10. Detouring around a pair of imaginary poles.

axis, arg $[F(j\omega)] \to -270°$. Letting $s = j\omega$ it follows, after some manipulations, that

$$F(j\omega) = \frac{-5\omega + j(\omega^2-6)}{\omega[(\omega^2-6)^2+25\omega^2]} = \underbrace{\frac{-5}{\omega^4+13\omega^2+36}}_{\text{Re }[F]} + j\underbrace{\frac{\omega^2-6}{\omega^5+13\omega^3+36\omega}}_{\text{Im }[F]} \tag{2}$$

Thus the only values of ω for which the imaginary part of $F(j\omega)$ will vanish are $\omega = \pm\sqrt{6}$, in which case $F(j\omega) = -\frac{1}{30}$. Because of the factor of ω in the denominator of Im $[F(j\omega)]$, it will become infinite as $\omega \to 0$, behaving as $-1/6\omega$ in the process. On the other hand Re $[F(j\omega)]$ remains finite as $\omega \to 0$, approaching the limit $-\frac{5}{36}$.

As for the behavior near the pole at $s = 0$, $F(s)$ can be defined as $\tilde{F}(s) = 1/[(s+2)(s+3)]$ which gives $\tilde{F}(0) = \frac{1}{6}$. From (1) it follows that for either infinitesimal semicircle around $s = 0$ the approximation $F(s) \approx (1/6\rho)e^{-j\theta}$ holds, where θ is the argument of s along the small semicircle. Putting the various pieces together, the two possible F-plane contours shown in parts (*b*) and (*c*) of Figure 11.11 result. The contour in part (*b*), denoted by Γ_{F_R}, corresponds to taking the small semicircle to the right of the s-plane origin (Figure 11.9*a*) and $n_p = 0$ since both of the poles of $\tilde{F}(s)$ are in the LHP, i.e., there are no poles of $F(s)$ within the s-plane contour. Finally, noting that the F-plane contour Γ_{F_R} crosses the negative real axis at $-\frac{1}{30}$ and the positive real axis at infinity, K must be restricted

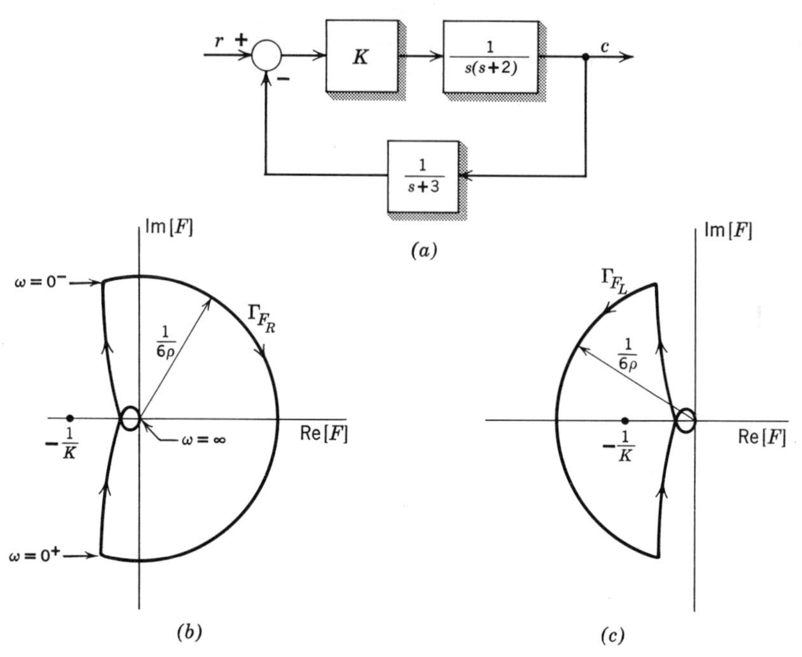

Figure 11.11. A third-order system with a single pole at the origin. (*a*) Block diagram. (*b*) Nyquist plot using Γ_{s_R}. (*c*) Nyquist plot using Γ_{s_L}.

to the interval $0 < K < 30$ in order to avoid an encirclement of the point $-1/K$ if the system is to be stable.

Using the alternate contour (Figure 11.9*b*) it follows that $n_p = 1$, thereby requiring one counterclockwise encirclement of the point $-1/K$ for stability. Reference to Figure 11.11*c* indicates that such is the case when the point $-1/K$ lies between $-\infty$ and $-\frac{1}{30}$, resulting in the same stability range as before.

11.3 DERIVATION OF THE NYQUIST STABILITY TEST

The Nyquist stability test is basically an application of a result from complex-variable theory known as the *argument principle*, the proof of which is carried out in Appendix A for the case of distinct poles and zeros. According to the argument principle, if $V(s)$ is a rational function of s, *any* closed contour in the s-plane that encloses n_z zeros and n_p poles of $V(s)$ will map into a V-plane contour that encircles the point $V(s) = 0$ exactly

$n_z - n_p$ times in the same sense as the s-plane contour. In order to apply this theorem as a stability test, we can *1*. choose the s-plane contour so as to enclose the entire RHP, and *2*. select $V(s)$ to be a rational function whose poles are known and whose zeros are the roots of the characteristic equation.

Doing this, the number of RHP zeros of $V(s)$, which is the number of RHP roots of the closed-loop characteristic equation, will be readily apparent from the number of encirclements of the origin in the V-plane. It was Nyquist† who showed that by using the contour Γ_s in Figure 11.5*a* and the function

$$V(s) \triangleq 1 + G(s)H(s) \tag{1}$$

the argument principle would provide an effective stability test. Before stating Nyquist's result, known as the *Nyquist criterion*, and going through the steps that lead to the specialized form referred to here as the Nyquist stability test, we shall present a geometric interpretation of the argument principle.

Geometric interpretation

Restricting $V(s)$ to be a rational function, it can be described to within a multiplicative constant (taken as positive) by its poles and zeros. A typical pole-zero pattern and test point s_0 are shown in Figure 11.12*a* and the point into which s_0 is mapped is denoted by the vector $V(s_0)$ in part (*b*), which is a representation of the V-plane. As indicated in Section 10.1, the argument of $V(s_0)$ is obtained by adding the angles of the vectors from the zeros to s_0 and subtracting the angles of the vectors from the poles. For the pole-zero set shown

$$\arg [V(s_0)] = \theta_1 + \theta_2 - (\phi_1 + \phi_2) = 46°$$

which is the angle made by the vector $V(s_0)$ in Figure 11.12*b*.

Any closed contour in the s-plane, denoted by C_s, will map into a closed contour in the V-plane, denoted by C_V. If C_s is clockwise and encloses a pole or zero of $V(s)$, the angle of the vector from that pole or zero to the test point will undergo a decrease of 360° as the test point moves around the contour. On the other hand, the angles to s_0 from any poles or zeros that are not enclosed will undergo no net change after the test point has

†Nyquist (1932).

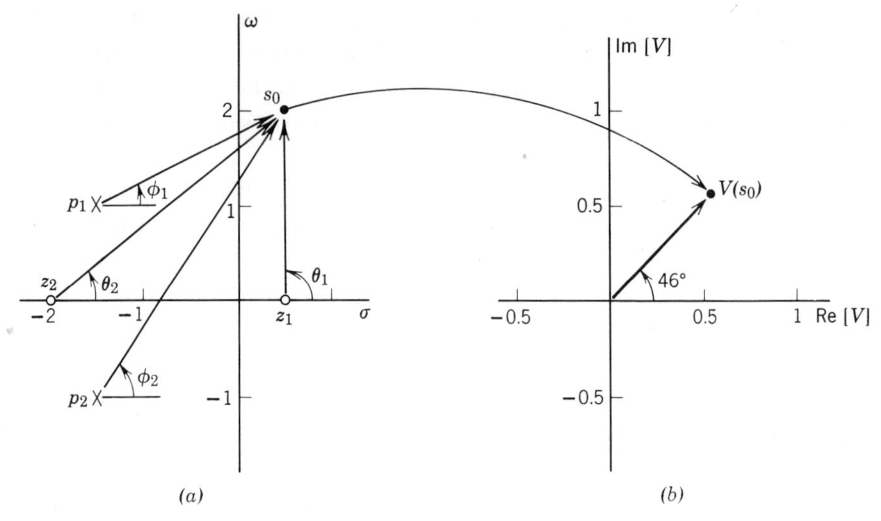

Figure 11.12. $V(s)$ corresponding to Figure 10.2.

returned to its starting position. Correspondingly, a change of 360° in arg $[V(s)]$ requires an encirclement of the V-plane origin by the contour C_V, with an increase in argument implying a counterclockwise encirclement and a decrease a clockwise encirclement.

In Figure 11.13 several sets of s-plane contours are sketched for the pole-zero pattern of Figure 11.12*a*, along with V-plane contours which exhibit the corresponding number of encirclements of the point $V = 0$. In Figure 11.13*a* none of the poles or zeros lies within C_s so none of the four angles (θ_1, θ_2, ϕ_1, or ϕ_2) undergoes a net change as s_0 moves around the contour. Hence, arg $[V(s)]$ does not undergo a net change, which is another way of saying that C_V does not encircle the origin of the V-plane. In part (*b*), θ_1 experiences a decrease of 360°, with no net change in the other angles. Thus, arg $[V(s)]$ has a net decrease of 360° which, in geometric terms, amounts to one clockwise encirclement of the V-plane origin. In part (*c*), C_V is in the counterclockwise direction and involves a net of two encirclements because both poles of $V(s)$ and neither of its zeros are within C_s. Part (*d*) again results in no encirclement of $V(s) = 0$ because one pole and one zero are enclosed by the s-plane contour, thereby canceling one another's contributions to the net change in arg $[V(s)]$.

With all of the preliminaries out of the way, we proceed at last to the proof of the Nyquist criterion.

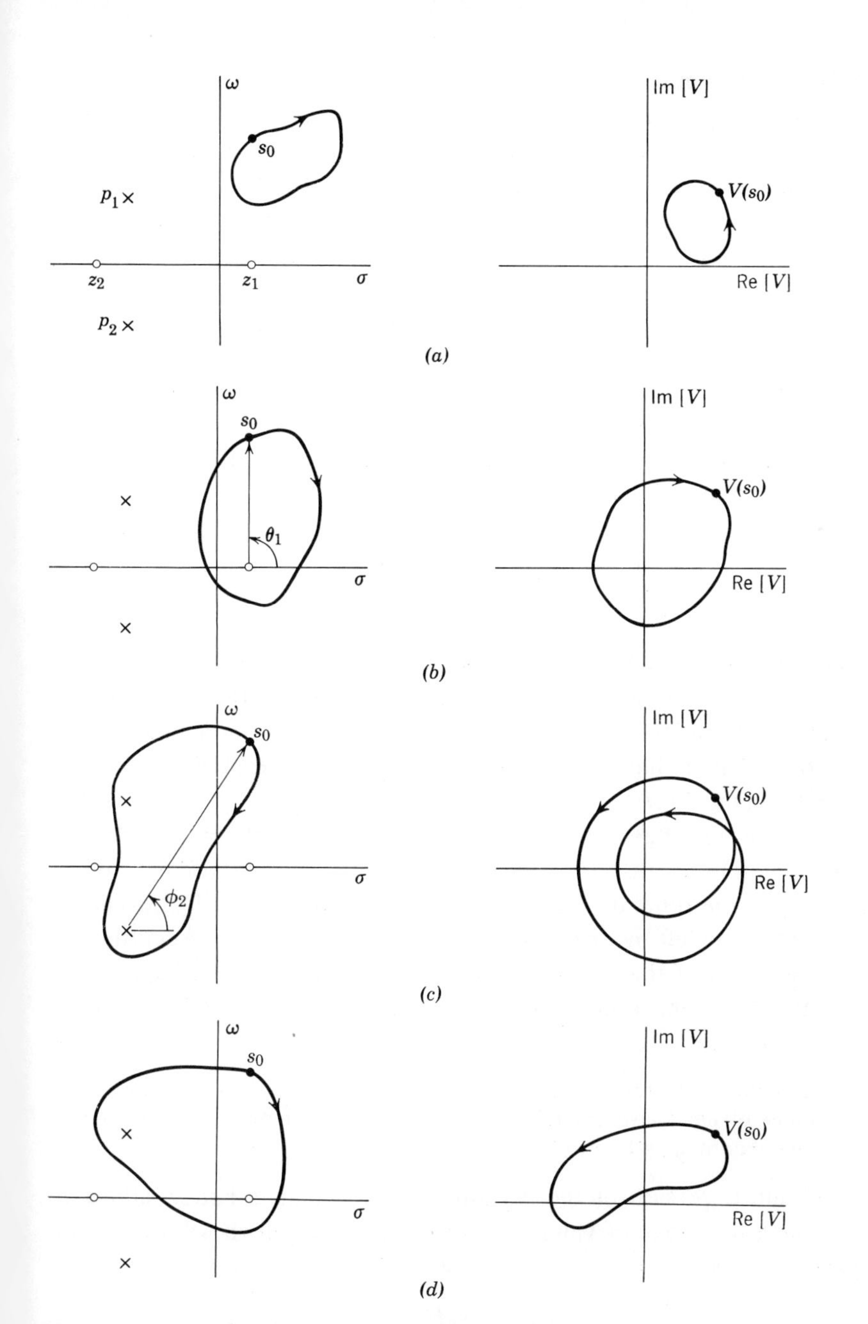

Figure 11.13. s- and V-plane contours. (*a*) No encirclements. (*b*) Encircling one zero. (*c*) Encircling two poles. (*d*) Encircling one pole and one zero.

The Nyquist criterion

If $V(s)$ is defined as in (1), i.e., $V(s) = 1 + G(s)H(s)$, its zeros will be the roots of the characteristic equation and its poles are the combined poles of $G(s)$ and $H(s)$, i.e., the open-loop poles. If the s-plane contour is taken as Γ_s (Figure 11.5*a*), with appropriate modifications made if either $G(s)$ or $H(s)$ should have a pole on the imaginary axis, the entire RHP will be enclosed as $r \to \infty$.

Thus, if the closed contour into which Γ_s is mapped (Γ_V) is plotted in the V-plane, the net number of *clockwise* encirclements of the origin will be equal to the difference between the number of zeros (n_z) and poles (n_p) of $V(s)$ inside Γ_s. But the poles of $V(s)$ are the system's open-loop poles and we should certainly know how many of these lie in the RHP. From the argument principle, it follows that the number of clockwise encirclements of the origin (n_{cw}) is given by $n_{cw} = n_z - n_p$ or, solving for the number of zeros of $V(s)$ in the RHP,

$$n_z = n_{cw} + n_p \tag{2}$$

Because the zeros of $V(s)$ are the characteristic-equation roots, we have shown that the number of closed-loop poles in the RHP is equal to the sum of: *1*. the number of *clockwise* encirclements of the V-plane origin made by the mapping $V(s) = 1 + G(s)H(s)$; and *2*. the number of open-loop poles in the RHP.

Because $V(s)$ differs from the product $G(s)H(s)$ only by the addition of unity, it follows that if Γ_s were mapped into another complex plane according to the function $GH(s) \triangleq G(s)H(s)$ the contour Γ_{GH} would be Γ_V shifted to the left by one unit, as in Figure 11.14*a* and *b*. Accordingly, since the origin of the V-plane corresponds to the point -1 in the GH-plane, the argument principle may be restated in terms of $GH(s)$ as the *Nyquist criterion*:

> The number of closed-loop poles in the RHP is equal to the sum of the number of *clockwise* encirclements of the point -1 made by $GH(s)$ plus the number of open-loop poles in the RHP.

To obtain a version of the Nyquist criterion that is better suited to ascertaining the range of values for a gain factor K which will result in a stable closed-loop system, it is convenient to plot the function

$$F(s) \triangleq \frac{GH(s)}{K}$$

yielding Figure 11.14*c*. Having done this we can investigate encirclements of the point $-1/K$ made by the contour Γ_F which in the previous section was defined to be the Nyquist plot. As depicted in Figure 11.14 the point $F(s) = -1/K$ corresponds to $GH(s) = -1$ or $V(s) = 0$. One advantage of this formulation is that changing the gain K amounts to only shifting the point $-1/K$ rather than replotting the contour Γ_{GH}. Furthermore, since there can be no roots of the characteristic equation in the RHP if the system is to be stable, n_z can be set to zero in (2). Hence, for a stable system it is necessary and sufficient that $n_{cw} + n_p = 0$ or, treating positive clock-

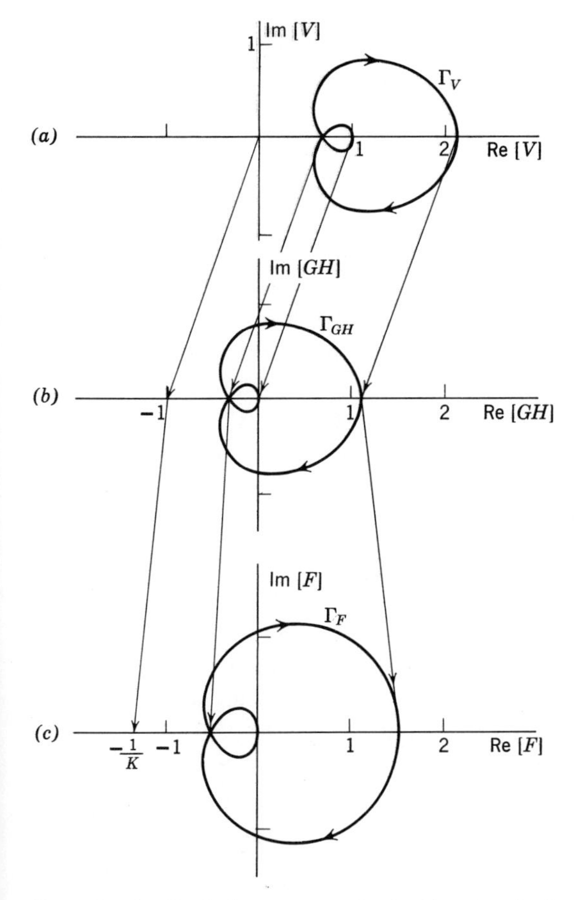

Figure 11.14. Equivalent contours for the Nyquist criterion. (*a*) V-plane. (*b*) GH-plane. (*c*) F-plane.

wise encirclements as negative *counterclockwise* encirclements,

$$n_{ccw} = n_p \tag{3}$$

In words, the system will be stable if and only if the number of counterclockwise encirclements by Γ_F of the point $-1/K$ is equal to the number of poles of $F(s)$ within Γ_s.

11.4 FREQUENCY-RESPONSE ANALYSIS

In addition to forming the basis for the Nyquist stability test, the open-loop frequency response function $F(j\omega)$ can be used to obtain both quantitative information and insight into the transient response of the closed-loop system. Several topics that have proved to be useful analysis tools will be discussed, although the interested reader should consult a control-oriented text for a more complete treatment.

Gain and phase margins

For the sake of discussion, let us assume that a particular gain value, say $\bar{K}$, has been selected so that the closed-loop system is known to be stable. If the positive-frequency portion of the Nyquist plot near the point $-1/\bar{K}$ appears as in Figure 11.15, we know from the arguments of Section 11.1 that the system will possess an undamped oscillatory mode if K is increased to the point where it equals K_{co}. Provided that the system is stable for $0 \leq K < K_{co}$ and unstable for $K > K_{co}$, the system's *gain margin* can be defined as

$$k_{mar} \triangleq \frac{K_{co}}{\bar{K}} \tag{1}$$

Hence, where applicable, the gain margin is the factor by which the gain $\bar{K}$ can be increased before the transition from a stable to an unstable system takes place. This factor, which must be greater than unity for a stable system, is often used as one of the design criteria that a closed-loop system must satisfy when working with frequency-domain methods. Unfortunately, its use is restricted to conventional single-loop systems for which the above conditions are satisfied. Systems that are stable for some gain $\bar{K}$ but unstable for $0 < K < \bar{K}$ are referred to as being *conditionally stable* and represent a situation where the gain-margin concept is invalid.

A graphical interpretation of the gain margin is shown in Figure 11.15.

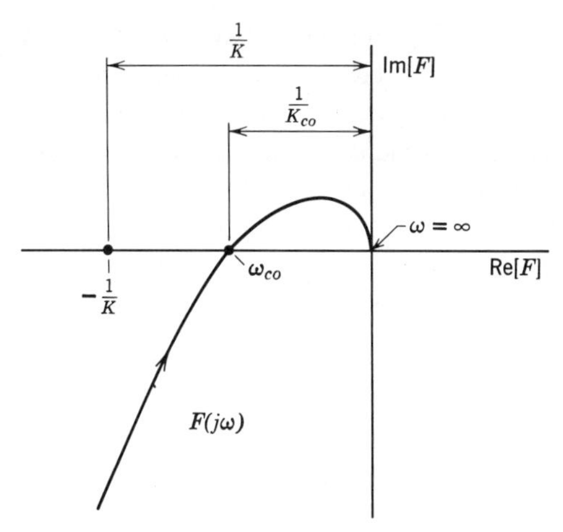

Figure 11.15. Interpretation of gain margin k_{mar}.

Because we have assumed that the actual gain $\bar{K}$ is less than K_{co}, the point $-1/\bar{K}$ will lie to the left of the point $-1/K_{co}$ and the ratio of the distances to these two points from the origin, namely

$$\frac{1/\bar{K}}{1/K_{co}} = \frac{K_{co}}{\bar{K}} > 1$$

is the gain margin. The gain margin gives a quantitative measure of the increase in gain that can be tolerated without causing the system to become unstable, in a sense, a measure of the "relative stability." But it contains no information regarding the phase-shift properties of $F(j\omega)$ which are all-important in determining the closed-loop dynamic behavior. For example, the two feedback systems represented in Figure 11.16 by

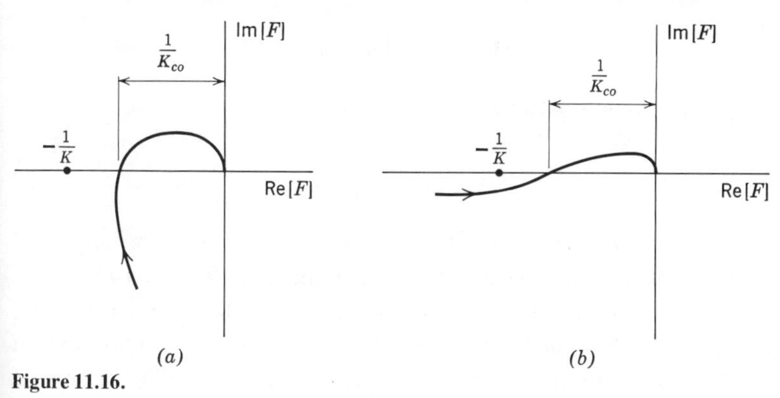

Figure 11.16.

the positive-frequency portions of their Nyquist plots have identical gain margins but can be shown to have rather different response characteristics. In either of the cases depicted the system could be made unstable without any change in K by introducing sufficient phase lag in $F(j\omega)$ such that the Nyquist plot encircled the point $-1/\bar{K}$. The additional phase lag that will make the curve of $F(j\omega)$ pass exactly through the point $-1/\bar{K}$ is known as the *phase margin*, denoted by ϕ_{mar}, and can be measured by drawing the arc of radius $1/\bar{K}$ shown in Figure 11.17, where the frequency

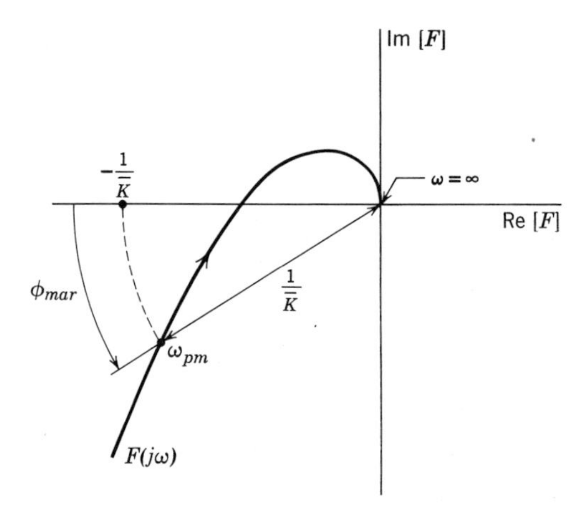

Figure 11.17. Phase margin ϕ_{mar}.

at which the arc intersects $F(j\omega)$ is denoted by ω_{pm}.

The phase margin can be expressed analytically as

$$\phi_{mar} = 180° + \arg [F(j\omega_{pm})] \tag{2}$$

where ω_{pm} satisfies the relationship

$$|F(j\omega_{pm})| = \frac{1}{\bar{K}} \tag{3}$$

Without being too quantitative, it seems reasonable that the gain and phase margins might be useful criteria for gauging the extent to which the closed-loop system will behave as an undamped oscillator. Clearly both a gain margin greater than unity (positive, if gains are expressed in decibels) and a positive phase margin are required for stability. Thus, as

$\phi_{mar} \to 0$ or $k_{mar} \to 1$ the system's zero-input response will become increasingly oscillatory, implying the presence of a dominant pair of complex roots for which the damping ratio $\zeta \to 0$. If one is willing to assume a particular form for the structure of the open-loop system, e.g., second order with one pole at the origin, approximate relationships between ϕ_{mar}, k_{mar}, and ζ are available which may be useful in estimating the effects of changes in K or $F(j\omega)$ upon time-domain responses.

Example 11.6

Demonstrating the calculations of gain and phase margins, consider the system studied in Example 11.5 when the gain is fixed at $\bar{K} = 30$, which is within the interval $-6 < K < 60$ known to yield a stable system. The relevant portion of the Nyquist plot is shown in Figure 11.18, along with

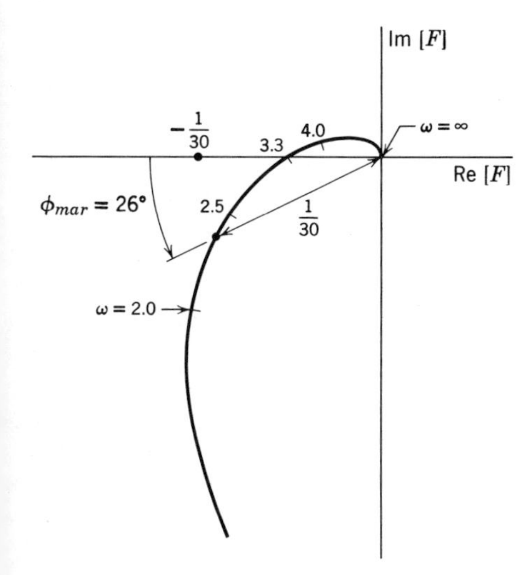

Figure 11.18.

the point $-1/\bar{K}$. Because $F(j\omega) = \frac{1}{60}$ when the plot crosses the negative real axis, it follows that $K_{co} = 60$ and, from (1), that $k_{mar} = \frac{60}{30} = 2.0$. By drawing the arc of radius $\frac{1}{30}$ between the point $-\frac{1}{30}$ and its intersection with $F(j\omega)$, and measuring the corresponding angle, we see that $\phi_{mar} = 26°$. Interpolating between the frequency points given in the figure, $\omega_{pm} \approx 2.4$ rad/sec.

Relationship between root-locus and Nyquist plots

Recall that the 180° root locus is a plot of all those points in the s-plane that satisfy the angle criterion $\arg[F(s)] = q180°$ where q is any odd integer. An alternate definition would be to say that the 180° root locus is the set of all points in the s-plane that are mapped into the negative real axis of the F-plane by the function $F(s)$. Similarly, the 0° locus is the set of all points in the s-plane that map into the positive real axis of the F-plane. In the same vein, the imaginary axis of the s-plane maps into the contour $F(j\omega)$, which constitutes the major portion of the Nyquist plot in the F-plane. The mapping from the s-plane to the F-plane is necessarily unique; however, the reverse mapping is not unique. For example, the negative real axis of the F-plane maps back into each of the n branches of the 180° root locus.

Both the Nyquist and root-locus plots can be thought of as containing two sets of loci with one member of the set being straight lines and the other curves – the choice of plot dictates which of the two sets of loci is to be straight. When the loci intersect in one plane they must necessarily intersect in the other plane. Furthermore, because $F(s)$ has been restricted to rational functions, the mapping has the property that it is comformal, implying that the angle of intersection of two lines in the s-plane is preserved when the mappings of the same two lines intersect in the F-plane. The manner in which this property can be used to understand the relationship between the phase and gain margins and the damping ratio and natural frequency of dominant closed-loop poles is illustrated in the following example.

Example 11.7

The 180° locus and the positive-frequency portion of the Nyquist plot for the system discussed in Example 11.4 are shown in Figure 11.19. The intersection of the upper branch of the root locus with the imaginary axis occurs at the point $s_0 = j3.32$ and corresponds to $K = 60$. The angle ψ between the locus and the imaginary axis at the intersection is, to a good approximation, 30°. In the F-plane this intersection has mapped into the intersection of the Nyquist plot with the real axis, which occurs at $F_0 = -\frac{1}{60}$ and $\omega = 3.32$. Because of the conformal nature of the mapping, the angle ψ appears in the F-plane as the angle formed by the real axis and the tangent to $F(j\omega)$ at $\omega = 3.32$, as indicated in part (b) of the figure. The points s_1 and s_2 and their counterparts F_1 and F_2 in the F-plane are also

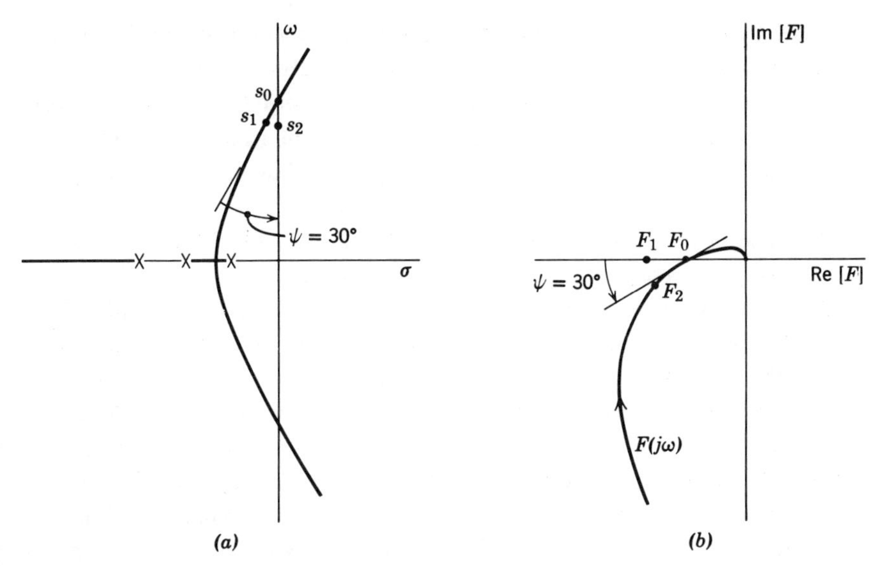

Figure 11.19. (*a*) Root locus. (*b*) Nyquist plot.

shown. Thus, if the gain K were selected so as to make the point $-1/K$ fall on F_1 in part (*b*), a closed-loop root would necessarily lie at s_1 in part (*a*).

11.5 BODE PLOTS

The Nyquist criterion provides the basis for the analysis of feedback systems using frequency-response information, and it has been demonstrated that plots of $F(j\omega)$ or $GH(j\omega)$, with ω treated as a parameter, can be used effectively for design and analysis purposes. However, there are other means of graphically representing the open-loop frequency response that are widely used in practice. The most common of these is the Bode plot which henceforth will refer to $|GH(j\omega)|$ in decibels and arg $[GH(j\omega)]$ in degrees, both plotted against log ω.

We shall assume that the reader has had prior experience with the straight-line approximations, which greatly simplify the construction of a Bode plot† and concentrate on the *interpretation* of the Bode plot as

†The definition of the decibel as a measure of magnitude and the approximations useful for drawing the Bode plot of a first-order dynamic element were given in Section 6.2. For a comprehensive treatment, including complex poles and zeros, consult almost any introductory circuit-theory or control-systems text; e.g., Close (1966, Sect. 6.5).

applied to feedback systems. It is essential to keep in mind that the Nyquist and Bode plots—as well as the related Nichols plot†—are merely different ways of presenting open-loop frequency-response information, and that the theoretical underpinning of the rules used to interpret these various plots is the Nyquist criterion, which is in turn based on the argument principle of complex-variable theory.

Because of the relative ease with which a Bode plot of $GH(j\omega)$ can be drawn and because the gain and phase margins are readily displayed, it is widely used when working with feedback systems. The Bode plot does have the disadvantage that the closed-loop frequency response cannot be found directly therefrom, except by certain approximations that usually hold only for high and low frequencies. Furthermore, it is of necessity restricted to finite, positive frequencies because of its logarithmic nature, and does not contain any information regarding encirclements of the point $-1/K$ as does the complete Nyquist plot. In most situations the open-loop system is stable, in which case $F(s)$ does not have any poles within the contour Γ_s. Thus the closed-loop system will be stable provided the positive-frequency portion of the Nyquist plot passes *below* the point $-1/K$ so as to avoid encirclements, thereby avoiding the need for a complete Nyquist plot.

Given the two situations depicted in Figure 11.20, we distinguish between their stability properties by using the gain and phase margins. Because of the manner in which these terms were defined in the preceding section, satisfying $k_{mar} > 1$ and $\phi_{mar} > 0$ implies that the closed-loop system will be stable. In most situations, either of the above inequalities

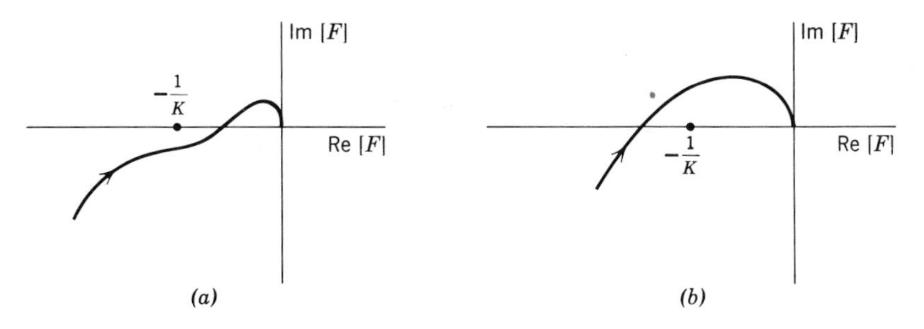

Figure 11.20. (*a*) Stable. (*b*) Unstable.

†A Nichols plot consists of $|GH(j\omega)|$ in decibels plotted against arg $[GH(j\omega)]$ in degrees, with frequency as a parameter.

will imply the other; if the system's characteristics are such that the gain and phase margin concept is not valid, then one should look to the Nyquist criterion for the correct conditions for ensuring stability. Systems for which a reduction of gain will cause instability (conditionally stable) or those having open-loop poles in the RHP are examples of situations where the Nyquist criterion should be used to guide the interpretation of the Bode plot.

Example 11.8

To illustrate the manner in which the Bode plot can be used to determine stability in terms of the gain and phase margins, we investigate the system having $G(s) = K/[s(s+1)(s+10)]$ and $H(s) = 1$. Setting $K = 50$, the magnitude and phase plots of $G(j\omega)$ can be drawn versus log ω from the data in Table 11.1, supplemented with the usual straight-line approximations

$$G(j\omega) \approx \frac{50}{j10\omega} = \frac{5}{j\omega} \qquad \omega << 1$$

and

$$G(j\omega) \approx \frac{50}{(j\omega)^3} = \left(\frac{3.68}{j\omega}\right)^3 \qquad \omega >> 10$$

The resulting plot is shown in Figure 11.21a, where the argument (phase) curve has been shown only in the vicinity of $-180°$ in order to obtain greater resolution.

Identifying the frequency at which $\phi = -180°$ as ω_{co}, we obtain the gain margin in decibels by measuring *down* from the 0 db axis on the magnitude

Table 11.1

ω (rad/sec)	arg $[G]$ (degrees)	$K = 50$		$K = 75$	
		$\lvert G\rvert$	$\lvert G\rvert_{db}$	$\lvert G\rvert$	$\lvert G\rvert_{db}$
1	−141	3.53	+11.0	5.30	+14.5
2	−165	1.10	+0.8	1.65	+4.3
$\sqrt{10}$	−180	0.46	−6.7	0.69	−3.2
5	−196	0.18	−14.9	0.27	−11.4

plot to the $|G(j\omega)|$ curve. If $|G(j\omega)| < 0$ db at this frequency, then the gain margin (k_{mar}) is greater than unity or 0 db; in this case, $k_{mar} = +6.7$ db. If the magnitude plot had been above the 0 db line at $\omega = \omega_{co}$, the gain margin would have been less than unity (< 0 db), implying that the closed-loop system would be unstable.

Denoting the frequency at which the magnitude plot crosses the 0 db axis as ω_{pm}, the phase margin can be measured as the number of degrees by which arg $[G(j\omega_{pm})]$ falls short of $-180°$; from the figure, $\phi_{mar} = 13°$. Thus the phase curve must lie above the $-180°$ line at $\omega = \omega_{pm}$ if the phase margin is to be positive, which is taken as the requirement for stability.

When the gain K is increased from 50 to 75, the only modification required in the Bode plot is to shift the magnitude plot upwards by 3.5 db which is the factor $\frac{75}{50}$ expressed in decibels. Figure 11.21*b* results, in which the gain margin is reduced by 3.5 db and is now $(6.7 - 3.5) = 3.2$ db. Likewise, the phase margin is also reduced from 13° to 6° because the frequency ω_{pm} at which $|G(j\omega)| = 0$ db is increased from 2.15 to 2.65 rad/sec, thereby reducing the difference between arg $[G(j\omega_{pm})]$ and $-180°$.

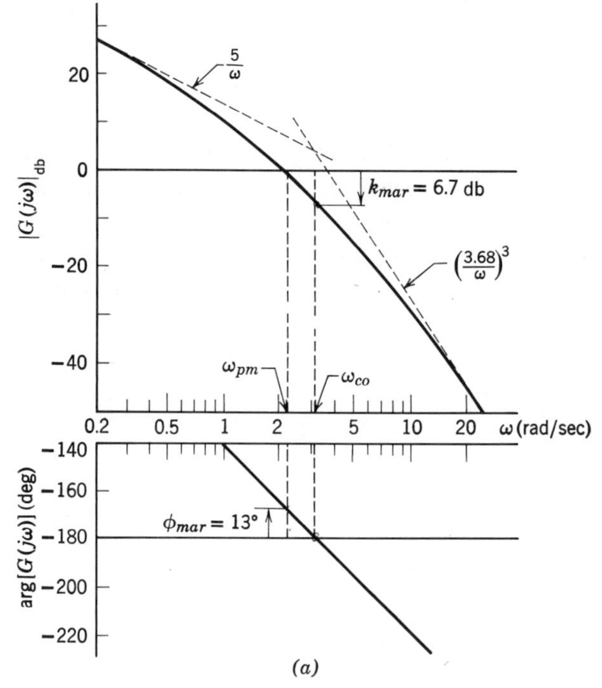

Figure 11.21. (*a*) $K = 50$.

The magnitude of the closed-loop frequency response, $|T_R(j\omega)|$, is plotted in Figure 11.22 for $K = 50$ and $K = 75$. In each case the low-frequency behavior is asymptotic to 0 db (unity gain) and for high frequencies the gain falls off at a rate of 40 db per decade. The peaking of $|T_R(j\omega)|$ in the intermediate frequency range results from the fact that the phase and gain margins are relatively low, which is another way of saying that the Nyquist plot approaches close to the point $F(j\omega) = -1/K$. The frequencies at which the peaks of $|T_R(j\omega)|$ occur are between the corresponding frequencies ω_{pm} and ω_{co} at which the phase and gain margins are measured.

It is apparent that the lower phase and gain margins for $K = 75$ result in a closed-loop system having a more pronounced resonant peak than for $K = 50$. The corresponding effect upon the step or impulse response would obviously be a reduction of the damping ratio as K is increased, until $\zeta = 0$ when the gain and phase margins reach zero. Graphical and approximate methods for obtaining the closed-loop frequency response

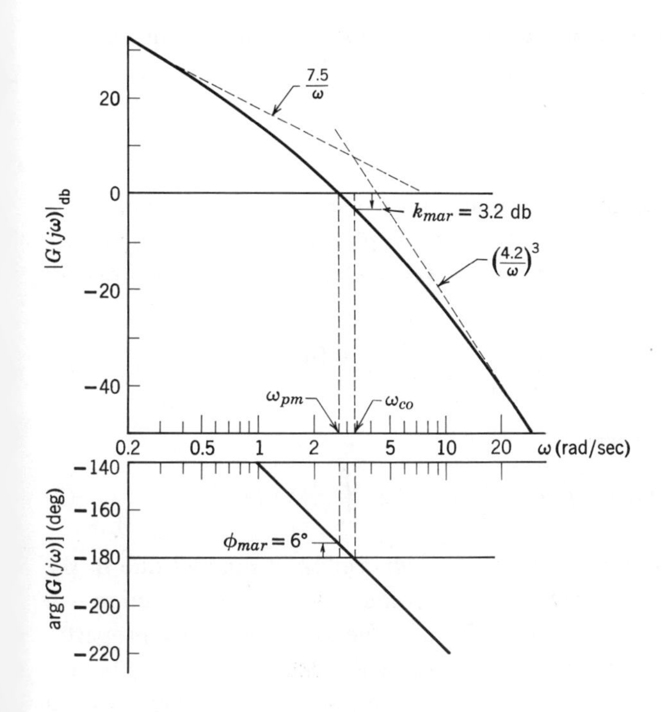

Figure 11.21. (*b*) $K = 75$.

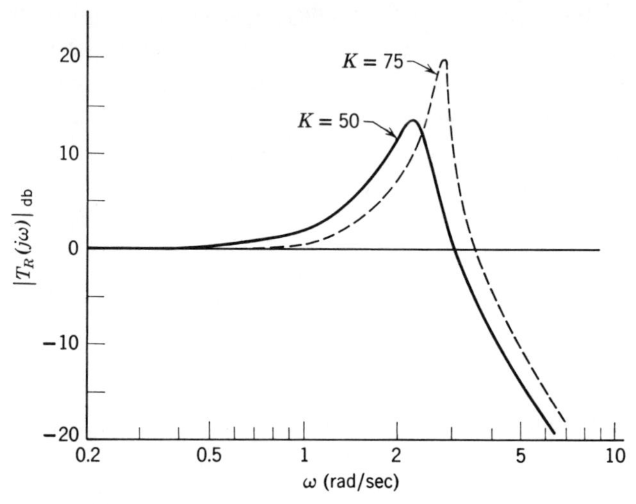

Figure 11.22. Closed-loop frequency response.

directly from the open-loop frequency response are discussed in detail in most introductory control-system texts.

11.6 APPLICATION

The controller designed in Section 10.5 using a P+I+D control law suffers from one major drawback: it is not physically realizable. Specifically, expressing the derived control law either in terms of its transfer function

$$G_c(s) = \frac{Y(s)}{X(s)} = 1 + 20s + \frac{0.0123}{s} \tag{1}$$

or by its input-output relationship

$$y(t) = x(t) + 20\frac{dx}{dt} + 0.0123\int_0^t x(\lambda)\,d\lambda \tag{2}$$

we can see that its impulse response will contain a doublet due to the differentiator and the magnitude of the frequency-response function will approach infinity as $\omega \to \infty$. In addition to the nonrealizability property, the amplification of high-frequency noise by the differentiator would be a severe practical drawback, as would the integration amplifier drift over a sufficiently long period of time.

However, rather than discard the existing design and start anew, we shall show that the P+I+D function can be approximated with a second-order filter. Basically, the filter's frequency-response function will be selected so as to closely approximate that of the P+I+D controller over the range of frequencies for which $|GH(j\omega)| \approx 1$ and $\arg[GH(j\omega)] \approx -180°$. The realizable system that results will be shown by simulation to perform virtually as well as the ideal P+I+D controller with regard to the response specifications used in Section 10.5

The network in question is known as a *lead-lag* filter – sometimes referred to as a notch filter because of the shape of its magnitude function, or as a lead-lag compensating network because in a sense it compensates for the undesirable aspects of the existing process (assuming it is properly designed). Although the design of such a network is often carried out using Bode plots, we shall only demonstrate via the frequency-response methods that the filter will do the job, leaving the task of presenting synthesis methods to the books that deal exclusively with control systems. Keeping all of the system parameters that are not involved in the P+I+D controller fixed at their nominal values and replacing the ideal controller with the lead-lag network whose transfer function is denoted by $G_L(s)$ leads to Figure 11.23. We shall assume that any loading effects present when the network is inserted in the loop have been incorporated in the transfer function.

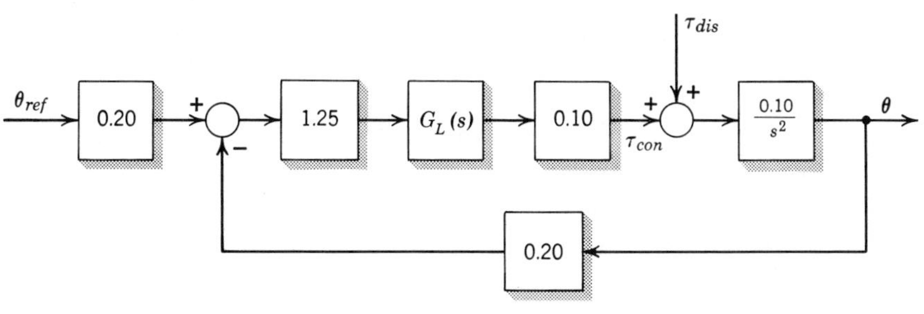

Figure 11.23. Satellite attitude system with compensation.

The lead-lag filter

We shall start by rewriting $G_c(s)$ as given by (1) in factored form so as to be able to identify its poles and zeros. Thus,

$$G_c(s) = \frac{20s^2 + s + 0.0123}{s} = 20\frac{(s+0.0222)(s+0.0278)}{s}$$

from which it is apparent that $G_c(s)$ has zeros at $z_1 = -0.0222$ and $z_2 = -0.0278$ and a single pole at the origin – the fact that the number of zeros exceeds that of the poles is another manifestation of the unrealizability of the P+I+D controller. The plots of the magnitude and phase of $G_c(j\omega)$ shown in Figure 11.24*a* can be drawn using the usual Bode-plot construction rules. Out of the frequency range shown in the figure one can use the high- and low-frequency approximations

$$G_c(j\omega) \approx \frac{0.0123}{j\omega} \qquad \omega \ll 0.0222$$

$$G_c(j\omega) \approx j20\omega \qquad \omega \gg 0.0278$$

Because the phase of the loop frequency-response function $G_cG_pH(j\omega)$ is close to $-180°$ only in the interval $0.01 < \omega < 0.1$, we shall endeavor to find a realizable network with frequency response $G_L(j\omega)$ which closely approximates $G_c(j\omega)$ over that interval and we shall not worry for the moment if the approximation is poor for much lower or much higher frequencies. Having found such a network, we can use our various analytical tools and computer simulation to determine whether the performance of the closed-loop system meets our specifications.

First, to rectify the fact that $G_c(s)$ has more zeros than poles we shall have $G_L(s)$ retain the zeros of $G_c(s)$ and add a second pole which has a 3-db corner frequency well to the right of those corresponding to the zeros z_1 and z_2 in Figure 11.24*a* – a separation of one decade is easily achieved in practice. Similarly, the pole of $G_c(s)$ at the origin will appear in $G_L(s)$ such that its corner frequency lies a decade below that of the zero z_1. With these changes the poles of $G_L(s)$ will lie at $p_1 = z_1/10 = -0.00222$ and $p_2 = 10z_2 = -0.278$. Since the zeros are unchanged, the transfer function of the new network will be

$$G_L(s) = k\frac{(s+0.0222)(s+0.0278)}{(s+0.00222)(s+0.278)} \tag{3}$$

where the gain k will be selected so as to make $G_L(s) \approx G_c(s)$ in the vicinity of $\omega = 0.025$. Figure 11.24*b* shows the result of drawing the magnitude plot of the frequency-dependent part of $G_L(s)$ and adjusting it vertically to coincide with $G_c(j\omega)$ for $\omega \approx 0.025$, thereby indicating that $k = +14.8\text{ db} = 5.52$. Using this value in (3) gives the lead-lag filter trans-

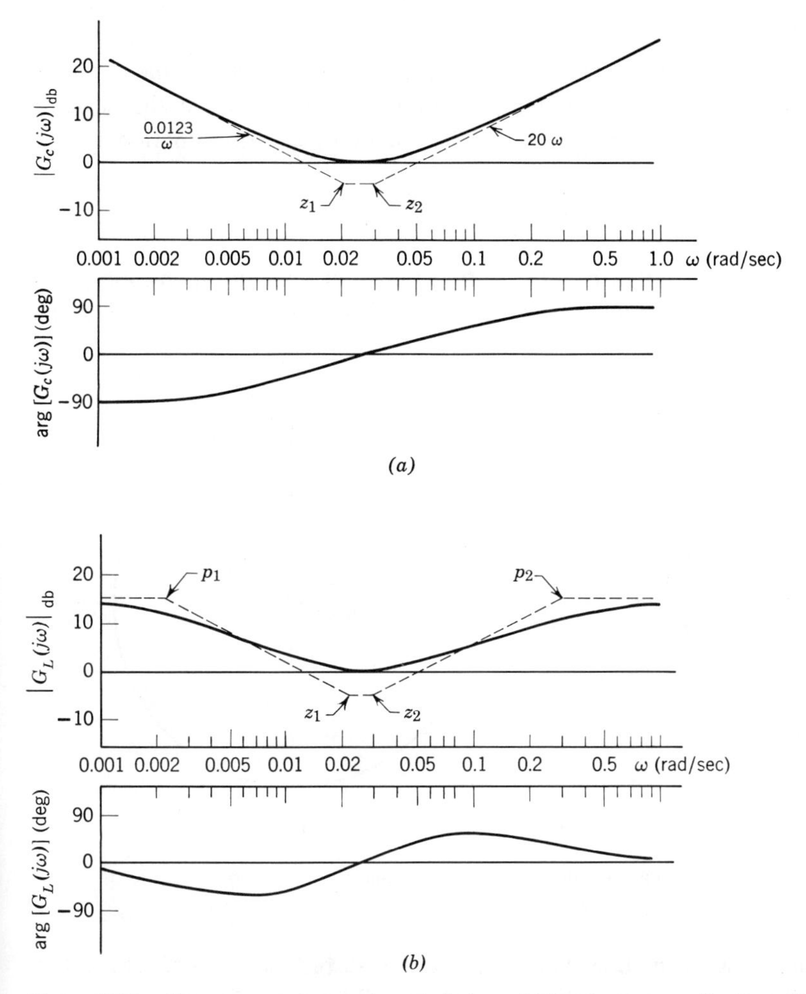

Figure 11.24. Frequency-response characteristics. (*a*) P + I + D controller (unrealizable). (*b*) Lead-lag filter (realizable).

fer function as

$$G_L(s) = 5.52\,\frac{(s+0.0222)\,(s+0.0278)}{(s+0.00222)\,(s+0.278)} \tag{4}$$

The magnitude and phase functions are shown in Figure 11.24*b* and, when compared with those of the ideal filter in part (*a*), are seen to agree rather well in both magnitude and phase in the interval $0.01 < \omega < 0.1$.

Stability

Having decided on $G_L(s)$ we can investigate the stability of the resulting system by examining the Nyquist plot and its accompanying s-plane contour sketched in Figure 11.25. The open-loop transfer function is

$$GH(j\omega) = \frac{1}{400s^2} G_L(s)\bigg|_{s=j\omega} \tag{5}$$

where $G_L(s)$ is given by (4). Noting that there are no open-loop poles within Γ_s and that there are no net encirclements of the point -1 in the

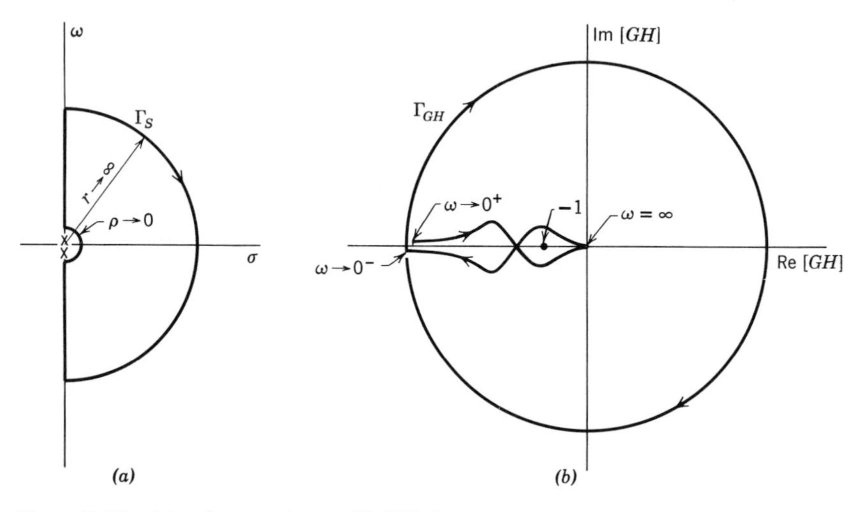

Figure 11.25. (*a*) s-plane contour. (*b*) GH-plane contour.

GH-plane, we conclude that the system is stable for $k = 5.52$. It is interesting to note, however, that the system will be stable for any increase in the loop gain but will become unstable if the gain is decreased to the point where the Nyquist plot encircles the point -1. The reason for this potential instability is the phase lag introduced at low frequencies by the lead-lag network. In order to have a stable system the phase lead resulting from the two zeros of $G_L(s)$ must reduce the total phase lag of $GH(j\omega)$ to less than 180° at the frequency for which $|GH(j\omega)| = 0$ db, thereby avoiding encirclement of the point -1.

Whereas the requirements for stability are clear from the Nyquist plot, such is not the case with the Bode plot, as evidenced by the fact that in Figure 11.26 the behavior of $GH(j\omega)$ in the range of frequencies for which $|GH(j\omega)| \approx 0$ db and arg $[GH(j\omega)] \approx -180°$ does not fall into the pattern for which the gain and phase margin concepts were defined; hence, they had best not be used. While the Bode plot may be more convenient for calculations than the Nyquist plot, it is essential that at least a qualitative sketch of the latter be made in order to ensure that the Bode plot is being interpreted correctly.

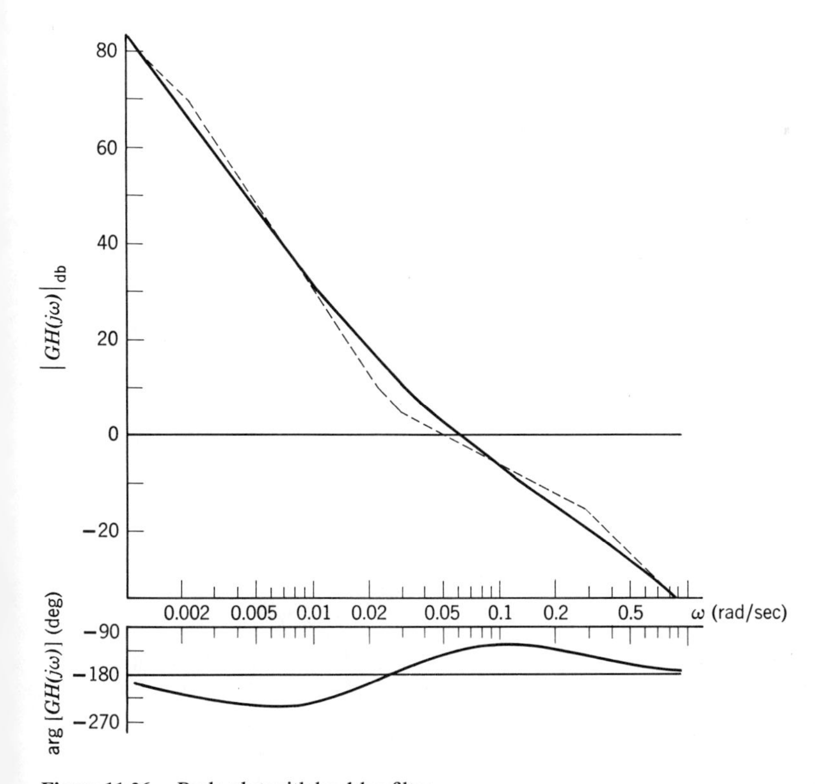

Figure 11.26. Bode plot with lead-lag filter.

Closed-loop roots

To assess the effects of the lead-lag network upon the dynamic behavior of the system, we can draw the root locus where the gain k associated with the compensating network is left free. Writing the open-loop transfer function in the root-locus form

$$GH(s) = \underbrace{\frac{k}{400}}_{K} \underbrace{\frac{(s+0.0222)(s+0.0278)}{s^2(s+0.00222)(s+0.278)}}_{F(s)}$$

The 180° locus is shown in Figure 11.27*a* with the roots corresponding to the design value $k = 5.52$ denoted by triangles. For comparative purposes, part (*b*) of the figure shows the root locus of the P + I + D version with the closed-loop roots shown for the nominal loop gain.

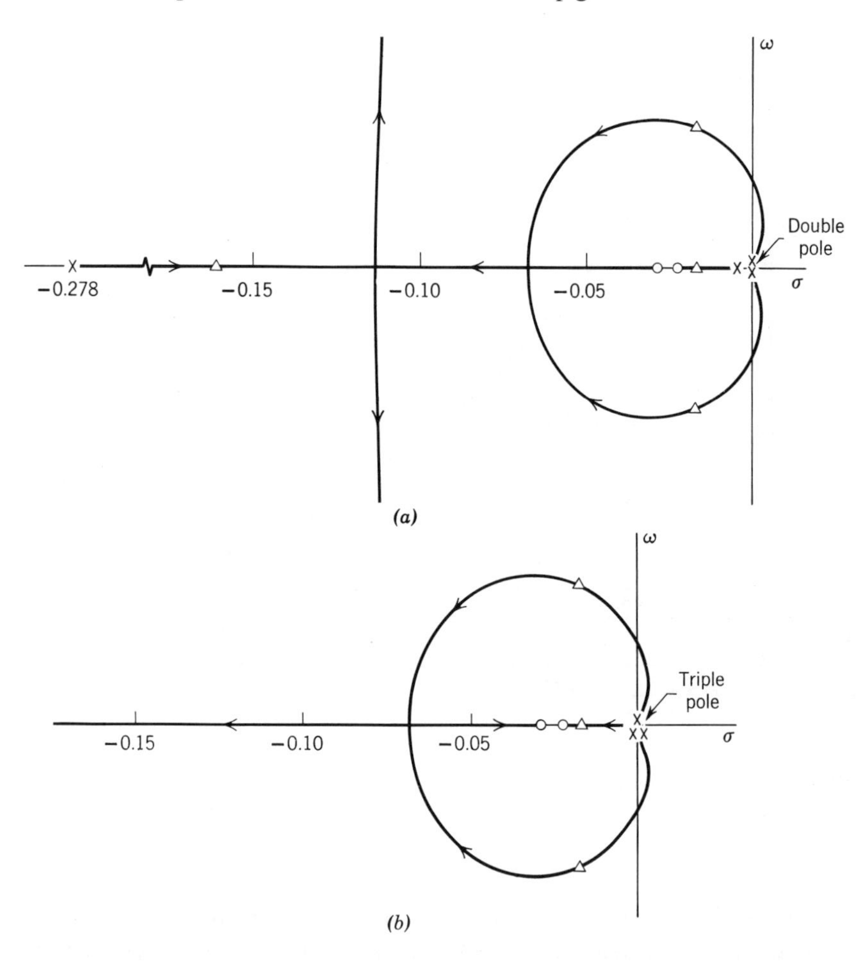

Figure 11.27. Root locus. (*a*) With lead-lag filter. (*b*) With P + I + D controller.

While there are major differences between the two plots – the plot for the lead-lag case has one more branch, and thus one more large-gain asymptote and closed-loop root than the other – they are quite similar in the vicinity of the origin, say $|s| < 0.05$. Although the system shown in part (a) is fourth-order, the dominant roots (the three that are closest to the imaginary axis) are quite close to the three roots of the system with the P+I+D controller. For $k = 5.52$ the fourth root, which was introduced by the lead-lag network, is more than ten times as far to the left in the s-plane as any of the three dominant roots; hence its contribution to the transient response is likely to be negligible. Because the zeros of $G_L(s)$ were selected to be those of $G_c(s)$, the closed-loop zeros of $T_R(s)$ must be identical for the two systems.

Disturbance torques

In Section 9.3 it was shown that a step or constant disturbance torque would result in a constant steady-state pointing error if a P+D controller was used and in zero error if an integral term was added. As a basis for comparison, the steady-state pointing error for the system with the P+D controller can be found by constructing the steady-state model shown in Figure 11.28a. The steady-state controller gain of unity was obtained by noting that its transfer function is $(1+20s)$ and its input will be constant in the steady state provided only that θ approaches a constant value.

Because of the form of the vehicle dynamics, Eq. (1), Sect. 9.3, the total torque $\tau_{tot} \triangleq \tau_{con} + \tau_{dis}$ must vanish if θ_{ss} is to be a constant. The steady-state model of Figure 11.28a indicates that for this to happen it is necessary that

$$(0.20)(1.25)(0.10)\theta_{ss} = \tau_{dis}$$

Thus,

$$\theta_{ss} = 40.0\,\tau_{dis} \tag{6}$$

which is to say that a constant disturbance torque of 1/10,000 ft-lb will cause a steady-state error of 0.004 rad (0.23°). Actually, this relationship between τ_{dis} and θ_{ss} is not as severe as (6) might indicate because the magnitudes of typical disturbance torques are exceedingly small by earthbound standards. In any event, we are only interested in obtaining a relative comparison between the three types of controllers in question.

When using the lead-lag network described by (4), the low-frequency

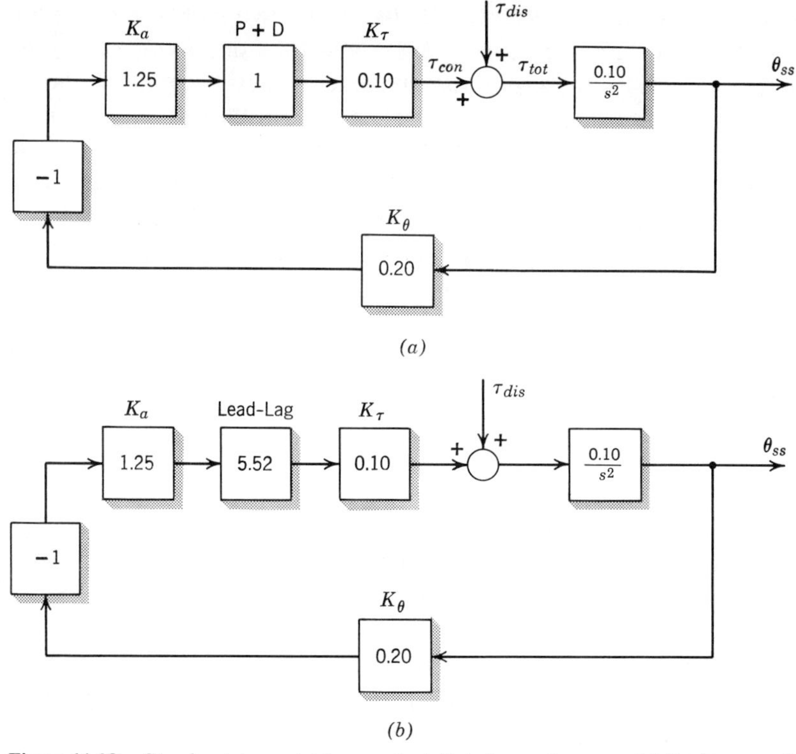

Figure 11.28. Steady-state model for constant disturbance torque. (*a*) P + D controller. (*b*) Lead-lag filter.

gain of the controller is

$$\lim_{s \to 0} G_L(s) = 5.52\left(\frac{0.0222}{0.00222}\right)\left(\frac{0.0278}{0.278}\right) = 5.52$$

and the steady-state model is as shown in Figure 11.28*b*. Solving for the value of θ_{ss} that will cause the torque τ_{dis} to be exactly canceled by the applied torque, in the steady-state,

$$\theta_{ss} = \frac{40.0}{5.52}\tau_{dis} = 7.25\tau_{dis}$$

Thus, the error has been reduced by a factor of 5.52 compared with that resulting from the P + D system, although not to the value of zero achievable with the P + I + D controller.

If a further reduction in the ratio θ_{ss}/τ_{dis} is necessary, the parameters of the lead-lag network could be modified by trial and error so as to increase the low-frequency gain of the compensating network, probably at the expense of the transient response of the system. It is up to the designer to assess the available tradeoffs between dynamic response, steady-state errors, reliability, cost, etc., so as to arrive at a satisfactory design.

The responses of the system to a change in reference pointing angle and to a step disturbance torque are shown in Figure 11.29. A comparison of

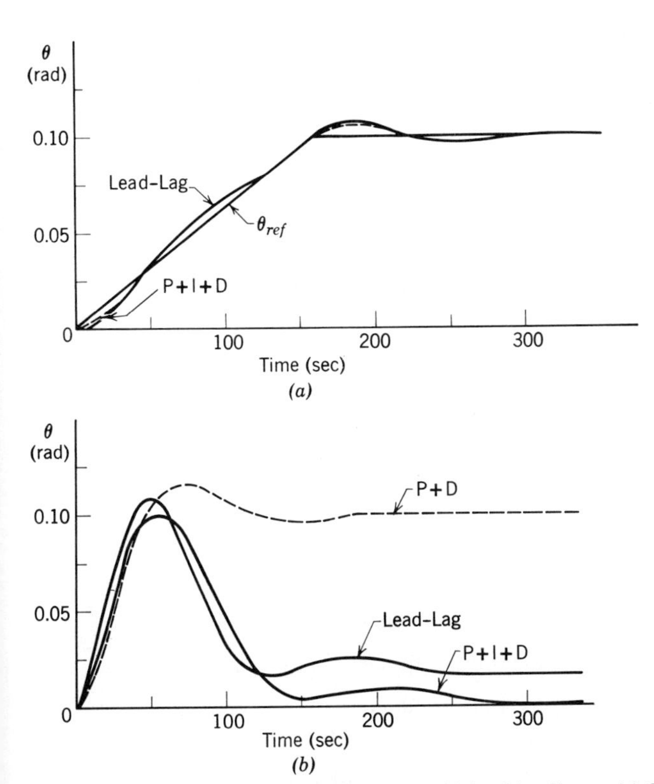

Figure 11.29. Simulated system responses with lead-lag filter. (*a*) Change of reference input. (*b*) Step disturbance input.

these plots with those in Figure 10.27, which were obtained using the ideal P + I + D controller, will substantiate the assertion that the lead-lag compensation has provided a suitable substitute for the P + I + D controller, at least for the response curves shown. We could readily extend the above assertion to quite arbitrary inputs, within the system's range of linearity.

Problems

11.1 Determine whether each of the systems described below is capable of producing a constant-amplitude oscillation for some value of the gain K. If it can, determine the values of K and ω_{osc}.

(a) $GH(s) = K/s(\tau s+1)$

(b) $GH(s) = K/s(\tau_1 s+1)(\tau_2 s+1)$

(c) $GH(s) = K/s^2(\tau s+1)$

(d) $GH(s) = Ke^{-st_d}$

(e) $GH(s) = Ke^{-st_d}/s$

11.2 For each of the open-loop frequency-response plots in Figure P11.1 state: *1*. the minimum number of open-loop poles, *2*. the minimum number of open-loop zeros, and *3*. the difference between the number of poles and zeros, that is, $n-m$. Also sketch a pole-zero pattern in the s-plane that could correspond to the frequency response in a qualitative sense.

11.3 Sketch Nyquist and Bode plots for the systems whose transfer functions have the pole-zero patterns shown in Figure P10.1 (take $K=1$).

11.4 Draw Bode and Nyquist plots for the open-loop transfer function of Problem 10.2. Use these plots to estimate the range of $K>0$ for which the closed-loop system will be stable.

11.5 Repeat Problem 11.4 for the transfer function of Problem 10.3.

11.6 Repeat Problem 11.4 for the transfer function of Problem 10.4.

11.7 Repeat Problem 11.4 for the transfer function of Problem 10.5.

11.8 Repeat Problem 11.4 for the transfer function of Problem 10.6.

11.9 Find the range of K, both positive and negative, for which the system shown in Figure P11.2 is stable. Do so by applying (a) the Nyquist stability test, (b) Routh's criterion, (c) a root-locus plot.

11.10 Sketch a Nyquist plot and use it to determine the range of $K>0$ for which the system having $G(s) = K(s+100)^2/s^2(s+10)$ and $H(s)=1$ is stable. *Answer:* $K>40$.

11.11 Use a Nyquist plot to find the range of K, both positive and negative, for which the system having $G(s) = K(s+10)/(s-1)$ and $H(s)=1$ is stable.

11.12 (a) Use the Nyquist stability test to show that the system for which

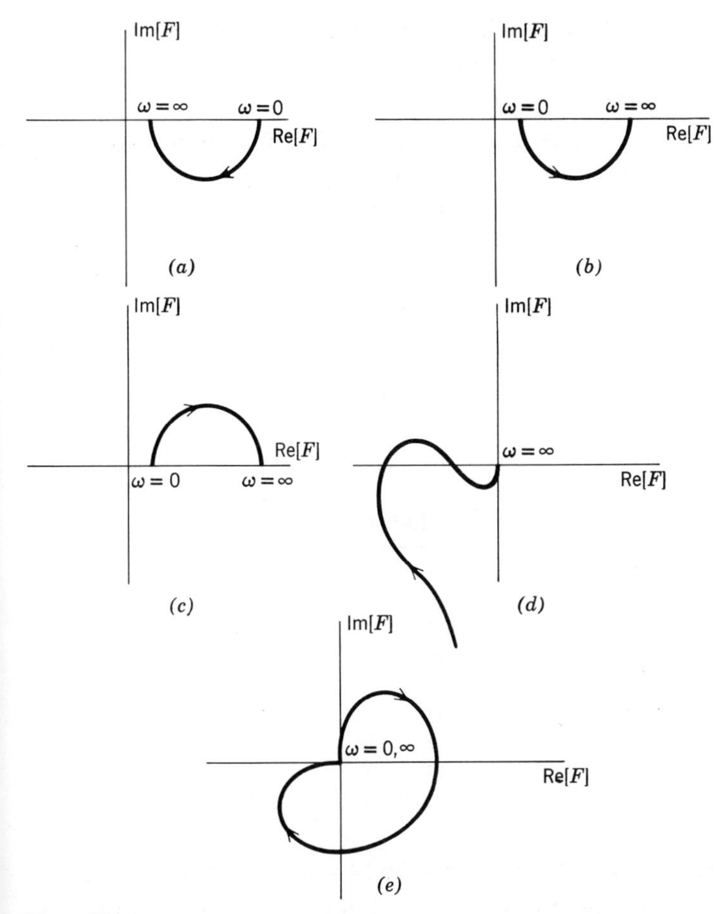

Figure P11.1

$G(s) = K/s(s+1)$ and $H(s) = 1$ is stable for all $K > 0$. Do the problem twice, using s-plane contours that pass to the left and to the right of the pole of $G(s)$ at the origin of the s-plane. Indicate clearly how the Nyquist plot is interpreted in each case.

(b) Construct and interpret the corresponding Bode plot.

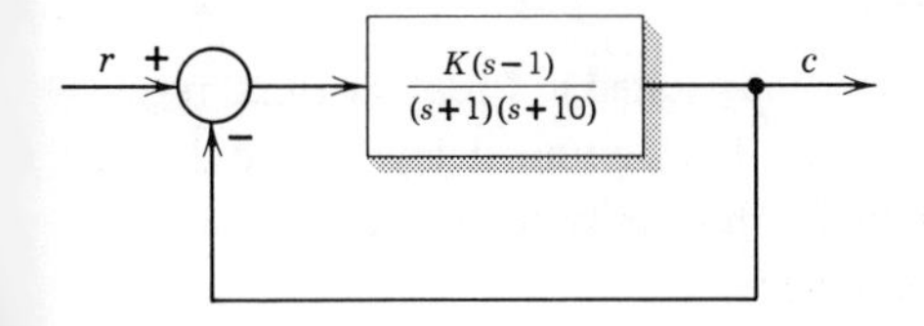

Figure P11.2

11.13 Use Nyquist and Bode plots to show that the system having $F(s) = 1/s(s+10)^2$ will be unstable for $K > 2000$ and that the frequency of oscillations for marginal stability will be 10 rad/sec.

11.14 The data in Table P11.1 represent the open-loop frequency response of a feedback system when its gain parameter K is unity.

(a) Evaluate the gain and phase margins when $K = 1$.

(b) Determine the value of K necessary to meet the following criteria: 1. $\phi_{mar} = 20°$, (what will k_{mar} be?); 2. $k_{mar} = 9$ db, (what will ϕ_{mar} be?).

(c) Find the rational function of s that corresponds to the tabulated points and estimate the damping ratio of the dominant closed-loop roots for the first case in part (b).

Table P11.1

ω(rad/sec)	$\lvert F(j\omega)\rvert_{db}$	arg $[F(j\omega)]$(deg)
0.1	15	−95
0.3	5	−110
0.5	0	−122
1.0	−8	−146
2.0	−19	−175
3.0	−26	−192
5.0	−37	−214
10.0	−52	−237

11.15 Repeat parts (a) and (b) of Problem 10.24 using a Bode plot in place of the root locus. In part (b), select the gain K so as to achieve a 50° phase margin rather than a specific damping ratio for the dominant roots.

11.16 A feedback system has $GH(s) = K/s(\tau s+1)^2$ with nominal values of $\bar{K} = 10$ and $\bar{\tau} = 0.10$. Give the gain and phase margins for the following conditions:

1. K and τ at their nominal values,
2. $\pm 50\%$ changes in K, with $\tau = \bar{\tau}$,
3. 50% increase in τ, with $K = \bar{K}$.

11.17 Use Nyquist plots to verify the assertions stated in Problem 10.20 regarding the stability of the feedback system having time delays. Note that for $s = j\omega$, the time-delay transfer function has unity magnitude and a phase lag that increases linearly with frequency.

11.18 Show that for a unity-feedback system, that is, $H(s) = 1$, the ratio (a/b) of the lengths of the two vectors drawn to the point $G(j\omega_0)$ in Figure P11.3 is the magnitude of the closed-loop transfer function $C(j\omega)/R(j\omega)$. Give an expression for arg $[C/R]$ in terms of the angles of these two vectors, ϕ_1 and ϕ_2. Describe how these properties can be used to obtain the closed-loop frequency response $C(j\omega)/R(j\omega)$ graphically from the open-loop frequency response $G(j\omega)$.

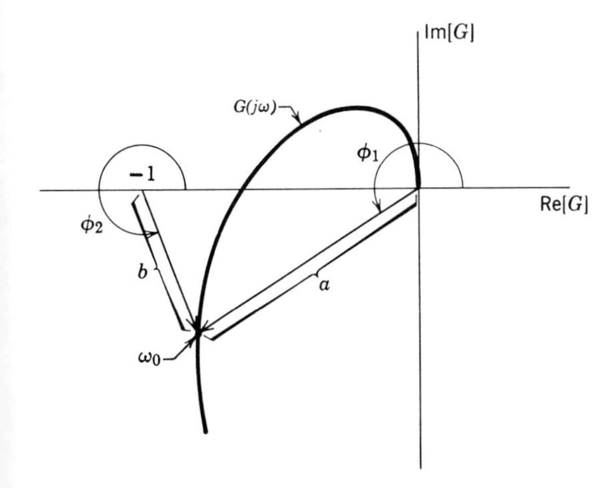

Figure P11.3

11.19 Repeat Problem 10.11, replacing the P + I controller by a lag network with transfer function $(\alpha s + 1)/(\tau s + 1)$, $\tau > \alpha$, in part (*a*), and the P + I + D controller by a lead-lag network in part (*b*). Use sketches of both root-locus and Nyquist plots to support your answers.

11.20 For the system having the open-loop transfer function $GH(s) = K/s(s+1)(0.10s+1)$, plot the following quantities versus K: *1*. gain margin, in decibels, *2*. phase margin, *3*. damping ratio (ζ) and natural frequency (ω_n) of the pair of dominant closed-loop roots. Values of gain yielding dominant roots with greater than critical damping need not be considered.

11.21 Draw the root locus for the system treated in Example 11.8 and locate the closed-loop roots for $K = 50$ and 75. Then give an interpretation of the frequency-domain characteristics, for example, gain margin, phase margin, closed-loop frequency-response magnitude, in terms of the root locus.

11.22 Show that the inverted stick treated in Problems 9.19 and 10.28 can be stabilized by passing the signal $\theta(t)$ through a filter having the transfer function $K(As+1)/(0.1As+1)$ and using the output to drive the motor.

(a) Use the Nyquist stability test to find values of the gain K and time constant A that will stabilize the system.

(b) This filter is often referred to as a "lead filter" and can be considered to be a realizable approximation to the P+D control equation. Explain the basis for this statement.

12

Sampling and discrete signals

Previous chapters have dealt with systems whose associated signals are *continuous*, i.e., functions of the continuous variable time. Furthermore, save for isolated discontinuities, these signals take the form of "smooth" curves typically illustrated in Figure 12.1*a*. Such time functions are commonly referred to by the designation *analog signals*, and most physical signals inherently belong to this category – with a few exceptions. One exception is the telegraph wave that has only two possible values, "on" and "off," although it is still a function of t as a continuous variable. Signals restricted in this fashion are said to be *discrete-amplitude* or *quantized*; Figure 12.1*b* shows a multilevel quantized version of Figure 12.1*a*.

In contrast to analog and quantized signals, *discrete-time signals* (hereafter called *discrete* signals, for short) are defined only at a sequence of separated points in time, i.e., they are functions of a discrete variable, say t_m. Often, but not always, the t_m are uniformly spaced and the values of the discrete signal may correspond to samples of a continuous signal, Figure 12.1*c*. Viewed more abstractly, however, a discrete signal is just an ordered sequence of numbers. *Digital signals*, a subclass of discrete signals, are simultaneously discrete in time and quantized in value, Figure 12.1*d*. As the name suggests, they arise quite naturally in digital systems, e.g., digital computers, regardless of whether sampling and quantizing of a continuous signal is involved.

The point of this discussion is that situations do come up when the systems engineer must work with other than continuous signals, at least insofar as mathematical representations are concerned. And, as we have tried to stress before, our analysis and design tools should be chosen to fit

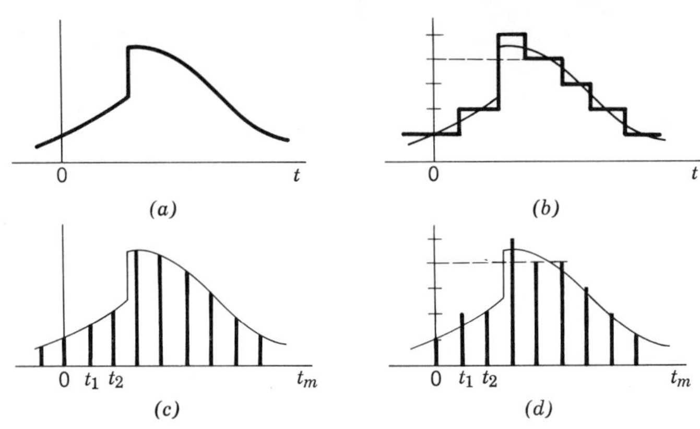

Figure 12.1. Signal classifications. (*a*) Analog. (*b*) Quantized. (*c*) Time-discrete. (*d*) Digital.

whatever problem is at hand. Those special techniques for handling discrete signals and systems will be developed here and in the following chapter.

The present chapter begins with an introduction to the theory of sampling, the process whereby a continuous signal is converted to discrete form, and reconstruction, by which the continuous signal is recovered from its sample values. This theory is then applied to such topics of interest as time-division multiplexing, analog-digital conversion, and the discrete Fourier transform. A closing section gives the signal-space interpretations.

12.1 UNIFORM SIGNAL SAMPLING

Every engineer has, at one time or another, drawn a smooth curve through a series of data points with the assumption that the data points are sufficient to completely define the entire curve. Sampling theory says essentially the same thing for functions of time, i.e., subject to certain conditions, a continuous signal is completely defined by a set of instantaneous sample values. That theory is covered below for the case of *periodic* or *uniform* sampling.

Choppers and sampling

As a simple but informative introduction to sampling, consider the switching circuit or *chopper* diagrammed in Figure 12.2*a*. The switch periodically shifts between the input-signal contact and the grounded contact, so the output signal $x_s(t)$ consists of short pieces or *samples* of the input waveform $x(t)$ but is otherwise zero. Figure 12.2*b* shows typical input and output waveforms; T_s is the switching period, corresponding to a switching rate or *sampling frequency* $f_s = 1/T_s$, and τ is the dwell time

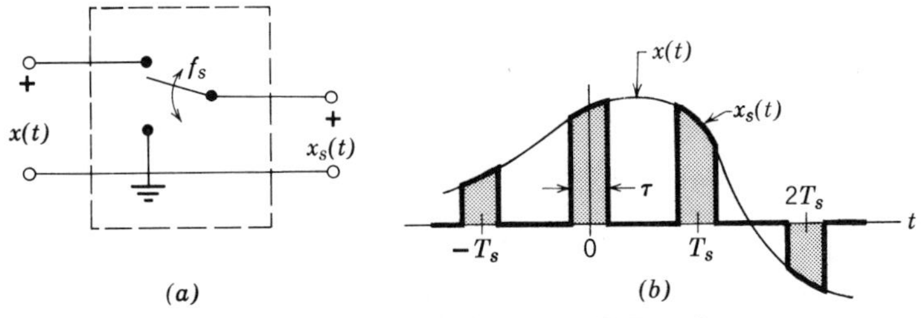

Figure 12.2. Chopper sampling. (*a*) Diagram. (*b*) Waveforms.

on the input contact, i.e., the *sample duration.* In view of the fact that most of the input wave has been chopped out by the switch, it would appear that one could not recover the original signal $x(t)$ from the sampled waveform $x_s(t)$. But, surprising as it may seem, recovery of $x(t)$ is possible providing certain conditions hold. To show this fact we turn to the frequency domain and examine $X_s(f) = \mathfrak{F}[x_s(t)]$, the spectrum of the sampled wave.

The necessary frequency-domain analysis will be simplified by introducing a *switching function* $s(t)$, the unit-amplitude rectangular pulse train with period T_s and pulse duration τ plotted in Figure 12.3*a*. Then, as a little thought will show, the switching operation can be mathematically modeled as *multiplication* of the input signal and the switching function, i.e.,

$$x_s(t) = x(t)s(t) \tag{1}$$

so we have the equivalent system diagram of Figure 12.3*b*. Since $s(t)$ is

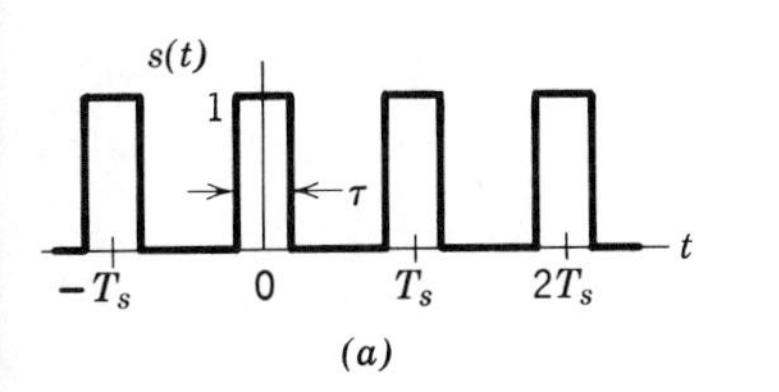

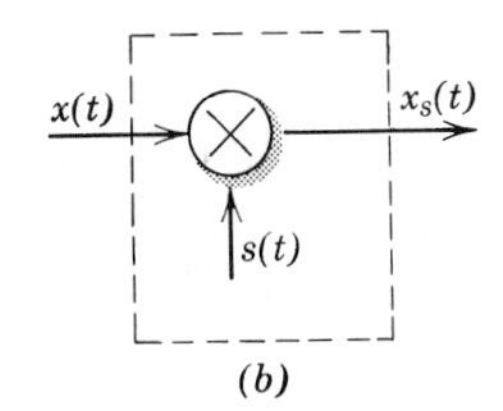

Figure 12.3. Model for chopper sampling. (*a*) Switching function. (*b*) Equivalent diagram.

periodic and real, it can be expressed as a trigonometric Fourier series

$$s(t) = \frac{\tau}{T_s}\left(1 + \sum_{n=1}^{\infty} 2A_n \cos n\omega_s t\right) \qquad \omega_s = 2\pi f_s \tag{2}$$

where we have used the results of Example 6.2 and, for compactness, introduced the new coefficients $A_n = \text{sinc}\, n\tau/T_s$. Combining (1) and (2) gives the term-by-term expansion of the sampled signal:

$$\begin{aligned} x_s(t) = \frac{\tau}{T_s}[x(t) &+ 2A_1\, x(t) \cos \omega_s t \\ &+ 2A_2\, x(t) \cos 2\omega_s t \\ &+ \cdots] \end{aligned} \tag{3}$$

Thus, if the input spectrum is $X(f) = \mathcal{F}[x(t)]$, the output spectrum is

$$\begin{aligned} X_s(f) = \frac{\tau}{T_s}\{X(f) &+ A_1[X(f-f_s) + X(f+f_s)] \\ &+ A_2[X(f-2f_s) + X(f+2f_s)] \\ &+ \cdots\} \end{aligned} \tag{4}$$

which follows directly from the Fourier-transform superposition and modulation theorems or, less directly, from the multiplication-convolution relationship.

Although (4) appears cumbersome to deal with at first glance, the spectrum actually is quite easily sketched – providing that the various spectral components do not overlap. Suppose, therefore, that $X(f) = 0$ for $|f| \geq W$ and that $f_s \geq 2W$. Figure 12.4 shows a typical input spectrum $X(f)$ and the corresponding output spectrum $X_s(f)$ under these assumptions. The significant feature to be observed here is that the sampling operation has left the original spectrum *completely intact*, merely repeating it regularly in the frequency domain with spacing f_s. Note, in particular, that $X_s(f)$ includes $X(f)$, multiplied by the scale factor τ/T_s but at its normal frequency position – corresponding to the first term in (3) or (4).

Drawing on this last observation, it follows that $X(f)$ can be extracted from $X_s(f)$ by *ideal lowpass filtering*. And if $X(f)$ is separated out from $X_s(f)$, then we have actually recovered $x(t)$ from the sampled signal $x_s(t)$. The basic condition for this to be possible is that the "sidebands" of $X_s(f)$ are *nonoverlapping in frequency*; this, in turn, requires that *1*. the

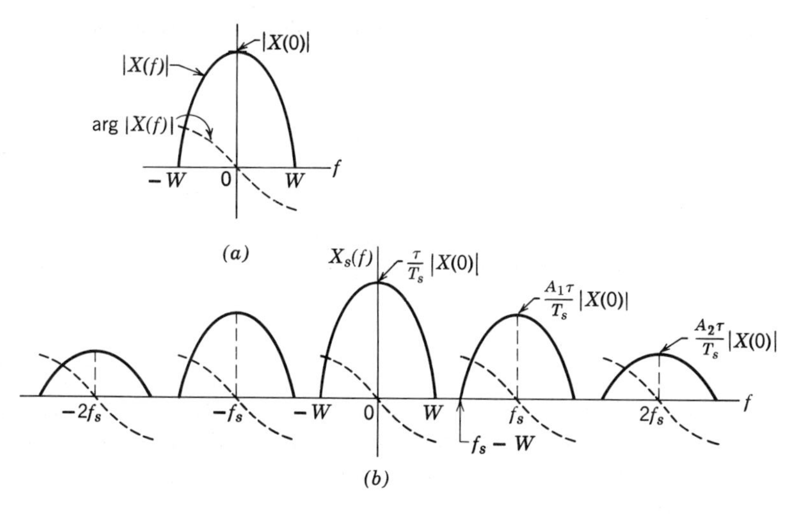

Figure 12.4. (*a*) Bandlimited input spectrum. (*b*) Spectrum of the sampled signal.

input signal must be *bandlimited*, say in W, and 2. the sampling frequency is sufficiently great that $f_s - W \geq W$ (see Figure 12.4*b*), i.e.,

$$X(f) = 0 \qquad |f| \geq W \tag{5a}$$

$$f_s \geq 2W \tag{5b}$$

If (5) is not satisfied then perfect recovery by lowpass filtering is impossible due to the presence of other frequency components in the filtered signal, a topic we shall return to later. The minimum sampling frequency for a given bandlimited signal is $f_{s_{\min}} = 2W$, known as the *Nyquist rate.*

To summarize here, our analysis has indicated that if a bandlimited signal is sampled periodically at the Nyquist rate or greater, it can be *completely reconstructed* from the sampled signal, the reconstruction being accomplished simply by lowpass filtering. These conclusions may be difficult to swallow at first exposure; they certainly test our faith in spectral analysis. Nevertheless, they are quite correct and are formalized in the theory of ideal sampling and reconstruction.

Ideal sampling

A quick review of the above results will show that they are essentially independent of the sample duration τ and should also apply to the case of *instantaneous* sampling where $\tau \to 0$. The only hitch is that if τ goes to zero, so does $x_s(t)$. Conceptually, we overcome this difficulty by defining the *instantaneous* or *ideal sampling wave*

$$s_\delta(t) \triangleq \lim_{\tau \to 0} \frac{1}{\tau} s(t)$$

with $s(t)$ the switching function previously introduced. Now $(1/\tau)s(t)$ would be a train of rectangular pulses each having duration τ and amplitude $1/\tau$. Thus, letting $\tau \to 0$ yields $s_\delta(t)$ as a *train of unit impulses*

$$s_\delta(t) = \sum_{m=-\infty}^{\infty} \delta(t - mT_s) \tag{6}$$

where T_s is the spacing. As graphically depicted in Figure 12.5, the ideal sampling wave looks like a "picket fence." We shall soon need the spectrum of this peculiar time function, so let us take a short digression to obtain $S_\delta(f) = \mathcal{F}[s_\delta(t)]$.

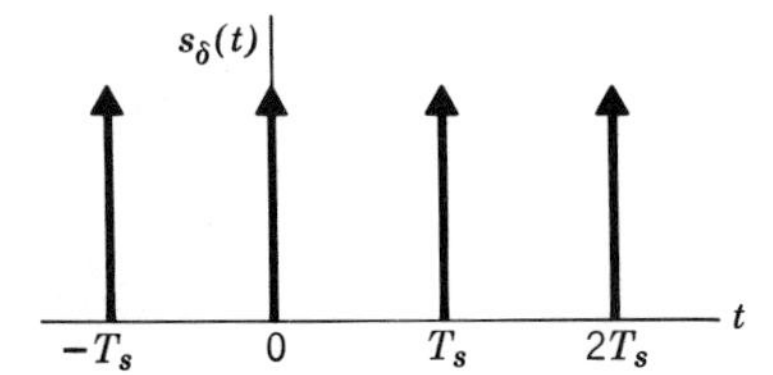

Figure 12.5. The ideal sampling wave.

Strictly speaking, $s_\delta(t)$ is not Fourier transformable. It has an infinite number of undefined values, neither its energy nor its power is well defined, and it violates all the usual conditions for the existence of a frequency-domain representation. Nonetheless, bearing in mind the abstract nature of $s_\delta(t)$, formally one can find a transform in the limiting sense by the following procedure. Since $s_\delta(t)$ is periodic with one impulse in each period, its exponential Fourier series coefficients are

$$c_n = \frac{1}{T_s} \int_{T_s} \left[\sum_{m=-\infty}^{\infty} \delta(t - mT_s) \right] e^{-jn\omega_s t}\, dt = \frac{1}{T_s} \int_{-T_s/2}^{T_s/2} \delta(t)\, e^{-jn\omega_s t}\, dt$$

$$= \frac{1}{T_s} e^{-jn\omega_s t} \bigg|_{t=0} = \frac{1}{T_s} = f_s$$

and hence

$$s_\delta(t) = \sum_{n=-\infty}^{\infty} f_s e^{jn\omega_s t} \tag{7}$$

Then, invoking superposition and the known transform pair

$$Ae^{j\omega_c t} \leftrightarrow A\delta(f-f_c)$$

yields

$$S_\delta(f) = \sum_{n=-\infty}^{\infty} f_s \delta(f-nf_s) \tag{8}$$

which says that the spectrum of the ideal sampling wave is a picket fence of impulses in the frequency domain with spacing f_s.

Returning to the main subject, we define *ideal instantaneous* sampling as the process that produces the sampled signal

$$x_\delta(t) = x(t)s_\delta(t) \tag{9a}$$

$$= x(t) \sum_{m=-\infty}^{\infty} \delta(t-mT_s) \tag{9b}$$

$$= \sum_{m=-\infty}^{\infty} x(mT_s)\delta(t-mT_s) \tag{9c}$$

where $x(mT_s)$ is the value of $x(t)$ at $t = mT_s$. Therefore, an ideally sampled signal is a train of impulses whose weights equal the instantaneous sample values $x(mT_s)$.

For $X_\delta(f)$, the spectrum of $x_\delta(t)$, we note that (9a) is multiplication in the time domain, which becomes convolution in the frequency domain. Hence

$$X_\delta(f) = X * S_\delta(f) = [X(f)] * \Big[\sum_{n=-\infty}^{\infty} f_s\delta(f-nf_s)\Big]$$

but, recalling the replication property of convolution with impulses, it follows that

$$X_\delta(f) = \sum_{n=-\infty}^{\infty} f_s X(f-nf_s) \tag{10}$$

as illustrated in Figure 12.6 taking $X(f)$ as the spectrum of Figure 12.4*a*. This spectrum differs from $X_s(f)$, Figure 12.4*b*, only in that all spectral components are multiplied by the same scale factor f_s rather than having a different scale factor for each sideband pair. Accordingly, if we invoke the same conditions as before, given by (5), then lowpass filtering will reconstruct $x(t)$ from its instantaneous samples $x(mT_s)$ as contained in $x_\delta(t)$.

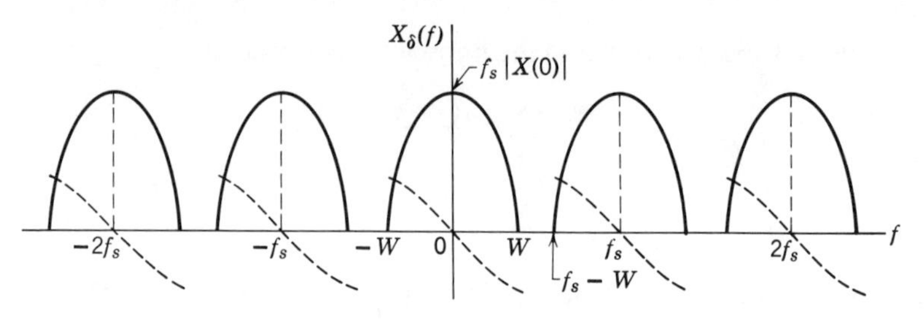

Figure 12.6. Output spectrum with ideal sampling.

The filter in question must be an ideal lowpass filter cutting off between W and $f_s - W$, i.e.,

$$H(f) = Ce^{-j\omega t_d}\Pi\left(\frac{f}{B}\right) \qquad W \leq B \leq f_s - W \tag{11}$$

where C is the gain and t_d the time delay. Since all frequency components in $X_\delta(f)$ outside the range $|f| \leq B$ are totally rejected, the filtered spectrum will be

$$Y(f) = Cf_s X(f) e^{-j\omega t_d}$$

and the corresponding output time function becomes

$$y(t) = \mathcal{F}^{-1}[Y(f)] = Cf_s x(t - t_d) \tag{12}$$

Thus, $y(t)$ equals the original signal $x(t)$ amplified by Cf_s and delayed by t_d, so ideal sampling and reconstruction is equivalent to *distortionless transmission.*

The *uniform sampling theorem* expresses the salient points of this section in the following statement:

> If a signal contains no frequency components for $|f| \geq W$, it is completely defined by instantaneous sample values uniformly spaced in time with period $T_s = 1/f_s \leq 1/2W$. If these samples are represented as weighted impulses, the signal can be exactly reconstructed by passing the impulse train through an ideal low-pass filter having bandwidth $W \leq B \leq f_s - W$.

The analysis leading to (10) and (12) constitutes our proof of the theorem; for an alternate derivation not based on frequency-domain arguments, see Carlson (1968, pp. 279–281).

Reconstruction

Further insight to the sampling process (and hopefully greater confidence in the theory) may be gained by viewing reconstruction in the *time domain.* For simplicity of discussion, let the filter specified by (11) have $B = f_s/2$, $C = 1/f_s$, and $t_d = 0$, so our frequency-domain result (12) reduces to $y(t) = x(t)$. We check this directly in the time domain using the filter's impulse response $h(t) = \mathcal{F}^{-1}[H(f)] = \text{sinc}\, f_s t$. Thus, since the input $x_\delta(t)$ is a train of impulses weighted by the sample values, as given in (9c), the output is just a train of impulse responses, namely,

$$y(t) = \sum_{m=-\infty}^{\infty} x(mT_s)h(t - mT_s)$$

$$= \sum_{m=-\infty}^{\infty} x(mT_s) \text{ sinc}\, f_s(t - mT_s) \qquad (13)$$

as plotted in Figure 12.7. Clearly, the correct values of $x(t)$ are recovered at the sampling instants $t = mT_s$, for at those times all the sinc functions are zero except for one whose peak is $x(mT_s)$. Between sample times, $x(t)$ is *interpolated* from the tails of all the sinc functions; that is to say that *all* the terms in (13) add up to yield the curve $x(t)$ when $t \neq mT_s$. For this reason the ideal lowpass filter may be called an *interpolating filter* and its impulse response is known as the *interpolation function.*

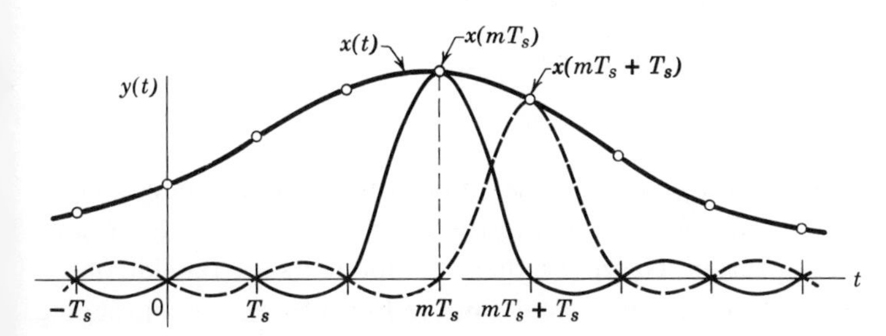

Figure 12.7. Ideal reconstruction, $y(t) = x(t)$.

But now we should give some attention to the practical implementation of signal sampling as contrasted with the ideal theory. In practice, there are two obvious departures: *1.* the sampled wave surely cannot be a train of *impulses*, and *2. ideal* lowpass filters are physically unrealizable.

With regard to pulse-shape considerations, our analysis of signal chopping has shown that rectangular sampling pulses are just as good as impulses insofar as recovery is concerned. Moreover, similar investigations support this conclusion for other finite pulse shapes. There is a minor complication when instantaneous sample values are extracted and "held" for a nonzero duration, but the resulting *aperture effect* can be corrected by equalization filtering at the final output; see Problem 12.7.

With regard to reconstruction, the principal difference between real and ideal filters is that the former do not have an abrupt change from passband to rejection nor do they have identically zero response outside the passband. But if we allow for a *transition region* of sufficient width and there are no frequency components present in this region then, to all intents and purposes, real filters will closely achieve the function of ideal filters. In the context of sampling this implies that the sampling rate f_s should be substantially greater than the theoretic minimum to allow a "gap" in the spectrum of the sampled wave into which the filter's transition region can be fitted. Figure 12.8 illustrates this strategy, and the

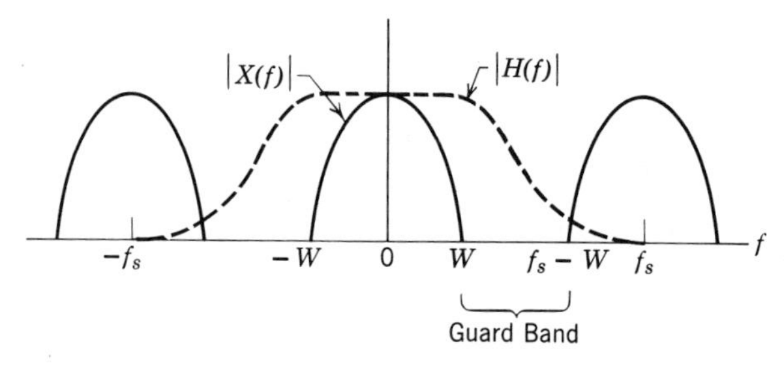

Figure 12.8.

spectral gaps are called *guard bands*.† Note that some frequency components at $|f| > W$ will get through to the output since a real filter does not give perfect rejection. However, these components are outside of the signal's bandwidth and they may be further reduced by either sharpening the cutoff characteristics of the filter or increasing f_s. In fact, the system designer commonly is faced with a trade-off between precisely these two factors.

Sometimes, especially in sampled-data control systems, a device known as the *zero-order hold* (also called a *box-car circuit*) is used for reconstruction purposes. This device makes a staircase approximation to the original signal by taking the instantaneous sample values and "holding" them for the entire sampling interval T_s, as shown in Figure 12.9.

†The use of guard bands is not unique to sampling systems, for they are also necessary in frequency-division multiplexing. After all, ideal *bandpass* filters are just as unrealizable as ideal lowpass filters. See Figure 7.23.

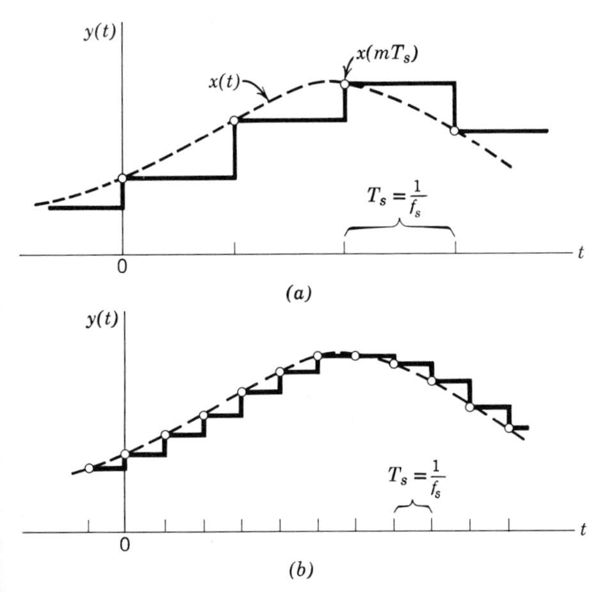

Figure 12.9. Approximate reconstruction using zero-order hold circuit. (*a*) "Low" sampling rate. (*b*) "High" sampling rate.

Analytically, its output would be

$$y(t) = \sum_{m=-\infty}^{\infty} x(mT_s)[u(t-mT_s) - u(t-mT_s-T_s)]$$

assuming that the input approximates the ideal sampled wave $x_\delta(t)$. It therefore follows that the impulse response of the zero-order hold is similar to the finite-time integrator, Eq. (16), Sect. 4.3, i.e.,

$$h(t) = u(t) - u(t-T_s) = \Pi\left[\frac{t-(T_s/2)}{T_s}\right] \tag{14a}$$

corresponding to the frequency response characteristic

$$H(f) = T_s \operatorname{sinc} fT_s e^{-j\omega T_s/2} \tag{14b}$$

which is sketched in Figure 12.10 (for positive f only) along with a typical input spectrum $X_\delta(f)$ and the resulting output spectrum $Y(f)$. This figure shows that additional lowpass filtering should substantially improve the recovery process; on the other hand, referring to Figure 12.9 indicates that additional filtering may not be necessary if T_s is sufficiently small, i.e., a *high sampling rate*.

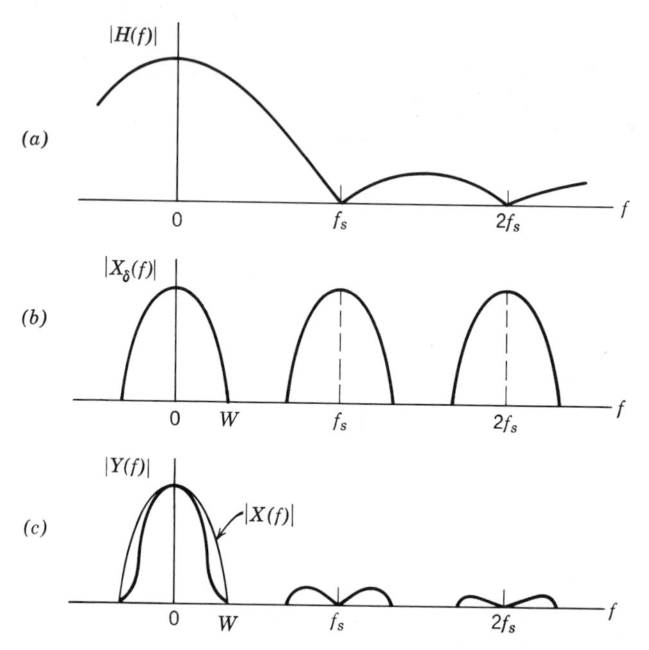

Figure 12.10. Frequency-domain analysis of the zero-order hold. (*a*) Amplitude ratio. (*b*) Input amplitude spectrum. (*c*) Output amplitude spectrum.

Bandlimiting, timelimiting, and aliasing

Although impulsive sampled waves and ideal lowpass filters are not essential for perfect recovery of a sampled signal, there is yet another assumption in ideal sampling theory that cannot be satisfied in practice and may cause serious problems. This assumption is that the original signal $x(t)$ has a *bandlimited* spectrum, i.e., there must be some finite constant W such that

$$X(f) = 0 \qquad f \leq -W \quad \text{and} \quad f \geq W \tag{15}$$

Consider, on the other hand, a *timelimited* signal defined as being zero everywhere outside a certain time interval, say $t_1 < t < t_2$, i.e.,

$$x(t) = 0 \qquad t \leq t_1 \quad \text{and} \quad t \geq t_2 \tag{16}$$

We see that (15) and (16) are dual time-domain and frequency-domain concepts. And it turns out that *timelimiting and bandlimiting are mutually exclusive*: a bandlimited signal cannot be timelimited, and vice versa. Proving this assertion is rather involved – see Wozencraft and Jacobs

(1965, pp. 352–353) – but its plausibility is strengthened by reviewing some of our known transforms of bandlimited and timelimited signals.

As to the significance for sampling, one must first observe that actual physical signals are *timelimited*; they start at some particular instant and eventually they will stop. Therefore, physical signals do not have bandlimited spectra and our theory of sampling does not hold in the strict sense. Specifically, regardless of the sampling rate, there will be *overlapping frequency components* in the spectrum of the sampled wave; this, in turn, produces a phenomenon known as *aliasing*.

By way of explanation, consider the signal spectrum $X(f)$ in Figure 12.11*a*, a spectrum which is not strictly bandlimited but does fall off rapidly with increasing frequency. It can thus be said that there exists some finite bandwidth W such that the range $|f| < W$ contains all the *significant frequency components* of $X(f)$. Of course, the choice of W is more or less arbitrary and usually must be made on the basis of experimental studies. Tests have shown, for example, that intelligible and natural reproduction of the human voice requires a bandwidth in the neighborhood of 3 to 4 kHz, while high-fidelity music systems need 10 to 20 kHz, depending on the acuity of the listener.

Suppose then that W has been chosen on some reasonable basis, the signal is sampled at $f_s = 2W$, and the recovery filter is essentially ideal with $B = f_s/2$. Figure 12.11*b* gives the spectrum of the sampled wave, and we see that after lowpass filtering there will be unwanted frequency components in the recovered signal which, in the original spectrum, were above $|f| = f_s/2$ but now, due to spectral overlap, appear in the range $|f| < f_s/2$. For instance, a frequency component such as $W + \Delta$, which previously was beyond the nominal signal bandwidth W, comes out as $f_s - (W + \Delta) = W - \Delta$, which is within W.

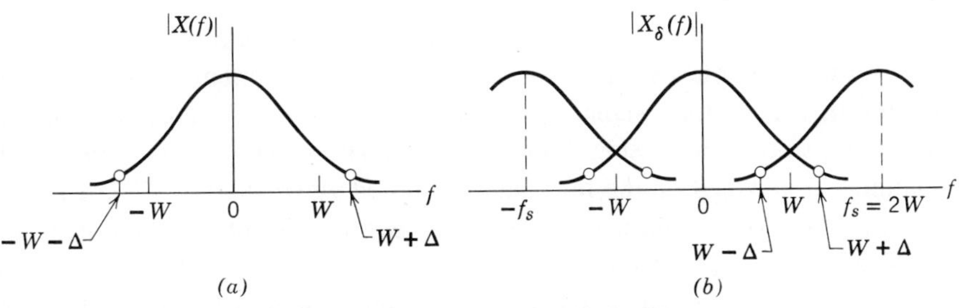

Figure 12.11. The aliasing phenomenon. (*a*) Nonbandlimited input spectrum. (*b*) Spectral overlap.

This transmutation of high-frequency to low-frequency components in the course of sampling and recovery is aliasing. Clearly, it will occur in all sampling systems because the input spectra are not bandlimited. However, through the use of lowpass filtering *before* sampling, together with a sufficiently high sampling rate, these aliased components can be rendered insignificant in most cases.

Summarizing in the light of the above considerations, we finally return to the ideal sampling theorem and paraphrase it in less precise form as a *guide for practical sampling*:

> If a signal has negligible frequency content for $|f| \geq W$ and is periodically sampled at a rate $f_s \geq 2W$, then the original waveform can be recovered with reasonable accuracy by passing the sampled wave through a lowpass filter of bandwidth W. Highly accurate recovery may entail lowpass filtering at the input, sampling at a rate much greater than $2W$, and, possibly, equalization filtering at the output.

With this guide in mind, let us look at some of the important applications.

12.2 APPLICATIONS OF SAMPLED SIGNALS

Three of the numerous applications of sampling theory in systems engineering have been selected for discussion below, namely time-division multiplexing, pulse-code modulation, and analog-digital conversion. Each represents a situation where the properties of sampled signals are exploited to achieve some desired goal. Our coverage here is necessarily rather abbreviated; for more complete information, consult the list of Supplementary Reading for source material.

Time-division multiplexing

Frequency-division multiplexing was introduced in Section 7.5 as a means of sending two or more signals on one transmission system, the trick being to assign the various signals to disjoint slots in the frequency domain. *Time*-division multiplexing (TDM) accomplishes the same task in a dual manner. It is based on the fact that a sampled wave such as Figure 12.2*b* is "off" most of the time, and therefore the intervals between samples can be filled with samples from other signals.

The rudiments of a TDM system are diagrammed in Figure 12.12*a* for the case of three input signals assumed to have the same nominal bandwidth W. A switch or *commutator*† rotating at the rate $f_s \geq 2W$ sequentially extracts samples from each input to produce the waveform $x(t)$, Figure 12.12*b*. Note that $x(t)$ contains one sample every $T_s = 1/f_s$ seconds for each input signal, giving a total of kf_s samples per second if there are k inputs. At the receiving end a similar switch, called a *distributor*, separates the interleaved samples and distributes them to a bank of lowpass filters for individual reconstruction. Thus, the essence of time-division multiplexing is assigning the signal samples to disjoint periodic slots in the time domain.‡

TDM offers several advantages (and disadvantages) as compared with FDM. For one thing, TDM instrumentation is simpler thanks to technological advances in integrated switching circuits; however, it does

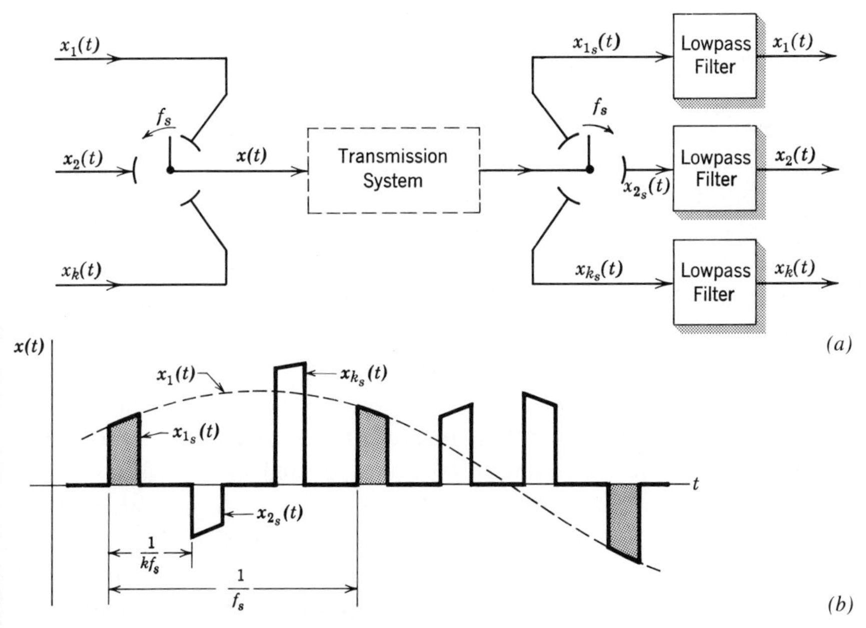

Figure 12.12. A three-channel TDM system. (*a*) Block diagram. (*b*) Multiplexed waveform.

†Figure 12.12*a* implies that the commutator and distributor are mechanical or electromechanical switches. This is often the case when the sampling rate is sufficiently low – as is true, for instance, in certain control and telemetry systems. Higher sampling frequencies are achieved by electronic switching.

‡And, consequently, the individual components of the TDM signal are mutually *orthogonal*.

require precise *synchronization* between commutator and distributor whereas some FDM systems do not. For another, TDM is immune to the usual sources of interchannel *cross talk* that afflict FDM, but the price of immunity is very large transmission bandwidth to accommodate the short-duration pulses. In brief, the choice between TDM and FDM for a given multiplexing system must be made by considering many factors, and there is no pat answer.

Pulse-code modulation

Pulse-code modulation (PCM) is a signal-transmission method in which an *analog* input signal is converted to *digital* form prior to sending. Figure 12.13*a* diagrams the operations at the transmitter while part (*b*) shows typical results.

The input $x(t)$ is first sampled at an appropriate rate. Then the sample values are *quantized* (or digitized) so that each $x(mT_s)$ is rounded off to be one of Q numbers, called the *quantization levels*. The result, $x_{sq}(mT_s)$, consists of a train of quantized numbers, one for each sample, and therefore is a *digital* signal. [Note that $x_{sq}(mT_s)$ may be viewed as the output of a digital voltmeter operating in a repetitive mode on $x(t)$.] The final step, *encoding*, transforms each quantized sample into a block of ν discrete-amplitude pulses – a pulse-code group – with each pulse limited to have one of μ possible amplitudes. Figure 12.13*b* has $Q = 8$ quantization levels and *binary* encoding, i.e., $\mu = 2$, represented by pulses of positive and negative polarity. Binary encoding, we might add, is the type most often employed.

To analyze PCM generation, one must determine the relationship between Q, ν, and μ. Assuming Q has been specified, as discussed shortly, the encoding parameters are found by the following argument. Since a pulse-code group consists of ν pulses with μ possible amplitudes, there are exactly μ^ν *different groups* to represent the Q possible quantized sample values. Therefore, it is necessary that $\mu^\nu \geq Q$ or, for a given μ,

$$\nu \geq \log_\mu Q \tag{1}$$

For instance, in correspondence to Figure 12.13*b* with $Q = 8$ and $\mu = 2$, the number of pulses per block becomes $\nu \geq \log_2 8 = \log_2 2^3 = 3$. Examining (1) also indicates that efficient operation, i.e., minimum ν, calls for the number of quantization levels to be chosen such that $\log_\mu Q$ is an integer.

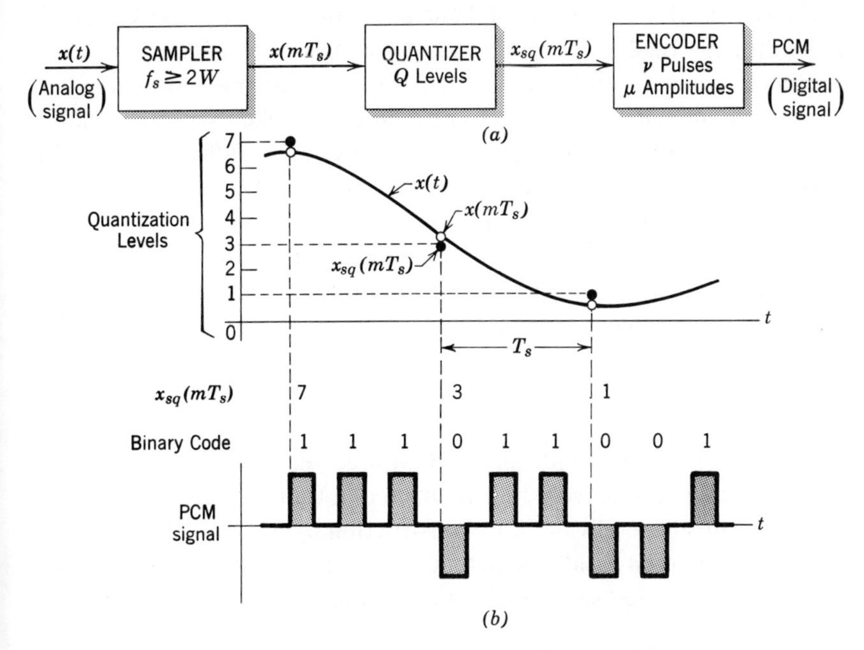

Figure 12.13. PCM transmitter (or A/D converter). (*a*) Block diagram. (*b*) Waveforms.

The fact that the encoded signal contains several pulses for each sample value has an important *bandwidth* implication. Recall from Example 7.1 that the spectrum of a rectangular pulse has most of its content in the range $-1/\tau < f < 1/\tau$, so for pulse transmission (when the exact output pulse shape is not critical) one can get by with a bandwidth $B \geq 1/2\tau$. Thus, to calculate the PCM bandwidth requirement, we need the maximum allowed duration of the coded pulses. That value is found by noting that there are νf_s coded pulses per second, where $f_s \geq 2W$; eliminating ν via (1) yields $\nu f_s \geq 2W \log_\mu Q$ and, for any one coded pulse,

$$\tau = \frac{1}{\nu f_s} \leq \frac{1}{2W \log_\mu Q} \tag{2}$$

Hence

$$B \geq \frac{1}{2\tau} \geq W \log_\mu Q \tag{3}$$

which shows that the PCM bandwidth B is generally much greater than the analog signal's bandwidth W, and it increases logarithmically with Q.

By now, in view of the complicated instrumentation and excessive bandwidth, the reader is probably questioning the practicality of pulse-code modulation. Unfortunately, a simple explanation is impossible here because the advantages of PCM are inherently linked with the relative merits of digital versus analog transmission systems. Suffice it to say that PCM gives the communications engineer another degree of freedom in tailoring his system design for a particular application. One of the flexibilities that PCM allows can be appreciated, at least qualitatively, from a discussion of the recovery process at the receiving end where the transmitter's digitizing comes home to roost in the guise of quantization noise.

Quantization noise

Figure 12.14 presents an idealized PCM receiver. The digital pulse train is first decoded and delivered in the form of a train of weighted impulses – or the realizable equivalent. Reconstruction by lowpass filtering or the zero-order hold then yields the output analog signal. Thus, it might appear that the reconstructed signal is identical to the original analog signal $x(t)$,

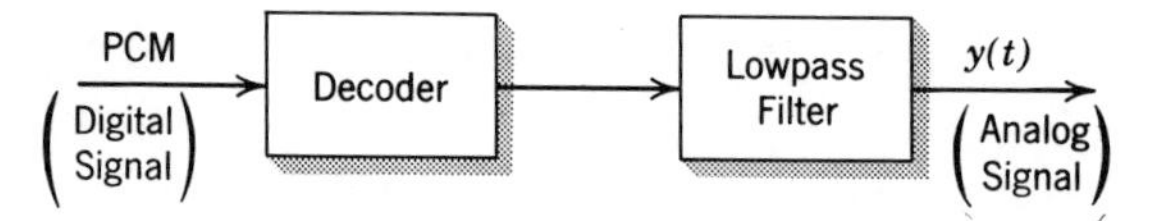

Figure 12.14. PCM receiver (or D/A converter).

subject only to the limitations of practical sampling and recovery. But this overlooks the fact that it is the *quantized* sample values $x_{sq}(mT_s)$, rather than $x(mT_s)$, emerging from the decoder. And no operation on the quantized data will ever restore the exact sample values; they have been irrevocably lost in the process of quantizing.

From a slightly different perspective, it is said that the recovered signal is contaminated by *quantization noise*. To clarify this point, we note that the order of sampling and quantizing at the transmitter is interchangeable, so one can think of the quantized signal $x_q(t)$ prior to sampling as being

$$x_q(t) = x(t) - \varepsilon(t)$$

where $\varepsilon(t)$ is the *error* waveform, Figure 12.15. If the quantizing levels are equally spaced, say by α units, and if the quantization consists of rounding-off to the nearest level, either up or down, then it follows that

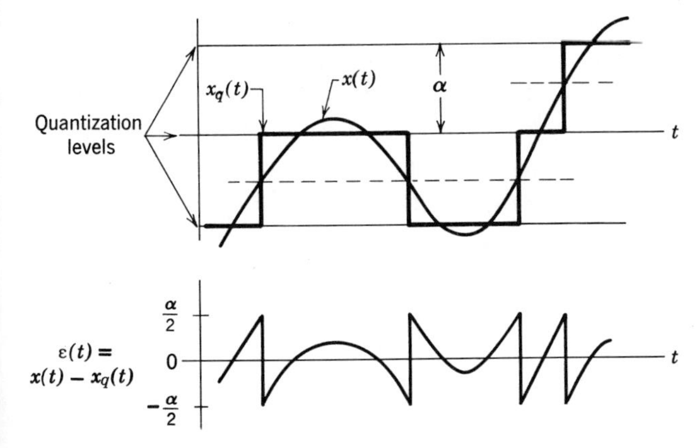

Figure 12.15. Waveforms in quantizing.

$\varepsilon(t)$ is bounded by

$$|\varepsilon(t)| \leq \frac{\alpha}{2} \tag{4}$$

as compared with $|x(t)| \leq Q\alpha/2$. Periodic sampling then gives

$$x_{sq}(mT_s) = x(mT_s) - \varepsilon(mT_s) \tag{5}$$

for the digital signal. Assuming now that $x(t)$ is bandlimited in W and we have perfect reconstruction based on the sample values given by (5), the recovered analog signal will be

$$y(t) = x(t) + n(t) \tag{6}$$

where $n(t)$ is the quantization noise corresponding to the samples of $-\varepsilon(t)$. Note that $n(t)$ does not equal $-\varepsilon(t)$ because the latter, in general, is not bandlimited.

Since (6) has the form of desired signal plus undesired "noise," a meaningful measure of system quality is the ratio of average signal power to noise power. Denoting P_x and P_n as the average powers in $x(t)$ and $n(t)$, respectively, it can be shown that the signal-to-noise ratio is†

$$\frac{P_x}{P_n} = KQ^2 \tag{7}$$

†See, for example, Carlson (1968, pp. 307–309).

where K is a constant falling in the range $0.1 < K < 10$ but whose exact value depends on the specific properties of $x(t)$. For most applications, the quantization noise may be deemed negligible if P_x/P_n is at least 30 to 50 db (power ratios of 10^3 to 10^5), which obviously means that the number of quantization levels must be quite large. In voice telephone PCM transmission, for example, it has been found that $Q = 128$ gives good intelligibility. (Incidentally, with the usual binary encoding, that value of Q requires $\nu = \log_2 128 = 7$ coded pulses per block, so, taking $W = 4$ kHz as the nominal voice bandwidth, the transmission bandwidth must be at least $B = 4 \times 10^3 \times 7 = 28$ kHz.)

Finally, relative to design flexibility, we point out that PCM provides a trade-off between transmission bandwidth B and signal-to-noise ratio P_x/P_n. This is demonstrated by taking (3) as an equality and solving for Q to get $Q = \mu^{B/W}$; whence, inserting in (7),

$$\frac{P_x}{P_n} = K\mu^{2B/W} \tag{8}$$

which shows that for fixed values of μ and W, the signal-to-noise ratio increases exponentially with B. This possible exchange of increased transmission bandwidth for improved system performance, called *wideband noise reduction*, is one of the more important features of pulse-code modulation. All signal transmission systems suffer from noise contamination; in analog systems it is random noise largely beyond the designer's control. But PCM is relatively unaffected by random noise and its own internal quantization noise is controllable in the sense of (8).

Analog-digital conversion

A common task of the systems engineer – be it in communication, control, or the like – is to design a signal processing system that operates on a continuous input signal $x(t)$ to produce a continuous output signal $y(t)$ that meets some set of specifications. Because both signals are continuous, it is only natural to think in terms of a continuous processing system, i.e., an *analog* system, at least as a first cut at the problem. But the possibilities of discrete or *digital* processing should not be overlooked; for, drawing upon modern microelectronics technology, switching theory, etc., certain operations are now being accomplished digitally that before had been completely impractical by analog methods.

Of course, appropriate *interfacing* is required when digital signal processing is used in conjunction with analog signals. Thus, one is led to the system configuration diagrammed in Figure 12.16, where the digital processor is sandwiched between two interfaces, an *analog-to-digital* (A/D) *converter* on the input side and a *digital-to-analog* (D/A) *converter* at the output. The functions of A/D and D/A conversion are virtually the same as take place in a PCM transmitter and receiver (Figures 12.13 and

Figure 12.16. Interfacing a digital signal processor with analog-to-digital (A/D) and digital-to-analog (D/A) converters.

12.14), respectively; Hoeschele (1968) may be consulted for further information. As for the processor itself, it is a *discrete system* and will be discussed in Chapter 13.

12.3 DISCRETE FOURIER TRANSFORMS

Just as the Fourier transform of a continuous signal leads to its description in the frequency domain, it is possible to define a *discrete Fourier transform* (DFT) as the frequency-domain representation in the case of discrete signals. Actually, as a mathematical operation, the DFT can be applied to any sequence of numbers as well as to discrete signals. More often than not, however, it is used in conjunction with sample values from a continuous signal; this is the context in which the frequency-domain interpretation becomes meaningful and it will be adopted when introducing the subject below.

Consider the problem of spectral analysis based on experimental observation of a continuous signal. While it might be possible to devise an expression for the observed time function and from that calculate the Fourier transform, a more practical approach would be to take sample values and use numerical methods to evaluate the frequency-domain representation. This latter course, obviously lending itself to implementation via digital computer, is the motivation behind the DFT. The recent development of highly efficient algorithms for computing DFT's – notably the so-called *fast Fourier transform* or FFT – has given further impetus in this direction.†

†Oppenheim (1970) describes an interesting application in the spectral analysis of speech.

Suppose, therefore, that M samples are extracted from a continuous signal $v(t)$ at the rate of one every T seconds.† For convenience we temporarily assume that this process is centered around $t = 0$, as in Figure 12.17*a*, so M is an odd number, say $M = 2N + 1$. This gives as our working base the sequence of values

$$v(mT) \qquad m = 0, \pm 1, \pm 2, \ldots \pm N \tag{1}$$

Because we have a restricted amount of data – restricted in the sense that a *finite* number of samples cannot completely describe a continuous signal, bandlimited or not – any spectral analysis derived therefrom must itself be of a restricted nature. Specifically, in view of the time-domain data, the frequency-domain representation can consist of at most M *independent terms*, meaning that it is inherently *discrete in frequency*.

To better appreciate this point, note that the minimum spacing between samples is T seconds; hence, from the sampling theorem we know that frequency components in $v(t)$ with periods shorter than $2T$ cannot be resolved uniquely due to the aliasing effect. At the other extreme, frequency components with periods longer than the observation interval‡ MT surely will cause troubles because we do not see a full cycle; hence, aside from a possible DC term, the lowest frequency that can be resolved is $1/MT$. Putting these two factors together, the time-domain data of (1) limits us to the set of *resolvable frequencies*

$$kf_0 = \frac{k}{MT} \qquad k = 0, \pm 1, \ldots \pm N \tag{2}$$

so that

$$|f|_{\max} = \frac{N}{MT} = \frac{1 - (1/M)}{2T} < \frac{1}{2T}$$

Thus, regardless of its specific details, the general structure of a DFT will be as indicated in Figure 12.17*b*, namely a *line spectrum* covering the range $|f| < 1/2T$ with $f_0 = 1/MT$ being the frequency increment or "fundamental." Despite the similarity between this figure and the line spectrum of a periodic signal, one should not conclude that frequencies other than kf_0 are not present in $v(t)$. What it means, simply, is that our data is insufficient to resolve unambiguously any other frequencies.

†The subscript s has been dropped from T_s for simplicity.

‡Although the time between the first and last sample is $(M - 1)T$ seconds, the observation interval is effectively MT seconds since we have M samples and each sample represents T seconds of $v(t)$.

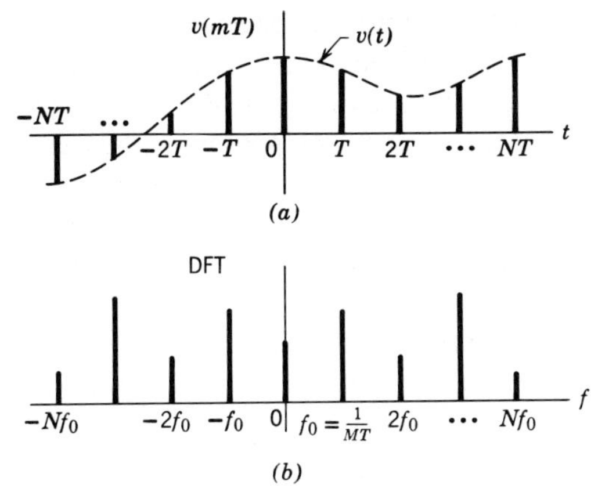

Figure 12.17. (*a*) The discrete signal $v(mT)$, $|m| \leq N$. (*b*) General structure of the discrete Fourier transform.

Definition and properties of the DFT

With the above ideas in mind, we proceed to define the discrete Fourier transform of $v(mT)$ to be

$$c(nf_0) = \frac{1}{M} \sum_{m=-N}^{N} v(mT) e^{-j2\pi n f_0 mT} \qquad n = 0, \pm 1 \ldots \pm N \tag{3}$$

where

$$f_0 = \frac{1}{MT} \qquad M = 2N + 1$$

The inverse DFT takes one back to the time-domain values via

$$v(mT) = \sum_{n=-N}^{N} c(nf_0) e^{j2\pi n f_0 mT} \qquad m = 0, \pm 1, \ldots \pm N \tag{4}$$

whose form is the same as an exponential Fourier series with t set equal to mT. Drawing upon this parallel, we can interpret the coefficients of (4) just as before, i.e., each complex number $c(nf_0)$ gives the amplitude and phase of the frequency component of $v(mT)$ at $f = nf_0$.

To clarify one matter immediately, the difference between the DFT and the Fourier series must be pointed out. The Fourier series represents a continuous periodic signal which, of course, lasts for $-\infty < t < \infty$, and the resulting line spectrum generally covers $-\infty < f < \infty$. The DFT represents a finite number of sample values in the finite observation interval $-MT/2 < t < MT/2$, and the resulting line spectrum is limited to $-1/2T < f < 1/2T$.† Nevertheless, the DFT has many properties identical or equivalent to those discussed in Section 6.4. For instance, the superposition theorem certainly holds here, and corresponding theorems can be developed covering time delay, scale change, etc. But perhaps the most convincing arguments in support of (3) as a frequency-domain representation are based on its frequency selectivity and its agreement with sampling theory.

The selectivity or "frequency response" of the DFT is best investigated by considering a continuous signal having only one frequency component, say

$$v(t) = Ae^{j(\omega_c t+\theta)} = Ae^{j\theta}e^{j2\pi f_c t} \tag{5}$$

Inserting the sample values into (3) gives

$$\begin{aligned} c(nf_0) &= \frac{1}{M}\sum_{m=-N}^{N} \overbrace{(Ae^{j\theta}e^{j2\pi f_c mT})}^{v(mT)} e^{-j2\pi nf_0 mT} \\ &= Ae^{j\theta}\underbrace{\left\{\frac{1}{M}\sum_{m=-N}^{N}[e^{-j2\pi(nf_0-f_c)T}]^m\right\}}_{F_M(nf_0-f_c)} \end{aligned}$$

which can be expressed in closed form by first letting $nf_0 - f_c = f$ and introducing

$$\begin{aligned} F_M(f) &\triangleq \frac{1}{M}\sum_{m=-N}^{N}(e^{-j2\pi fT})^m \\ &= \frac{1}{M}(e^{j2\pi fT})^{-N}\sum_{m=0}^{2N}(e^{j2\pi fT})^m \end{aligned} \tag{6}$$

We then note that $2N = M-1$ and invoke the summation formula for geometric progressions:

$$\sum_{m=0}^{M-1} w^m = \frac{w^M - 1}{w-1} = w^{(M-1)/2}\frac{w^{M/2}-w^{-M/2}}{w^{1/2}-w^{-1/2}}$$

†Actually, the DFT does predict values outside these ranges if one takes $|m| > N$ or $|n| > N$. Indeed, both (3) and (4) are *periodic* with periods $\Delta f = 1/T$ and $\Delta t = MT$, respectively.

where, in (6), $w = e^{j2\pi fT}$ and $(M-1)/2 = N$. Therefore,

$$F_M(f) = \frac{1}{M}\frac{e^{j\pi MfT} - e^{-j\pi MfT}}{e^{j\pi fT} - e^{-j\pi fT}}$$

$$= \frac{\sin \pi MfT}{M \sin \pi fT} \tag{7}$$

and, finally,

$$c(nf_0) = Ae^{j\theta}F_M(f-f_c)\big|_{f=nf_0} \tag{8}$$

Equation (8) states that the *shifted* function $|F_M(f-f_c)|$ defines the *envelope* of $|c(nf_0)|$ when $v(t)$ is given by (5). So an understanding of this situation entails looking at the envelope per se as a continuous function of f. Figure 12.18 shows plots of $F_M(f)$ versus f for two values of M. It is seen to be rather like a *periodically repeated sinc function*; the fundamental period is $1/T$, there is even symmetry, and there are zero-crossings at all integer multiples of $f_0 = 1/MT$ save for $f = 0, \pm 1/T, \pm 2/T, \ldots$, where $F_M = 1$. Applying this information to (8), it follows that we have three different cases to consider, depending on the exact value of f_c.

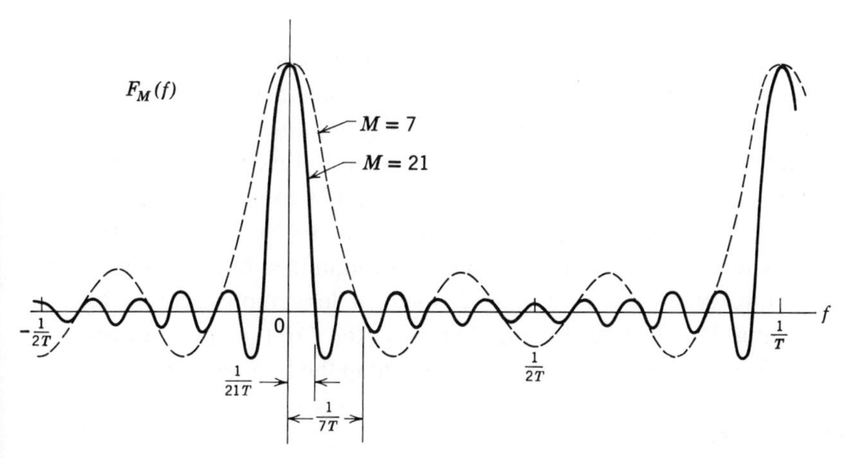

Figure 12.18. Plot of $F_M(f) = (\sin \pi MfT)/(M \sin \pi fT)$.

Case (a): $f_c = kf_0$, $|k| \leq N$. Since $F_M(nf_0 - kf_c) = F_M[(n-k)/MT] = 0$ except for $k = n$, (8) gives just one nonzero term, i.e.,

$$c(nf_0) = \begin{cases} Ae^{j\theta} & n = k \\ 0 & n \neq k \end{cases}$$

Therefore, the correct amplitude and phase has been detected from the sample values in this case. The amplitude spectrum is plotted in Figure 12.19*a* with the envelope indicated by a dashed line.

Case (*b*): $f_c \neq kf_0$, $|f_c| < 1/2T$. Now, instead of one line at f_c, the DFT yields *many* frequency components, predominantly in the neighborhood of f_c but none exactly at f_c, as illustrated in Figure 12.19*b*. The inability of the DFT to properly resolve the frequency content here stems directly from the limited working data. Indeed, referring back to Figure 12.18 reveals that the resolution will improve with increasing M, i.e., more samples, since increasing M causes $F_M(nf_0 - f_c)$ to narrow-in around $f = f_c$.

Case (*c*): $|f_c| \geq 1/2T$. Here f_c is beyond the allowed spectral range and $c(nf_0)$ should be zero for all $|n| \leq N$. But, because of the periodicity of $F_M(f)$, the DFT actually gives one or more frequency components within the range $|f| < 1/2T$. This effect, depicted in Figure 12.19*c*, is the aliasing phenomenon of the discrete Fourier transform.

Combining Case (*a*) with the superposition theorem, we can say that the DFT exactly detects the amplitude and phase of all frequency components in $v(t)$ that belong to the set of resolvable frequencies listed in (2). But any other components that may be present in the signal will result in false or *spurious* spectral terms, as in Cases (*b*) and (*c*), thereby implying that $v(t)$ contains frequencies that are not really there. Such spurious terms are eliminated or at least minimized by: *1*. sampling at a higher rate (decreasing T), and *2*. using a longer observation interval (increasing M). Putting this another way, the DFT cannot squeeze out more information than is inherently present in the time-domain data.

DFT's and sampling theory

Pursuing this last remark, let us show that the DFT does give an accurate frequency-domain representation when one has "sufficient" data. For a continuous signal $v(t)$ to be described with reasonable accuracy by a finite number of samples, say $v(mT)$ with $|m| \leq N$, it is necessary that $v(t)$ be bandlimited in $W < 1/2T$ and be negligibly small† outside the observation

†But $v(t)$ cannot be zero everywhere outside the observation interval because time-limiting and bandlimiting are mutually exclusive.

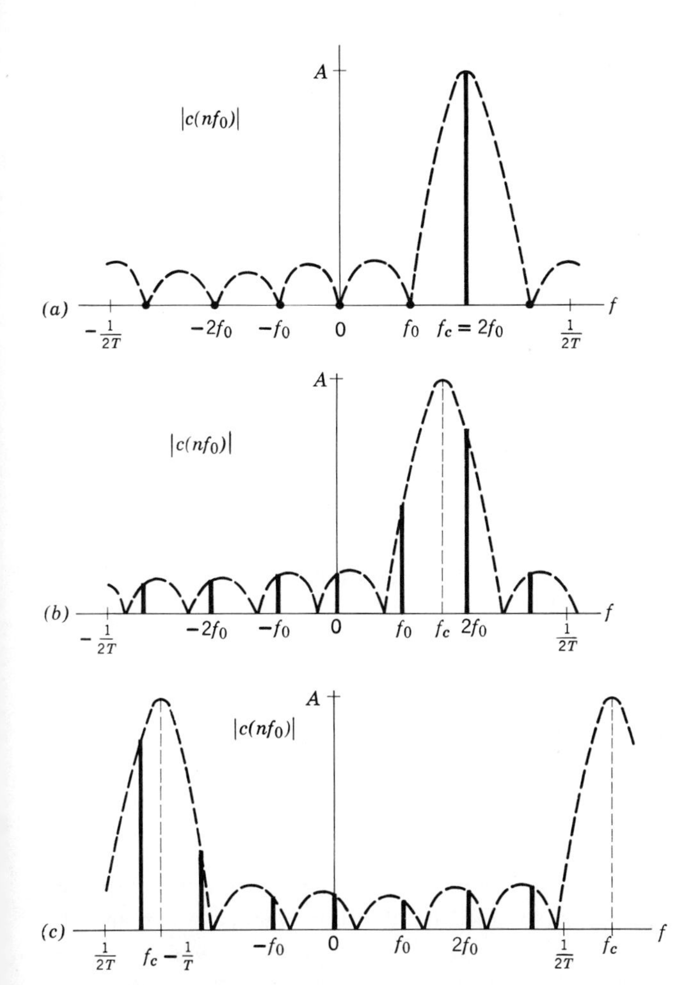

Figure 12.19. DFT of $v(mT) = Ae^{j\theta}e^{j2\pi f_c mT}$. (a) $f_c = kf_0$, $|k| \le N$. (b) $f_c \ne kf_0$, $|f| < 1/2T$. (c) $|f_c| \ge 1/2T$.

interval so that $v(mT) \approx 0$ for $|m| > N$. Under these conditions, sampling theory tells us that $v(t)$ may be expressed in terms of its sample values in the form

$$v(t) \approx \sum_{m=-N}^{N} v(mT)\,\text{sinc}\left(\frac{t-mT}{T}\right)$$

which is Eq. (13), Sect. 12.1, modified for the present context. Fourier transformation then gives the continuous spectrum as

$$V(f) = \mathfrak{F}[v(t)] = \sum_{m=-N}^{N} v(mT)\,\mathfrak{F}\left[\text{sinc}\left(\frac{t-mT}{T}\right)\right]$$

$$= \sum_{m=-N}^{N} v(mT)[T\Pi(fT)e^{-j\omega mT}]$$

$$= T\sum_{m=-N}^{N} v(mT)e^{-j2\pi fmT} \qquad |f| < 1/2T$$

Now, comparing the above with $c(nf_0)$ as defined in (3) shows that $V(nf_0) = MTc(nf_0)$, or

$$c(nf_0) = \frac{1}{MT}V(nf_0) \tag{9}$$

which is interpreted graphically in Figure 12.20. Thus, to within a constant of proportionality, the discrete spectrum obtained from the DFT equals the corresponding sample values of the continuous spectrum when the time-domain samples are sufficient.

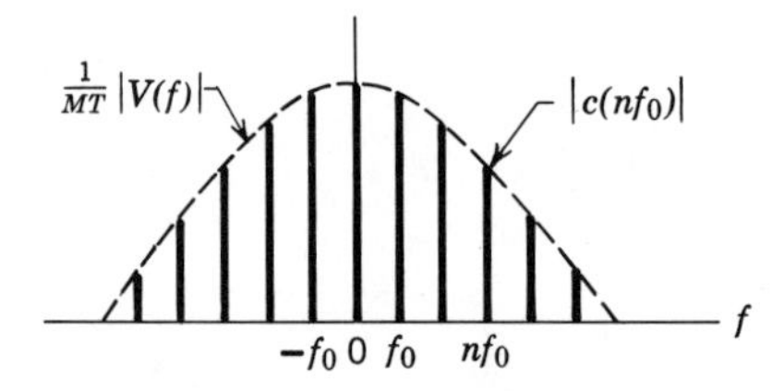

Figure 12.20.

But, lest there be any confusion on this score, we hasten to repeat that the DFT is not restricted to sampled bandlimited signals. As an analytic tool, it may be applied to any sequence of numbers for which a transformed representation is desired. Reflecting this broader view, DFT algorithms are commonly based on writing (3) and (4) as

$$\begin{aligned} c(n) &= \frac{1}{M}\sum_{m=0}^{M-1} v(m)e^{-j2\pi nm/M} \qquad & n = 0, 1, \ldots M-1 \\ v(m) &= \sum_{n=0}^{M-1} c(n)e^{j2\pi nm/M} \qquad & m = 0, 1, \ldots M-1 \end{aligned} \tag{10}$$

where M now is either odd or even†; in this format, the same computer

†However, the FFT algorithm requires M to be a power of 2; see Brigham and Morrow (1967) or Bergland (1969) for further details.

program can be used to calculate both sequences $c(n)$ and $v(m)$. However, for the frequency-domain interpretation, the values of $c(n)$ for $|n| > M/2$ actually represent frequency components at $f = -(M-n)/MT$.

12.4 SIGNAL-SPACE INTERPRETATIONS*

As a retrospective finale to this chapter we shall apply the vector concepts and notation of Chapter 5 to sampled and discrete signals. Specifically, it will be shown that bandlimited signals may be represented in a signal space whose basis functions are sinc functions, rather than exponentials, a representation of considerable importance in communication theory. Then the discrete Fourier transform will be related to the signal-space representation of discrete signals. For clarity in both developments it is necessary to relable some of the indices used in prior sections; this should cause little confusion if one concentrates on the meaning of those indices instead of their symbols.

Bandlimited energy signals

Let $v(t)$ be an energy signal whose spectrum $V(f)$ is bandlimited in W so, via the inverse Fourier transform,

$$v(t) = \int_{-W}^{W} V(f) e^{j2\pi ft} df \tag{1}$$

the finite range of integration reflecting the fact that $V(f) = 0$ for $|f| \geq W$. Equation (1) is the Fourier representation of $v(t)$ in the sense of Section 5.4. But another representation, keyed to the bandlimited property, is possible; its derivation starts with the *approximation*

$$\hat{v}(t) = \sum_{n=-N}^{N} c_n \phi_n(t) \tag{2}$$

where

$$\phi_n(t) = \sqrt{2W} \operatorname{sinc}(2Wt - n) \tag{3}$$

Note that these basis functions, like $v(t)$, are bandlimited in W since

$$\Phi_n(f) = \mathcal{F}[\phi_n(t)] = \frac{1}{\sqrt{2W}} \Pi\left(\frac{f}{2W}\right) e^{-j\pi fn/W} \tag{4}$$

a result the reader can easily confirm.

Providing that the $\phi_n(t)$ are *orthonormal* $\hat{v}(t)$ will be a *least-square-error* approximation if

$$c_n = \langle v(t), \phi_n(t) \rangle = \int_{-\infty}^{\infty} v(t)\phi_n^*(t)\,dt \tag{5}$$

which is Eq. (6), Sect. 5.3, with the energy-signal definition of the scalar product. Therefore, our work here entails showing that

$$\langle \phi_n(t), \phi_m(t) \rangle = \delta_{nm}$$

and then evaluating (5). Both of these tasks are expedited by drawing upon Eq. (6), Sect. 7.5,

$$\int_{-\infty}^{\infty} v(t)w^*(t)\,dt = \int_{-\infty}^{\infty} V(f)W^*(f)\,df$$

Hence, testing for orthonormality,

$$\begin{aligned}
\langle \phi_n(t), \phi_m(t) \rangle &= \int_{-\infty}^{\infty} \phi_n(t)\phi_m^*(t)\,dt \\
&= \int_{-\infty}^{\infty} \overbrace{\frac{1}{\sqrt{2W}}\Pi\left(\frac{f}{2W}\right)e^{-j\pi fn/W}}^{\Phi_n(f)} \overbrace{\frac{1}{\sqrt{2W}}\Pi\left(\frac{f}{2W}\right)e^{+j\pi fm/W}}^{\Phi_m^*(f)}\,df \\
&= \frac{1}{2W}\int_{-W}^{W} e^{j\pi(m-n)f/W}\,df \\
&= \operatorname{sinc}(m-n) = \delta_{nm}
\end{aligned}$$

so the basis functions are orthonormal, as required. By similar procedure,

$$\begin{aligned}
c_n &= \int_{-\infty}^{\infty} V(f)\Phi_n^*(f)\,df \\
&= \frac{1}{\sqrt{2W}}\int_{-W}^{W} V(f)e^{j2\pi f(n/2W)}\,df \\
&= \frac{1}{\sqrt{2W}}v\left(\frac{n}{2W}\right)
\end{aligned} \tag{6}$$

the last step being accomplished from comparison with (1).

Putting these results in (2) we have

$$\hat{v}(t) = \sum_{n=-N}^{N} \frac{1}{\sqrt{2W}}v\left(\frac{n}{2W}\right)\sqrt{2W}\operatorname{sinc}(2Wt-n)$$

which shows that a bandlimited energy signal may be approximated by a linear combination of sinc functions whose coefficients are proportional to instantaneous sample values of $v(t)$ taken every $1/2W$ seconds. The uniform sampling theorem then goes on to say, in effect, that letting $N \to \infty$ gives $\hat{v}_{\infty}(t) = v(t)$, as follows from Eq. (13), Sect. 12.1, with sampling at the Nyquist rate ($f_s = 2W$, $T_s = 1/2W$) and m replaced by n, i.e.,

$$v(t) = \sum_{n=-\infty}^{\infty} v\left(\frac{n}{2W}\right) \operatorname{sinc} 2W\left(t - \frac{n}{2W}\right) \tag{7}$$

Although we derived this expression with specific reference to sampling and reconstruction, it can be viewed equally well simply as a *series representation* for any energy signal $v(t)$ that is bandlimited in W – just as a periodic signal is represented by a Fourier series. The Fourier series, recall, consists of exponential functions, while here the series consists of sinc functions. In both cases the series functions are orthogonal and the coefficients are derived from least-square-error considerations, i.e., the *projection theorem.*

Turning to the vector interpretation of (7), we have the space containing all vectors of the form

$$v = \sum_{n=-\infty}^{\infty} c_n \phi_n$$

which is the set of all energy signals bandlimited in W. And, from (6), the vector "coordinates" are directly proportional to $v(n/2W)$, so the vector picture is more readily visualized than those encountered in Chapter 5. For instance, the norm or "length" of v is found from

$$\|v\|^2 = \sum_{n=-\infty}^{\infty} |c_n|^2 = \frac{1}{2W} \sum_{n=-\infty}^{\infty} \left| v\left(\frac{n}{2W}\right)\right|^2$$

But $\|v\|^2$ equals the signal's *energy* E, so

$$\sum_{n=-\infty}^{\infty} \left| v\left(\frac{n}{2W}\right)\right|^2 = 2WE \tag{8}$$

a surprisingly simple relationship between the energy and the sample values.

Needless to say, we have barely touched on these ideas. Their real payoff comes in the *mathematical theory of communication* (or *information theory*) pioneered by Shannon. A short but lucid introduction to that theory is given by Raisbeck (1964).

Discrete signals

Finally, consider the representation of a *discrete signal* $v(kT)$ based on the inverse DFT expression

$$v(kT) = \sum_{n=-N}^{N} c(nf_0)e^{j2\pi nf_0kT} \qquad k = 0, \pm 1, \ldots \pm N \tag{9}$$

with $f_0 = 1/MT$ and $M = 2N+1$. The vector interpretation of (9) is quite simple: $v(kT)$ is a vector having coordinates $c(nf_0)$ in an M-dimensional space whose basis functions are the *discrete exponentials*

$$\phi_n(kT) = e^{j2\pi nf_0kT} \qquad n = 0, \pm 1, \ldots \pm N \tag{10}$$

But justifying the interpretation requires a scalar-product formula for this space.

Since *summation* is the discrete equivalent of integration, we define the scalar product as

$$\langle v(kT), w(kT)\rangle \triangleq \frac{1}{M} \sum_{k=-N}^{N} v(kT)w^*(kT) \tag{11}$$

when $v(kT)$ and $w(kT)$ are specified for $|k| \leq N$. [A quick check shows that (11) satisfies all the basic properties, Eqs. (2) to (4), Sect. 5.2, and its form certainly agrees with our earlier definitions.] Putting (11) immediately to work, the orthonormality of the basis functions is established as follows:

$$\begin{aligned} \langle \phi_n(kT), \phi_m(kT)\rangle &= \frac{1}{M} \sum_{k=-N}^{N} [e^{-j2\pi(mf_0-nf_0)T}]^k \\ &= F_M(mf_0 - nf_0) = F_M\left(\frac{m-n}{MT}\right) \\ &= \delta_{nm} \end{aligned} \tag{12}$$

where we have drawn upon Eq. (6), Sect. 12.3, and Figure 12.18.

Presumably, then, a least-square-error *approximation* for $v(kT)$ can be written as

$$\hat{v}(kT) = \sum_{n=-N}^{N} c_n\phi_n(kT) \tag{13}$$

where, according to the projection theorem,

$$c_n = \langle v(kT), \phi_n(kT) \rangle$$
$$= \frac{1}{M} \sum_{k=-N}^{N} v(kT) e^{\,j2\pi n f_0 kT} \tag{14}$$

so c_n is identical to $c(nf_0)$, Eq. (3), Sect. 12.3. Now, in point of fact, $\hat{v}(kT)$ is an *exact* representation with zero error, not just an approximation. We demonstrate that $\hat{v}(kT) = v(kT)$ by substituting (14) – with a new index m – into the right-hand side of (13), i.e.,

$$\sum_{n=-N}^{N} c_n \phi_n(kT) = \sum_{n=-N}^{N} \left[\frac{1}{M} \sum_{m=-N}^{N} v(mT) e^{-j2\pi n f_0 mT} \right] e^{j2\pi n f_0 kT}$$
$$= \sum_{m=-N}^{N} v(mT) \underbrace{\left\{ \frac{1}{M} \sum_{n=-N}^{N} [e^{-j2\pi(mf_0 - kf_0)T}]^n \right\}}_{F_M\left(\frac{m-k}{MT}\right)}$$
$$= \sum_{m=-N}^{N} v(mT) \delta_{mk} = v(kT)$$

Therefore, besides justifying the vector interpretation, we have also *proved* the validity of the discrete Fourier transform.

Problems

12.1 Describe what happens in Figure 12.4b when T_s/τ is an integer, say k. Consider, in particular, the case $T_s/\tau = 1$.

12.2 Let $x(t)$ be processed by a *bipolar chopper*, producing $x_s(t)$ as shown in Figure P12.1. (a) Find and sketch the corresponding spectrum $X_s(f)$, taking $X(f)$ per Figure 12.4*a*. *Hint*: The switching function is a *square wave* (Problem 6.13). (b) Based on your results, discuss possible applications of

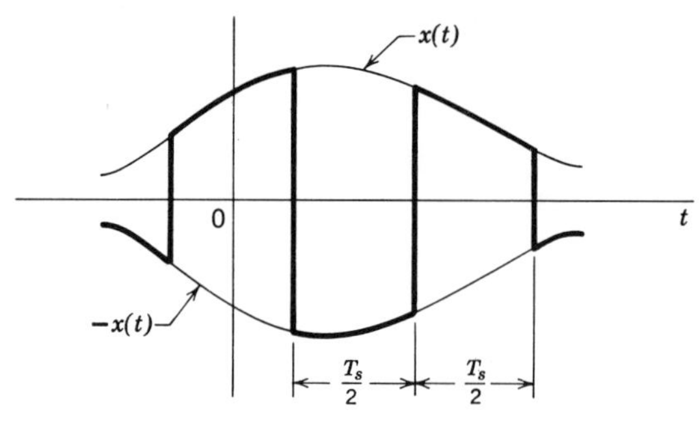

Figure P12.1

bipolar choppers. *Answer*: $X_s(f) = (2/\pi)[X(f-f_s)+X(f+f_s)] - (2/3\pi)[X(f-3f_s)+X(f+3f_s)] + (2/5\pi)[X(f-5f_s)+X(f+5f_s)] - \cdots$.

12.3 Prior to carrier modulation, the multiplexed signal in FM stereo is $x(t) = C_1[x_L(t)+x_R(t)] + C_2[x_L(t)-x_R(t)]\cos 2\pi f_s t$, where $x_L(t)$ and $x_R(t)$ are the left and right channels, respectively, each bandlimited in $W = 15$ kHz, and where the subcarrier is $f_s = 38$ kHz. Show that the system in Figure P12.2 can be used to generate $x(t)$ when the switch alternately dwells for $\tau = T_s/2$ on each of the two input contacts. *Hint*: Note that there are two switching functions involved, one delayed by $T_s/2$ with respect to the other.

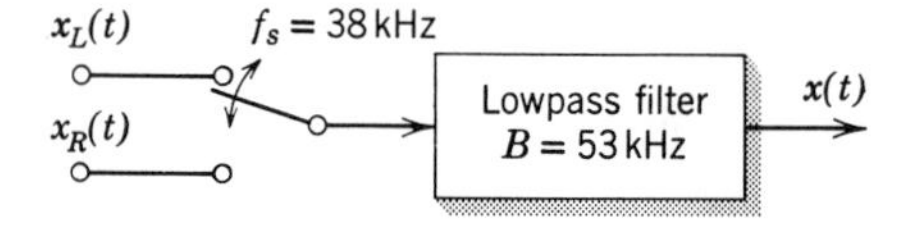

Figure P12.2

12.4 Plot the sample values when each of the following signals is instantaneously sampled at $t = 0, \pm\frac{1}{4}, \pm\frac{2}{4}, \ldots$: (a) $\cos 2\pi t$; (b) $\sin 2\pi(7/4)t$; (c) $\sin 2\pi 2t$. In the first two cases, convince yourself that no other waveform, bandlimited in 2 Hz, can be interpolated from these points. In the third case, explain why the sampling theorem apparently fails.

12.5 The signal $x(t) = \text{sinc}^2\, Wt$ is ideally sampled at $t = 0, \pm 1/2W, \pm 2/2W, \ldots,$ and reconstructed by an ideal lowpass filter having bandwidth W, unit gain, and zero delay. Carry out the reconstruction graphically, similar to Figure 12.7, for $|t| \leq 1/W$.

12.6 Consider a practical sampling system in which $x_s(t) = x(t) \sum_{m=-\infty}^{\infty} p(t-mT_s)$, where $p(t)$ is an arbitrarily shaped sampling pulse. By expanding

$\sum_{m=-\infty}^{\infty} p(t-mT_s)$ in a trigonometric Fourier series, demonstrate that $X_s(f)$ is similar to Eq. (4), Sect. 12.1.

12.7 *Aperture effect* arises when the sampled signal is of the form $x_s(t) = \sum_{m=-\infty}^{\infty} x(mT_s)p(t-mT_s)$, where $p(t) \neq \delta(t)$ is the sampling pulse. Show that $X_s(f) = P(f)X_\delta(f)$, where $P(f) = \mathcal{F}[p(t)]$, and describe how the reconstruction process must be modified to recover $x(t)$ without distortion, assuming Eq. (5), Sect. 12.1, holds. *Hint*: Use the fact that $p(t-mT_s) = [p(t)] * [\delta(t-mT_s)]$.

12.8 If $\cos 2\pi 50t$ and $\cos 2\pi 350t$ are ideally sampled at $f_s = 400$, the reconstructed waveforms will be identical because of aliasing. Confirm this assertion by plotting the sample values and examining the spectra of the sampled waves.

12.9 A 1-millisec rectangular pulse is sampled and ideally reconstructed. Sketch the resulting output waveform when T_s equals: (a) 0.4 millisec; (b) 0.2 millisec.

12.10 Given the choice between time-division and frequency-division multiplexing, state which you would use, and why, for each of the following situations: (a) The signals to be multiplexed have drastically different bandwidths; (b) the transmission system has nonlinear distortion such that the output is of the form $x(t)+0.1x^2(t)$; (c) the transmission system has a nonlinear phase shift; (d) the transmission medium is subject to narrow-band (frequency-selective) fading; (e) the transmission medium is subject to across-the-band fading.

12.11 *Hyperquantization* is the process whereby M successive quantized samples are represented by *one* number having Q^M possible values. Show that hyperquantized PCM can yield *bandwidth compression*, that is, $B < W$.

12.12 A certain PCM system has $B = 20$ kHz and is to be used for an input signal bandlimited in $W = 3$ kHz. Taking $K = 1$ in Eq. (7), Sect. 12.2, specify values for μ, ν, Q, and f_s such that $P_x/P_n \geq 6000$. *Answer*: $\mu = 3$, $\nu = 4$, $Q = 81$, $f_s = 10$ kHz.

12.13 Using the DFT definition, Eq. (3), Sect. 12.3, prove the following: (a) If $z(mT) = \alpha v(mT) + \beta w(mT)$, then $c_z(nf_0) = \alpha c_v(nf_0) + \beta c_w(nf_0)$; (b) If $v(mT)$ is real, then $c(-nf_0) = c^*(nf_0)$; (c) If $v(mT)$ is real and has even symmetry, then $c(nf_0)$ is real and even.

12.14 When dealing with a finite sequence $v(mT)$, $|m| \leq N$, it is often convenient to introduce the *periodic continuation* $\tilde{v}(kT)$ defined for any integer k such that $\tilde{v}(kT)$ is periodic in k with period $M = 2N+1$, and equals $v(kT)$ for $|k| \leq N$. Similarly, for $|m| \leq N$, $\tilde{v}[(m-m_d)T]$ equals $v(mT)$ delayed by m_dT and with the last m_d values moved to the beginning of the sequence. Show that if $z(mT) = \tilde{v}[(m-m_d)T]$, $|m| \leq N$, then $c_z(nf_0) = c_v(nf_0)$ $\exp(-j2\pi nf_0m_dT)$, which is the discrete version of the *time-shift theorem.*

12.15 Given two sequences $v(mT)$ and $w(mT)$, $|m| \leq N$, their *periodic convolution* is defined as

$$v * w(mT) \triangleq \frac{1}{M} \sum_{k=-N}^{N} v(kT)\tilde{w}[(m-k)T]$$

where $M = 2N+1$ and $\tilde{w}$ is as defined in the previous problem. (a) Taking $v(mT) = m^2$ and $w(mT) = 2-m$, $-2 \leq m \leq 2$, calculate the sequence $v * w(mT)$ and plot $v(mT)$, $w(mT)$, and $v * w(mT)$. (b) Let $z(mT) = v * w(mT)$, where $v(mT)$ and $w(mT)$ are arbitrary sequences; obtain a general expression for $c_z(nf_0)$ in terms of $c_v(nf_0)$ and $c_w(nf_0)$. *Hint*: Use the discrete time-shift theorem from Problem 12.14.

12.16 A fixed linear system having transfer function $H(f)$ is driven by an input signal $x(t)$ and produces the zero-state response $y(t)$. Assuming there is a value of f_0 such that Eq. (9), Sect. 12.3, holds with reasonable accuracy for both $X(f)$ and $Y(f)$, show that $y(t)$ may be computed numerically via

$$y(mT) = \sum_{n=-N}^{N} H(nf_0)c_x(nf_0)e^{j2\pi nf_0mT} \qquad |m| \leq N$$

where $c_x(nf_0)$ is the DFT of $x(mT)$.

12.17 Suppose an available computer subroutine takes an input sequence $\alpha(k)$, $k = 0, 1, \ldots M-1$, and produces the output sequence

$$\beta(i) = \sum_{k=0}^{M-1} \alpha(k)e^{-j2\pi ik/M} \qquad i = 0, 1, \ldots M-1$$

Either or both sequences may be complex and, if so, each complex number is handled as a pair of numbers, the real and imaginary parts. Utilizing this subroutine, outline computer programs for the following tasks: (a) Given $v(mT)$ for $m = 0, 1, \ldots M-1$, calculate $|c_v(nf_0)|$ and $\arg[c_v(nf_0)]$ for $-(M-1)/2 \leq n \leq (M-1)/2$; (b) Given $c_v(nf_0)$ as in part (a), calculate $v(mT)$ for $m = 0, 1, \ldots M-1$.

12.18 When a signal is *timelimited*, such that $v(t)=0$ for $|t| \geq T_0/2$, its Fourier transform is completely specified by the *frequency-domain sample values* $V(nf_0)$, $n=0, \pm1, \pm2, \ldots,$ where $f_0 = 1/T_0$. Prove this assertion by the following procedure: (a) Define the periodic continuation $\tilde{v}(t) \triangleq \sum_{m=-\infty}^{\infty} v(t-mT_0)$ and show that its exponential Fourier-series coefficients are $c_n = (1/T_0)V(nf_0)$; (b) Noting that $V(f) = \mathfrak{F}[\tilde{v}(t)\Pi(2t/T_0)]$, derive the relationship

$$V(f) = \sum_{n=-\infty}^{\infty} V(nf_0) \operatorname{sinc}(T_0 f - n)$$

(c) Compare this result with Eq. (7), Sect. 12.4, and comment.

12.19 With the help of Rayleigh's energy theorem and Eq. (8), Sect. 12.4, evaluate the sum $\sum_{n=-\infty}^{\infty} \operatorname{sinc}^2 n$.

12.20 Based on Eq. (11), Sect. 12.4, develop summation expressions analogous to Eqs. (8) and (9a), Sect. 5.2.

13

Discrete-time systems: z-transforms

Having studied signals reduced to a sequence of numbers via sampling, we now take up the analysis of systems that contain discrete-time signals either exclusively or within a subsystem. By expanding the scope of his analytical tools to include discrete-time concepts, the engineer is in a position to take profitable advantage of the many advances in digital technology. In a communication system, for example, the use of *digital filtering* facilitates highly sophisticated signal processing that would be impractical with conventional filters. Likewise, *sampled-data* control systems can be designed to achieve almost any desired control law, compensation, etc. In either case, depending on the circumstances, the discrete-time function may be carried out by specialized digital components (hardware) or digital-computer programs (software).

These points will be illustrated in the course of our work. First, however, we must establish suitable mathematical models for dynamic discrete-time systems and then introduce methods for analyzing their response. The modeling process leads to the use of *difference* equations, whose solution can be treated both in the "time" domain and, by introducing the appropriate transform, in a "frequency" domain. Throughout this chapter, our attention will focus on discrete systems that are linear and time-invariant. We shall draw heavily upon the concepts and analytical techniques developed earlier for continuous systems, thereby allowing a compact treatment that still conveys the fundamentals.

13.1 DISCRETE MODELS

Sampling a continuous signal $x(t)$ every T seconds gives the sequence of sample values $\ldots, x(-2T), x(-T), x(0), x(T), x(2T), \ldots$, or

$$x(kT) \qquad k = \ldots, -1, 0, 1, 2, \ldots$$

This is a discrete signal, a function of the discrete variable kT. Noting that T is a constant, we can remove the association with time by defining a discrete signal more abstractly as

$$x(k) \qquad k = \ldots, -1, 0, 1, 2, \ldots$$

In other words, a discrete signal is simply an *ordered sequence of numbers*; the ordering index k is restricted to integer values and is the independent variable. Figure 13.1 is a graphical illustration.

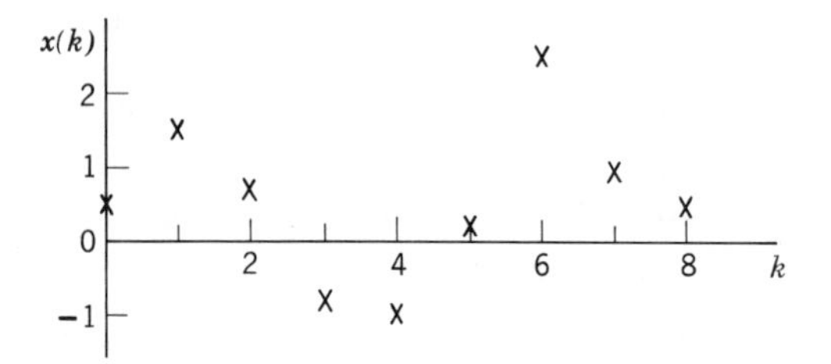

Figure 13.1. A discrete signal.

The values of $x(k)$ may or may not themselves be discrete, i.e., $x(k)$ is not necessarily quantized. Of course, if the signal in question comes from a digital computer, then it must be both time-discrete and quantized. Having discussed quantization effects in the preceding chapter, we shall limit our attention to time-discrete signals and the systems within which they occur.

Now we define a *discrete system* (more accurately, a discrete-time system) as a system in which all the physical variables are *discrete signals*. Thus, similar to Eq. (8), Sect. 2.3, the general input-output relation for an SIO discrete system is

$$y(k) = S[\mathbf{q}_0; x(k)] \qquad k \geq k_0$$

where $x(k)$ is the input, $\mathbf{q}_0$ is the initial state at $k = k_0$, and $y(k)$ the response. A fixed linear discrete system then obeys *decomposability*

$$y(k) = \underbrace{S[\mathbf{q}_0; 0]}_{y_{zi}(k)} + \underbrace{S[\mathbf{0}; x(k)]}_{y_{zs}(k)}$$

and *superposition* and *stationarity*, i.e., for the zero-state response,

$$S[\mathbf{0}; \alpha v(k) + \beta w(k)] = \alpha S[\mathbf{0}; v(k)] + \beta S[\mathbf{0}; w(k)]$$
$$S[\mathbf{0}; x(k-K)] = y_{zs}(k-K)$$

where α and β are constants and K is a fixed integer – and similarly for $y_{zi}(k)$.

For fixed linear continuous systems, we found the four primative building blocks to be scalors, integrators, differentiators, and delayors. Since integration and differentiation are inherently continuous operations, they are not defined for discrete signals. A discrete fixed linear system has only the two primative operations, *scalors* and *unit delayors*, depicted in Figure 13.2. Note that the output of the delayor is the previous value of the input; it is the only *dynamic* element of discrete systems.

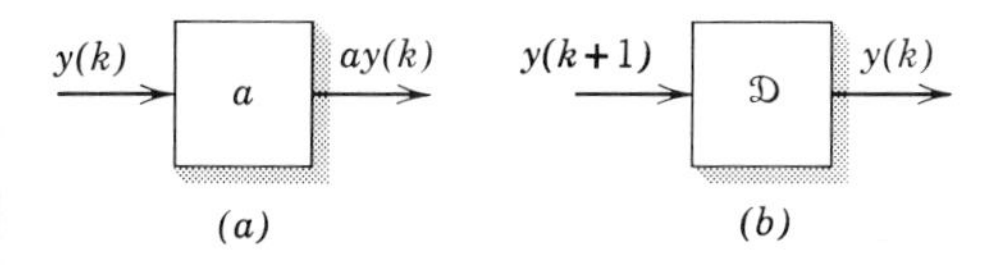

Figure 13.2. (*a*) Scalor. (*b*) Delayor.

Example 13.1

To show how a discrete system can be used for signal processing, suppose that it is desired to process the continuous signal $x(t)$ so as to obtain its running time-average over the previous τ seconds, i.e., we want to implement the finite-time integrator, first introduced in Chapter 4. Mathematically, the desired input-output relationship is

$$y(t) = \frac{1}{\tau} \int_{t-\tau}^{t} x(\lambda)\, d\lambda$$

While this operation can be done (approximately) with continuous systems, it may be more satisfactory to use a discrete system based on the following.

Splitting up the integration into N equal parts of duration $T = \tau/N$, we have

$$y(t) = \frac{1}{NT}\left[\int_{t-T}^{t} + \int_{t-2T}^{t-T} + \cdots + \int_{t-\tau}^{t-(N-1)T} x(\lambda)\, d\lambda\right]$$

Then, if T is sufficiently small such that $x(t)$ is essentially constant over each interval, the running average can be approximated by

$$\hat{y}(t) \triangleq \frac{1}{N}[x(t) + x(t-T) + \cdots + x(t - NT + T)]$$

and, at any $t = kT$,

$$\hat{y}(k) = \frac{1}{N}[x(k) + x(k-1) + \cdots + x(k-N+1)]$$

$$= \frac{1}{N}\sum_{j=0}^{N-1} x(k-j)$$

where T has been dropped from the arguments of $\hat{y}(kT)$ and $x(kT)$.

The desired task can therefore be accomplished, to any reasonable degree of accuracy, by sampling $x(t)$ every T seconds and using a discrete system to process the sequence $x(k)$. The block diagram for the system is shown in Figure 13.3; it involves $N-1$ delayors in a cascade configuration, i.e., the tapped delay line encountered in Section 7.5.

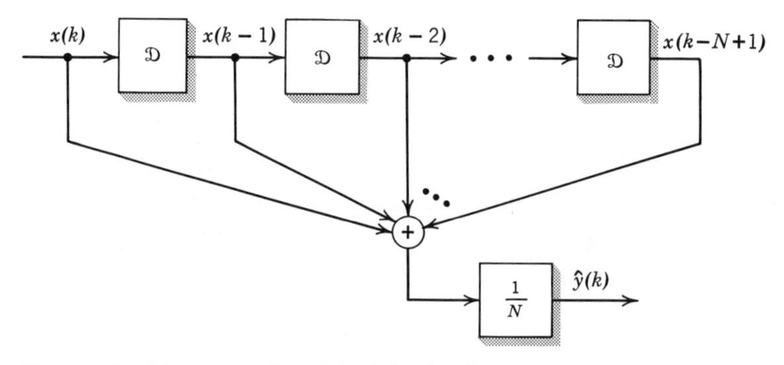

Figure 13.3. Discrete version of the finite-time integrator.

Simulation diagrams and difference equations

Because delayors are the only dynamic elements involved, the mathematical models of discrete systems contain terms such as $y(k)$, $y(k+1)$, etc., and are called *difference equations*. As a simple illustration, consider the feedback configuration shown in Figure 13.4.

As with continuous systems, the equation that the system obeys is found by equating the summing-junction output to the sum of its inputs. Thus,

$$y(k+1) = -ay(k) + x(k) \tag{1a}$$

or

$$y(k+1) + ay(k) = x(k) \tag{1b}$$

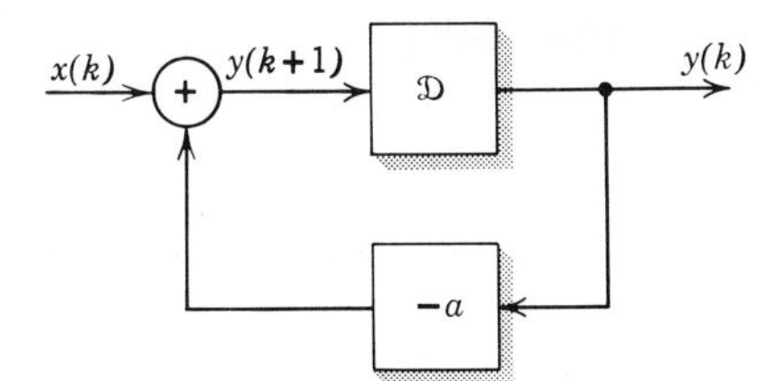

Figure 13.4. A first-order feedback system.

which is a first-order, nonhomogeneous linear difference equation. As a point of comparison, replacing the unit delayor in the block diagram with an integrator would result in the differential equation $\dot{y} + ay = x(t)$ for the continuous signal $y(t)$. Clearly, difference equations play the same role in discrete systems that differential equations do in continuous systems. Moreover, the simulation diagrams have the same structure as that for continuous systems, the only change being the replacement of integrators by unit-delay elements.

To generalize these points, consider the simulation diagram of the nth-order SIO discrete system depicted in Figure 13.5. Denoting the output of the last delayor as $q(k)$ imposes the designations $q(k+1)$,

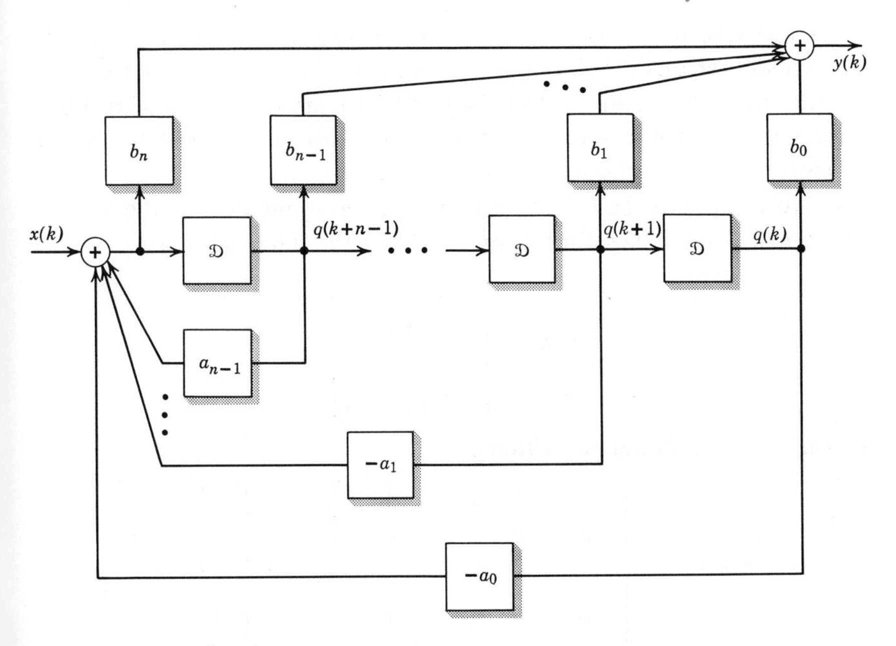

Figure 13.5. An nth-order system.

$q(k+2), \ldots, q(k+n-1)$ as outputs of preceding delayors, the entire set comprising the *state* of the system. Then, forming the equation of the input summing junction, we have

$$q(k+n)+a_{n-1}q(k+n-1)+\cdots+a_0q(k)=x(k) \tag{2}$$

which is the state equation.

Similarly, the equation of the output summing junction is

$$y(k)=b_0q(k)+b_1q(k+1)+\cdots+b_mq(k+m) \tag{3}$$

where $m \leq n$ (the figure is shown for $m=n$). Eliminating $q(k)$ gives the input-output difference equation

$$y(k+n)+a_{n-1}y(k+n-1)+\cdots+a_0y(k)=b_mx(k+m)+\cdots+b_0x(k)$$

or

$$\sum_{i=0}^{n} a_i y(k+i)=\sum_{j=0}^{m} b_j x(k+j) \tag{4}$$

analogous to the differential equation

$$\sum_{i=0}^{n} a_i y^{(i)}=\sum_{j=0}^{m} b_j x^{(j)}$$

As was true for continuous systems, we shall not allow $m>n$ in (4). The reason for this restriction is very simple; if $m>n$ the system is *not causal*, the current output depending on future input values. To illustrate, if $m=n+1$, then (4) implies that $y(i)$ depends on $x(i+1)$, which happens to be one time interval in the future.

If we desired to have a vector-matrix representation of the system equations, it would be rather natural to define the state vector as

$$\mathbf{q}(k)=\begin{bmatrix} q(k) \\ q(k+1) \\ \vdots \\ q(k+n-1) \end{bmatrix}$$

in which case (2) could be written as

$$\mathbf{q}(k+1)=\begin{bmatrix} 0 & 1 & 0 \cdots & 0 \\ 0 & 0 & 1 \cdots & 0 \\ \vdots & \vdots & \vdots & \vdots \\ 0 & 0 & 0 \cdots & 1 \\ -a_0 & -a_1 & -a_2 \cdots & -a_{n-1} \end{bmatrix}\mathbf{q}(k)+\begin{bmatrix} 0 \\ 0 \\ \vdots \\ 0 \\ 1 \end{bmatrix}x(k)$$

By using (2) to substitute for $q(k+n)$ in (3) the output variable can be written as a linear combination of the state variables and the input, to give

$$
\begin{aligned}
y(k) &= b_0 q(k) + \cdots + b_{n-1} q(k+n-1) \\
&\quad + b_n[-a_0 q(k) \cdots - a_{n-1} q(k+n-1) + x(k)] \\
&= [b_0 - b_n a_0] q(k) + \cdots + [b_{n-1} - b_n a_{n-1}] q(k+n-1) + b_n x(k)
\end{aligned}
$$

Making the obvious definitions for the matrix **A**, the vectors **b** and **c**, and the scalar d, we can write the state and output equations in the compact form

$$
\begin{aligned}
\mathbf{q}(k+1) &= \mathbf{A}\mathbf{q}(k) + \mathbf{b}x(k) \\
y(k) &= \mathbf{c}^T\mathbf{q}(k) + dx(k)
\end{aligned}
$$

From this point, the extension to multi-input-output systems, where x and y are represented by vectors, is almost immediate.

Hybrid systems

Discrete and continuous systems are often used together by interfacing the two with analog-to-digital (A/D) and digital-to-analog (D/A) converters.† The combination is referred to as a hybrid system, or, in the context of control, a *sampled-data* system.

A single-loop hybrid sampled-data control system is shown in Figure 13.6, which has continuous input and output signals $r(t)$ and $c(t)$, respectively. The analog error signal $\varepsilon(t)$ is sampled periodically and converted to the sequence of numbers, $\varepsilon(kT)$. In this example the control law, whatever it may be, is implemented by performing calculations in the digital computer upon receipt of a new value for $\varepsilon(kT)$. Assuming that the time required to compute $y(kT)$ is very short compared with the sampling interval T and that the converters are synchronized, the output of the

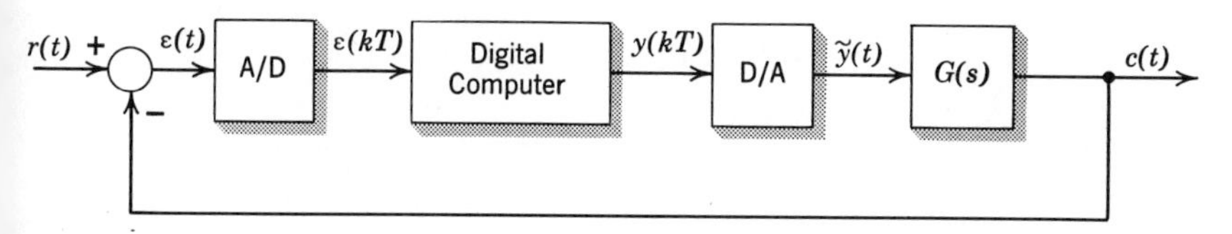

Figure 13.6. A hybrid feedback system.

†Here we shall not be concerned with the encoding and decoding operations mentioned in Chapter 12, relegating them to the digital computer.

D/A converter – assumed to be a zero-order hold as defined in Chapter 12 – will switch to $y(kT)$ and then maintain that value throughout the following T seconds, at which time the digital computer will put out a new value, $y[(k+1)T]$.

The piecewise-constant output of the D/A converter, denoted by $\tilde{y}(t)$, is the input to the continuous subsystem whose transfer function is $G(s)$ and whose output is $c(t)$. In order to construct a mathematical model of a hybrid system we shall obtain a discrete model of the continuous elements, thereby yielding a discrete model of the entire system. To achieve this, it is necessary to ignore the behavior of the continuous elements *between* the sampling times $t = kT$ by considering only the values $r(kT)$, $c(kT)$, and $\varepsilon(kT)$.

Generally, the output of the D/A converter is a staircase or piecewise-constant waveform such that

$$\tilde{y}(t) = y(kT) \qquad kT < t \leq (k+1)T \tag{5}$$

as sketched in Figure 13.7. This figure also shows a typical process output response $c(t)$ over the same interval. Just prior to the stepwise change in

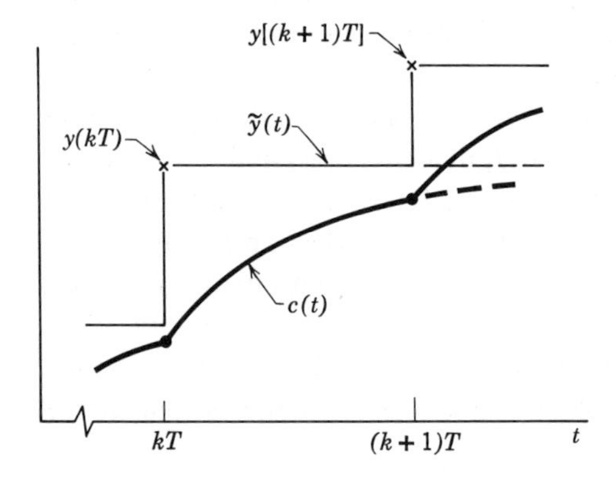

Figure 13.7.

$\tilde{y}(t)$ the process has the *state* $\mathbf{q}(kT)$. Therefore, assuming the process is linear, time-invariant, and causal, we can express $c(t)$ over the interval $kT < t \leq (k+1)T$ by breaking it up into two parts: *1.* the *zero-input* response to a step of height $y(kT)$ applied at $t = kT$, and *2.* the *zero-state* response resulting from $\mathbf{q}(kT)$. This technique is demonstrated in the following example.

Example 13.2

The continuous first-order process in Figure 13.8 has as its zero-input response

$$c_{zi}(t) = e^{-\alpha(t-kT)}c(kT)$$

for $t \geq kT$. Similarly, during the interval $kT \leq t \leq (k+1)T$ the zero-state response is

$$c_{zs}(t) = [1 - e^{-\alpha(t-kT)}]y(kT)$$

Setting $t = (k+1)T$ and adding the two components of the output signal gives the desired recursive relationship, namely,

$$c[(k+1)T] = e^{-\alpha T}c(kT) + [1 - e^{-\alpha T}]y(kT)$$

The process response over the entire interval $kT \leq t \leq (k+1)T$ is depicted in Figure 13.7.

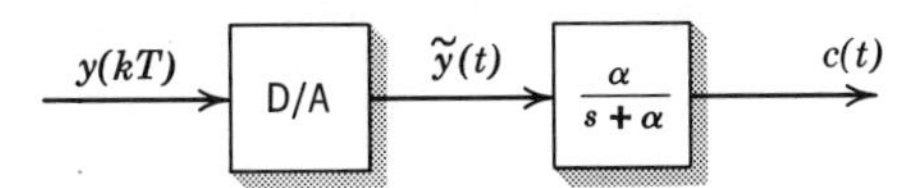

Figure 13.8.

13.2 RESPONSE OF FIRST-ORDER SYSTEMS

In order to establish a feeling for the behavior of discrete systems we shall consider the feedback system described by the first-order equation

$$y(k+1) + a_0 y(k) = b_0 x(k) \tag{1}$$

which leads to the simulation diagram in Figure 13.9. As was done in

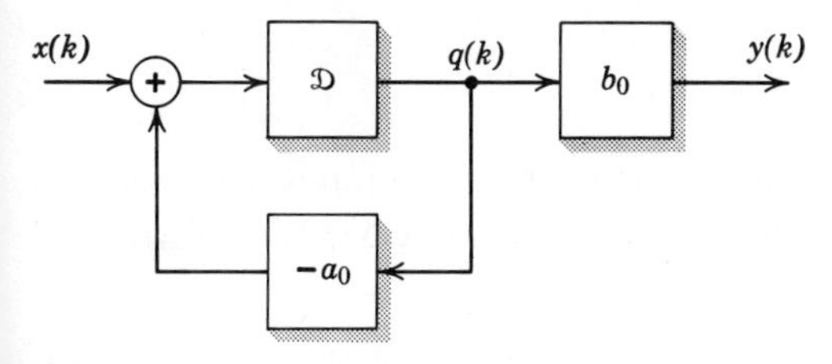

Figure 13.9. A first-order system $(b_1 = 0)$.

Chapter 2, we shall solve this equation by what amounts to a brute-force method, treating the zero-input and zero-state responses separately.†

Zero-input response

First, let us find the zero-input response, denoted by $y_{zi}(k)$, subject to the initial condition $y(0) = y_0$. Setting $x(k) = 0$ for all k, we can rewrite (1) as

$$y(k+1) = -a_0 y(k) \qquad y(0) = y_0$$

which indicates that $y(k+1)$ depends only on the previous value $y(k)$, hence the designation "first-order." Since we know $y(0)$, we can solve successively for $y(k)$ in the following manner:

$$y(1) = -a_0 y_0$$

$$y(2) = -a_0 y(1) = (-a_0)^2 y_0$$

$$y(3) = -a_0 y(2) = (-a_0)^3 y_0$$

and so forth. In this case we can easily deduce the value of the solution for any $k \geq 0$ to be

$$y(k) = (-a_0)^k y_0 \tag{2}$$

In Figure 13.10, the normalized response $y(k)/y_0$ is plotted for $a_0 = \pm 0.9$ and $a_0 = \pm 1.1$.

The question of whether or not the solution decays is more closely related to the *magnitude* of a_0 than to its *sign*. In particular, for $|a_0| > 1$, $|y(k)|$ grows with increasing k while it decays when $|a_0| < 1$. The growth or decay is monotonic when a_0 is negative and alternates in sign when a_0 is positive.

Rather than working directly with the output $y(k)$, we could have solved for the response of the state variable $q(k)$ in terms of the initial state $q(0)$. Since the state and output are related by $y(k) = b_0 q(k)$, the method is the same.

Zero-state response

Because we took $k = 0$ as the initial time for the zero-input response, we shall now calculate the zero-state response assuming $x(k) = 0$ for $k < 0$

†Henceforth, our attention will be restricted to causal systems and signals, unless explicitly stated to the contrary.

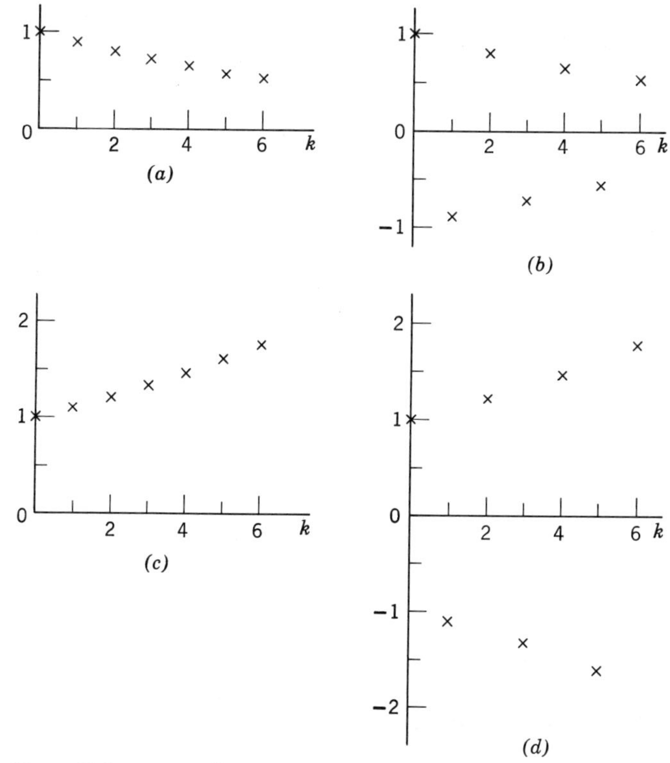

Figure 13.10. Normalized zero-input response $y(k)/y_0$. (*a*) $a_0 = +0.9$. (*b*) $a_0 = -0.9$. (*c*) $a_0 = +1.1$. (*d*) $a_0 = -1.1$.

i.e., a *causal* input. Therefore, (1) becomes

$$y(k+1) = -a_0 y(k) + b_0 x(k) \qquad k \geq 0$$

with a zero-state of $q(0) = y(0) = 0$. This equation can be solved repetitively to yield

$$y(0) = 0$$

$$y(1) = b_0 x(0)$$

$$y(2) = -a_0 b_0 x(0) + b_0 x(1)$$

$$y(3) = (-a_0)^2 b_0 x(0) + (-a_0) b_0 x(1) + b_0 x(2)$$

et cetera. The general term becomes

$$y(k) = (-a_0)^{k-1} b_0 x(0) + (-a_0)^{k-2} b_0 x(1) + \cdots + b_0 x(k-1) \qquad (3)$$

Examining (3), we note that the zero-state response is a linear combination of the input sequence $x(0), x(1), \ldots, x(k-1)$ and there appears to be a definite structure to the various weights.

To be more specific, let us define the *unit delta* function as

$$d(k) \triangleq \begin{cases} 1 & k=0 \\ 0 & k \neq 0 \end{cases} \tag{4}$$

and denote the corresponding response, with zero initial state, as $h(k)$, i.e., $h(k) \triangleq S[\mathbf{0}; d(k)]$. It then follows from (3) that, for the system of Figure 13.9,

$$h(k) = \begin{cases} 0 & k=0 \\ (-a_0)^{k-1} b_0 & k \geq 1 \end{cases}$$

Rewriting the zero-state response to an arbitrary causal input in terms of the unit-delta response $h(k)$ gives

$$y(k) = h(k)x(0) + h(k-1)x(1) + \cdots + h(1)x(k-1) + h(0)x(k) \tag{5}$$

where the term $h(0)x(k)$ has been included for completeness, although in this particular instance it is equal to zero.

When (5) is written in either of the equivalent forms

$$y(k) = \sum_{i=0}^{k} h(k-i)x(i) \tag{6}$$

$$= \sum_{j=0}^{k} h(j)x(k-j) \tag{7}$$

the reader should recognize the two expressions as *convolution summations*,† where $h(k)$ is the response of the system to the unit-delta function defined by (4).

Although our development has been restricted to first-order systems – not even the most general first-order system, because $b_1 = 0$ in (1) – the general features of the response of discrete systems are beginning to emerge. The most important result is that a direct parallel exists between the responses of discrete and continuous systems. It is not surprising that this similarity in structure has resulted in the development of similar

†More generally, the limits in the convolution summation are $+\infty$ and $-\infty$. Here, the limits of 0 and k are due to the fact that both $h(k)$ and $x(k)$ are causal functions.

methods of analysis, e.g., the simulation diagram with only the dynamic element changed from an integrator to a delayor, and the convolution summation versus the convolution integral. In fact, we shall soon introduce a transform method – known as the z-transform and tailored expecially for fixed linear discrete systems – which bears many similarities to the Laplace transform.

The general case ($b_1 \neq 0$)

Before moving on to a discussion of higher-order systems, we state the corresponding results for the general causal first-order system, namely that described by

$$y(k+1) + a_0 y(k) = b_1 x(k+1) + b_0 x(k) \tag{8}$$

and represented in Figure 13.11. If the initial state is specified in terms of $q(0)$, we can easily solve for the corresponding value of $y_{zi}(0)$. Specifically, setting $x(k) = 0$, one sees immediately from the simulation diagram that $y(k) = b_0 q(k) - a_0 b_1 q(k)$ where $q(k)$ obeys (2), namely,

$$q(k) = (-a_0)^k q(0) \qquad k = 0, 1, 2, \ldots$$

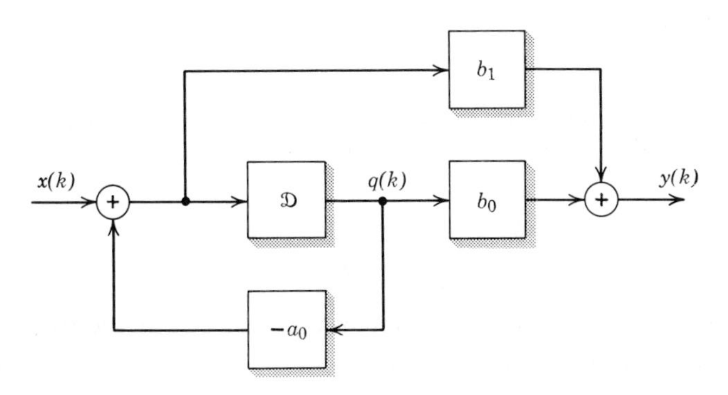

Figure 13.11. The general first-order system ($b_1 \neq 0$).

The zero-state response can be found by first obtaining the delta response $h(k)$, which is

$$h(k) = \begin{cases} b_1 & k = 0 \\ (-a_0)^{k-1}(b_0 - a_0 b_1) & k \geq 1 \end{cases} \tag{9}$$

Because $b_1 \neq 0$ there is a direct path, without a delayor, on the simulation diagram from the input to output; hence $h(0) \neq 0$. As before, the zero-state response can be expressed as the convolution of $h(k)$ with $x(k)$ giving, upon substitution of (9),

$$y(k) = b_1 x(k) + (b_0 - a_0 b_1)\left[\sum_{i=0}^{k-1} (-a_0)^{k-1-i} x(i)\right]$$

which is merely a linear combination of the input sequence $x(k)$, $k = 0, 1, 2, \ldots$, with weights which depend upon the system parameters a_0, b_0, and b_1.

The complete response

Although we have considered the zero-input and zero-state responses separately, based on our thorough treatment of continuous systems and the property of superposition, we would expect that the complete response would be

$$y(k) = y_{zs}(k) + y_{zi}(k)$$

Such is indeed true. However, it is worth stressing the point that one should be careful to distinguish between the initial state $q(0)$, which is independently specified and the initial *output* $y(0)$ which, in general, will depend on both $q(0)$ and $x(0)$.

Example 13.3

To demonstrate the steps in solving for the complete response, consider the first-order discrete system shown in Figure 13.12 with the input

$$x(k) = \begin{cases} 0 & k = 0 \\ 1 & k \geq 0 \end{cases}$$

and the initial state $q(0) = 2$. From the simulation diagram we see that the state and output equations are, respectively,

$$q(k+1) + \tfrac{1}{2}q(k) = x(k)$$

$$y(k) = -q(k+1) + 2q(k)$$

which leads directly to the input-output relationship

$$y(k+1) + \tfrac{1}{2}y(k) = -x(k+1) + 2x(k)$$

To obtain $y_{zi}(k)$, we solve for $y_{zi}(0)$ in terms of $q(0)$, getting

$$y_{zi}(0) = [2 + (-\tfrac{1}{2})(-1)]2 = 5$$

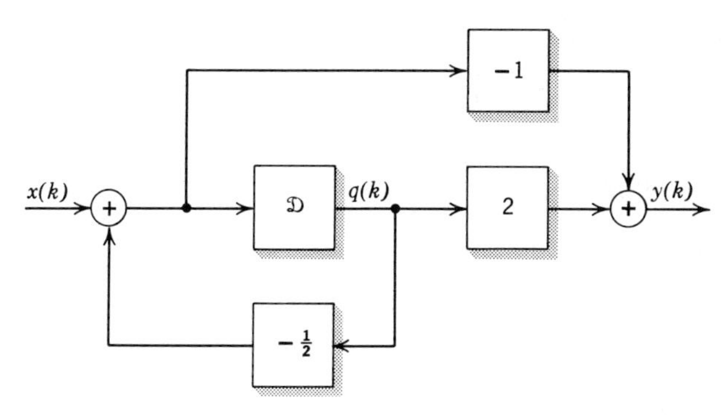

Figure 13.12.

Thus, from (2),

$$y_{zi}(k) = 5(-\tfrac{1}{2})^k \qquad k = 0, 1, 2, \ldots \tag{10}$$

The zero-state response will be computed by first finding the response to the unit-delta function, with $q(0) = 0$, and then convolving the result with the input $x(k)$. Direct calculation of $h(k)$ or the use of (9) leads to the sequence

k	0	1	2	3	$\cdots$
$h(k)$	-1	$\frac{5}{2}$	$-\frac{5}{4}$	$\frac{5}{8}$	$\cdots$

which has as its general term

$$h(k) = \tfrac{5}{2}(-\tfrac{1}{2})^{k-1} \qquad k \geq 1$$

Because $x(k-i) = 1$ for all $i \leq k$, the convolution summation for $y_{zs}(k)$ yields

$$\begin{aligned} y_{zs}(k) &= \sum_{i=0}^{k} h(i)x(k-i) = \sum_{i=0}^{k} h(i) \\ &= \sum_{i=0}^{k} \tfrac{5}{2}(-\tfrac{1}{2})^{i-1} \end{aligned} \tag{11}$$

With a little work (11) can be converted to the closed-form expression

$$y_{zs}(k) = \begin{cases} -1 & k = 0 \\ \tfrac{2}{3} - \tfrac{5}{3}(-\tfrac{1}{2})^k & k \geq 1 \end{cases}$$

The complete solution is obtained by adding $y_{zi}(k)$ as given by (10) to

$y_{zs}(k)$, yielding, after some simplification,

$$y(k) = \tfrac{2}{3}[1+5(-\tfrac{1}{2})^k] \qquad k \geq 0$$

13.3 DIFFERENCE EQUATIONS

To avoid having to pursue the solution of difference equations of arbitrary order with the brute-force methods used in the previous section, we shall proceed by drawing upon the notion of the characteristic equation and by extending the use of the convolution summation. In the work to follow, we shall consider input-output equations of the form

$$y(k+n)+a_{n-1}y(k+n-1)+\cdots+a_0y(k) = b_mx(k+m)+\cdots+b_0x(k) \tag{1}$$

where $m \leq n$ and $b_m \neq 0$. As it will be essential to distinguish between the state and output variables, we establish an equivalent set of equations consisting of the *state equation*

$$q(k+n)+a_{n-1}q(k+n-1)+\cdots+a_0q(k) = x(k) \tag{2}$$

and the *output equation*

$$y(k) = b_mq(k+m)+\cdots+b_0q(k) \tag{3}$$

with the initial state of the system being the n numbers

$$q(0), q(1), \ldots, q(n-1)$$

The characteristic equation: Mode functions

When the input is zero but the initial state is not, the state variable $q(k)$ must satisfy the homogeneous state equation

$$q(k+n)+a_{n-1}q(k+n-1)+\cdots+a_0q(k) = 0 \tag{4}$$

In order to obtain a unique solution, the initial values of the outputs of the delay elements in Figure 13.5 must be specified: $q(0), q(1), \ldots, q(n-1)$.

In the previous section we found the solution of a first-order homogenous difference equation to be of the form $y(k) = Az^k$, where A and z are constants. Therefore, recalling our analysis of nth-order differential equations, we might be inclined to look for a solution to (4) having the form

$$q(k) = A_1z_1^{\,k}+A_2z_2^{\,k}+\cdots+A_nz_n^{\,k}$$

where the function z_i^k is the ith *mode* of the zero-input state-variable response and is a property of the system, independent of the initial state. Likewise, the constant A_i is the weighting of the ith mode and its value will depend on the initial state.

To determine whether such a solution will satisfy (4) and, if so, what constraints must be placed up the numbers z_i, we assume the trial solution $q(k) = Az^k$. Because of the form assumed for $y(k)$, it is particularly easy to compute the delayed values of $y(k)$, which are

$$q(k+1) = Az^{k+1} = zq(k)$$
$$\vdots$$
$$q(k+n) = Az^{k+n} = z^n q(k)$$

Substituting these into (4) gives

$$z^n q(k) + a_{n-1} z^{n-1} q(k) + \cdots + a_0 q(k) = 0$$

or, since $q(k)$ can be factored from each term,

$$(z^n + a_{n-1} z^{n-1} + \cdots + a_0) q(k) = 0$$

which will yield a nontrivial solution if and only if z is a root of the *characteristic equation*

$$z^n + a_{n-1} z^{n-1} + \cdots + a_0 = 0 \tag{5}$$

The polynomial on the left is known as the *characteristic polynomial* and will be denoted by $P(z)$. Provided its roots z_i are distinct there will be exactly n such modes: $z_1^k, z_2^k, \ldots, z_n^k$.

We observe that (5) has the same form as the characteristic equation for a continuous system – the only difference being that the complex number z is used rather than s. However, the assumed solution was of the form z^k rather than $e^{st}|_{t=kT} = (e^{sT})^k$. This distinction notwithstanding, there are many similarities between the solutions of difference and differential equations that stem from the fact that we can associate a characteristic polynomial with the discrete system.

For example, any complex roots of $P(z) = 0$ will occur in complex-conjugate pairs provided only that the system parameters $a_0, a_1, \ldots, a_{n-1}$ are real, i.e., that the system is real. If $z_1 = Me^{j\phi}$ is a complex root, then its complex-conjugate $z_2 = z_1^* = Me^{-j\phi}$ is also a root and the weighting factors A_1 and A_2 in the zero-input response will likewise be complex conjugates. Therefore, the contribution of the pair of roots to the response would be

$$A_1 z_1^k + A_2 z_2^k = 2|A_1|M^k \cos(k\phi + \arg[A_1])$$

since $|A_2| = |A_1|$ and $\arg[A_2] = -\arg[A_1]$. Another point of similarity to continuous systems is that if z_i is a repeated root of multiplicity two the zero-input response will contain the terms z_i^k and kz_i^k, analogous to the terms $e^{s_i t}$ and $te^{s_i t}$, which would result from a repeated root in the characteristic polynomial of a continuous system.

In Figure 13.13, the complex z-plane is shown with the mode functions corresponding to a variety of roots of the characteristic equation, in the same fashion as Figure 3.4 for continuous systems and the s-plane. As suggested in the previous section, the most important feature of a root of the characteristic equation is its location relative to the *unit circle* in the z-plane.

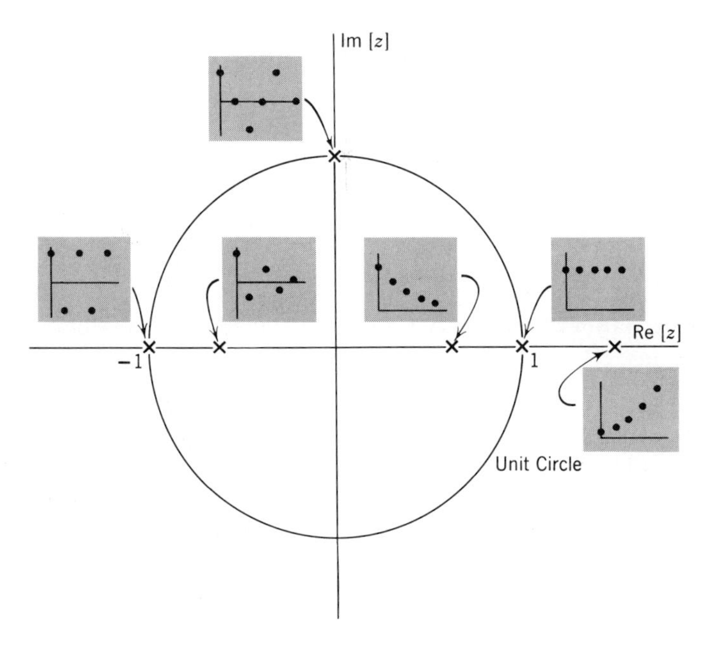

Figure 13.13. Discrete-time functions associated with points in the z-plane.

Zero-input response

Having established the behavior of the state variable $q(k)$ when $x(k) = 0$, it is a simple matter to use the output equation (3) to express the zero-input response as

$$y_{zi}(k) = b_m q(k+m) + \cdots + b_0 q(k)$$

where $q(k)$ satisfies the homogeneous state equation and the initial state.

Example 13.4

The system described by

$$y(k+2) - 3y(k+1) + 2y(k) = x(k+1) - 2x(k)$$

has as its characteristic equation

$$z^2 - 3z + 2 = (z-1)(z-2) = 0$$

the roots of which are $z_1 = 1$ and $z_2 = 2$. Thus the response of the state variable $q(k)$ must be of the form

$$q(k) = A_1 1^k + A_2 2^k = A_1 + A_2 2^k \qquad k \geq 0$$

To satisfy the initial state $q(0) = 0$, $q(1) = 1$, we require that

$$q(0) = A_1 + A_2 = 0$$
$$q(1) = A_1 + 2A_2 = 1$$

which gives $A_1 = -1$ and $A_2 = 1$. Hence, the state variable $q(k)$ behaves as $q(k) = -1 + 2^k$, $k \geq 0$, which is composed of one constant mode and another mode which doubles its value at each increment in k.

Next, the output equation $y(k) = q(k+1) - 2q(k)$ is invoked, yielding

$$y_{zi}(k) = (-1 + 2^{k+1}) - 2(-1 + 2^k)$$
$$= 1 \qquad k \geq 0$$

which hardly appears to be the unforced response of a second-order system. However, the absence of the mode 2^k in the response results from the fact that the particular combination of state variables specified by the output equation is such that that mode is unobservable and thus will never appear in the output signal $y(k)$, no matter what the initial state – hence the necessity for distinguishing between a system's state and output variables.

Zero-state response

The zero-state response of the system of arbitrary order described by (1) can be obtained in exactly the same manner as was developed in the previous section for treating first-order systems. The result is the convolution summation

$$y_{zs}(k) = \sum_{i=0}^{k} h(i)x(k-i) \tag{6}$$

$$= \sum_{j=0}^{k} h(k-j)x(j) \tag{7}$$

where $h(k)$ is the response to the unit-delta function $d(k)$. A demonstration of the technique follows.

Example 13.5

In the previous example we showed that when there was no input the general response of the state variable $q(k)$ was $q(k) = A_1 + A_2 2^k$ which, using the output equation, leads to the unit-delta response

$$\begin{aligned} h(k) &= (A_1 + A_2 2^{k+1}) - 2(A_1 + A_2 2^k) \\ &= -A_1 \qquad k \geq 1 \end{aligned}$$

By solving for $h(1)$ and $h(2)$ directly from the difference equation we find that $h(1) = h(2) = 1$, which indicates that $A_1 = -1$, so

$$h(k) = \begin{cases} 0 & k = 0 \\ 1 & k \geq 1 \end{cases}$$

The response to any causal input function can now be found by evaluating the convolution summation

$$y_{zs}(k) = \sum_{i=0}^{k} h(i)x(k-i)$$

Specifically, for the input $x(k) = 1$, $k \geq 0$, substitution for $h(i)$ yields

$$y_{zs}(k) = \sum_{i=0}^{k-1} 1 = k$$

Again, the output $y(k)$ does not exhibit the unobservable mode that grows as 2^k. Such a situation might correspond to a chemical process that is "running away" but does not contain instrumentation to indicate the fact – obviously not a desirable situation.

13.4 z-TRANSFORMS

Having developed the time-domain solution of fixed linear difference equations along the same lines used for differential equations, it follows naturally that frequency-domain methods exist that are analogous to the Laplace transform. The particular transform that will convert difference equations into algebraic equations is called the z-transform and as its properties are developed the similarity to the Laplace transform will be apparent.

The z-transform

Although a bilateral or two-sided z-transform can be defined for functions of a discrete variable that ranges from $-\infty$ to $+\infty$, we shall restrict our attention to the one-sided (unilateral) transform defined for causal functions of the discrete integer variable k, i.e., for functions that vanish for $k < 0$. For the function $v(k)$, which corresponds to the sequence $v(0)$, $v(1), v(2), \ldots$, we define its z-transform as the infinite *summation*

$$V(z) = \sum_{k=0}^{\infty} v(k) z^{-k} \tag{1}$$

$$= v(0) + v(1)z^{-1} + v(2)z^{-2} + \cdots$$

which will be denoted on occasion as $V(z) = \mathfrak{Z}[v(k)]$ or by the transform pair

$$v(k) \leftrightarrow V(z)$$

We shall show in Section 13.5 that if the transform $V(z)$ is known, the sequence $v(k)$ can be found by evaluating the inversion integral

$$v(k) = \frac{1}{2\pi j} \oint_C V(z) z^{k-1} dz \tag{2}$$

where C is any contour that encloses all of the poles of the integrand – usually taken as a circle centered at the origin of the z-plane.

Superposition

Like the Fourier and Laplace transforms, the z-transform obeys superposition. Thus,

$$\alpha v(k) + \beta w(k) \leftrightarrow \alpha V(z) + \beta W(z) \tag{3}$$

where $v(k) \leftrightarrow V(z)$, etc., for all constants α and β and transformable functions $v(k)$ and $w(k)$.

Advance (Forward Shift)

As we have written difference equations in terms of advanced versions of the input, output, and state variables, e.g., $y(k+1)$, $y(k+n)$, it is important that we be able to relate the transforms of these terms to $Y(z)$, the transform of $y(k)$. However, because we are shifting values ahead in k, some points will be shifted to negative values of k, in which case the one-sided transform is no longer defined.

Hence, before $v(k)$ is shifted we must delete from the sequence the terms that will be moved to the left of the origin, so as to avoid negative arguments. For example, if $v(k)$ is advanced by one interval we must drop $v(0)$ from the sequence, thereby creating the new function

$$w(k) = \begin{cases} 0 & k < 0 \\ v(k+1) & k \geq 0 \end{cases}$$

which is causal and has as its transform

$$W(z) = \sum_{k=0}^{\infty} w(k) z^{-k} = \sum_{k=0}^{\infty} v(k+1) z^{-k}$$

$$= z \sum_{k=0}^{\infty} v(k+1) z^{-(k+1)}$$

Letting $i = k+1$ and denoting the result as $\mathscr{Z}[v(k+1)]$ – although, as pointed out above, this notation is not strictly correct – the summation becomes

$$\mathscr{Z}[v(k+1)] = z \sum_{i=1}^{\infty} v(i) z^{-i} = z\Big[\underbrace{\sum_{i=0}^{\infty} v(i) z^{-i}}_{V(z)} - v(0)\Big]$$

yielding the transform pair

$$v(k+1) \leftrightarrow zV(z) - zv(0) \tag{4}$$

Similarly, the relationship for an advance of m intervals, where m is any nonnegative integer, is

$$v(k+m) \leftrightarrow z^m V(z) - z^m v(0) - z^{m-1} v(1) - \cdots - zv(m-1) \tag{5}$$

Delay (backward shift)

Having discussed the forward shift, the backward shift or delay by n units is easily handled since there are no problems about the shifted function having a nonzero value for a negative value of k. The resulting transform pair is simply

$$v(k-n) \leftrightarrow z^{-n}V(z) \qquad n \geq 0 \tag{6}$$

Summation

Analogous to the operation of integration for which Laplace and Fourier-transform relationships exist, we can define the summation operation

$$w(k) = \sum_{i=0}^{k} v(i)$$

where w is a function of k due to the upper limit. In the course of deriving an expression for $W(z)$ in terms of $V(z)$ we shall need the well-known infinite series sum

$$S_z \triangleq 1+z^{-1}+z^{-2}+\cdots = \frac{1}{1-z^{-1}} = \frac{z}{z-1} \tag{7}$$

which converges provided that $|z| > 1$.

Rather than derive the result for $W(z)$ by the more compact method of manipulating the double-summation expression resulting from a direct use of the transform summation (1), we note that the successive values of $w(k)$ are

$$\begin{aligned} w(0) &= v(0) \\ w(1) &= v(0)+v(1) \\ &\vdots \\ w(k) &= v(0)+v(1)+\cdots+v(k) \end{aligned}$$

Thus, $W(z)$ is the infinite sum

$$\begin{aligned} W(z) &= w(0)+z^{-1}w(1)+\cdots+z^{-k}w(k)+\cdots \\ &= v(0)+z^{-1}[v(0)+v(1)]+\cdots+z^{-k}[v(0)+\cdots+v(k)]+\cdots \\ &= v(0)[1+z^{-1}+z^{-2}+\cdots]+v(1)[z^{-1}+z^{-2}+\cdots] \\ &\quad +\cdots+v(k)[z^{-k}+z^{-k-1}+\cdots]+\cdots \end{aligned}$$

However, the coefficient of $v(0)$ is just the infinite series S_z which converges to $z/(z-1)$. Furthermore, by factoring out z^{-k} from the term involving $v(k)$, for all $k > 0$, the same series appears in all subsequent terms. Thus

$$W(z) = \left(\frac{z}{z-1}\right)\underbrace{[v(0) + v(1)z^{-1} + v(2)z^{-2} + \cdots]}_{V(z)}$$

so

$$\sum_{i=0}^{k} v(i) \leftrightarrow \left(\frac{z}{z-1}\right)V(z) \tag{8}$$

Discrete singularity functions

The unit-delta function, $d(k)$, has been defined as

$$d(k) = \begin{cases} 1 & k = 0 \\ 0 & k \neq 0 \end{cases} \tag{9}$$

Clearly, its transform is unity, since all terms in the series for the transform vanish, except for $d(0) = 1$. Thus

$$d(k) \leftrightarrow 1 \tag{10}$$

analogous to the unit impulse for continuous systems.

The *discrete step function*, defined as

$$u(k) = \begin{cases} 0 & k < 0 \\ 1 & k \geq 0 \end{cases} \tag{11}$$

has the transform

$$\mathfrak{Z}[u(k)] = \sum_{k=0}^{\infty} u(k)z^{-k} = \sum_{k=0}^{\infty} z^{-k} = S_z$$

which converges† to $z/(z-1)$ for $|z| > 1$. Thus,

$$u(k) \leftrightarrow \frac{z}{z-1} \tag{12}$$

†z-transforms have a region within the z-plane for which the transform summation converges. For those functions with which we shall be dealing the region of convergence will be the entire z-plane outside the smallest circle centered at the origin which encloses all of the poles of $V(z)$. Just as with Laplace transforms, the contour for the inversion integral in (2) must lie within the region of convergence. See Schwarz and Friedland (1965, Sect. 8.2) for details.

The transforms of several other commonly used discrete functions are tabulated in Appendix B.

Convolution

The convolution of two causal, discrete functions $v(k)$ and $w(k)$ has been defined to be

$$v * w(k) \triangleq \sum_{i=0}^{k} v(i)w(k-i) = \sum_{i=0}^{\infty} v(i)w(k-i)$$

where the upper limit of the summation may be taken as either k or ∞ as a result of the restriction that both functions are causal. Using an upper limit of infinity and writing the z-transform series gives

$$\mathscr{Z}[v * w(k)] = \sum_{k=0}^{\infty}\Big[\overbrace{\sum_{i=0}^{\infty} v(i)w(k-i)}^{v * w}\Big]z^{-k}$$

$$= \sum_{i=0}^{\infty} v(i)z^{-i} \sum_{k=0}^{\infty} w(k-i)z^{-(k-i)} \tag{13}$$

provided, of course, that z is restricted to the appropriate region of convergence and the summations can be interchanged.

Because $w(k-i) = 0$ for $k < i$, we can start the second summation in (13) at $k = i$. Then, defining the index $j = k - i$, we can write

$$\mathscr{Z}[v * w(k)] = \underbrace{\Big[\sum_{i=0}^{\infty} v(i)z^{-i}\Big]}_{V(z)}\underbrace{\Big[\sum_{j=0}^{\infty} w(j)z^{-j}\Big]}_{W(z)}$$

where the summations in brackets can be recognized as the transforms $V(z)$ and $W(z)$, respectively. Thus, we have

$$[v * w(k)] \leftrightarrow V(z)W(z) \tag{14}$$

which is the foundation for the extensive use of transfer functions and pole-zero concepts in the z-plane.

The transfer function

The transfer function of a fixed linear SIO discrete system is the ratio of the output and input z-transforms when the initial state is zero. Transforming the state equation, Eq. (2), Sect. 13.3, with $q(0)$,

$q(1), \ldots, q(n-1)$ all set to zero and a_n shown explicitly gives

$$Q(z) = \frac{X(z)}{a_n z^n + a_{n-1} z^{n-1} + \cdots + a_0}$$

Likewise, transforming the output equation, Eq. (3), Sect. 13.3 gives

$$Y(z) = (b_m z^m + b_{m-1} z^{m-1} + \cdots + b_0) Q(z)$$

and combining the two results in

$$Y(z) = \underbrace{\left(\frac{b_m z^m + \cdots + b_0}{a_n z^n + a_{n-1} z^{n-1} + \cdots + a_0} \right)}_{H(z)} X(z) \tag{15}$$

from which the transfer function for the system model represented by Eqs. (2) and (3), Sect. 13.3, can be identified as

$$H(z) = \frac{b_m z^m + \cdots + b_0}{a_n z^n + a_{n-1} z^{n-1} + \cdots + a_0} \tag{16}$$

As a case of particular interest, when the input is the unit-delta function $d(k)$ whose transform is unity, the output is $h(k)$ whose transform is the transfer function $H(z)$.

Initial-value theorem

Given the z-transform $V(z)$, the initial value $v(0)$ can be found as

$$v(0) = \lim_{z \to \infty} V(z) \tag{17}$$

To prove this result, we merely note that

$$V(z) = v(0) + v(1) z^{-1} + v(2) z^{-2} + \cdots.$$

Thus, as $z \to \infty$ each term in the series approaches zero, except the first. Following similar reasoning, we can see that

$$v(1) = \lim_{z \to \infty} z[V(z) - v(0)]$$

Final-value theorem

The final-value theorem states that

$$\lim_{k \to \infty} v(k) = \lim_{z \to 1} (z-1) V(z) \tag{18}$$

provided that $(z-1)V(z)$ is analytic for $|z| \geq 1$. The condition on $V(z)$ is necessary to ensure that the sequence $v(k)$ does indeed have a limit as $k \to \infty$.

Relationship to Laplace transforms

The situation often arises that the Laplace transform of a continuous signal is known and the z-transform of the sample values of that signal is desired. Rather than inverting the Laplace transform and then taking the z-transform of the sample values, it is usually possible to go directly from the Laplace transform $V(s)$ to the z-transform $V(z)$. The theorem that provides the transition can be obtained by starting with

$$v(t) = \frac{1}{2\pi j} \int_{\sigma-j\infty}^{\sigma+j\infty} V(s) e^{st} ds$$

which is the Laplace-transform inversion integral. Thus, with a sampling interval of T the sample values will be

$$v(kT) = \frac{1}{2\pi j} \int_{\sigma-j\infty}^{\sigma+j\infty} V(s) e^{skT} ds \qquad k = 0, 1, 2, \ldots$$

The z-transform of the sampled signal is

$$V(z) = \sum_{k=0}^{\infty} v(kT) z^{-k}$$

which, after interchanging the order of integration and summation, is

$$V(z) = \frac{1}{2\pi j} \int_{\sigma-j\infty}^{\sigma+j\infty} V(s) \sum_{k=0}^{\infty} (e^{sT} z^{-1})^k ds$$

Provided that $|z| > |e^{sT}|$ so the summation converges to $1/(1-e^{sT}z^{-1})$, the z-transform is

$$V(z) = \frac{1}{2\pi j} \int_{\sigma-j\infty}^{\sigma+j\infty} \frac{V(s)}{1-e^{sT}z^{-1}} ds$$

which, when written with positive powers of z, becomes

$$V(z) = \frac{1}{2\pi j} \int_{\sigma-j\infty}^{\sigma+j\infty} \frac{zV(s)}{z-e^{sT}} ds \qquad (19)$$

For almost all Laplace transforms of interest, a more useful form of (19) can be obtained by using the residue theorem (see Appendix A) to obtain

$$V(z) = \sum \text{Res} \left\{ \frac{zV(s)}{z - e^{sT}} \right\}_{\text{poles of } V(s)} \tag{20}$$

To demonstrate, if s_1 is a simple pole of $V(s)$, then

$$\text{Res} \left\{ \frac{zV(s)}{z - e^{sT}} \right\}_{s_1} = \frac{z(s - s_1)}{z - e^{sT}} V(s) \bigg|_{s=s_1} = \frac{C_1 z}{z - e^{s_1 T}} = \frac{C_1 z}{z - z_1}$$

where C_1 is the residue of $V(s)$ at its pole s_1 and $z_1 = e^{s_1 T}$ is the pole in the z-plane of the term contributed by the pole of $V(s)$ at $s = s_1$. Thus, if $V(s)$ has n poles, all of which are distinct, the corresponding z-transform is

$$V(z) = \sum_{i=1}^{n} \frac{C_i z}{z - e^{s_i T}}$$

which has n poles in the z-plane at the points $z_i = e^{s_1 T}, e^{s_2 T}, \ldots, e^{s_n T}$.

The relationship between the s-plane and z-plane can be better seen by substituting $s_i = \sigma_i + j\omega_i$ into $z_i = e^{s_i T}$ and then solving for the magnitude and argument of z_i, namely $z_i = e^{(\sigma_i + j\omega_i)T} = e^{\sigma_i T} e^{j\omega_i T}$. Hence,

$$|z_i| = e^{\sigma_i T}$$

$$\arg [z_i] = \omega_i T$$

For example, poles of $V(s)$ lying on the imaginary axis of the s-plane will lie on the unit circle of the z-plane since $|e^{j\omega T}| = 1$ for any combination of ω and T. In particular, poles of $V(s)$ at $s = 0$ become poles of $V(z)$ at $z = 1$. It is interesting to note, however, that distinct poles in the s-plane do not necessarily imply distinct z-plane poles since two values of s that have the same real parts but imaginary parts that differ by $2\pi/T$ radians will result in the *same* pole location in the z-plane. The reason for this situation is that the continuous functions corresponding to s-plane poles that differ by exactly $2\pi/T$ (or any integer multiple thereof) in their imaginary parts can both yield precisely the same sample values when sampled every T seconds.

13.5 z-TRANSFORM INVERSION

When $V(z) \triangleq \mathfrak{Z}[v(k)]$ is available as a rational function, essentially two methods exist for finding the corresponding sequence $v(0), v(1), \ldots$. These methods involve either *1.* an expansion of $V(z)$ into a form from which $v(k)$ is apparent or 2. a contour integration in the z-plane, usually evaluated by applying the residue theorem.

Expansion methods

Within the category of expansion methods, two types are commonly used. First, if its poles are known, $V(z)$ may be written as a *partial-fraction* expansion in essentially the same manner as was done for Laplace transforms. The objective, of course, is to obtain $V(z)$ as the *sum* of easily recognizable transforms such as those of the exponential, sine, cosine, step, and delta functions.

The only essential difference in the two cases is that the form used for the transforms of the elementary functions (see the table of z-transforms in Appendix B) contains a factor of z in the numerator, e.g., $u(k) \leftrightarrow z/(z-1)$. To insure that the partial-fraction expansion will yield terms corresponding to those tabulated, it is customary to first expand the function $V(z)/z$ and then multiply the resulting expansion by z.

For instance, if $V(z)$ has the n distinct poles $z_1, z_2, \ldots, z_n$, we are justified in writing

$$\frac{V(z)}{z} = \frac{A_0}{z} + \frac{A_1}{z-z_1} + \cdots + \frac{A_n}{z-z_n}$$

where the constants $A_0, A_1, \ldots, A_n$ can be evaluated by applying the conventional partial-fraction expansion rules. Multiplying by z,

$$V(z) = A_0 + A_1 \frac{z}{z-z_1} + \cdots + A_n \frac{z}{z-z_n}$$

and, using the transform pair $\alpha^k \leftrightarrow z/(z-\alpha)$ from Appendix B,

$$v(k) = A_0 d(k) + A_1 z_1{}^k + \cdots + A_n z_n{}^k$$

When only the first few numbers in the sequence $v(0), v(1), \ldots$ are desired, rather than the function $v(k)$, it is often convenient to expand $V(z)$ in the *power series*

$$V(z) = B_0 + B_1 z^{-1} + B_2 z^{-2} + \cdots$$

Referring back to Eq. (1), Sect. 13.4, we note that the coefficients of such a series must be the desired sequence values, i.e., $v(i) = B_i$. To obtain the first few numbers in the sequence, we can perform a long division of the numerator of $V(z)$ by its denominator, where the divisor and dividend polynomials are arranged in descending powers of z.

Example 13.6

As an illustration of the above techniques, we shall use both to find $v(k)$, or at least its first few values, from the transform

$$V(z) = \frac{2z^2 - 1.5z}{z^2 - 1.5z + 0.5}$$

To apply the partial-fraction expansion we first form

$$\frac{V(z)}{z} = \frac{2z - 1.5}{z^2 - 1.5z + 0.5} = \frac{2z - 1.5}{(z - 0.5)(z - 1)}$$

$$= \frac{1}{z - 0.5} + \frac{1}{z - 1}$$

Multiplying both sides by z,

$$V(z) = \frac{z}{z - 0.5} + \frac{z}{z - 1}$$

whence

$$v(k) = 0.5^k + u(k) = 0.5^k + 1 \qquad k \geq 0$$

Alternately, the powers-series expansion of $V(z)$ is the quotient of the long-division process

$$\begin{array}{r|l}
 & 2 + 1.5z^{-1} + 1.25z^{-2} + 1.125z^{-3} + \cdots \\
\hline
z^2 - 1.5z + 0.5 & 2z^2 - 1.5z \\
 & \underline{2z^2 - 3z \quad + 1} \\
 & \qquad 1.5z - 1 \\
 & \qquad \underline{1.5z - 2.25 + 0.75z^{-1}} \\
 & \qquad\qquad 1.25 - 0.75z^{-1} \\
 & \qquad\qquad \underline{1.25 - 1.875z^{-1} + 0.625z^{-2}} \\
 & \qquad\qquad\qquad 1.125z^{-1} - 0.625z^{-2}
\end{array}$$

Thus, the first four values in the sequence are 2.0, 1.50, 1.250, and 1.125, in agreement with the function $v(k)$ found above. Note that it is essential that the components of both the divisor and the dividend be placed in the order shown if the quotient is to have the required form. Other forms yield valid expansions of $V(z)$ but do not have the numbers $v(k)$ as their coefficients and hence are of no interest to us.

The inversion integral

At the beginning of the previous section it was asserted that the inverse transform could be obtained by evaluating the integral

$$v(k) = \frac{1}{2\pi j} \oint_C V(z) z^{k-1} dz \tag{1}$$

where the contour C must enclose all the poles of $V(z)$ and is usually taken to be a circle centered at the origin of the z-plane. To prove this assertion we rewrite the integral with $V(z)$ in its power-series form:

$$\oint_C V(z) z^{k-1} dz = \oint_C [v(0) + v(1)z^{-1} + \cdots + v(i)z^{-i} + \cdots] z^{k-1} dz$$

Multiplying each term of the series by the factor z^{k-1} and interchanging the order of summation and integration gives

$$\oint_C V(z) z^{k-1} dz = v(0) \oint_C z^{k-1} dz + \cdots + v(i) \oint_C z^{k-i-1} dz + \cdots.$$

However, as shown in Appendix A, every integral on the right-hand side vanishes except the single one for which $i = k$, in which case the integrand is z^{-1} and

$$\oint_C z^{k-i-1} dz \bigg|_{i=k} = \oint_C z^{-1} dz = 2\pi j$$

Thus, the infinite summation of integrals reduces to the single term

$$\oint_C V(z) z^{k-1} dz = 2\pi j\, v(k)$$

which becomes (1) upon division of both sides by $2\pi j$.

Having expressed $v(k)$ as a contour integral we can employ the residue theorem as a convenient method for evaluating the integral in most problems of interest. In particular

$$\oint V(z) z^{k-1} dz = 2\pi j \sum \text{Res}\, \{V(z) z^{k-1}\}_{\text{poles inside } C}$$

so

$$v(k) = \sum \text{Res}\, \{V(z) z^{k-1}\}_{\text{poles inside } C} \tag{2}$$

where, except for $k = 0$, the only poles of $V(z) z^{k-1}$ inside the contour C will be the poles of $V(z)$.

Example 13.7

To find the function $v(k)$ corresponding to the transform $V(z)$ of the previous example by using the inversion integral, we start by writing the integrand in factored form as

$$V(z)z^{k-1} = \frac{(2z-1.5)z^k}{(z-0.5)(z-1)}$$

Noting that for $k \geq 0$ the integrand has only the two poles $z = 0.5$ and $z = 1$, the appropriate residues are

$$\text{Res}\,\{V(z)z^{k-1}\}_{z=0.5} = \left.\frac{(2z-1.5)z^k}{z-1}\right|_{z=0.5} = 0.5^k$$

and

$$\text{Res}\,\{V(z)z^{k-1}\}_{z=1} = \left.\frac{(2z-1.5)z^k}{z-0.5}\right|_{z=1} = 1^k = 1$$

Substituting into (2) gives the closed-form expression

$$v(k) = 0.5^k + 1 \qquad k \geq 0$$

which agrees with the result obtained by partial-fraction expansion in the previous example.

13.6 DIGITAL-FILTERING APPLICATION

To demonstrate the application of many of the foregoing analytical tools and, hopefully, to arouse the interest of the reader in the possibilities of discrete-time filters, we shall design and discuss the implementation of a simple lowpass filter. Somewhat arbitrarily, the digital filter will be specified by requiring that its response to the unit step function $u(k)$ coincide at sampling instants with the continuous unit step response† of the second-order Butterworth filter and have a DC gain of unity. By making the discrete and continuous filter outputs agree at $t = kT$ for a step input we shall be assuring that the respective frequency-response functions agree at least for zero frequency.

†This requirement is an arbitrary choice. In fact, it is more common to match up impulse responses or frequency-response functions rather than step responses, as described in Gold and Rader (1969, Chap. 3).

Filter transfer function

The analog filter meeting these requirements with a 3-db frequency of ω_n rad/sec is known to have the transfer function

$$H_a(s) = \frac{\omega_n^{\,2}}{s^2 + \sqrt{2}\,\omega_n s + \omega_n^{\,2}} \tag{1}$$

and the pole-zero pattern shown in Figure 13.14a. The response of the analog filter to a unit step is

$$w_a(t) = 1 - \sqrt{2}\exp\left(-\frac{\omega_n}{\sqrt{2}}t\right)\sin\left(\frac{\omega_n}{\sqrt{2}}t + \frac{\pi}{4}\right) \tag{2}$$

which is plotted in part (b) of the figure.

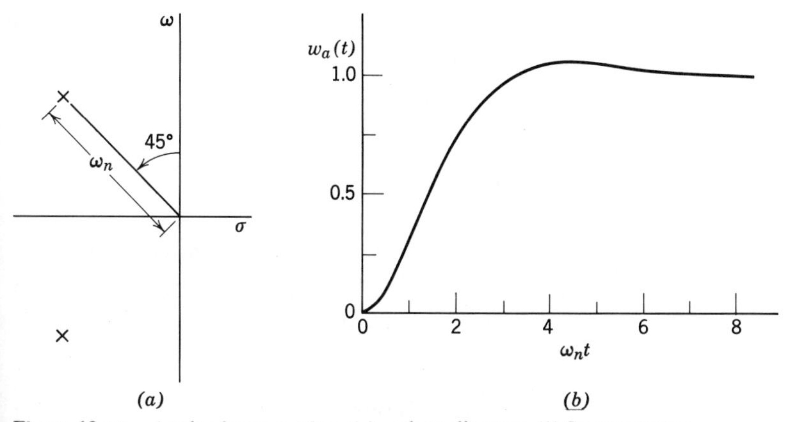

Figure 13.14. Analog lowpass filter. (a) s-plane diagram. (b) Step response.

Denoting the unit-step response of the digital filter as $w_d(k)$ and leaving the sampling period T and the continuous-filter bandwidth ω_n as free parameters for the moment, it follows from (2) that

$$\begin{aligned} w_d(k) &= w_a(kT) \\ &= 1 - \sqrt{2}\exp\left(-\frac{\omega_n T}{\sqrt{2}}k\right)\sin\left(\frac{\omega_n T}{\sqrt{2}}k + \frac{\pi}{4}\right) \end{aligned} \tag{3}$$

One way to proceed would be to take the z-transform of $w_d(k)$ and divide it by $\mathcal{Z}\,[u(k)] = z/(z-1)$, thereby obtaining the transfer function $H_d(z)$ of the digital filter. Once the transfer function is known, it is a simple matter to write the difference equation that the filter must satisfy and, thus, the computer-program instructions for a software mechanization or the interconnection of delayors, gains, and summers for a hardware mechanization.

However, since the prototype analog filter is more likely to be specified in the frequency domain by its transfer function $H_a(s)$, rather than its step response $w_a(t)$, we shall use Eq. (20), Sect. 13.4, to express $W_d(z)$, the z-transform of the digital filter's step response, directly in terms of $W_a(s)$, the Laplace transform of the analog filter's step response. Making use of the identity $W_a(s) = H_a(s)/s$,

$$W_d(z) = \sum \text{Res}\left\{\frac{\omega_n{}^2 z}{s(s^2+\sqrt{2}\,\omega_n s+\omega_n{}^2)(z-e^{sT})}\right\}_{\text{poles of } W_a(s)}$$

where the three poles of $W_a(s)$ are $s_1 = 0$, $s_2 = -(\omega_n/\sqrt{2})(1-j)$, and $s_3 = -(\omega_n/\sqrt{2})(1+j)$. Using the notation $\phi = \omega_n T/\sqrt{2}$ it follows that the corresponding residues at s_1 and s_2 are

$$\frac{z}{z-1} \qquad \text{and} \qquad \frac{-z}{(1+j)(z-e^{-\phi(1-j)})}$$

with the residue at s_3 being the complex conjugate of that at s_2. Adding the three residues to obtain $W_d(z)$ and dividing the result by $z/(z-1)$ to obtain the discrete transfer function yields, after a bit of manipulation,

$$H_d(z) = \frac{Az+B}{z^2-(2e^{-\phi}\cos\phi)z+e^{-2\phi}} \tag{4a}$$

where

$$A = 1-e^{-\phi}(\sin\phi+\cos\phi)$$

$$B = e^{-\phi}(e^{-\phi}+\sin\phi-\cos\phi) \tag{4b}$$

$$\phi = \omega_n T/\sqrt{2}$$

As a quick check on our work before considering the mechanization and response of the filter, the steady-state value of the unit-step response is, from the final-value theorem,

$$\lim_{k\to\infty} w_d(k) = \lim_{z\to 1}(z-1)W_d(z) = \lim_{z\to 1}(z-1)\left(\frac{z}{z-1}\right)H_d(z)$$

$$= H_d(1) = \frac{A+B}{1-2e^{-\phi}\cos\phi+e^{-2\phi}} = 1$$

Software realization

For a second-order discrete system obeying the input-output relationship

$$a_2 y(k+2) + a_1 y(k+1) + a_0 y(k) = b_2 x(k+2) + b_1 x(k+1) + b_0 x(k)$$

the transfer function $Y(z)/X(z)$ is

$$H(z) = \frac{b_2 z^2 + b_1 z + b_0}{a_2 z^2 + a_1 z + a_0}$$

Upon comparison with (4a) we see that the difference equation of the filter is

$$y(k+2) - 2e^{-\phi} \cos \phi \, y(k+1) + e^{-2\phi} y(k) = Ax(k+1) + Bx(k) \tag{5}$$

where the parameters A, B, and ϕ are defined in (4b). Since none of the filter's parameters depend on the independent variable k, i.e., it is a fixed filter, we can rewrite (5) with the argument of each variable reduced or increased by any fixed amount. In particular, it will be helpful to decrease all arguments by two and transpose the two lower-order terms from the left-hand side so as to express the difference equation in the form

$$y(k) = Cy(k-1) + Dy(k-2) + Ax(k-1) + Bx(k-2) \tag{6}$$

where $C = 2e^{-\phi} \cos \phi$ and $D = -e^{-2\phi}$. Thus the calculation of the current filter output $y(k)$ makes use of the two most recent outputs, $y(k-1)$ and $y(k-2)$, and the inputs $x(k-1)$ and $x(k-2)$. We note that in computing $y(k)$, the corresponding input $x(k)$ is not used; rather, $x(k)$ is stored for one sample interval and then used in forming $y(k+1)$.

If the sampling interval happens to be $T = 10^{-4}$ sec and the bandwidth of the analog prototype filter is $\omega_n = 1000$ rad/sec, it follows that $\phi = 0.0707107$ and (6) becomes

$$\begin{aligned} y(k) = {} & 1.858806 y(k-1) - 0.8681235 y(k-2) \\ & + 0.004768465 x(k-1) + 0.004548856 x(k-2) \end{aligned}$$

The following segment of computer code, written in FORTRAN, will implement the filter calculations. Note that the input value X is not used until the next pass through the loop, after it has been shifted to XM1.

```
┌→READ X
│ Y = 1.858806 * YM1 − 0.8681235 * YM2
│     + 4.768465E-3 * XM1 + 4.548856E-3 * XM2
│ ØUTPUT Y
│ YM2 = YM1
│ YM1 = Y
│ XM2 = XM1
└ XM1 = X
```

If the filter is running in "real time," the input samples occur every 10^{-4} sec and it is obvious that the time required for one circuit of the program loop must not exceed 10^{-4} sec. Also, if the computational time for the loop is appreciably less than 10^{-4} sec, the computer can perform other calculations while it is waiting for the next value of x to be accepted. On the other hand, if the filter is processing stored data in an "off-line" mode, it is possible for the sampling interval of the original data to be less than the time required for each loop through the filter calculations.

The response to a unit-step function agreed with the computed values of $w_a(kT)$ to within $\pm 10^{-5}$ when carried out on a digital computer.

Hardware realization

If one desires to construct the filter in the form of digital hardware, a simulation diagram will provide a basis for interconnecting the various elements. The diagram in the form of Figure 13.5 can be drawn directly from the difference equation and is shown in Figure 13.15 for $\omega_n = 1000$ rad/sec and $T = 10^{-4}$ sec.

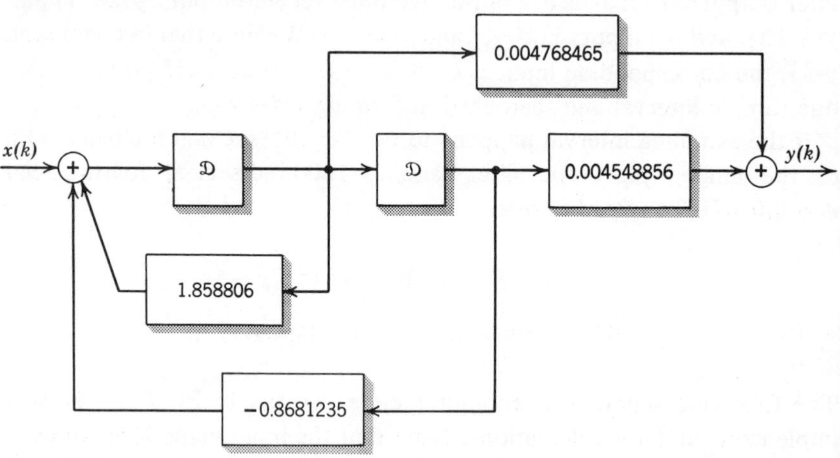

Figure 13.15.

Frequency response

As we have done so often for analog (continuous) filters, the digital filter can be characterized in terms of its frequency response. When sampled every T seconds, the signal $\cos \omega t$ yields the input sequence $\cos \omega Tk$, which can be thought of as existing over the infinite interval $-\infty < k < \infty$ since we are dealing with the sinusoidal steady state. Following a line of reasoning similar to that used in Chapter 6 in establishing the frequency-response function for continuous systems, we let

$$x(k) = e^{j\omega Tk} = 2 \text{ Re } [\cos \omega Tk]$$
$$y(k) = H_d(j\omega) e^{j\omega Tk}$$

The as-yet unknown function $H_d(j\omega)$ is the *frequency-response function* of the discrete filter, a function of the *continuous* frequency variable ω and not dependent upon k.

Substituting for $x(k)$ and $y(k)$ and the appropriately shifted values in the filter's difference equation, (5), it follows that

$$(e^{j2\omega T} - 2e^{-\phi} \cos \phi \; e^{j\omega T} + e^{-2\phi}) H_d(j\omega) e^{j\omega Tk} = (Ae^{j\omega T} + B) e^{j\omega Tk}$$

where $\phi = \omega_n T/\sqrt{2}$, a dimensionless parameter of the filter, and A and B depend upon ϕ per (4). Solving for the frequency-response function, we obtain

$$H_d(j\omega) = \frac{Ae^{j\omega T} + B}{e^{j2\omega T} - 2e^{-\phi} \cos \phi \; e^{j\omega T} + e^{-2\phi}}$$

An alternate method of obtaining $H_d(j\omega)$ that lends itself to a useful graphical interpretation is to set $z = e^{j\omega T}$ in the discrete filter's transfer function, (4). The basis for this assertion is the fact that $\cos \omega Tk$ has both of its poles on the unit circle at $z = e^{\pm j\omega T}$ and it is these poles in the output transform $Y(z)$ that account for the sinusoidal-steady-state behavior. The partial-fraction coefficient in question will involve

$$H_d(j\omega) = H_d(z)\Big|_{z=e^{j\omega T}}$$

which can be represented by drawing the vectors from the poles and zero of $H_d(z)$ to the point $z = e^{j\omega T}$ on the unit circle. Figure 13.16 depicts the vectors corresponding to $\omega T = 0.85$ rad, or $\omega = 8500$ rad/sec. As ωT increases beyond 2π, the point at which the vectors converge begins another full circuit around the unit circle, hence the periodicity of $H_d(j\omega)$. The

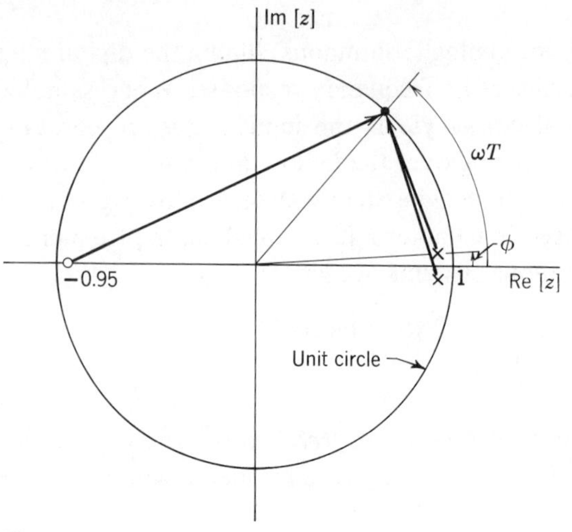

Figure 13.16.

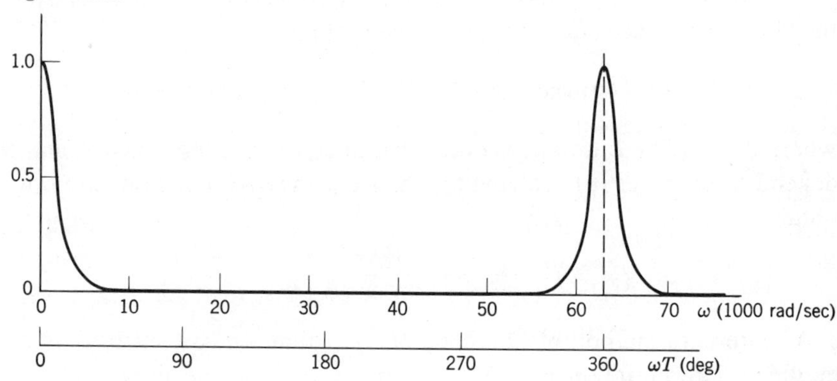

Figure 13.17. Amplitude ratio $|H_d(j\omega)|$.

magnitude of $H_d(j\omega)$, plotted in Figure 13.17, is the length of the vector from the zero divided by the product of the lengths of the vectors from the poles multiplied by the factor A. Likewise, arg $[H_d(j\omega)]$ can be obtained by adding the angles of the three vectors with the appropriate signs.

13.7 SAMPLED-DATA CONTROL APPLICATION

The feedback system for controlling the pointing direction of a satellite that was studied in Chapters 9, 10, and 11 can also be implemented by the introduction of a digital computing device with analog-to-digital (A/D)

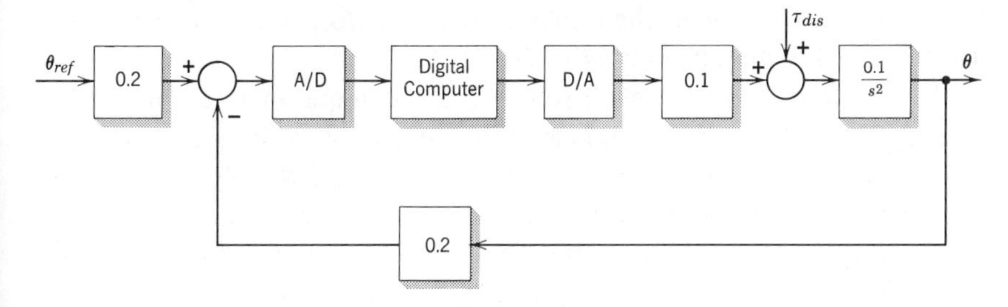

Figure 13.18. Satellite attitude system with digital controller.

and digital-to-analog (D/A) conversion equipment. Recasting the system block diagram (Figure 10.16*a*) in terms of the hybrid system shown in Figure 13.6 results in Figure 13.18.

Discrete model

Before proceeding to the analysis of the system and the determination of a suitable control algorithm, a completely discrete model must be constructed per the discussion of hybrid systems in Section 13.1. First, the disturbance torque is assumed to be zero in order that we may restrict our attention to the stability of the closed-loop system and its response to the reference input $\theta_{ref}(t)$. Then, moving the two gains of 0.20 (input and feedback transducers) inside the loop and combining the two factors of 0.10 (the torque constant and reciprocal of the moment of inertia) the system block diagram reduces to Figure 13.19, where the control algorithm has been represented by the discrete transfer function $F(z)$, as yet unspecified.

It remains to find the relationship between the discrete signal $y(k)$ and the sample values of the continuous output variable, $\theta(kT)$. Two methods present themselves. First, the D/A converter may be modeled by the discrete transfer function $(1-z^{-1})$ and the continuous elements by the Laplace transform of the unit-step response, $H(s)/s$, whereupon Eq. (20),

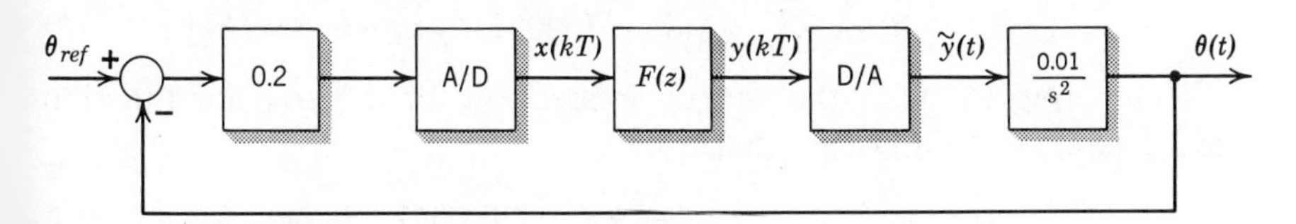

Figure 13.19. Hybrid model without disturbance input.

Sect. 13.4, which relates the Laplace and z-transforms, can be used to find the transfer function from $y(kT)$ to $\theta(kT)$.

As an alternative, we shall follow the technique of Example 13.2 in finding the difference equation relating successive values of $\theta(kT)$ and $y(kT)$, from which the transfer function will be immediately apparent. We start by writing the differential equation for $\theta(t)$, which, because of the piecewise-constant output of the D/A converter, is valid only over the interval $kT < t \leq (k+1)T$. Specifically, we write

$$\ddot{\theta}(t) = 0.010y(kT) \tag{1}$$

The appropriate initial conditions are $\theta(kT)$ and $\dot{\theta}(kT)$, the differential equation being second-order. The solution of (1) is

$$\theta(t) = \theta(kT) + (t-kT)\dot{\theta}(kT) + 0.010\frac{(t-kT)^2}{2}y(kT) \tag{2}$$

Setting $t = (k+1)T$ and dropping the sampling interval T from the arguments of the time functions, (2) reduces to

$$\theta(k+1) = \theta(k) + T\dot{\theta}(k) + 0.005T^2y(k) \tag{3}$$

To eliminate $\dot{\theta}(k)$ so as to obtain a difference equation in terms of $\theta(k)$ and its shifted values $\theta(k+i)$, we start by advancing the index in (3) one unit and subtracting (3) from the result, obtaining the second-order equation

$$\theta(k+2) - \theta(k+1) = \theta(k+1) - \theta(k) + T[\dot{\theta}(k+1) - \dot{\theta}(k)] + 0.005T^2[y(k+1) - y(k)] \tag{4}$$

which still involves $\dot{\theta}(k)$. However, because $\dot{\theta}(t) = 0.010 \int y \, dt$, $\dot{\theta}(k)$ obeys the first-order difference equation

$$\dot{\theta}(k+1) = \dot{\theta}(k) + 0.010Ty(k)$$

Substituting into (4) yields the desired relationship

$$\theta(k+2) - 2\theta(k+1) + \theta(k) = 0.005T^2[y(k+1) + y(k)] \tag{5}$$

Comparing (5) with the general expression for the transfer function of a discrete system, we see that

$$\frac{\theta(z)}{Y(z)} = 0.005T^2\left(\frac{z+1}{z^2 - 2z + 1}\right) = \frac{0.005T^2(z+1)}{(z-1)^2} \tag{6}$$

which has a double pole at $z = 1$ and a single zero at $z = -1$. Changing the input signal to $\theta_{ref}(k)$ – values of $\theta_{ref}(t)$ *between* sample points are ignored by the A/D converter anyway, so no information is lost – and no longer showing the A/D and D/A converters explicitly, the discrete mathematical model of the system is as shown in Figure 13.20. Finally we are prepared to examine the behavior of the closed-loop discrete model, keeping in mind the fact that $\theta(k)$ represents the continuous output $\theta(t)$ only at the sample points $t = kT$, $k = 0, 1, 2, \ldots$.

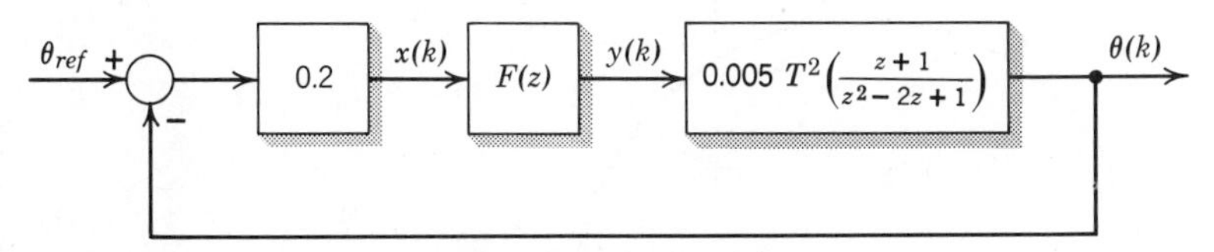

Figure 13.20. Discrete model without disturbance input.

Proportional control

The simplest control algorithm is to make the control input be proportional to the current error signal. Stated mathematically the proportional control law is

$$y(k) = K_a x(k) \tag{7}$$

where $x(k) = 0.20[\theta_{ref}(k) - \theta(k)]$. Transforming (7),

$$F(z) = \frac{Y(z)}{X(z)} = K_a$$

and the system block diagram reduces to Figure 13.21*a*.

From this point on, we can use block-diagram algebra to find the closed-loop transfer function which is

$$\frac{\theta(z)}{\theta_{ref}(z)} = \frac{0.001T^2K_a(z+1)}{z^2 - (2 - 0.001T^2K_a)z + (1 + 0.001T^2K_a)}$$

Treating T^2K_a as the gain parameter, the root-locus plot of Figure 13.21*b* can be drawn for $K_a > 0$. Inspection of the plot leads to the conclusion that the system will be unstable for any values of T and K_a (T is inherently positive and making $K_a < 0$ is of no help). It is interesting to recall that in Chapter 9 we found that the same proportional control law resulted in an oscillatory closed-loop system when used with a continuous rather

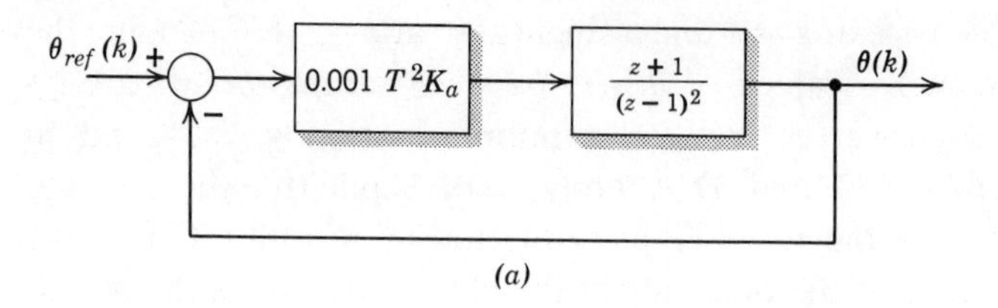

(a)

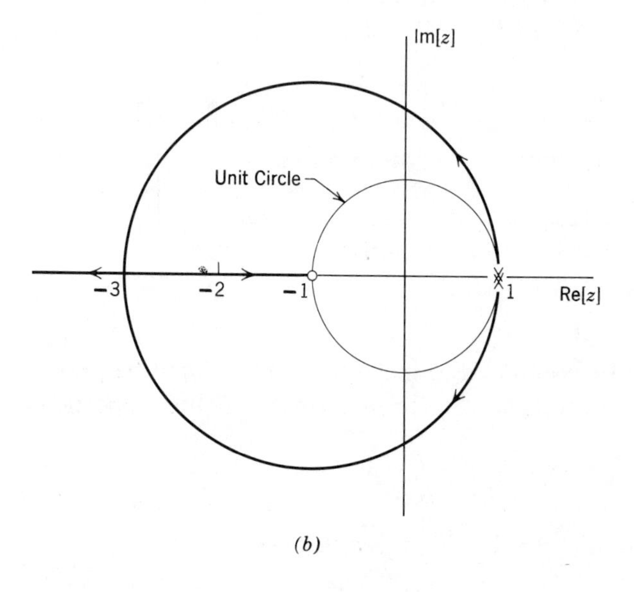

(b)

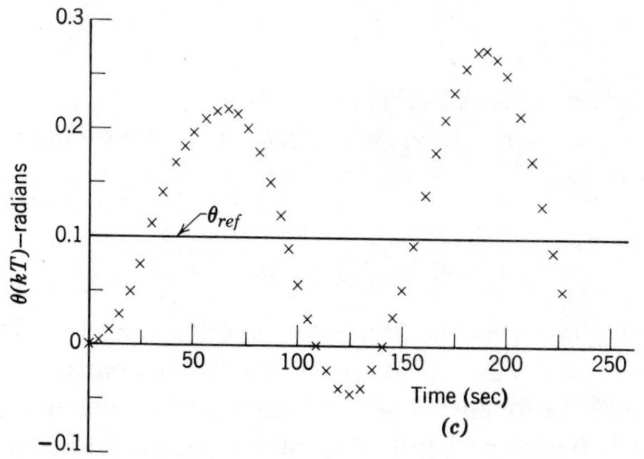

(c)

Figure 13.21. Proportional control. (*a*) System diagram. (*b*) Root locus. (*c*) Sample values of step response, $K_a = 1.25$.

than discrete controller. Evidently the introduction of sampling has caused the marginally stable continuous system to become an unstable sampled system.

The transient response resulting from a step input of 0.1 rad for θ_{ref} is shown in Figure 13.21*c* for $T = 1$ sec and $K_a = 1.25$. It is no accident that the period of the growing oscillation is very close to $40\pi = 125.7$ sec, the period of the continuous system's undamped oscillation when $K_a = 1.25$. Setting $K_a = 1.25$ and applying the root-locus magnitude criterion indicates that the closed-loop poles will be near the points $z = e^{\pm j0.05}$, which correspond to the sample values of the function sin $0.05\,kT$.

Because proportional control is clearly unsatisfactory for this application and because of the favorable results obtained with a proportional-plus-derivative control law for the continuous controller, we shall next consider adding a "difference" term to (7).

Proportional-plus-difference control

Making $y(k)$ be a linear combination of $x(k)$ and the difference $x(k) - x(k-1)$ is equivalent to the control law

$$y(k) = K_a[x(k) - \tau x(k-1)] \tag{8}$$

where the parameters K_a and τ are to be chosen so as to yield (hopefully) a stable closed-loop system. Transforming (8) using the delay theorem,

$$Y(z) = K_a[X(z) - \tau z^{-1}X(z)]$$

whence

$$F(z) = \frac{Y(z)}{X(z)} = K_a(1 - \tau z^{-1}) = K_a\left(\frac{z-\tau}{z}\right) \tag{9}$$

Substituting (9) for $F(z)$ in Figure 13.20 leads to the block diagram of Figure 13.22*a* and, for $\tau = 0.75$, the root locus shown in part (*b*).

From the root-locus plot it is apparent that the addition of the zero at $z = 0.75$ due to the presence of $x(k-1)$ in the control law has made it possible to stabilize the system by attracting the two branches emanating from $z = 1$ inside the unit circle. While the pole at the origin which was introduced along with the zero has a destabilizing effect, its presence can be tolerated by placing the zero close to $z = 1$ and keeping the gain K_a sufficiently low. The sample values of the step response corresponding to several values of τ resulting in a stable system are plotted in Figure 13.22*c*. Although not shown, the combination of $\tau = 0.25$ and $K_a = 1.25$ resulted

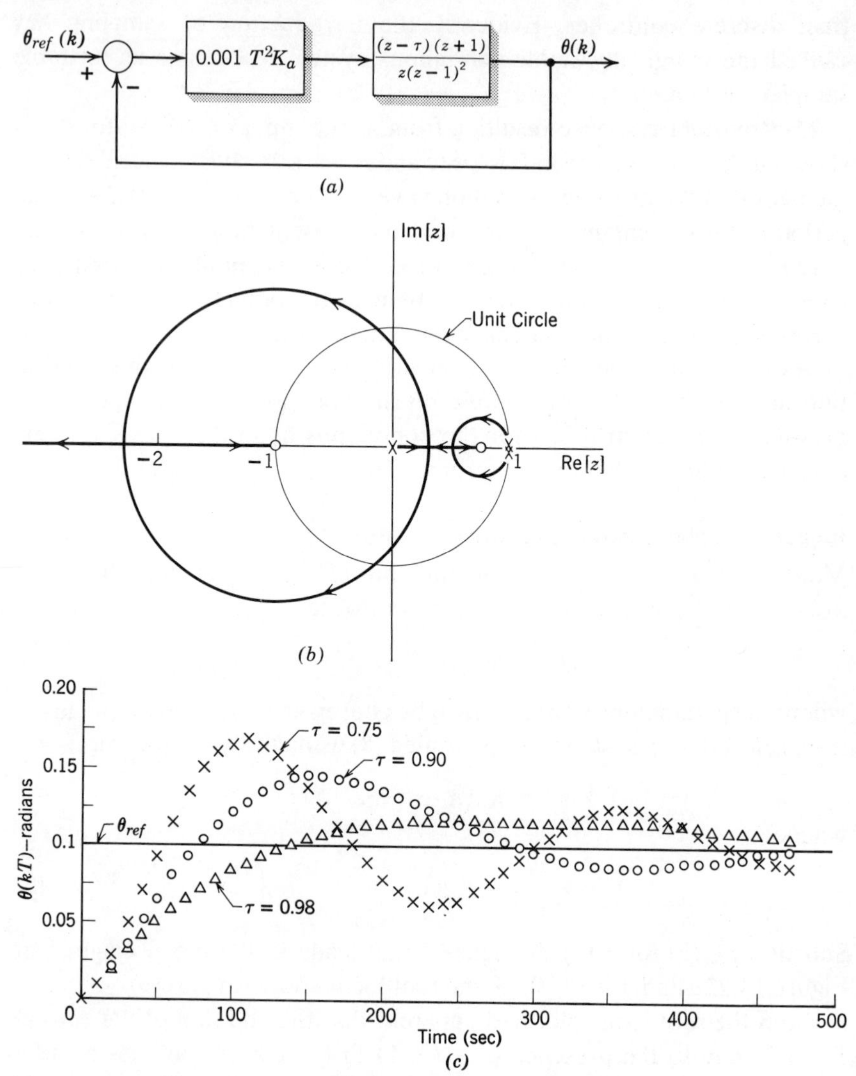

Figure 13.22. Proportional-plus-difference control. (*a*) System diagram. (*b*) Root locus, $\tau = 0.75$. (*c*) Sample values of step response (every other point shown), $T = 5$ sec, $K_a = 1.25$.

in slightly growing oscillations. As should be expected, making $\tau = 0.98$ has greatly reduced the overshoot at the expense of having a slowly decaying mode, since the zero at $z = 0.98$ insures that a closed-loop pole will lie close to $z = 1.0$.

Other control laws

Having come this far, we leave the reader on his own to explore the possibilities of other more complicated control laws. For example, to achieve the counterpart of the continuous P + I + D controller one need only add the summation process to (8), yielding the algorithm

$$y(k) = K_a\left[x(k) - \tau x(k-1) + \beta \sum_{i=0}^{k} x(i)\right] \tag{10}$$

Another possibility is to incorporate a digital filter as part of $F(z)$ – in fact, one can view the above control algorithms as merely specific examples of digital filters.

Problems

13.1 For the fluid reservoir described in Problem 2.2, derive the discrete incremental model relating the sample values of the flows in and out of the tank. Assume that the incremental inflow $x(t)$ will remain constant between sample times because of a digitally controlled valve, the dynamics of which can be neglected. Draw a simulation diagram for the discrete process. *Answer*: $y(k+1) - e^{-T/\alpha}y(k) = (1 - e^{-T/\alpha})x(k)$, where $\alpha \triangleq 2A\sqrt{\bar{H}}/C$.

13.2 Obtain a discrete model of the tape-drive system described in Problem 2.3 when a digital controller is used, (a) assuming that $\omega_i = (\omega_i)_{des}$, $i = 1, 2$. (b) assuming that ω_i and $(\omega_i)_{des}$ are related by the differential equation $\tau\dot{\omega}_i = (\omega_i)_{des} - \omega_i$, $i = 1, 2$, where τ is the time constant of the drive motors.

13.3 Write the difference equation obeyed by a sum of money in a bank account at a yearly interest rate of R percent, compounded every T years. Denote the amount immediately following the kth interest period as $A(kT)$. Find the differential equation which results as $T \to 0$. *Answer*: $\dot{A} = (R/100)A$.

13.4 Numerically integrating a differential equation is done by constructing a related difference equation that can be solved in a recursive manner. One approximation to the first derivative of $y(t)$ is

$$\dot{y}(t)|_{t=kT} \approx \frac{y(kT+T)-y(kT)}{T}$$

for a step size of T units.

(a) Show that the corresponding approximation to $\ddot{y}$ is

$$\ddot{y}(t)|_{t=kT} \approx \frac{y(kT+2T)-2y(kT+T)+y(kT)}{T^2}$$

(b) Find the difference equation whose solution should closely approximate the solution to $\ddot{y}(t)+a\dot{y}(t)+by(t)=x(t)$ and determine the conditions on $y(k)$ corresponding to the initial conditions $y(0)=C, \dot{y}(0)=D$.

13.5 Figure P13.1 shows an RC circuit driven by a D/A converter whose input is the sequence $x(kT)$. Derive the difference equation relating $x(kT)$ and $y(kT)$ and sketch a typical response between sample points. *Answer*: $y(k+1)-e^{-T/\tau}y(k)=[(1-e^{-T/\tau})\tau/(R_1C)]x(k)$ where $\tau \triangleq R_1R_2C/(R_1+R_2)$.

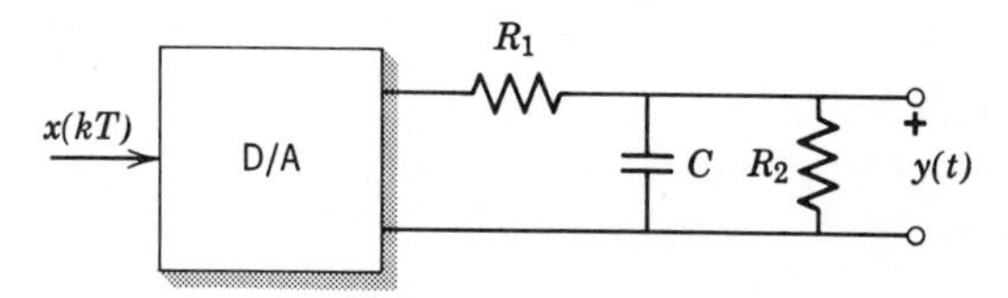

Figure P13.1

13.6 For the hybrid circuit modeled in Problem 13.5, find the unit delta response $h(k)$ and the sample values of the zero-input response to an initial charge on the capacitor. Let $R_1=R_2=1\ \text{M}\Omega$, $C=0.001\ \mu\text{f}$, and the sampling frequency be 4 KHz.

13.7 For the following difference equations, find the mode functions and the weighting of each mode in the unit delta response:

(a) $y(k+2)-0.6y(k+1)-0.16y(k)=x(k)$

(b) $y(k+3)-2\sqrt{2}y(k+2)+y(k+1)=x(k)$

(c) $y(k+2)-y(k+1)+0.25y(k)=x(k+2)-x(k+1)$

Answer: (a) $h(k)=1.25(0.80)^k+5.0(-0.20)^k, \quad k \geq 1$.

13.8 Compute and plot *1*. the unit delta responses and *2*. the responses to the causal input $x(k)=k$ over the interval $0 \leq k \leq 6$ for the following discrete systems:

(a) $y(k+2)+y(k)=x(k)$

(b) $y(k+2) - y(k) = x(k)$

(c) $y(k+2) - y(k) = x(k+1) - x(k)$

Answer: (a) Starting at $k = 0$, $h(k) = 0, 0, 1, 0, -1, 0, 1, \ldots$ and $y(k) = 0, 0, 0, 1, 2, 2, 2, \ldots$

13.9 Prove that the response of the system described by Eq. (1), Sect. 13.3, to the unit step $u(k)$ is the sum $y(k) = h(0) + h(1) + \ldots + h(k)$ where $h(k)$ is the unit delta response.

13.10 The differential equation $\dot{y} = -ay$, $y(0) = 1$ is to be numerically integrated using the algorithm $\hat{y}(kT+T) = \hat{y}(kT) - aT\hat{y}(kT)$ with the initial value $\hat{y}(0) = 1$. (a) Justify the appropriateness of the proposed algorithm. (b) Find the error of the integration process defined as $\varepsilon(kT) \triangleq y(kT) - \hat{y}(kT)$. Comment on the dependence of the error on a, T, and k.

13.11 (a) Present an argument as to why the convolution summations given by Eqs. (6) and (7), Sect. 13.3, yield the zero-state response of a causal system with a causal input. Show the corresponding form of each version for a noncausal system and a noncausal input.

(b) Prove that a discrete function convolved with the unit delta is the function itself, that is, that $v * d(k) = v(k)$.

13.12 Derive the z-transforms given in Appendix B for the causal functions: (a) k, (b) e^{aTk}, (c) α^k, (d) $\sin \omega Tk$, (e) $\cos \omega Tk$.

13.13 Use Eq. (20), Sect. 13.4 to derive the z-transforms of the functions: (a) e^{aTk}, (b) $\sin \omega Tk$, (c) $e^{aTk} \sin \omega Tk$.

13.14 Prove the final-value theorem for z-transforms and explain why $(z-1)V(z)$ must be analytic for $|z| \geq 1$ in order for the theorem to apply. *Hint*: Start with the relationship $\mathcal{Z}[v(k+1) - v(k)] = (z-1)V(z) - zv(0)$.

13.15 For the following z-transforms, find $v(0)$, $v(1)$, and $\lim_{k\to\infty} v(k)$. Also determine whether the corresponding system is asymptotically stable, marginally stable, or unstable if $V(z)$ represents the system's transfer function.

(a) $$V(z) = \frac{z^2 + 2z}{(z^2 - 1)(z + 0.5)}$$

(b) $$V(z) = \frac{2z^2 - 3z + 1}{z^2 - 4z - 5}$$

(c) $$V(z) = \frac{z^3 + 2z^2 - z + 1}{z^3 + z^2 + 0.5z}$$

Answer: (a) $v(0) = 0$, $v(1) = 1$, $v(\infty)$ does not exist, marginally stable.

13.16 With reference to the transfer function

$$H(z) = \frac{b_m z^m + \cdots + b_0}{a_n z^n + a_{n-1} z^{n-1} + \cdots + a_0}$$

show that the system is causal if and only if $m \leq n$, and that the delta response $d(k)$ remains zero until $k = n - m$ at which time it takes on the value b_m / a_n.

13.17 Show that the transfer function of the discrete model relating $y(kT)$ and $x(kT)$ in a continuous system having the transfer function $G(s)$ following a D/A converter (see part *a* of Figure P13.2) can be expressed as

$$H(z) = \frac{Y(z)}{X(z)} = \sum \text{Res}\left\{\frac{(z-1)G(s)}{(z-e^{sT})s}\right\}_{s=0 \text{ and poles of } G(s)}$$

Hint: Show that the model in part *b* of the figure is a valid representation of the original system. Then consider only the sample values and apply Eq. (20), Sect. 13.4, resulting in the model shown in part *c*.

13.18 Use the transfer-function relationship for $H(z)$ given in Problem 13.17 to compute the transfer function and difference equation of the hybrid system discussed in Example 13.2.

13.19 Find the transfer function $\theta(z)/Y(z)$ of the satellite discussed in Section 13.7 by using the theorem stated in Problem 13.17. *Answer*: See Eq. (6), Sect. 13.7.

13.20 (a) Verify that the *bilinear transformation* $z = (s+1)/(1-s)$ maps the exterior of the *z*-plane unit circle onto the right half of the *s*-plane, that is, $|z| > 1$ implies $\text{Re}[s] > 0$ and vice versa. Also show that this is the same transformation referred to in Problem 13.26.

(b) Using the bilinear transformation show that in order for $P(z) = a_2 z^2 + a_1 z + a_0$, $a_2 > 0$, to be the characteristic polynomial of an asymptotically stable discrete system, it is necessary and sufficient that each of the following conditions be satisfied:

1. $a_0 + a_1 + a_2 > 0$
2. $a_0 + a_2 - a_1 > 0$
3. $|a_0| < a_2$.

(c) Can you suggest a procedure that will allow one to test a characteristic polynomial of any order in z for stability?

13.21 A filter often used as a part of computer-control algorithms obeys the first-order difference equation $y(k) = \alpha y(k-1) + \beta x(k)$.

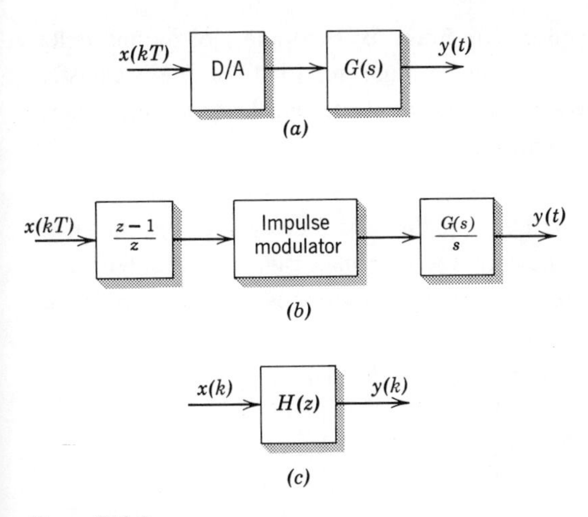

Figure P13.2

(a) Show that if the filter is to have unity gain for DC signals, it is necessary that $\beta = 1 - \alpha$.

(b) Draw a simulation diagram and find the filter's transfer function $H(z) = Y(z)/X(z)$.

(c) For $\alpha = 0.1$, 0.5, and 0.9 and $\beta = 1 - \alpha$, plot the magnitude of the frequency response over the interval $0 \leq \omega \leq 3\pi/T$, where T is the sampling interval.

(d) Find the transfer function of a continuous filter that can be considered as analogous to the digital filter with $\alpha = 0.5$.

13.22 Consider the tapped-delay-line filter shown in Figure P13.3. (a) Write the

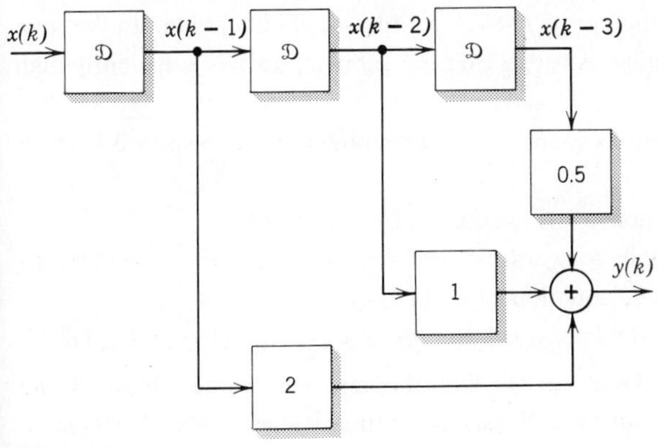

Figure P13.3

difference equation obeyed by the filter. (b) Find and plot its unit delta response; comment on any unusual feature. (c) Find the filter's transfer function and make a pole-zero plot. (d) Evaluate and sketch the magnitude of the frequency-response function.

Answer: (c) $H(z) = (2z^2 + z + 0.5)/z^3$.

13.23 Select the weights a_0, a_1, a_2, and a_3 in the tapped-delay-line filter shown in Figure P13.4 so as to provide unity DC gain and a gain of zero for both $\omega = \pi/2000$ and $\pi/1000$ rad/sec. The sampling interval is $T = 0.001$ sec. Find

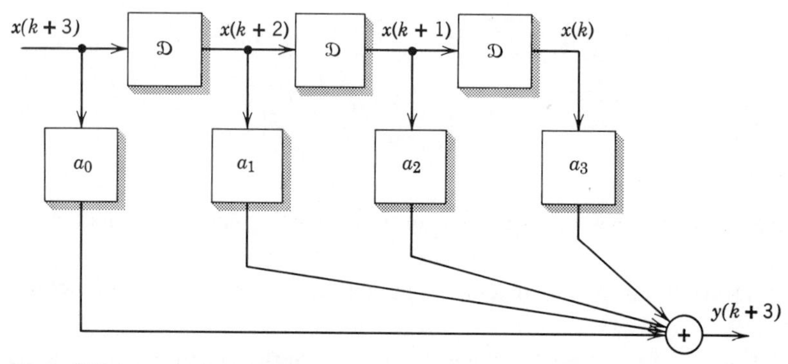

Figure P13.4

the filter's transfer function and plot the magnitude of its frequency response.

13.24 The analog (continuous) filter described by $\dot{y} + ay = ax$, $a > 0$, is to be used as the basis for designing an "equivalent" digital filter. Use each of the following methods and compare the resulting digital filters in both the time and frequency domains. Assume that the sampling rate is sufficiently high that $aT < 1$.

1. Let the unit delta response $h_d(k)$ equal the impulse response $h_a(t)$ at all $t = kT$.
2. Let the step response $w_d(k)$ equal $w_a(t)$ at all $t = kT$.
3. Replace $\dot{y}(t)$ by its forward-difference version $[y(k+1) - y(k)]/T$ to obtain a difference equation and set $t = kT$.
4. Repeat *3* using the backward-difference version $[y(k) - y(k-1)]/T$.

13.25 Given $H(z) = (\alpha - 1)/(\alpha - z^{-1})$, find the magnitude and phase of the frequency-response function $H(j\omega)$ by setting $z = e^{j\omega T}$. Sketch $|H(j\omega)|^2$ for $\alpha = 5$ and $\alpha = 2$.

13.26 Digital filters are sometimes designed using the bilinear transformation by starting with an analog transfer function $H_a(s)$ and replacing s by $(z-1)/(z+1)$. Use this approach for $H_a(s) = a/(s+a)$; draw the resulting digital simulation diagram and sketch $|H(j\omega)|^2$ for $a^2 >> 1$ and $a^2 << 1$. *Hint*: For the latter task note that $(z-1)/(z+1) = j \tan \omega T/2$ when $z = e^{j\omega T}$.

13.27 Repeat the digital filter design of Sect. 13.6, with $H(z)$ obtained by applying Eq. (20), Sect. 13.4, directly to $H_a(s)$ so that $h_d(k)$ will equal the sample values of the impulse response $h_a(t)$. Draw the digital filter's block diagram and pole-zero plot, find $\lim_{k\to\infty} w_d(k)$, and sketch $|H_d(j\omega)|$ taking $\omega_n = 10^3$ and $T = 10^{-4}$. Compare these with the original design and reconcile any differences.

13.28 Computer-control systems often calculate the increment in the controller setting from its value at the previous sample time instead of calculating the setting itself. Show that the P + I analog control law

$$y(t) = \alpha\left[x(t) + \beta \int_0^t x(\lambda)\, d\lambda\right]$$

becomes

$$\Delta y(kT) = \alpha[x(kT) - x(kT-T) + \beta T x(kT)]$$

$$y(kT) = \sum_{n=0}^{k} \Delta y(nT)$$

when implemented as an incremental digital control law with sampling interval T.

13.29 In practice, the incremental P + I control law described in Problem 13.28 may encounter small inaccuracies in the process of summing the computed increments. To model this situation the disturbance input $d(k)$ is introduced between the increment-calculation and the summation blocks. Letting $G(z)$ represent the transfer function of the plant and using unity feedback, Figure P13.5 results.

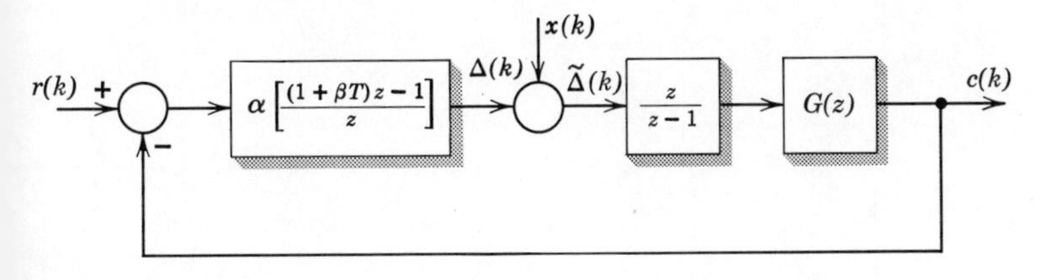

Figure P13.5

(a) Show that when strictly proportional control is used ($\beta = 0$) *1.* an error made in a single pass through the computation loop will cause a constant steady-state error in the output $c(k)$, and *2.* a constant error made on each pass results in an error that is a combination of ramp and constant terms. In both cases, express the errors in $c(k)$ in terms of the system parameters. (b) Show that if P + I control is used ($\beta > 0$) that *1.* a single-pass error in summing the increments will not cause a steady-state error in $c(k)$, and *2.* a repeated error will result in a constant steady-state error in $c(k)$. Find an expression for the error.

13.30 Design a sampled-data control system to balance the inverted stick treated in Problems 8.32, 9.19, 10.28 and 11.22. Assume that $\theta(kT)$ is available from an A/D converter and that the desired torque, constant between sample times, can be applied by driving the motor with a D/A converter. Consider, in order, proportional and proportional-plus-difference control laws.

Appendix A

Complex variables

Because so many of the analytic techniques used in the analysis of fixed linear systems have their foundations in the theory of functions of a complex variable, we shall state the properties and prove the theorems that are of greatest use in this book. However, a complete treatment of the subject is not attempted, and the reader is referred to any standard text for further coverage, e.g., Churchill (1948).

A.1 FUNCTIONS OF A COMPLEX VARIABLE

Letting s be the complex variable $s = \sigma + j\omega$, such that $\sigma = \text{Re}\,[s]$ and $\omega = \text{Im}\,[s]$, the function $V(s) = \text{Re}\,[V] + j\,\text{Im}\,[V]$ is a rule for mapping points in the s-plane into *unique* points in the V-plane, as indicated in Figure A.1. Although the mapping is one-to-one going from s to V, there may be more than one value of s that maps into a single value of V. Two equivalent forms for $V(s)$, referred to as the rectangular and polar forms, respectively, are

$$V(s) = \text{Re}\,[V(s)] + j\,\text{Im}\,[V(s)]$$
$$= |V(s)|e^{j\arg[V(s)]}$$

In general, our work will be limited to three classes of functions: *polynomials*, *rational functions* (ratios of polynomials), and the *exponential function* (e.g., e^{st}).

If the function $V(s)$ is defined in a domain D in the s-plane, it is said to be *analytic* in D if the derivative dV/ds is continuous everywhere in that domain. It can be proved that the sum and product of functions that are analytic in the same domain are also analytic. Because the function $V(s) = s$ is analytic everywhere in the finite s-plane, it follows that all polynomials in s are analytic everywhere in the finite s-plane.† Furthermore, it can be shown that the quotient of two analytic functions is analytic everywhere except at those points for which the denominator

†The point at *infinity* is defined as the limit of s as $|s| \to \infty$, regardless of arg $[s]$, i.e., regardless of the direction in which $s \to \infty$.

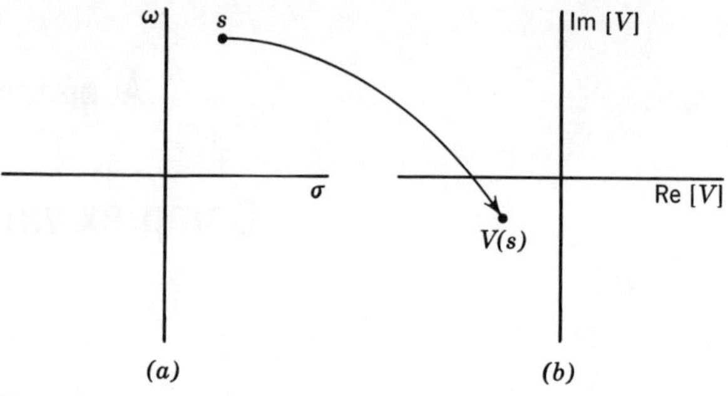

Figure A.1. (*a*) *s*-plane. (*b*) *V*-plane.

vanishes. Hence, any rational function of s is analytic everywhere except at the zeros of the denominator polynomial.

Points at which a function is not analytic are referred to as *singularities*. If $V(s)$ has a singularity at the point s_0 but is analytic everywhere else in a neighborhood of the point, then s_0 is said to be an *isolated singularity*. We shall concern ourselves exclusively with this type.

Contour integrals

If $V(s)$ is analytic in a domain D, and C_1 is a path between points s_1 and s_2 lying in D, the *contour integral* shown in Figure A.2, is defined to be

$$\underset{C_1}{\int_{s_1}^{s_2}} V(s)\,ds = \int_{\sigma_1+j\omega_1}^{\sigma_2+j\omega_2} (\text{Re}\,[V]+j\,\text{Im}\,[V])\,(d\sigma+jd\omega)$$

If we consider another integral having the same terminal points but a different

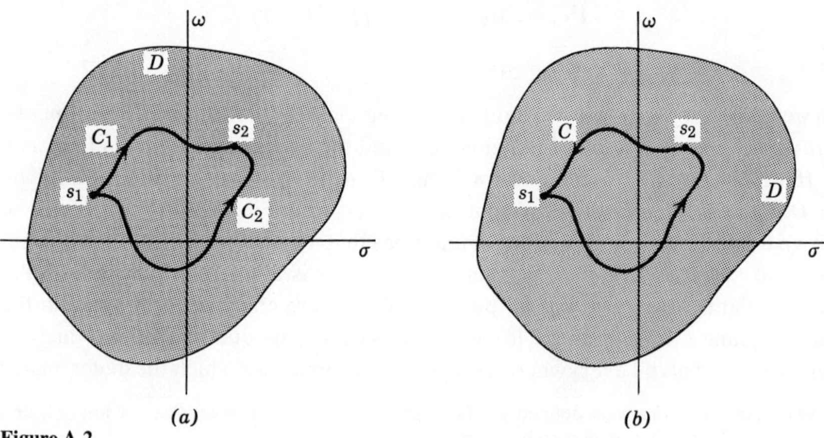

Figure A.2.

path, C_2, which also lies entirely within D, it follows from Cauchy's *integral theorem* that

$$\int_{\substack{s_1 \\ C_1}}^{s_2} V(s)\,ds = \int_{\substack{s_1 \\ C_2}}^{s_2} V(s)\,ds$$

Equivalently, the theorem can be stated in terms of an integral along the *closed contour* C as

$$\oint_C V(s)\,ds = 0$$

where C is any closed contour lying entirely within D, Figure A.2*b*. As indicated in the figure, when the closed contour C is traversed in the positive sense the "interior" is on the left-hand side of the path.

Power-series expansions

If $V(s)$ is analytic within the domain D, then it can be expanded about any point s_0 within D in the *Taylor series*

$$V(s) = V(s_0) + \left.\frac{dV}{ds}\right|_{s_0}(s-s_0) + \frac{1}{2!}\left.\frac{d^2V}{ds^2}\right|_{s_0}(s-s_0)^2 + \cdots$$

The series will converge at every point within the largest circle centered at s_0 that does not include any singularities of $V(s)$, referred to as the *circle of convergence.* Furthermore, the series will diverge at all points outside the circle of convergence.

If $V(s)$ is analytic within an annular domain $R_{min} < |s-s_0| < R_{max}$ centered at s_0, Figure A.3, it can be expanded in the *Laurent series*

$$V(s) = \sum_{n=-\infty}^{\infty} c_n(s-s_0)^n$$

where

$$c_n = \frac{1}{2\pi j}\oint_C \frac{V(s)}{(s-s_0)^{n+1}}\,ds \tag{1}$$

and the contour C lies within the annulus. Our primary use of the Laurent series is in evaluating the integral of $V(s)$ around a closed contour. Setting $n=-1$ in (1) gives

$$c_{-1} = \frac{1}{2\pi j}\oint_C V(s)\,ds \tag{2}$$

Thus, if we are able to find the coefficient of $(s-s_0)^{-1}$ in the Laurent-series expansion of $V(s)$, we immediately know the value of the integral of $V(s)$ around any closed contour in the annulus of convergence.

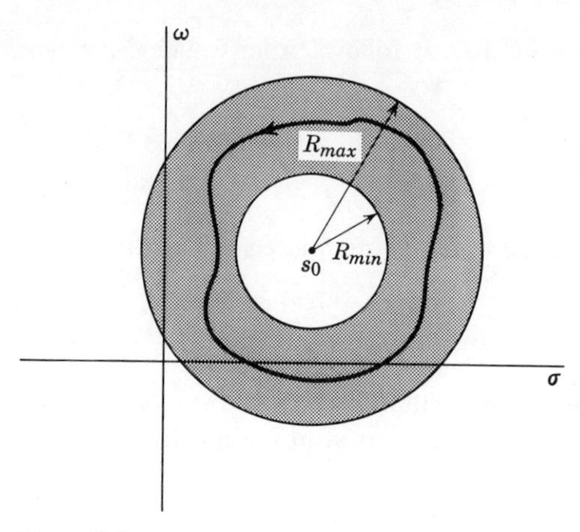

Figure A.3.

Poles and residues

If $V(s)$ is analytic in a domain D except at the point s_0 its Laurent-series expansion will be

$$V(s) = \frac{c_{-m}}{(s-s_0)^m} + \cdots + \frac{c_{-1}}{(s-s_0)} + c_0 + c_1(s-s_0) + \cdots$$

If $-m$, the index of the lowest-order term, is negative but finite, then the singularity of $V(s)$ at s_0 is said to be a *pole* of multiplicity m. The coefficient c_{-1} is known as the *residue* of $V(s)$ at the pole s_0 and can be easily evaluated from $V(s)$ when $m = 1$ or 2, which covers most cases of interest.

When s_0 is a pole with $m = 1$, referred to as a *simple pole*, the Laurent-series expansion about $s = s_0$ will be

$$V(s) = \frac{c_{-1}}{s-s_0} + c_0 + c_1(s-s_0) + \cdots$$

To evaluate the residue c_{-1}, denoted by Res $\{V(s), s_0\}$ we merely multiply $V(s)$ by $(s-s_0)$, giving

$$(s-s_0)V(s) = c_{-1} + c_0(s-s_0) + c_1(s-s_0)^2 + \cdots$$

and then set $s = s_0$. Hence,

$$\text{Res}\,\{V(s), s_0\} = (s-s_0)V(s)\big|_{s=s_0} \qquad m = 1$$

If s_0 is a pole of multiplicity two, the Laurent series is

$$V(s) = \frac{c_{-2}}{(s-s_0)^2} + \frac{c_{-1}}{(s-s_0)} + c_0 + \cdots$$

Thus

$$(s-s_0)^2V(s) = c_{-2}+c_{-1}(s-s_0)+c_0(s-s_0)^2+\cdots$$

and

$$\frac{d}{ds}[(s-s_0)^2V(s)] = c_{-1}+2c_0(s-s_0)+\cdots$$

from which it follows that

$$\text{Res}\{V(s), s_0\} = \frac{d}{ds}[(s-s_0)^2V(s)]\Big|_{s=s_0} \qquad m=2$$

This process can be extended for greater multiplicities but it is seldom that $m > 2$.

A.2 THEOREMS

Several important complex-variable theorems are given below, with the chapters in which they are used indicated in parentheses.

Residue theorem (Chapter 13)

When the contour C encloses k isolated singularities, then (2) can be extended to yield Cauchy's *residue theorem*, namely

$$\oint_C V(s)ds = 2\pi j \sum_{i=1}^{k} \text{Res}\{V(s), s_i\} \tag{3}$$

where the s_i are the isolated singularities of $V(s)$ within the contour C. Almost always, these singularities are poles. If the direction of integration is reversed, the sign of right-hand side of (3) must be reversed.

As a corollary to (3), consider $V(s) = 1/(s-s_0)^n$, where n is any integer. Since $V(s)$ is already in the form of its Laurent-series expansion, its residue is 1 if $n=1$ and zero for all other n. As a consequence,

$$\oint_C \frac{1}{(s-s_0)^n}ds = \begin{cases} 2\pi j & n=1 \\ 0 & n \neq 1 \end{cases}$$

where the contour C encloses the point $s = s_0$. This result is used in proving the inverse z-transform theorem in Section 13.5.

Partial-fraction expansion theorem (Chapter 8)

We shall assume that $V(s)$ is a rational function of s which has m zeros, z_i, and n

distinct poles, p_j, and thus can be written as

$$V(s) = \left(\frac{b_m}{a_n}\right) \frac{\prod_{i=1}^{m} (s - z_i)}{\prod_{j=1}^{n} (s - p_j)} = \frac{b_m s^m + \cdots + b_0}{a_n s^n + a_{n-1} s^{n-1} + \cdots + a_0} \tag{4}$$

where $m < n$. The derivation can be extended to handle repeated poles without difficulty but at the expense of more complicated notation.

Expanding $V(s)$ in a Laurent series about the pole p_1,

$$V(s) = \frac{A_1}{s - p_1} + \underbrace{\sum_{i=0}^{\infty} c_i (s - s_0)^i}_{g_1(s)}$$

where A_1 is the residue and $g_1(s)$ is a polynomial in s. Similarly, the residues A_j and polynomials $g_j(s)$ can be defined for $j = 2, 3, \ldots, n$ by considering expansions about the remaining poles. Now, we construct the function

$$W(s) \triangleq V(s) - \sum_{j=1}^{n} \frac{A_j}{s - p_j} \tag{5}$$

and proceed to show that $W(s)$ must be identically zero. Because of the manner in which it has been defined, the Laurent-series expansion of $W(s)$ about any of the poles of $V(s)$ will result in *nonnegative* powers of s; $W(s)$ is said to have a removable singularity at these points. Furthermore, $W(s)$ will be analytic at all other points in the finite s-plane.

Examining the behavior at infinity we see that $V(s) \to 0$ as $s \to \infty$ because $m < n$ and the summation on the right-hand side of (5) obviously vanishes. Thus, $W(s)$ must be a polynomial that vanishes at infinity, which can happen only if it is identically zero, resulting in the widely used partial-fraction expansion

$$V(s) = \sum_{j=1}^{n} \frac{A_j}{s - p_j} \tag{6}$$

The argument principle (Chapter 11)

Assume that $V(s)$ can be written as in (4), with the further restriction that its zeros be distinct. The first step is to show that

$$\frac{d}{ds} [\ln V(s)] = \sum_{i=1}^{m} \frac{1}{s - z_i} + \sum_{j=1}^{n} \frac{-1}{s - p_j} \tag{7}$$

which is to say that the derivative of the logarithm of $V(s)$ has $m + n$ poles consisting of the zeros and poles of $V(s)$ and the residues are either $+1$ or -1. To obtain (7), we take the logarithm of $V(s)$ which, from (4), is

$$\ln V(s) = \ln\left(\frac{b_m}{a_n}\right) + \sum_{i=1}^{m} \ln(s - z_i) - \sum_{j=1}^{n} \ln(s - p_i) \tag{8}$$

Recalling that

$$\frac{d}{ds}[\ln V(s)] = \frac{1}{V}\frac{dV}{ds}$$

whence, for the terms on the right-hand side of (8),

$$\frac{d}{ds}[\ln(s - z_i)] = \frac{1}{s - z_i}$$

and

$$\frac{d}{ds}[\ln(s - p_j)] = \frac{1}{s - p_j}$$

Equation (7) follows directly when (8) is differentiated.

Next we multiply both sides of (7) by ds and integrate around any closed contour Γ, arbitrarily taking the clockwise direction as positive, yielding

$$\oint_\Gamma d[\ln V(s)] = \oint_\Gamma \left\{\sum_{i=1}^{m} \frac{1}{s - z_i} + \sum_{j=1}^{n} \frac{-1}{s - p_j}\right\} ds \tag{9}$$

The right-hand side of (9) can be integrated using the residue theorem to give

$$\oint d[\ln V(s)] = 2\pi j(n_z - n_p) \tag{10}$$

where the number of zeros and poles of $V(s)$ which are inside the s-plane contour Γ are denoted by n_z and n_p, respectively. The left-hand side of (10) can be integrated by writing $V(s)$ in polar form before taking the logarithm. Thus

$$\begin{aligned}\oint_\Gamma d[\ln V(s)] &= \oint_\Gamma d\{\ln|V(s)| + j\arg[V(s)]\} \\ &= \ln|V|\Big|_{s_1}^{s_2} + j\arg[V]\Big|_{s_1}^{s_2}\end{aligned}$$

where s_1 and s_2 denote the beginning and end of the contour Γ. Because Γ is closed, $s_1 = s_2$ and $|V(s_1)| = |V(s_2)|$. However, arg $[V(s_2)]$ will differ from arg $[V(s_1)]$ by an integer multiple of 2π radians if the contour in the V-plane encircles the origin, and the two arguments will be equal if the origin is not encircled. Because of the assumed positive direction of integration the integral reduces to

$$\oint d[\ln V(s)] = 2\pi j\, n_{cw} \tag{11}$$

where n_{cw} is the number of clockwise encirclements of the origin by the V-plane contour.

Finally the argument principle results if the right sides of (10) and (11) are equated to give

$$n_{cw} = n_z - n_p \tag{12}$$

where the s-plane contour is traversed in a clockwise direction. Actually the basic result is independent of the sense assumed as positive provided only that the *same* sense is used in both the s- and V-planes. Thus, if Γ is a counterclockwise contour, (12) becomes

$$n_{ccw} = n_z - n_p$$

Appendix B

Transform tables

For convenient reference, this appendix gives tabulated summaries of definitions, theorems, and common transform pairs for the Fourier, Laplace, and z-transforms. Details and restrictions will be found in the corresponding sections of the text proper, while Appendix C defines the symbolic notation.

B.1 FOURIER TRANSFORMS (Sections 7.1 to 7.3)

Definitions

Transform $\quad V(f) = \mathfrak{F}[v(t)] = \int_{-\infty}^{\infty} v(t) e^{-j2\pi ft} dt$

Inversion $\quad v(t) = \mathfrak{F}^{-1}[V(f)] = \int_{-\infty}^{\infty} V(f) e^{j2\pi ft} df$

Theorems

	Time Function	*Fourier Transform*
Superposition	$\alpha v(t) + \beta w(t)$	$\alpha V(f) + \beta W(f)$
Time shift	$v(t - t_d)$	$V(f) e^{-j2\pi f t_d}$
Time scaling	$v(ct)$	$\frac{1}{\lvert c \rvert} V\left(\frac{f}{c}\right)$
Duality	$V(t)$	$v(-f)$
Frequency translation	$v(t) e^{j2\pi f_c t}$	$V(f - f_c)$
Modulation	$v(t) \cos 2\pi f_c t$	$\frac{1}{2}[V(f - f_c) + V(f + f_c)]$
Differentiation	$v^{(n)}(t)$	$(j2\pi f)^n V(f)$

Integration	$v^{(-n)}(t)$	$(j2\pi f)^{-n}V(f)$
Convolution	$v * w(t)$	$V(f)W(f)$
Multiplication	$v(t)w(t)$	$V * W(f)$

Integral theorems

$$\int_{-\infty}^{\infty} |v(t)|^2 dt = \int_{-\infty}^{\infty} |V(f)|^2 df$$

$$\int_{-\infty}^{\infty} v(t)w^*(t)dt = \int_{-\infty}^{\infty} V(f)W^*(f)df$$

Transform pairs

	$v(t)$	$V(f)$
Rectangular pulse	$\Pi(t/\tau)$	$\tau \operatorname{sinc} f\tau$
Triangular pulse	$\Lambda(t/\tau)$	$\tau \operatorname{sinc}^2 f\tau$
Sinc pulse	$\operatorname{sinc} 2Wt$	$\frac{1}{2W}\Pi\left(\frac{f}{2W}\right)$
Exponential	$e^{-\lvert at\rvert}u(t)$	$\frac{1}{\lvert a\rvert + j2\pi f}$
Gaussian	$e^{-\pi(t/\tau)^2}$	$\tau e^{-\pi(f\tau)^2}$
Constant	1	$\delta(f)$
Phasor	$e^{j(2\pi f_c t+\theta)}$	$e^{j\theta}\delta(f-f_c)$
Sinusoid	$\cos(2\pi f_c t+\theta)$	$\frac{1}{2}[e^{j\theta}\delta(f-f_c)+e^{-j\theta}\delta(f+f_c)]$
Impulse	$\delta(t-t_d)$	$e^{-j2\pi f t_d}$
Step	$u(t)$	$\frac{1}{2}\delta(f)+\frac{1}{j2\pi f}$

B.2 LAPLACE TRANSFORMS (Section 8.1)

Definitions

Transform $\quad V(s) = \mathcal{L}[v(t)] = \int_0^{\infty} v(t)e^{-st}dt$

Inversion $\quad v(t) = \mathcal{L}^{-1}[V(s)] = \frac{1}{2\pi j}\int_{\sigma_c - j\infty}^{\sigma_c + j\infty} V(s)e^{st}ds$

Theorems

	Time Function	*Laplace Transform*
Superposition	$\alpha v(t)+\beta w(t)$	$\alpha V(s)+\beta W(s)$
Time delay ($t_d \geq 0$)	$v(t-t_d)$	$V(s)e^{-st_d}$
Time scaling ($c > 0$)	$v(ct)$	$\frac{1}{c}V\left(\frac{s}{c}\right)$
Differentiation	$v^{(n)}(t)$	$s^nV(s)-s^{n-1}v(0)$ $-s^{n-2}\dot{v}(0)-\cdots-v^{(n)}(0)$
Integration	$\int_0^t v(\lambda)d\lambda$	$\frac{1}{s}V(s)$
Convolution	$v * w(t)$	$V(s)W(s)$
Multiplication by an exponential	$e^{at}v(t)$	$V(s-a)$

Initial-value and final-value theorems

$$v(0+) = \lim_{s\to\infty} sV(s)$$

$$\lim_{t\to\infty} v(t) = \lim_{s\to\infty} sV(s)$$

Transform pairs

	$v(t)$	$V(s)$
Impulse	$\delta(t)$	1
Step	$u(t)$	$\frac{1}{s}$
Ramp	t	$\frac{1}{s^2}$
Exponential	e^{at}	$\frac{1}{s-a}$
Sine	$\sin \omega t$	$\frac{\omega}{s^2+\omega^2}$
Cosine	$\cos \omega t$	$\frac{s}{s^2+\omega^2}$

B.3 z-TRANSFORMS (Section 13.4)

Definitions

Transform $$V(z) = \mathfrak{Z}[v(k)] = \sum_{k=0}^{\infty} v(k)z^{-k}$$

Inversion $$v(k) = \mathfrak{Z}^{-1}[V(z)] = \frac{1}{2\pi j} \oint V(z)z^{k-1}\,dz$$

Theorems

	Discrete-Time Function	*z-Transform*
Superposition	$\alpha v(k) + \beta w(k)$	$\alpha V(z) + \beta W(z)$
Advance ($m \geq 0$)	$v(k+m)$	$z^m V(z) - z^m v(0) - z^{m-1}v(1) - \cdots - zv(m-1)$
Delay ($n \geq 0$)	$v(k-n)$	$z^{-n}V(z)$
Summation	$\sum_{i=0}^{k} v(i)$	$\frac{z}{z-1}V(z)$
Convolution	$v * w(k)$	$V(z)W(z)$
Multiplication by an exponential	$e^{kaT}v(k)$	$V(e^{-aT}z)$

Initial-value and final-value theorems

$$v(0) = \lim_{z\to\infty} V(z)$$

$$\lim_{k\to\infty} v(k) = \lim_{z\to 1} (z-1)V(z)$$

Transform pairs

	$v(k)$	$V(z)$
Delta	$d(k)$	1
Step	$u(k)$	$\frac{z}{z-1}$
Ramp	k	$\frac{z}{(z-1)^2}$
Exponential	e^{kaT}	$\frac{z}{z-e^{aT}}$

Power	α^k	$\dfrac{z}{z-\alpha}$
Sine	$\sin k\omega T$	$\dfrac{z \sin \omega T}{z^2 - 2z \cos \omega T + 1}$
Cosine	$\cos k\omega T$	$\dfrac{z(z - \cos \omega T)}{z^2 - 2z \cos \omega T + 1}$

Appendix C

Symbolic notation

C.1 FUNCTIONS

Impulse $\delta(t)$ (see Sections 4.2 and 8.1)

Step $$u(t) = \int_{-\infty}^{t} \delta(\lambda)\, d\lambda = \begin{cases} 1 & t > 0 \\ 0 & t < 0 \end{cases}$$

Doublet $$\delta'(t) = \frac{d\delta(t)}{dt}$$

Rectangle $$\Pi\left(\frac{t}{\tau}\right) = \begin{cases} 1 & |t| < \tau/2 \\ 0 & |t| > \tau/2 \end{cases}$$

Triangle $$\Lambda\left(\frac{t}{\tau}\right) = \begin{cases} 1 - |t|/\tau & |t| < \tau \\ 0 & |t| > \tau \end{cases}$$

Sinc $$\operatorname{sinc} t = \frac{\sin \pi t}{\pi t}$$

C.2 COMPLEX NUMBERS

Imaginary unit $$j = \sqrt{-1}$$

Rectangular form $$z = \operatorname{Re}[z] + j \operatorname{Im}[z]$$

Polar form $$z = |z|\, e^{j \arg[z]}$$

Magnitude $$|z| = \sqrt{\operatorname{Re}^2[z] + \operatorname{Im}^2[z]}$$

Angle $$\arg[z] = \arctan \frac{\operatorname{Im}[z]}{\operatorname{Re}[z]}$$

Conjugate $$z^* = \operatorname{Re}[z] - j \operatorname{Im}[z] = |z|\, e^{-j \arg[z]}$$

C.3 OPERATIONS†

Differentiation

$$\dot{v}(t) = \frac{dv(t)}{dt}, \quad \ddot{v}(t) = \frac{d^2v(t)}{dt}, \qquad \text{etc.}$$

$$v^{(k)}(t) = \frac{d^kv(t)}{dt^k}$$

Integration

$$v^{(-k)}(t) = \int_{-\infty}^{t} \int_{-\infty}^{\lambda_k} \cdots \int_{-\infty}^{\lambda_2} v(\lambda_1)\,d\lambda_1 d\lambda_2 \cdots d\lambda_k$$

Convolution‡

$$v * w(t) = \int_{-\infty}^{\infty} v(\lambda)w(t-\lambda)\,d\lambda$$

Scalar Product

$$\langle v, w\rangle = \begin{cases} \lim\limits_{T\to\infty} \int_{-T}^{T} v(t)w^*(t)\,dt \\ \lim\limits_{T\to\infty} \frac{1}{2T} \int_{-T}^{T} v(t)w^*(t)\,dt \end{cases}$$

Norm

$$\|v\| = \langle v, v\rangle^{1/2}$$

C.4 MISCELLANEOUS SYMBOLS

$\approx$ "Approximately equals"
$\triangleq$ "Equals by definition", or "denotes"
$\leftrightarrow$ Denoting a transform pair

$\int_{T_0}$ Denoting $\int_{t_1}^{t_1+T_0}$ where t_1 is arbitrary

†See Appendix B for transform operations.
‡Also written as $v * w$ or $[v(t)] * [w(t)]$.

Supplementary reading

A selected and annotated bibliography is given below for the reader who wishes to pursue the subject of systems analysis and design. These books and papers are readily available in a technical library of moderate size and, unless otherwise indicated, they are understandable to anyone who has completed the corresponding sections of this text. Complete citations are listed under the title "References."

GENERAL SYSTEMS ANALYSIS

Several textbooks more or less parallel our coverage of the *analysis of fixed linear systems*. Among them are Cheng (1959), Martens and Allen (1969), Lathi (1965, Chaps. 1–10), and Cooper and McGillem (1967, Chaps. 1–8), the latter two concentrating on electrical systems. Cannon (1967) deserves special mention for its extensive discussion of *mathematical modeling* of physical systems, including numerous devices of practical interest. LePage (1961) gives a thorough treatment of the s-plane and the Laplace transform, while Aseltine (1958) is a handy reference on the various *transform methods*.

Somewhat more advanced treatments of linear systems analysis will be found in Schwarz and Friedland (1965, Chaps. 1–8) or Gupta (1966). Kaplan (1962) is recommended to those interested in the mathematical niceties we have omitted. Rosenbrock (1970) relates the frequency-domain and state-space methods in a mathematical vein.

State equations and the state-space approach are fully developed in such works as DeRusso, Roy, and Close (1965), Timothy and Bona (1968), and Brockett (1970) – any of which would be a natural sequel to this text. Zadeh and Desoer (1963) is substantially more sophisticated and abstract, while Rosenbrock and Storey (1970) present the mathematics of the state-space approach from an advanced point of view. Shilov (1961) covers the supporting mathematical theory of linear spaces.

For the student going beyond the confines we have assumed, the following are good starting points: D'Angelo (1970) for *time-varying* systems; Šiljak (1969) for *nonlinear* systems; and for systems with *random* inputs, Bendat (1958) or Papoulis (1965). Modeling and analysis of urban and world dynamics, two of the newer applications of the system concept, are discussed by Forrester (1969), (1971).

COMMUNICATION SYSTEMS

The basic principles and techniques of *modulation systems* are covered by Stein and Jones (1967), Carlson (1968), Lathi (1968), and Schwartz (1970), to name a few recent texts. All of these are written at about the same level as this book, and they consider both analog and digital modulation. An extended treatment of digital data communication will be found in Lucky, Salz, and Weldon (1968), while Pierce (1968) gives an informative survey of some of the practical problems in digital signal transmission.

Since the *Fourier transform* underlies much of modulation theory, Papoulis (1962) or Bracewell (1965) is strongly suggested for further study of Fourier analysis along with an introduction to impulses as generalized functions. Lighthill (1958) has the same scope from the mathematician's viewpoint.

Statistical communication theory, concerned with optimizing signal transmission in the presence of noise, is another aspect of communications engineering. Two reasonably self-contained introductions to this subject are Sakrison (1968) and Thomas (1969). Mention should also be made of Pierce (1961), a fascinating book intended for laymen (i.e., no mathematics!) but nonetheless of great conceptual value.

CONTROL SYSTEMS

It has been our intent to concentrate on the *analysis* of control systems, without delving into systematic methods for their *design* or *synthesis*. The design of single-loop control systems is covered in a multitude of senior-level texts of which Savant (1964), D'Azzo and Houpis (1965) and Dorf (1967) are examples.

Multi-input-output control systems are discussed in Schwarz and Friedland (1965), DeRusso, Roy, and Close (1965), Timothy and Bona (1968), and Gupta (1966), all of which were cited above. Perkins and Cruz (1969) and Melsa and Schultz (1969) are recent undergraduate-level texts that bridge the gap between the traditional SIO and the more modern MIO approaches to feedback systems.

Buckley (1964) and Coughanowr and Koppel (1965) cover both the basic analytical methods and the physical aspects of chemical process control. Finally, the monthly publication *Control Engineering* is a valuable source of practical applications and information on control-system components.

DISCRETE-TIME SYSTEMS

The study of z-transforms and the use of discrete-time systems for control purposes is usually covered under the heading of sampled-data control systems. Examples of texts that are devoted exclusively to sampled-data systems are Ragazzini and Franklin (1958), Kuo (1963), Jury (1964), and Lindorff (1965). Also, recent control systems texts generally include treatments of discrete-time systems in parallel with continuous systems.

Insofar as the use of discrete-time systems for the filtering of signals (digital filtering) is concerned, the literature is primarily in the form of journal papers, and is just reaching the textbook stage of development. However, Gold and Rader (1969) provides an excellent introduction to the subject, the first few chapters of which can be read following Chapter 13.

References

Angelo, E. J., Jr.
Electronic Circuits, 2d ed., McGraw-Hill, New York, 1964.

Aseltine, J. A.
Transform Method in Linear System Analysis, McGraw-Hill, New York, 1958.

Ash, R. H., and G. R. Ash
"Numerical computation of root loci using the Newton-Raphson technique," *IEEE Transactions on Automatic Control*, **AC-13**, 576–582 (October, 1968).

Bendat, J. S.
Principles and Applications of Random Noise Theory, Wiley, New York, 1958.

Bergland, G. D.
"A guided tour of the fast Fourier transform," *IEEE Spectrum*, **6**, 41–52 (July, 1969).

Blum, J. J.
Introduction to Analog Computation, Harcourt, Brace and World, New York, 1969.

Bode, H. W.
Network Analysis and Feedback Amplifier Design, Van Nostrand, Princeton, N.J., 1945.

Brigham, E. O., and R. E. Morrow
"The fast Fourier transform," *IEEE Spectrum*, **4**, 63–70 (December, 1967).

Bracewell, R.
The Fourier Transform and Its Applications, McGraw-Hill, New York, 1965.

Brockett, R. W.
Finite Dimensional Linear Systems, Wiley, New York, 1970.

Buckley, P. S.
Linear Control Systems, McGraw-Hill, New York, 1969.

Cannon, R. H., Jr.
Dynamics of Physical Systems, McGraw-Hill, New York, 1967.

Carlson, A. B.
Communication Systems, McGraw-Hill, New York, 1968.

Cheng, D. K.
Analysis of Linear Systems, Addison-Wesley, Reading, Mass., 1959.

Christian, E., and E. Eisenmann
Filter Design Tables and Graphs, Wiley, New York, 1966.

Churchill, R. V.
Complex Variables and Applications, McGraw-Hill, New York, 1948.

Close, C. M.
The Analysis of Linear Circuits, Harcourt, Brace and World, New York, 1966.

Cooper, G. R., and C. D. McGillem
Methods of Signal and System Analysis, Holt, Rinehart and Winston, New York, 1967.

Coughanowr, D. R., and L. B. Koppel
Process Systems Analysis and Control, McGraw-Hill, New York, 1965.

D'Angelo, H.
Linear Time-Varying Systems, Allyn and Bacon, Boston, 1970.

D'Azzo, J. J., and C. H. Houpis
Feedback Control System Analysis and Synthesis, 2d ed., McGraw-Hill, New York, 1965.

DeRusso, P. M., R. J. Roy, and C. M. Close
State Variables for Engineers, Wiley, New York, 1965.

Desoer, C. A., and E. S. Kuh
Basic Circuit Theory, McGraw-Hill, New York, 1969.

Dorf, R. C.
Modern Control Systems, Addison-Wesley, Reading, Mass, 1967.
Matrix Algebra, Wiley, New York, 1969.

Forrester, J. W.
Urban Dynamics, M.I.T. Press, Cambridge, Mass., 1969.
World Dynamics, Wright-Allen Press, Cambridge, Mass., 1971.

Gleason, A. M.
Fundamentals of Abstract Analysis, Addison-Wesley, Reading, Mass., 1966.

Gold, B., and C. M. Rader
Digital Processing of Signals, McGraw-Hill, New York, 1969.

Golomb, S. W.
"Mathematical models – uses and limitations," *Astronautics and Aeronautics*, **6**, 57–59 (January, 1968).

Gupta, S. C.
Transform and State Variable Methods in Linear Systems, Wiley, New York, 1966.

Harmuth, H. F.
"Applications of Walsh functions in communications," *IEEE Spectrum*, **6**, 82–91 (November, 1969).

Healy, T. J.
"Convolution revisited," *IEEE Spectrum*, **6**, 87–93 (April, 1969).

Hoeschele, D. F., Jr.
Analog-to-Digital/Digital-to-Analog Conversion Techniques, Wiley, New York, 1968.

Jury, E. I.,
Theory and Applications of the z-Transform Method, Wiley, New York, 1964.

Kaplan, W.
Operational Methods for Linear Systems, Addison-Wesley, Reading, Mass., 1962.

Kuo, B. C.
Analysis and Synthesis of Sampled-Data Control Systems, Prentice-Hall, Englewood Cliffs, N.J., 1963.

LaSalle, J., and S. Lefschetz
Stability by Liapunov's Direct Method, Academic Press, New York, 1961.

Lathi, B. P.
Signals, Systems and Communications, Wiley, New York, 1965.
Communication Systems, Wiley, New York, 1968.

LePage, W. R.
Complex Variables and the Laplace Transform for Engineers, McGraw-Hill, New York, 1961.

Lighthill, M. J.
An Introduction to Fourier Analysis and Generalized Functions, Cambridge, New York, 1958.

Lindorff, D. P.
Theory of Sampled-Data Control Systems, Wiley, New York, 1965.

Lucky, R. W., J. Salz, and E. J. Weldon, Jr.
Principles of Data Communication, McGraw-Hill, New York, 1968.

Martens, H. R., and D. R. Allen
Introduction to Systems Theory, Merrill, Columbus, Ohio, 1969.

Melsa, J. L., and D. G. Schultz
Linear Control Systems, McGraw-Hill, New York, 1969.

Nyquist, H.
"Regeneration theory," *Bell System Technical Journal* **11**, 126–147 (January, 1932).

Oppenheim, A. V.
"Speech spectrograms using the fast Fourier transform," *IEEE Spectrum*, **7**, 57–66 (August, 1970).

Papoulis, A.
The Fourier Integral and Its Applications, McGraw-Hill, New York, 1962.
Probability, Random Variables, and Stochastic Processes, McGraw-Hill, New York, 1965.

Perkins, W. R., and J. B. Cruz, Jr.
Engineering of Dynamic Systems, Wiley, New York, 1969.

Pierce, J. R.
Symbols, Signals and Noise, Harper and Row, New York, 1961.
"Some practical aspects of digital transmission," *IEEE Spectrum*, 5, 63–70 (November, 1968).

Ragazzini, J. R., and G. F. Franklin
Sampled Data Control Systems, McGraw-Hill, New York, 1958.

Raisbeck, G.
Information Theory – An Introduction for Scientists and Engineers, M.I.T. Press, Cambridge, Mass., 1964.

Rosenbrock, H. H.
State-space and Multivariable Theory, Wiley, New York, 1970.

Rosenbrock, H. H., and C. Storey
Mathematics of Dynamical Systems, Wiley, New York, 1970.

Sakrison, D. J.
Communication Theory, Wiley, New York, 1968.

Savant, C. J.
Control System Design, 2d ed., McGraw-Hill, New York, 1964.

Schwartz, M.
Information Transmission, Modulation, and Noise, 2d ed., McGraw-Hill, New York, 1970.

Schwarz, R. J., and B. Friedland
Linear Systems, McGraw-Hill, New York, 1965.

Shilov, G. E.
Theory of Linear Spaces, translated by R. A. Silverman, Prentice-Hall, Englewood Cliffs, N.J., 1961.

Šiljak, D.
Nonlinear Systems, Wiley, New York, 1969.

Stein, S., and J. J. Jones
Modern Communication Principles with Application to Digital Signaling, McGraw-Hill, New York, 1967.

Thomas, J. B.
Statistical Communication Theory, Wiley, New York, 1969.

Timothy, L. K., and B. E. Bona
State Space Analysis, McGraw-Hill, New York, 1968.

Wozencraft, J. M. and I. M. Jacobs
Principles of Communication Engineering, Wiley, New York, 1965.

Zadeh, L. A., and C. A. Desoer
Linear System Theory: The State Space Approach, McGraw-Hill, New York, 1963.

INDEX

Boldface page numbers signify the location of principle definitions.